GENES IN CONFLICT

GENES IN CONFLICT

The Biology of Selfish Genetic Elements

AUSTIN BURT *and* ROBERT TRIVERS

THE BELKNAP PRESS OF
HARVARD UNIVERSITY PRESS
Cambridge, Massachusetts
London, England
2006

Library of Congress Cataloging-in-Publication Data

Burt, Austin.
Genes in conflict : the biology of selfish genetic elements / Austin Burt
and Robert Trivers.
p. cm.
Includes bibliographical references and index.
ISBN 0-674-01713-7 (cloth : alk. paper)
1. Selfish genetic elements. I. Trivers, Robert. II. Title.
QH447.8.S45B87 2006
572.8'6–dc22   2005051219

Contents

SEVEN

Plates follow p. 262

Preface

IT HAS BEEN CLEAR FOR SOME TIME that within-individual conflict would become an important topic in evolutionary genetics. By logic, situations of conflict, in which natural selection acts in opposing directions simultaneously, are likely to generate strong coadaptations on each side. At the level of individuals interacting within (or between) species, this conflict is seen to generate a wide range of behaviors and structures. Conflict within an individual might be expected to generate similar kinds of genetic complexity. Empirically it has become apparent that some categories of selfish genetic elements (e.g., transposable elements) are very widespread, if not universal, and that other categories (e.g., B chromosomes, genomic imprinting) are important in a substantial minority of species.

In the spirit of exploring this emerging world together, we embarked in the early 1990s on a book that would cover all aspects of selfish genetic elements, those stretches of DNA that spread in spite of being injurious to the individuals they occupy. We taught a course together on the subject at the University of California at Santa Cruz in 1992 and we have taught the material since then at Rutgers University and Imperial College London. Throughout the book we have been guided by two principles: we have tried to include all types of selfish elements in all species (except bacteria and viruses) and we have tried to organize these examples logically by reference to the way in which natural selection acts on the selfish elements and, in turn, on the larger genomes they inhabit.

In doing this work we have had substantial help from others. We begin by expressing our gratitude to those who have provided the financial support that has permitted this work. We are grateful to the Harry Frank Guggenheim Foundation, NSERC (Canada), the Ann and Gordon Getty Foundation, the Biosocial Research Foundation, Imperial College London, JK May, NERC and BBSRC (UK), the John Simon Guggenheim Foundation, the College of Law, Arizona State University, the Wellcome Trust, the Center for Evolutionary Studies, Rutgers University, and the FNIH (Grand Challenges in Global Health initiative).

For reading the entire book and giving us their comments, we are most grateful to Jim Bull, Juan Pedro Camacho, Alan Grafen, and David Haig. For reading individual chapters we are grateful to Camille Barr, John Brookfield, Brian Charlesworth, Kelly Dawe, Tom Eickbush, Steve Frank, Jody Hey, Hopi Hoekstra, John Jaenike, Neil Jones, Vasso Koufopanou, Benjamin Normark, Sally Otto, Brian Palestis, Carmen Sapienza, Jonathan Swire, William Zimmerman, and Elefterios Zouros. We are also grateful to David Haig for sharing over many years his extensive knowledge of the subject and to Vasso Koufopanou for advice and support throughout. We are most grateful to Elizabeth Collins for her copyediting, Margaret Nelson for drawing all of the figures, Darine Zaatari for help with the bibliography, and Anne McGuire and Jennifer Snodgrass for help with final editing and production. And we are especially grateful to our editor Michael Fisher for encouragement and unfailing support over many years.

GENES IN CONFLICT

ONE

Selfish Genetic Elements

THE GENES IN AN ORGANISM sometimes "disagree" over what should happen. That is, they appear to have opposing effects. In animals, for example, some genes may want (or act as if they want) a male to produce lots of healthy sperm, but other genes in the same male want half the sperm to be defective. Some genes in a female want her to nourish all her embryos; others want her to abort half of them. Some genes in a fetus want it to grow quickly, others slowly, and yet others at an intermediate level. Some genes want it to become a male, others a female—and the reason they want it to be a female is so that a quarter of her fertilized eggs will be defective! In plants too there can be internal conflicts. Some genes want a plant to make both pollen and seeds, others only seeds. Some genes want the plant to allow a particular pollen grain to fertilize an ovule, and others want to kill that pollen grain.

And how should an organism manage its DNA? Some genes want to protect chromosomes from damage, while others want to break them. Some genes want the organism to snip out bits of DNA and insert them elsewhere in the genome; other genes want to stop this from happening. Some genes want to activate a particular gene, and others want to silence it. Indeed, in the extreme case, some genes want to inactivate half of the genome, while the targeted half prefers to remain active.

Some of these genes are known from only a few species, others from virtually all. In addition there are genes whose existence is predicted, but not yet confirmed, such as genes in fathers to favor their sons versus genes to favor their daughters. Or genes that want a female to judge her mate as more attractive than he really is versus genes that want the opposite effect. And so on.

These conflicts arise because genes are able to spread in a population despite being harmful to the larger organism. Such genes give themselves a benefit but typically cause negative effects on other nonlinked genes in the same creature. In that sense, they are selfish. Indeed, we can define selfish genetic elements as stretches of DNA (genes, fragments of genes, noncoding DNA, portions of chromosomes, whole chromosomes, or sets of chromosomes) that act narrowly to advance their own interests—in other words, replication—at the expense of the larger organism. They, in turn, select for nonlinked genes that suppress their activity, and thereby mitigate the harm. That is, the evolution of selfish genetic elements inevitably leads to within-individual—or intragenomic—conflict. This occurs over evolutionary time, as genes at different locations within the genome are selected to have contradictory effects. It also occurs over developmental time as organisms experience these conflicting effects. In this sense, we speak of "genes in conflict," that is, genes within a single body that are in conflict over the appropriate development or action to be taken.

Genetic Cooperation and Conflict

We mammals have some 30,000 genes that together make an astonishingly complex creature. Most of these genes are beneficial to us, that is, they make a positive contribution to survival and reproduction most of the time. This is no accident: most genes have spread in populations precisely because they increase organismal fitness. All the complex adaptations that organisms show for survival and reproduction—that distinguish living from non-living things, biology from chemistry and physics—exist because of natural selection; and for most genes, the selection of alternative alleles is based on how they contribute to organismal function.

This cooperation arises in large part because most genes are transmitted from one generation to the next in a transparent, "fair" manner, with diploid individuals transmitting the 2 copies of each gene with equal likelihood, 50:50. Thus the gametes produced by an individual are usually a faithful reflection of the gametes from which it was derived, and the transmission of genes from one generation to the next does not in itself lead to a change in gene frequencies. This genetic fairness is beneficial to the organism (Crow 1979). It avoids the introduction of an arbitrary bias into the genome, one that would warp gene frequencies away from what is optimal for phenotypic function. Ignoring mutation and random effects, gene frequencies will change only if individuals of different genetic constitution reproduce to different degrees. This is conventional natural selection. Because different genes in the same organism are selected in the same direction (to increase survival and reproduction), they can evolve to cooperate in, say, the construction of an eye or leg.

But some genes have discovered ways to spread and persist without contributing to organismal fitness. At times, this means encoding actions that are diametrically opposed to those of the majority of genes. As a consequence, most organisms are not completely harmonious wholes and the individual is, in fact, divisible. This book is about these alternative routes to genetic posterity, and the diverse array of adaptations and counteradaptations that such selection has produced. As we shall see, selfish genetic elements are a universal feature of life, with pervasive effects on the genetic system and the larger phenotype, diversified into many different forms, with adaptations and counteradaptations on both sides, a truly subterranean world of sociogenetic interactions, usually hidden completely from sight.

Three Ways to Achieve "Drive"

Most selfish genetic elements act by somehow contriving to be transmitted to a disproportionate fraction of the organism's progeny. While most genes in a sexual organism are transmitted to the Mendelian 50% of progeny, these selfish genetic elements manage to get into 60%, or 99%, or 50.001%. Instead of being fair and transparent, the process of gene transmission in itself leads to an increase in gene frequencies. Genes inherited in a biased manner can spread in a population without doing anything good for the organism. Indeed, they can spread even if they are harmful. Such a gene is said to "drive."

So how does a length of DNA distort its own transmission? This book documents all the many answers to this question that have been discovered to date, at least in eukaryotic organisms (that is, we exclude bacteria and viruses). Wherever the information is known, we describe the underlying molecular structures, biochemical capabilities, and mechanisms of action, both for their intrinsic interest and for what they reveal about the evolutionary processes that created the selfish elements. Here we sketch in broad outline the 3 main strategies that have evolved over and over again.

1. *Interference.* This strategy is the most blatantly selfish: getting ahead by disrupting the transmission of the alternative allele. For example, a gene in a diploid organism might sabotage the 50% of the gametes to which it is not transmitted, thereby increasing its own success (Fig. 1.1). Or it might kill the 50% of offspring that do not carry a copy of it, thereby increasing survival of the 50% that do. This is the most diverse class of selfish genetic elements, and we shall see examples throughout the book (see Chaps. 2, 3, 5, and 10). They have been found on autosomes, sex chromosomes, and in organelles; and the targets of their action can be genes, chromosomal regions, whole chromosomes, or entire haploid sets of chromosomes. In the latter case, selfish genetic elements can radically alter the species' genetic system, changing rules of inheritance for major sections of the genome (see Chap. 10). Note that in all of these cases, a kind of "kinship" discrimination is being made, at least at a very local level. That is, the selfish element is able to direct behavior against others according to the chance it will be located in them.

2. *Overreplication.* Other selfish genetic elements bias their transmission to the next generation by getting themselves replicated more often than other genes in the same organism (Fig. 1.2). In the nucleus, complex mechanisms have evolved to ensure that most genes are replicated exactly once per cell cycle, but various classes of selfish element manage to circumvent these mechanisms. Some encode their own DNA polymerase, which they use to make extra copies of themselves (off an RNA template); others cause chromosomes to break and then get themselves replicated as part of the repair process; and yet others literally jump across the DNA replication complex as the complex travels along a chromosome, and so get copied twice when normal host genes are getting copied once. This general strategy of overreplication is used by the most abundant class of selfish genes, the transposable elements, as well as other less well known types such as homing endonuclease genes and group II introns (see Chaps. 6 and 7). And outside the nucleus, where DNA replication is less strictly controlled, entire mitochondrial and chloroplast genomes can replicate a variable number of times per cell cycle, and so can compete among each other to replicate faster and

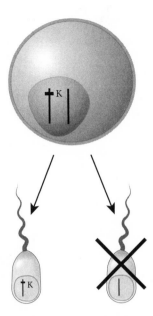

Figure 1.1 Drive by killing the opposition. A killer gene (K) is able to kill non-K bearing sperm. This should benefit the K gene's chance of fertilizing an egg.

faster in germinal tissue. Such selection may account in part for the small size of organelle genomes (see Chap. 5). Finally, genes can also increase in frequency within an organism by increasing the replication rate of the entire cell, if different cells are genetically distinct, thereby creating a selfish cell lineage (see Chap. 11).

3. *Gonotaxis*. The third strategy of drive is to move preferentially toward the germline, and away from somatic cells, when presented with the choice (Fig. 1.3). For example, female meiosis in most species gives rise to 1 functional egg or ovule and 2 or 3 nonfunctional "polar bodies." This asymmetry presents an opportunity for a selfish gene: any gene that can preferentially segregate to the egg or ovule and avoid the polar bodies will thereby be transmitted to more than 50% of the gametes. Competition to be included in the egg and avoid the polar bodies has led to the evolution of "knobs" on the chromosomes of maize, which act as centromeres during meiosis and pull themselves along the spindle (see Chap. 8). Indirect evidence suggests that this same asymmetry has been a crucial factor in the

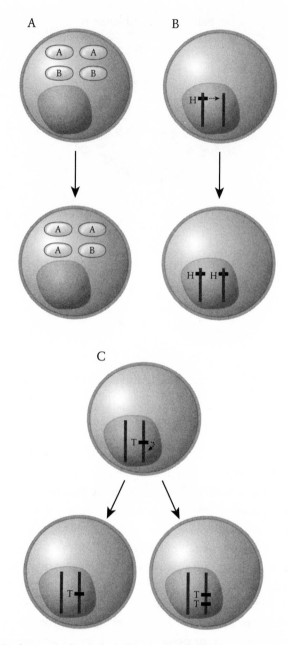

Figure 1.2 Drive by overreplication. A. The A mitochondria replicate faster than the B mitochondria. B. A homing endonuclease gene (H) copies itself onto the paired chromosome, thereby doubling in number. C. A transposable element (T) makes a second copy of itself and one daughter cell inherits 1 copy while the other inherits 2 copies. In all cases, genes have given themselves an advantage by replicating themselves more often than other genes in the cell.

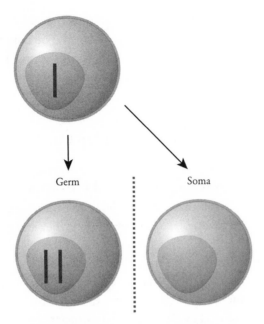

Figure 1.3 Drive by gonotaxis. A B chromosome replicates itself and both copies move to the germline and avoid the somatic cell.

rapid evolution of centromeric sequences and associated proteins. In some groups it may also have determined whether chromosomes have 1 arm or 2. Many species of plants and animals have parasitic "B" chromosomes that also exploit this opportunity, as well as analogous opportunities at mitosis when one daughter cell is destined to stay in the germline and the other to differentiate into a somatic cell (see Chap. 9). And some transposable elements are expressed in somatic cells from which they then invade nearby germline cells.

Thus a selfish gene can increase in frequency by interfering with the replication of the alternative allele, by itself replicating more often than normal, or by biased movement toward the germline. A gene using any one of these strategies is not increasing in frequency by increasing organismal quality; indeed, because it usually imposes some cost on the host organism, it may attract the evolution of suppressors. Note that other types of genes may be transmitted to more than 50% of the progeny of a diploid heterozygote but also help the organism. A gene expressed in pollen grains whose only effect is to make them grow faster toward the ovule would be transmitted to more than 50% of the progeny. But such a gene benefits the organism rather than harming it. It will not attract the evolution of suppressors, and we would not call it selfish.

Within-Individual Kinship Conflicts

Not all selfish genetic elements distort their own transmission. The other major route to selfishness is to distort how the organism behaves toward its relatives. It is well known that natural selection leads to organisms treating their relatives more favorably the more closely they are related, because closer relatedness means a higher probability that the relative carries copies of the organism's own genes. Crucially, this probability of sharing genes that are identical by descent may vary across the genome. As a consequence, the selection of genes affecting social behavior can also vary across the genome. The best empirical evidence for such intragenomic kinship conflicts comes from work on the evolution of genomic imprinting (see Chap. 4). In mice and humans (and probably most mammals), there are perhaps 100 genes in the fetus in which the maternally and paternally derived alleles are expressed at different levels (usually one allele is silent in some or all tissues). As expected from kinship considerations, genes in which the maternal allele is more highly expressed are likely to be growth suppressors (and therefore presumably reduce the burden of the fetus on the mother, who carries a copy of the gene), whereas genes in which the paternal allele is more active are more likely to be growth promoters (being less harmed by the costs of fetal demand on maternal fitness).

Kinship relations may differ not only between maternally and paternally derived genes, but also between autosomes, X-chromosomes, Y-chromosomes, and organelle genomes (Table 1.1). Sex-linked genes that distort the organism's social behavior have long been predicted, but have yet to be actually demonstrated. Unimprinted autosomal genes are much more numerous than the other classes of genes, and so are expected to dominate, but this may require the constant evolution of genes suppressing the activity of genes in the other classes. Of course, organisms differ enormously in the importance of posthatching kinship interactions, and we would expect on these grounds for species with the most extensive lifetime kin interactions

Table 1.1 Degrees of relatedness (*r*) between a boy and his maternal half-sister (plus approximate number of protein-encoding genes available) as a function of location in the genome

Location	*r*	No. of Genes
mtDNA	1	13
Y chromosome	0	70
X chromosome	½	1000
Autosome		
Unimprinted	¼	30,000
Mat. derived	½	100[a]
Pat. derived	0	100[a]

[a]One allele per locus.

to be the most internally conflicted. Humans come to mind as one such possibility (and ants as another).

This conflict due to within-individual variation in relatedness to others is not entirely distinct from conflict over drive or transmission to the next generation. For example, a gene that is active in a mother may cause her to favor progeny that inherited a copy of the gene over those that did not. An extreme version of this preference is to kill the progeny without the gene so as to free up resources for those with it. Here it is mostly a matter of convenience whether we call this a conflict over gene transmission or a conflict over behavior toward relatives. Similarly, nuclear genes that kill gametes to which they were not transmitted, and mitochondrial genes that disrupt pollen production (from which they are not transmitted), can be treated either as cases of drive or of kinship conflicts.

Rates of Spread

In the 1970s a strange genetic phenomenon was first reported in crosses between strains of *Drosophila melanogaster* that were being maintained in the lab (Engels 1989). This included high frequencies of sterile progeny, high mutation rates, and recombination in males (which normally does not occur in *Drosophila*). Subsequent research showed that this "hybrid dysgenesis" occurred because some strains carried a genetic factor not found in the other strains; and if males with the so-called *P* factor were mated to females without it, the progeny had these abnormal properties. Moreover, a most in-

Table 1.2   The invasion of *D. melanogaster* by *P* elements

Year(s) Collected	No. Strains with *P*s / Total Number Examined
1930–59	0/8
1960–64	0/3
1965–69	6/14
1970–74	8/11
1975–79	17/17
1980	13/13

Strains collected from France, Czechoslovakia, and USSR. From Anxolabéhère et al. (1988).

teresting fact emerged: The strains carrying the *P* factor had been collected from the wild relatively recently, whereas strains that had been collected decades previously lacked the *P* factor entirely (Table 1.2). Finally, a great deal of genetic detective work showed that the *P* factor was a transposable element that had spread worldwide through the *D. melanogaster* gene pool in the middle of the 20th century. It had been introduced (somehow) from the distantly related species *D. willistoni* (see Chap. 7). In other words, a selfish element arrived from outside the species during the 20th century and spread to fixation within it before the century was out.

For most evolutionary phenomena, a timespan of several decades is very short. At the vast majority of loci in the *D. melanogaster* gene pool, there was probably only very modest change in gene frequencies over this period. The invasion and spread of *P* elements thus illustrates the quantitative power of non-Mendelian transmission ratios to produce dramatic increases in gene frequency in a very short time span. Quantitative modeling shows that this is an expected, general property of selfish elements (Fig. 1.4). Even if the deviation from Mendelian inheritance is so small as to be undetectable in routine surveys, it can still cause a gene to increase in a population from undetectable to nearly universal in a trivial amount of time.

Drosophila melanogaster is unique in the extent to which geneticists have access to natural collections taken from all over the world for the better part of a century. For other species, we must usually infer change over time from the divergence of extant populations and species. The expectation is that when populations or species become isolated from each other, their selfish genetic elements and host suppressor genes may be among the first components of the genome to diverge. Consistent with this expectation, we review

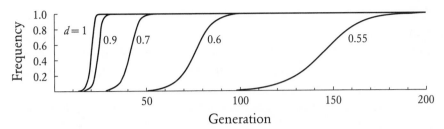

Figure 1.4 Spread of a driving gene through a population for different levels of drive. Frequency of a driving gene is plotted as a function of generation number for different levels of drive (d), where d is defined as the fraction of gametes produced by heterozygotes that carry the selfish element ($d = 0.5$ for Mendelian inheritance). Calculations assume a panmictic host population; no harm to the organism; drive in both sexes; and a starting frequency of 10^{-6}. Note that for rates of drive that would be undetectable in most studies, the driver can still spread to fixation in an evolutionarily trivial amount of time.

many examples of selfish gene activity in species hybrids. This evolutionary dynamism, still largely unexplored, is part of the appeal of selfish genetic elements.

Effects on the Host Population

To avoid one possible source of confusion, we emphasize that designating a gene as "selfish" does not necessarily imply anything about the effect of the gene on the host population, species, or clade. Indeed, it is entirely possible for a selfish gene that harms the host organism—and therefore attracts the evolution of suppressors—to be good for the host population. For example, the productivity of most populations depends much more on the number of females than the number of males, as long as there are enough males to fertilize the females. It follows that selfish genetic elements on sex chromosomes or in mitochondrial genomes that spread by biasing sex ratios toward female function may actually increase population productivity (see Chaps. 3 and 5). Though there is no direct evidence, it is plausible that populations infected with these selfish elements will be larger and persist for longer than equivalent uninfected populations. In addition, over longer evolutionary time spans, transposable elements may be important sources of mutational variability for the host, and it is possible (though as yet unproven) that some lineages have benefited substantially from their activity (see Chap. 7). But regardless of these possible benefits to the host lineage, driving X chromosomes, male-sterile mitochondria, and transposable elements are still harmful to the individual organism and will attract the evolution of suppressors. They are still, by definition, selfish genetic elements.

Other classes of selfish genetic element seem more likely to put the host population at risk. Two such examples are driving Y chromosomes that bias sex ratios toward males and an unusual class of selfish genetic elements that, when transmitted through sperm or pollen, cause the loss of all maternally transmitted chromosomes in the zygote ("androgenesis"; see Chap. 10). Interestingly, these 2 classes of selfish genes are relatively rare. One possible explanation (among several) for their rarity is that they frequently cause extinction soon after arising. In addition, transposable elements are known to inflate genome size in plants, and in plants increasing genome size is strongly associated with increased risk of extinction (see Chap. 7). Thus effects at the level of species or clade may affect the incidence of some selfish genetic elements.

There is also a practical reason to be interested in the effects of various selfish genetic elements on population productivity. Despite much effort, many species harmful to humans are beyond current methods of control. Selfish genetic elements seem, in many ways, ideal tools for population control, and our ever-more-detailed molecular understanding of how they work makes such control increasingly likely (see Chap. 6).

The Study of Selfish Genetic Elements

The history of the study of selfish genetic elements can be summarized in a few sentences. The existence of selfish genetic elements has been known since of the birth of genetics more than 100 years ago, but their selfish nature has been appreciated only more recently. What began as a trickle of information is now a flood. The elements best studied—with some fortunate exceptions—are those found in the genetic model organisms (*Drosophila*, mice, maize, and yeast). And, finally, only recently has it become generally appreciated that selfish elements are widespread and that they have had important coevolutionary effects on the larger genetic system.

Table 1.3 provides a thumbnail sketch of the field through the early 1990s, highlighting the first discovery of each class of selfish element and

Table 1.3   Selected milestones in the study of selfish genetic elements

Year	Scientist(s)	Event
1887	Boveri	Discovery of chromatin diminution (in nematodes)
1906	Correns	Discovery of cytoplasmic male sterility (in *Silene vulgaris*)
1906	Wilson	Discovery of B chromosomes (in bugs)
1926	Metz	Discovery of paternal genome loss (PGL; in *Sciara*)
1927	Dobrovolskaïa-Zavadskaïa and Kobozieff	Discovery of *t* haplotype in mice through isolation of lab mutant (T) that revealed phenotype of *t*
1928	Gershenson	Discovery of driving X chromosome (*SR* in *Drosophila obscura*) and clear statement of potential threat to species
1941	Lewis	Clear statement of cytoplasmic male sterility as an example of conflict between cytoplasmic and nuclear DNA
1942	Rhoades	Discovery of female meiotic drive (chromosomal knobs in maize)
1945	Östergren	Clear statement of the parasitic nature of B chromosomes
1952	McClintock	Discovery of transposable elements (*Ac* and *Ds* elements in maize)
1955	Ephrussi, deMargerie-Hottinguer and Roman	Discovery of overreplicating but defective mitochondria (in yeast)
1957	Sandler and Novitski	First review of meiotic drive as an evolutionary force
1959	Sandler, Hiraizumi and Sandler	Discovery of *Segregation Distorter* (*SD*), an autosomal killer on the second chromosome of *Drosophila melanogaster*
1961	Brown and Nelson-Rees	Radiation experiments first show that it is the paternal chromosomes in male scale insects that are inactivated (and later eliminated), hence PGL
1961	Schultz	Discovery of hybridogenetic reproduction (haploid genome from one species excludes that from another during gametogenesis; in topminnows)
1964	Brown	Clear statement that PGL can evolve by drive ("automatic frequency responses")
1966	Hickey and Craig	Discovery of driving Y chromosome (in mosquitoes) and investigation into its potential use for population control
1967	Hamilton	In-depth treatment of sex chromosome drive, with special reference to population structure
1972	Hamilton	Clear statement of internal conflict based on differential relatedness
1976	Fredga et al.	Discovery of X*Y females (in lemmings)
1977	Bengtsson	Clear statement that X* in lemmings drives
1977	Lyttle	Creation of a selfish element for population control (translocation of *SD* and Y chromosomes, in laboratory populations of *Drosophila*)
1980	Doolittle and Sapienza; Orgel and Crick	Clear criticisms of the phenotype paradigm as being undercut by drive and inadequate to explain genome parameters such as prevalence of noncoding DNA, transposable elements, and genome size

Table 1.3 (continued)

Year	Scientist(s)	Event
1980 –81	Eberhard; Cosmides and Tooby	Clear statements of organelle/nuclear and organelle/organelle conflict
1982	Rubin and Spradling	Transposable elements used to genetically engineer an animal (P elements and Drosophila)
1982	Jones and Rees	Comprehensive treatment of B chromosomes in all species (~1200)
1983	Charlesworth and Charlesworth	In-depth treatment of transposable element population biology
1984	Lyon	Drive of t caused by multiple distorters acting on responder locus
1986	Colleaux et al.	Characterization of a homing endonuclease (ω in yeast)
1988	Nur et al.	Discovery of unusually destructive B chromosome that kills all the As with which it is transmitted
1989	Haig and Westoby	First explanation of genomic imprinting in terms of kinship theory (in plants and mice)
1990	Daniels et al.	Discovery of horizontal transfer of a selfish element (P elements and Drosophila)
1991	Barlow et al.; DeChiara et al.; Haig and Graham	Discovery of first two imprinted genes in mice (and their explanation in terms of kinship theory)
1991	Lyttle, Sandler, Prout and Perkins (eds.)	Entire issue of The American Naturalist devoted to meiotic drive
1992	Beeman, Friesen and Denell	Discovery of maternal effect killers (in beetles)
1992	Tinti and Scali	Discovery of androgenesis (in stick insects)
2002	Craig et al. (eds.)	Recent edited volume on transposable elements, 30 review chapters, >1000 pages, thousands of references
2003	Montchamp-Moreau and Atlan (eds.)	Entire issue of Genetica devoted to intragenomic conflict
2004	Camacho (ed.)	Special volume of Cytogenetics and Genome Research devoted to Bs

the first appreciation of its selfish nature. We have also noted some important conceptual advances, broad syntheses, and, more recently, the devotion of entire journal issues to the topic, signaling the degree to which selfish elements are now seen as a subject of significance. Advances in the 1990s were so rapid and extensive that making a list of milestones after 1990 would be an increasingly arbitrary task.

The first selfish elements discovered were those that were easiest to observe. The failure of a plant to produce pollen is easily detected, and maternally inherited mutations with this effect (now called cytoplasmic male sterility, CMS) were first discovered in 1906, by Correns. Likewise, B chromosomes are easily seen under the microscope and they were first described in the same year, by Wilson. In both cases, ease of observation has led to them being described in hundreds (CMS) or thousands (Bs) of species since then. Other early discoveries were driving X chromosomes in *Drosophila* (due to some crosses producing almost all daughters) and the *t* haplotype of mice (observable due to the happy accident of creating tailless mice in a particular mutant background). More recently, degenerate PCR (polymerase chain reaction) primers and genome-sequencing projects are making transposable elements easy to detect, and information on their abundance is growing rapidly.

Some selfish genetic elements have been intensively studied because of their practical importance or medical relevance. Cytoplasmic male sterile systems, for example, have proven valuable to crop breeders in producing hybrid seeds without the need to hand-emasculate plants that otherwise might self-fertilize. Early work on driving Y chromosomes in mosquitoes was motivated in part by their potential use in population control. And much work on transposable elements has been stimulated by the discovery that they can be used to genetically engineer *Drosophila* and other species. Transposable elements have also been well studied because our own genomes contain 2 active families, which occasionally cause mutations and thus disease. In addition, some transposable elements are related to retroviruses such as HIV and use similar methods of proliferation. Much research into genomic imprinting has also been motivated by its involvement in several genetic disorders in humans. In the dozen years since the discovery of the first 2 imprinted genes in mice in 1990, more than 60 such genes have been described in mice (and almost as many in humans).

For much of the 20th century, conceptual advances in the field were only loosely correlated with the empirical discoveries. Perhaps the first discussion of genes increasing in frequency purely because of biased transmission involved the driving X chromosome of *Drosophila pseudoobscura* (Gershenson 1928; Sturtevant and Dobzhansky 1936). Here the realization was coincident with the discovery of the phenomenon. But for many other selfish genes, decades have passed between first discovery and the realization that the gene can spread purely through a transmission advantage. Cytoplasmic male sterility was discovered in 1906, but it was not until Lewis's paper in 1941 that such mutations were clearly recognized to have a transmission advantage. B chromosomes were first described in 1906, but their parasitic nature was not clearly appreciated until Östergren's far-seeing paper in 1945. Transposable elements were first described in 1952, but it was not until 1980 that papers appeared suggesting they were selfish. More recently, the interval between discovery and interpretation has shortened: For example, the first imprinted genes were discovered in mice in 1990, and their fit to kinship expectations noted only a year later.

As the study of selfish genetic elements has matured, a specialized vocabulary has developed–though, as usual, much of the terminology came long after the central ideas were first conceived. The term "selfish gene" was first introduced as the title of a popular book by Dawkins (1976) to capture the increasingly gene-centered view of evolution that developed during the 20th century; he used it to refer to all genes. But genes that affect other individuals can be cooperative, altruistic, and even spiteful, while within individuals most genes are cooperative. Thus, in the scientific literature "selfish gene" is now used almost exclusively to refer to the minority of genes that spread at a cost to the organism. To be more inclusive about the underlying molecular structure, which can range from a gene fragment to a complete chromosome, or set of chromosomes, we prefer the more general term "selfish genetic element" (Werren et al. 1988). Obviously, in using the term, no comment is being made on the morality of the DNA concerned. Other authors have used "outlaw gene" (Alexander and Borgia 1978), "ultraselfish gene" (Crow 1988) and "self-promoting genetic element" (Hurst et al. 1996a) to denote virtually identical concepts, and long ago Östergren (1945) referred to B chromosomes as "parasitic."

Another key term is "drive," which we use to denote the greater-than-Mendelian ("super-Mendelian") transmission of a selfish genetic element. "Drag" is the opposite: less-than-Mendelian inheritance ("sub-Mendelian"). Sandler and Novitski (1957) first introduced the term "meiotic drive" to refer to increased transmission at female meiosis due to polar body avoidance; it is now used more broadly, and for those cases in which the phenomenon does not occur at meiosis, it is appropriate to shorten it to just "drive." "Transmission ratio distortion" and "segregation distortion" are also commonly used, but these consume more space while failing to indicate whether the effect is positive or negative.

Design of This Book

The subject of selfish genetic elements has never been reviewed before in a book. It is now a very large subject, encompassing tens of thousands of scientific studies on thousands of species and touches every corner of genetics. Work on the various categories has largely proceeded independently, with rare review articles (for example, Werren et al. 1988; Hurst et al. 1996a; Hurst and Werren 2001) and some edited volumes (*American Naturalist, Genetica*). This book introduces the subject of selfish genetic elements in all its aspects:

Molecular
Cytogenetic
Genetic
Physiological
Behavioral
Comparative
Populational
Evolutionary

It would be very nice if all these lines of evidence were available for a single selfish element, but they never are. Consider B chromosomes. Much is known about their cytogenetics and population biology, but very little is known about their genetics or molecular structure. By contrast, a great deal is known about the molecular structure and behavior of homing endonuclease genes, but very little on their population biology or effects in nature.

Considering the entire subject together permits us to concentrate on particular elements in order to solve given problems. For example, it has long been appreciated that outbreeding and sexual reproduction are, in theory, congenial to the spread of driving elements. For most elements the evidence is weak or nonexistent; but for B chromosomes, a data set on some 250 species of flowering plants, characterized for breeding system and presence of Bs, can be used to demonstrate a strong positive association (see Chap. 9). This finding, in turn, can be linked with experimental work on homing endonuclease genes showing a similar correlation (see Chap. 6).

In reviewing the entire subject, the evidence always seems to be just short of what we need. Whether it is the molecular mechanism of the selfish element, the magnitude of its phenotypic cost (if any), the reasons for its particular frequency, or its long-term evolutionary effects on the larger genetic system, the evidence available is often inadequate to answer even some of the most elementary questions. This partially reflects the degree to which this subject is only now cohering as a unitary whole, with its own logic and interconnected questions. We review the evidence in more detail than we can make sense of and we describe logic beyond what the evidence will verify. Our aim is to strike a balance between what is known and what is not, the better to invite others to join in generating the missing logic and evidence.

Throughout the book we emphasize evolutionary logic coupled to knowledge of the underlying mechanisms of the selfish elements, whenever possible based on molecular evidence. For each kind of selfish element, we have been drawn to an interconnected set of questions, though for any given element many of the questions remain unanswered:

How does it gain its selfish advantage?
How did it arise?
How long ago?
What are its effects on the larger organism?
How fast is it expected to spread?
What is its frequency within a species?
What determines this?
Why is it found in some species and not others?
What counteradaptations has it stimulated in the rest of the genome?
What other effects has it had on the host lineage?

Each chapter typically deals with a major category of selfish element, though chapters may combine more than one. Each chapter is designed to stand on its own, so you can easily skip ahead to a topic of particular interest. This approach is facilitated by the fact that the literatures on these elements have often developed in isolation from each other. We try to draw attention to interconnections wherever possible, but the book can profitably be read in almost any order. The final chapter seeks to review the evidence topically for all elements with a special eye to the future.

TWO

Autosomal Killers

WE BEGIN OUR ACCOUNT with a detailed treatment of one selfish element, the *t* haplotype in mice. It is a killer—or, at least, a disabler—more specifically, a gamete killer. It disables (kills function in) sperm not containing it and thus gains an advantage (drives) in single inseminations of females. The *t* has existed for at least 3 million years. It has grown steadily in size since its origin to become more than 1% of the mouse genome. Its spread has generated adaptations on both sides, improvements in *t* action, as well as evolution of countermeasures by the rest of the genome. The *t* has been studied for the past 75 years and a great deal is known about it, although unfortunately on several key points the information is contradictory.

At the same time, the *t* is only one of many examples of killers. Another well-studied case is *Segregation Distorter* in *Drosophila,* and there are others in both fungi and plants (Table 2.1). Killing can also act against offspring lacking the killer, as in maternal-effects killers. Genes can discriminate against pollen grains not containing copies, and so on. We shall also consider the opposite possibility, that a gene may give a benefit preferentially to those with a copy of itself. This is perhaps especially likely in intimate relations such as between mother and fetus. Many of these examples have been discovered only recently (since 1990) so that evidence on them is fragmentary; but by one way of looking at the matter, the majority of species may harbor at least one of the kinds of killers described here, while there must be many

Table 2.1   Gamete killers

ANIMALS

Mus	*t* haplotype	In *t*-heterozygous males, wildtype sperm are incapacitated and *t* sperm fertilize ca. 90% of eggs. Transmission is normal in females. *t*/*t* homozygotes are invariably male sterile and often lethal in both sexes. Found in *M. musculus* s.l. (i. ssp. *musculus, domesticus, castaneus,* and *bactrianus*) at frequencies of about 5% (Lyon 1991, Ardlie and Silver 1996, 1998).
Drosophila	*Segregation Distorter*	In *SD*/+ heterozygous males, + sperm are incapacitated and *SD* sperm can fertilize 95–99% of eggs. Transmission is normal in females. *SD*/*SD* homozygotes may or may not be viable and fertile. Found in *D. melanogaster* at frequencies of 1–5% (Temin et al. 1991).

FUNGI

Neurospora, Giberella, Podospora, Cochliobolus	*Spore killers*	In *Sk*/+ heterozygous asci, + ascospores are killed and only *Sk* ascospores are viable. *Sk*/*Sk* homozygotes are normal. Some alleles are neutral, neither killing nor being killed. Multiple *Sk* genes have been found in *Neurospora* (i. *intermedia, sitophila,* and *celata*) and *Podospora anserina,* and 1 gene is known from each of *Giberella fujikuroi* (= *Fusarium moniliforme*) and *Cochliobolus heterostrophus.* Frequencies vary among species from <0.1% to 80% (Turner and Perkins 1991, Raju 1994, van der Gaag et al. 2000).

PLANTS

Lycopersicon	*Gamete eliminator*	*Ge* is a locus in the proximal heterochromatin of chromosome 4. In *Ge*^{p}/*Ge*^{c} heterozygotes, *Ge*^{c} pollen and ovules are killed and *Ge*^{p} is transmitted to >95% of viable pollen and >85% of viable ovules. *Ge*^{n} is neutral. Found in *L. esculentum* and *L. pimpinellifolium* (Rick 1966, 1971).
Lycopersicon	*X*	In *L. esculentum* × *pimpinellifolium* hybrids, *X*^{esculentum}–bearing pollen are killed and *X*^{pimpinellifolium} is transmitted to about 95% of progeny. Transmission is normal through females (Bohn and Tucker 1940, Alexander 1973).
Nicotiana	*Pollen killer*	In *N. tabacum* plants with a particular *N. plumbaginifolia* chromosome present as a supernumerary, pollen not containing the supernumerary is killed, and it is transmitted to ca. 95% of progeny. Transmission through females is about 15% due to loss of the supernumerary at meiosis (Cameron and Moav 1957).

Table 2.1   (continued)

Triticum	Ki	In *T. aestivum vulgare* var. Chinese Spring × var. Timstein hybrids, pollen containing the $Ki^{Timstein}$ allele are killed and $Ki^{Chinese\ Spring}$ is transmitted to >90% of progeny. Transmission is normal through ovules. Some other strains are neutral (Loegering and Sears 1963; see also Nyquist 1962).
Aegilops	Cuckoo chromosomes, gametocidal chromosomes	In *Triticum aestivum* plants with particular chromosomes from particular *Aegilops* species, either as substitutions or as additions (= supernumeraries), pollen and ovules not containing the *Aegilops* chromosome are killed by many chromosome breaks and the *Aegilops* chromosome is transmitted to 98–100% of progeny through both pollen and ovules. Possibly due to derepression of transposable elements (Finch et al. 1984, Endo 1990, Nasuda et al. 1998, Friebe et al. 2003).
Oryza	S_1 & S_2	*O. glabberima* appears to be fixed for a pollen killer gene S_1 (linkage group 1, chromosome 6), which acts upon the homolog found in *O. sativa*. Depending upon the genetic background, S_1 can also kill ovules. Recombination is suppressed near S_1 in the hybrid, perhaps indicating an inversion. Another locus on the 8th linkage group acts as a pollen killer. Finally, *O. sativa* appears to be fixed for a pollen killer at a different locus, S_2, which kills pollen carrying the homolog from *O. glabberima*. The result is a high degree of hybrid sterility (Sano et al. 1979, Sano 1983, Sano 1990).
Agropyron / Lophopyrum	Sd-1	A segregation distorter factor *Sd-1* on chromosome 7 of *A.* (=*L.*) *elongatum* and *L. ponticum* kills ovules without it, in a wheat background (Dvorak 1980, Scoles and Kibirge-Sebunya 1982, Dvorak and Appels 1986).

other kinds of killers undiscovered in the species we do cover. We concentrate here on how each element gains its selfish benefit and on the consequences of its spread—for itself and for the larger genome.

The *t* Haplotype

The *t* haplotype in mice is a variant form of chromosome 17 that shows drive in males but is transmitted normally through females (Fig. 2.1). That is, in single matings by males it is transmitted to more than one-half of offspring—in fact, to about 90% (reviewed in Silver 1993, Lyon 2003; see also Ardlie and Silver 1996). This effect was first thought to be caused by a single

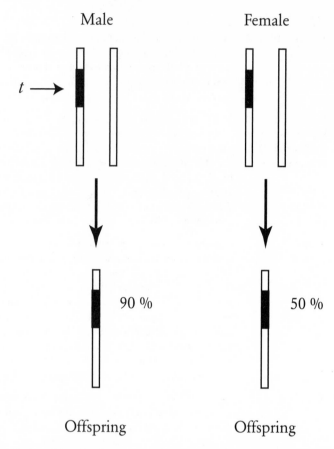

Male   Female

$t \longrightarrow$

90 %   50 %

Offspring   Offspring

Figure 2.1 Drive of the *t* haplotype in male mice. A male mouse heterozygous on his 17th chromosome for a *t* haplotype transmits the *t* to 90% of offspring. By contrast, *t* transmission is normal (50%) in female heterozygotes.

gene, hence the original term "*t* allele." The allele was believed to give superiority in swimming ability to *t*-bearing sperm. Later, it was imagined that in the face of strong drive, a simple form of group selection acted to keep *t* numbers low.

We now know that all of these notions are false. Far from being a single gene, the *t* haplotype spans more than one-third of chromosome 17. It gives its own sperm no special abilities but acts spitefully to damage sperm not containing itself, thereby gaining an advantage when thrown into competition with them. Finally, its numbers in nature are far too low to generate situations in which group selection is likely. Instead, *t* numbers are probably held in check in part by social forces (such as multiple mating and inbreeding) and by degenerative effects (for example, as caused by its very limited ability to recombine).

We begin with a detailed description of the *t* itself. How was it discovered? What does it consist of? How does such a long stretch of DNA inherit as a single unit? How does it act destructively to achieve drive, and how is it protected against its own destructiveness? What is the evolutionary history and current distribution of the *t*? What are the forces unleashed on itself— and on the larger organism—by its spread? And what exactly determines its numbers in nature?

Discovery

The *t* was discovered in the 1920s when a new mutation happened to be isolated in the laboratory that revealed whether a mouse was carrying a *t* (Dobrovolskaïa-Zavadskaïa and Kobozieff 1927). Called T, for Tail-factor, the mutant caused a normal non-*t* mouse (+/+) to develop a small tail, but caused a +/*t* heterozygous mouse to develop without any tail at all. If the tailless mouse was a male, he propagated his *t* chromosome to more than half his progeny—that is, the *t* showed drive.

A couple of features of this story are noteworthy. Just like the *t*, most selfish elements have little effect on the external phenotype. This in itself is an important observation. It suggests selection on both the element and the larger genome to reduce the negative effects at the individual (phenotypic) level. Because selfish elements are so often without obvious phenotypic effects, they are also usually uncovered only through careful genetic work in the laboratory and are, therefore, known preferentially from genetically well-studied species, such as the mouse. Indeed, as we shall see, 3 other possible killers (maternal effect) are also known from the mouse, not to mention a host of other selfish genetic elements.

The system had another feature that turned out to be useful in lab propagation. T/T individuals are embryonic lethals, as are *t*/*t* individuals (here the T mutant is on the + chromosome). Thus, breeding T/*t* mice with each other generated only more T/*t* mice, because both of the homozygotes perish. This is called a balanced lethal stock and permits a particular gene or haplotype to be maintained indefinitely, with little effort (though selection may continue to operate in the lab). Nevertheless, the absence of phenotypic effects meant that for decades an individual collected from nature could only be scored for the presence of *t* by breeding it to one of these tester stocks. Biases and errors in ascertainment were almost inevitable. Only recently could mice be scored by snipping a piece of the tail and searching for *t*-specific DNA (Schimenti and Hammer 1990, Ardlie 1995). Despite the difficulty in recognizing *t* mice, an enormous amount of work has been done on the *t*, almost all of it in the lab, with many interesting findings.

Structure of the *t* Haplotype

The *t* is about 1.2% of the entire mouse genome. Located near the centromere, it is 30–40Mb long and has scores, if not hundreds, of genes. (On the basis of length alone, it is expected to have more than 300 genes.) The *t* inherits as a single unit because it is almost completely covered by 4 nonoverlapping inversions that act to prevent recombination between the *t* and wildtype chromosomes (Fig. 2.2). They reduce recombination from an expected value of 20% to an actual value of about 0.1%. As we shall see in detail, their primary function is to lock together a series of distorting genes (each of which increases the rate of transmission) with a responder-insensitive allele that provides protection from the disabling effects of the distorters. Each of the 3 known drivers is located in a different inversion, and the responder-insensitive allele is located very close to the inversion that arose first.

Most *t* haplotypes carry 1 recessive embryonic lethal (and only 1) such that an individual homozygous for that *t* perishes *in utero*. Indeed, a *t* haplotype is characterized by the lethal it carries, and 16 complementing *t* haplotypes have been described in which complementing haplotypes have different recessive lethals, x and y, so that t^x/t^y individuals survive (Klein et al. 1984). But such males are invariably sterile. Females are fertile but have low fecundity, at least in the lab. When the *t* was thought to be a single locus, the existence of a series of lethals acting at different stages of embryonic life suggested that the *t* might be a master-control locus for early mouse development (and funding was sought on this basis).

The long-term consequences of linkage are presumably negative. The spread of the *t* complex is forcing one-third of chromosome 17 to inherit as

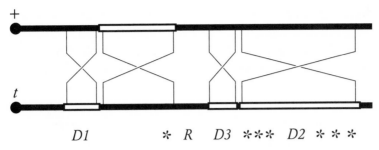

Figure 2.2 *t* haplotype and wildtype (+) forms of chromosome 17 in the mouse. The 4 inversions that define the *t* region are shown, along with the driving genes discovered to date *(D1–D3)* and the responder *(R)*. Asterisks denote *t*-associated lethals. Adapted from Silver (1993) and Lyon (2000).

a single asexual unit (recombination is possible between *t*'s in t^x/t^y females, but their numbers are few and fecundity low). We will discuss whether this high degree of linkage explains the frequency of lethals later; but, in general, selfish genetic elements almost invariably set in place forces that cause their own deterioration. This can come in a variety of ways, but the *t* haplotype illustrates one: increasing linkage (e.g., via inversions) provides a way for the selfish element to grow but robs the now-linked DNA of the benefits of recombination.

The evolutionary growth of the *t* has other implications. More loci become available to mutate in both advantageous and disadvantageous ways. Greater size means that a higher frequency of *t*'s will contain lethals each generation and there will be a greater chance that any given one may be randomly associated with a new driving haplotype and thus may be carried to high frequency. There are more loci available for beneficial effects, more places to appear that give greater drive, more targets for hostile action, and so on.

History and Distribution

The two major inversions of the *t* have been dated and these reveal an ancient origin. Comparing the degree of sequence divergence between DNA within each of these inversions and the comparable wildtype sequences suggests that the proximal inversion arose about 3 million years ago (mya), and the distal one about 1.5mya (Hammer and Silver 1993). If the inversions were selected to prevent recombination between killer and nonkiller chromosomes, the killer genotype must have arisen more than 3mya and within the proximal inversion. Note that the responder-insensitive allele is located very close to the proximal inversion (though not within it) so that this first inversion could have tightly linked a distorter allele with a responder-insensitive one. Because there has been time for substantial coevolution between the *t* and the rest of the genome, complex coadaptations are likely for *t*-bearing individuals and those interacting with them.

Currently, *t* complexes are found in all 4 subspecies of *Mus musculus: musculus, domesticus, castaneus,* and *bactrianus.* These latter are thought to have diverged from each other about 0.5–1mya. However, the *t* complexes found within them are exceedingly similar, and their most recent common ancestor is thought to have existed perhaps 0.1mya (Morita et al. 1992, Hammer and Silver 1993). This indicates that a *t*-complex variant has recently spread across subspecies boundaries, perhaps due to some beneficial mutation that has swept through the 4 subspecies.

Genetics of Drive

A key part of the *t* haplotype is the responder-insensitive allele *Tcr* that gives protection from the distorting effects of the drivers (Fig. 2.2). It is a novel protein-coding gene formed by the fusion of the front part of a *Smok* (sperm motility kinase) gene and the back part of an *Rsk3* (ribosome S6 kinase 3) gene (Herrmann et al. 1999); the resultant protein is 484 amino acids long. Transcription of this gene only occurs late in spermatogenesis, while the spermatids are elongating but still connected by cytoplasmic bridges allowing the passage of soluble components. Despite these cytoplasmic bridges, *Tcr* shows haploid-specific expression, that is, only sperm carrying it are rescued from the killing. The reason for this is not yet known, but one possibility is that the mRNA is compartmentalized and the protein quickly becomes associated with a membrane, perhaps even as it is being synthesized, as has been shown for other loci with haploid-specific expression (Zheng et al. 2001). An unusual feature of *Tcr* is that alone, in the absence of any drivers, it shows drag (~20% transmission) opposite a wildtype chromosome (Lyon 1991).

In addition to *Tcr*, there are 3 or more distorter loci that combine additively to cause *Tcr* to drive (Lyon 1991). These have yet to be unambiguously identified. Curiously, for at least 1 of the distorter loci, a chromosomal deletion behaves like the *t* allele, suggesting that the latter is itself a knockout (Lyon 1992, Lyon et al. 2000, Planchart et al. 2000). If so, this rules out perhaps the most obvious possible mechanism for drive, in which the drivers produce some toxin that is distributed to all sperm and the *Tcr* allele produces some antidote that remains associated with the sperm carrying it. Perhaps instead the distorter loci are members of a multigene family making partially or fully redundant products, of which the amounts produced are critical. Knockouts of these genes progressively impair sperm function, but *Tcr* rescues the knockouts in a haploid-specific manner, perhaps by making something similar itself. That is, the *t* complex may be failing to make an important protein at a time or in a form that is shared among meiotic products, but it makes a private supply only for itself. However, this "stash" is not sufficient by itself–perhaps it comes too late–and hence *t* homozygotes are sterile. Too much of the product is also harmful–hence the drag of *Tcr* in a wildtype background. Alternative mechanisms are suggested by Herrmann et al. (1999).

Whatever the molecular mechanism, the net result is sperm dysfunction. *t* heterozygotes produce and ejaculate equal numbers of the two types of sperm, but the +-bearing sperm are dysfunctional, probably due to a variety of mechanisms including motility defects (Olds-Clarke and Johnson 1993, Olds-Clarke 1997) and in some cases possibly a premature acrosome reaction, such that sperm reaching the egg no longer have the chemicals needed to dissolve its membrane (Brown et al. 1989, Mittal et al. 1989). There are hints that the degree of dysfunction depends on features of the female reproductive tract. For example, fertilization *in vitro* results in a normal transmission ratio (McGrath and Hillman 1980). Moreover, if copulation is experimentally delayed to occur just prior to ovulation (rather then being allowed to occur, as is normal, about 6 hours prior), then the rate of drive decreases substantially (Braden 1958). The sabotage must occur during spermiogenesis, as sperm from wildtype males is fully functional when mixed with sperm from *t*-heterozygous males prior to artificial insemination (Olds-Clarke and Peitz 1986).

Importance of Mating System and Gamete Competition

The strategy adopted by the *t* is one of spite: getting ahead by killing the competition. How effective the strategy is depends on the extent to which the benefits of killing accrue to the killers themselves, and not to the population as a whole. This in turn depends on the lifecycle and mating system of the host species (Hartl 1972). For example, in a completely monogamous rodent, the only sperm competition is between the sperm of a single male, and, in the ideal situation, any eggs not fertilized by wildtype sperm will be fertilized by killer sperm from that same male. In such a case, there is complete compensation for the killed sperm, the heterozygous male suffers no loss in fertility, all the benefits of killing accrue to the killers, and the strength of selection for the killer is maximized. But house mice are not completely monogamous: in a survey of 179 pregnant females from wild populations, 10 wildtype females were found to have mated with a *t*-heterozygous male; and of these, 3 had also mated with another male (Ardlie and Silver 1996). Though the numbers are small, they suggest multiple paternity is not a rare occurrence (~30%). Thus, some of the benefits of killing wildtype sperm are shared with sperm from a different male that do not carry *t* (and this sharing is even greater if the *t*-bearing sperm show any damage from the action of their distorters). The return benefits of killing are presumably even less for pollen killers, and lower still for spore killers, because spores from a single fungus are not usually in competition to fertilize.

Fate of Resistant Alleles

The degree to which the benefits of reduced competition accrue only to killer alleles is also important for the fate of resistant alleles that neither kill nor are killed. Such alleles have been observed for a number of gamete killer systems (Table 2.1), and have been recovered in the lab for the *t* complex. Even if they do not preexist when the killer first arises, we would expect resistant alleles to arise in all killer systems as degenerate forms of the killer that have one of the phenotypes but have lost the other. If the host organism compensates completely for the lost gametes and all the benefits of killing accrue to the killer itself, then resistant alleles would be selectively neutral and, if rare, could be lost by drift. But if the benefits of killing were shared more widely, resistant alleles would also benefit and would increase in frequency along with the killer. If the sensitive allele eventually disappears, killing would no longer be observed, and there would be a cryptic polymorphism of killer and resistant alleles. If the killer allele has additional costs for the host (for example, reducing fertility when homozygous), a resistant allele that is otherwise normal would increase in frequency and go to fixation, driving the killer to extinction.

It is essential, then, for the long-term persistence of a killer that there not be resistant alleles that are "otherwise normal," and the higher the cost of resistance, the better for the long-term fate of the killer. As noted earlier, alleles resistant to the *t* complex that have arisen in the lab by recombination (i.e., that carry *Tcr* but none of the drivers) show very low transmission rates in males when paired with wildtype chromosomes (Lyon 1984, 1991). Whatever the mechanistic basis for this, it presumably explains the low frequency of resistant alleles in natural populations and thus has probably contributed to the long-term persistence of the *t* complex in mouse populations. Note too that because resistant chromosomes are rare in the population, selection on unlinked genes to ameliorate the drag is weak.

Selection for Inversions

Inversions block recombination and, to understand their spread on the *t* haplotype, we must first consider the effects of recombination on killer systems. In the *t* (and other systems), drive is not due to a single locus, but rather to an interaction between at least 2 loci (killer and resistant). The killer complex will only become established in the population if its increase in frequency due to drive outweighs the loss in frequency due to recombination with wildtype chromosomes. This will act as a selective filter on the establishment of killer complexes. If the killer is too far away from the resistant allele, it will not be able to spread through a population. Mathematical modeling shows that in species with approximately equal recombination in the two sexes, the rate of recombination between killer and resistant must be less than 1/3 in order for the killer to spread (Charlesworth and Hartl 1978). These results are for the situation most favorable to the killer (resistance rare; no effect on host fitness), and in other cases the conditions on linkage are more restrictive. Thus, tight linkage facilitates the spread of a killer. This is presumably an important constraint on the evolution of 2-locus killers, with recombination creating an important "sieve" for the evolution of killer complexes (Charlesworth and Hartl 1978). Indeed, this presumably explains why most known killers are near a centromere, where recombination is often reduced (*t, SD, Sk-2^K, Sk-3^K, Ge^p;* see Table 2.1).

Once a 2-locus killer system becomes established in a population, the frequency of recombination between killer and target may itself evolve. If both killer-insensitive (*K, ins*) and wildtype-sensitive (+, *sen*) chromosomes are present in a population, recombination between them will produce killer-sensitive "suicide" chromosomes. The production of such chromosomes will select for reduced recombination between killer and target. If an inversion arises on either chromosome that prevents recombination between the two, the inversion can go to fixation (among that chromosome type). Moreover, there is a pleasant symmetry about which chromosome will evolve the inversion (Charlesworth and Hartl 1978). The selection coefficient for an inversion on a killer-resistant chromosome is proportional to the frequency of the wildtype-sensitive chromosome, and vice versa. Thus if (+, *sen*) chromosomes are, say, 10 times more frequent than (*K, ins*) chromosomes, an inversion is 10 times more likely to arise on a (+, *sen*) chromosome, but the selection coefficient will be one-tenth of that for a (*K, ins*) inversion. Because the probability that a beneficial mutation will be established is proportional to its selection coefficient, these differences cancel out, and the overall probability of an inversion being established is the same for the 2 chromosomes. In the case of the *t* haplotype, 3 inversions arose on the driving chromosome and 1 on the wildtype chromosome.

This selection for reduced recombination applies only to modifiers of recombination that are closely linked to the killer and target loci (of which inversions are a special case). For modifiers of recombination that are distantly linked to the killer complex, or on another chromosome, there will be selection for increased recombination, to break it up. Indeed, using a 3-locus model, Haig and Grafen (1991) argue that the breaking up of incipient killer systems is an important function of recombination, in the sense of selecting for observed levels of recombination. They do not, however, quantify the strength of selection for such modifiers for reasonable rates of origin of gametic killers. Nevertheless, the modeling does demonstrate that there is a conflict of interest over recombination, with linked genes selected to have less recombination and unlinked genes selected to have more. This conflict seems relatively easy for the linked genes to "win," just by having an inversion, for then even unlinked genes will not be selected to force recombination, as that would cause aneuploidy (either more than 2 copies of a gene at a given locus or fewer).

Recessive Lethals in *t* Complexes

Though inversions can be selected for because they bind together genes directly involved in the killing phenotype, as a side effect they also bind together many other genes that are wholly unrelated to killing. The resulting absence of recombination is expected to have the long-term effect of reducing the efficacy of natural selection on these genes, which may therefore degenerate over time (due to the accumulation of deleterious mutations and the failure to keep up with a changing environment). A reduced efficacy of selection may explain the high frequency of recessive lethal mutations in *t* complexes. As noted earlier, most *t* complexes recovered from natural populations contain a recessive lethal, though different haplotypes may have different lethals, and a total of 16 different recessive lethal mutations have been recovered from *t* complexes worldwide (Klein et al. 1984).

These lethals may be positively selected, either by group selection (because lethals are less likely to lead to population extinction than steriles; Dunn and Levine 1961), or by kin selection (because early death of homozygous offspring saves mothers from investing in sterile males; Charlesworth 1994). The conditions that allow group selection to have a strong effect on *t* frequencies do not seem to be satisfied in nature (Durand et al. 1997), but the effects of kin benefit are less clear. Charlesworth's (1994) model shows that there must be a high level of compensation (such that a lethal embryo is replaced by 85% of a new one) and near-complete sterility for a *t* lethal to invade and displace a nonlethal *t*. It is easy to see why: lethals strike both sons and daughters, but only sons are sterile. Even then the selection pressure is very weak (~0.1%) and a polymorphism results that invites the invasion of a new lethal, and so on (see also van Boven and Weissing 1999, 2001). But in Charlesworth's model, the homozygous *t* female is given normal fitness because she is not sterile, while her actual reproductive success is probably much lower, so that compensation is more likely to be beneficial.

Lethals differ greatly in their time of action during embryonic life and, therefore, the cost they impose on the mother (Bennett 1975). For example, t^{w5} acts in the blastocyst stage, before implantation, while t^{w12} acts late in gestation. By pairing heterozygotes with either the same *t* or complementing *t*'s, laboratory studies suggest that there is a 35% reduction in litter size when the *t* with a costly lethal mates with itself instead of with a complementing *t* male, but only a 5% reduction for the inexpensive lethal (Lenington et al. 1994). The best evidence comes from nature and shows that matings between heterozygotes with the same lethal produce litters that are about 30% smaller than matings between mice with complementing

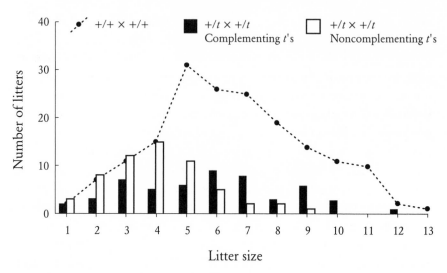

Figure 2.3 The relative proportion of different-sized litters produced by 3 crosses. Dotted line: cross between (mostly) wildtype mice. Black bars: cross between +/t individuals in which the t is complementing. White bars: cross between +/t mice in which the t is noncomplementing. Note that litters from noncomplementing t's show a sharp reduction in size. Adapted from Ardlie (1995)

lethals (Fig. 2.3). Also, females prefer to mate with males having a complementing t haplotype, suggesting no net gain from the lethals (Coopersmith and Lenington 1990). As expected, the mating preference of females is stronger against the costlier lethal but is still expressed against the early-acting one (Lenington et al. 1994). In North America, the t with the highest frequency (of 3) is t^{w5}, which has the earliest-acting lethal (Bennett 1975); but in Europe, lethals seem to be clumped by geographic region and are in some places absent from t's, without any known pattern reflecting stage of action (Klein et al. 1984). It is noteworthy that, in nature, the highest frequencies of t haplotypes are seen in populations with semilethal or complementing t haplotypes, which generate functional t/t females (Ardlie and Silver 1996).

Thus it is unclear at this time whether lethals have been positively selected in mice. The kin effect—along with the low frequency of t's—certainly lowers selection against recessive lethals, and the earlier acting the lethal is, the stronger the corresponding effect.

Enhancers and Suppressors

Intragenomic conflict is expected over the intensity of killing itself (Eshel 1985, Crow 1991). Alleles on killer chromosomes are selected to enhance the killing (because the killing favors their own drive), while alleles on the homologous wildtype chromosome are strongly selected to suppress the killing (so they do not suffer drag). Thus, if mutations affecting the extent of drive arise at loci linked to the killer, we would expect the enhancing allele to be found disproportionately on the killer chromosome and the suppressing allele on the wildtype chromosome. As before, this situation selects for inversions that reduce the frequency of recombination between the killer and the modifier. Thus, killer complexes can be expected to expand along a chromosome over evolutionary time.

Suppressors of killing can be expected to evolve not only on the target chromosome but also on nonhomologous chromosomes, the latter being selected (more weakly than the former) to ameliorate any negative effects of the killing on fertility and to reduce the probability of being cotransmitted with a killer. Suppressors of the t complex have evolved in balanced lethal laboratory populations, both on the homologous chromosome and on the other chromosomes, reducing drive from 90–95% to 65–70%, or even lower (Bennett et al. 1983, Gummere et al. 1986). Curiously, such suppressors appear to be rare in natural populations (Ardlie and Silver 1996).

Not all the coevolution between t complexes and the rest of the genome need be antagonistic. For example, both parties may be selected to compensate for the reduced quality of sperm produced by t-heterozygous males. This may explain the tendency of t males to suppress the reproduction of $+/+$ males in seminatural groups (Franks and Lenington 1986). We would also predict compensatory increases in sperm production in t-heterozygous males.

t and the Major Histocompatability Complex

One of the loci tied up in the distal inversion of the t complex is the major histocompatability complex (MHC). This is a series of tightly linked genes that are involved in the immune response and that also affect kin and self-recognition, as well as mate choice. They are highly polymorphic. Some loci have more than 100 alleles and all mice are typically heterozygous at these loci (Nadeau et al. 1988). *t* haplotypes show a striking reduction in their MHC diversity, with very few MHC haplotypes, all of them unique to *t*'s (Shin et al. 1982, Silver 1982, Nizetic et al. 1984). Indeed, in 2 separate global samples, of 13 and 20 different *t* haplotypes, only 5 and 4 different MHC haplotypes were found, while wildtype MHCs are essentially never expected to be identical (Hammerberg and Klein 1975, Sturm et al. 1982). This reduced diversity presumably reflects the fact that the distal inversion linking the MHC to the killer arose only once. This linkage with the MHC is probably unfortunate from the *t*'s perspective. The highly polymorphic nature of the MHC on wildtype chromosomes indicates strong selection for rare alleles, yet the *t* complex is stuck with a limited diversity, and so will suffer selection against their MHC types as *t*'s become common in a population. (On the other hand, for *t* the most important matings are between a *t*-heterozygous male and a wildtype female; and having a different MHC allele than is found in the female may ensure that the male is never discriminated against and, in small inbred populations, he may even be positively preferred.)

Linkage with the MHC could also explain some results of mate choice experiments, in which *t*-heterozygous females were observed to prefer to mate with males carrying a different *t* allele over males carrying the same *t* allele (Lenington et al. 1994). *t*-heterozygous females also prefer wildtype males over *t*-heterozygous males—a choice that may also be controlled by the MHC (but see Lenington et al. 1992). Wildtype females have no obvious preferences for or against *t*-heterozygous males (Lenington et al. 1992; Williams and Lenington 1993).

Heterozygous (+/*t*) Fitness Effects: Sex Antagonistic?

The *t* haplotype (like many selfish elements) shows drive through one sex only. This has a very interesting consequence. Because *t* genes that benefit a +/*t* male's survival or reproduction will find these effects nearly doubled by drive, they will be more strongly selected there than similar genes in females. Put another way, a sex-antagonistic gene that gives a +/*t* male a 10% gain in fitness will be positively selected even if it is associated with a 15% decline in +/*t* female fitness. Sex-biased drive causes a new category of sex-antagonistic genes to be favored, namely, those whose net effect is negative but not so negative as to outweigh the gains in drive.

Early effects are not expected to be sex biased (Rice and Chippindale 2001), nor are they found to be: in nature t males and females survive equally well into adulthood (Ardlie and Silver 1998, Huang et al. 2001). In adulthood, $+/t$ mice are more asymmetrical (= fluctuating asymmetry) than are pure wildtype (Leamy et al. 2001)–a trait often associated with lower phenotypic quality and reproductive success–but when mice are separated by sex, only females are more asymmetrical (L. Leamy, personal communication). It has long been noted that female t mice seem less fit in the lab than t males. $+/t$ females produce 30% fewer nestlings in the lab than do $+/+$ females (Johnson and Brown 1969), while $+/t$ males show greater fitness than do $+/+$ males (Dunn et al. 1958).

Early measures of dominance relied on attack latency and attack frequency in staged encounters between males, and in this situation $+/t$ males are more aggressive than $+/+$ ones (Lenington 1991, Lenington et al. 1996). In very small experimental groups, $+/t$ males do as well as $+/+$ ones, while $+/t$ females do only a third as well as $+/+$ females (Franks and Lenington 1986). In addition, when dominant, $+/t$ males act more spitefully than do $+/+$ dominants, often killing the subordinate and always suppressing his reproduction. This makes sense if multiple matings by a female with both a t- and a $+$-male sharply reduces t transmission. Here, the evidence is unclear. In 3 experimental multiple inseminations, t transmission averaged 22% (Olds-Clarke 1997), and 3 putative double matings from the wild showed t transmission of only 10–20% (Ardlie and Silver 1996). Both of these values are below the expected frequency if it had never driven in the first place (25%); but in 19 litters sired in seminatural conditions by at least one $+/t$ and $+/+$ male, the t was transmitted to 36.4% of progeny (Carroll et al. 2004).

Measures of reproductive success are missing from nature and the most comprehensive set of data on heterozygous fitness effects comes from a recent study of 10 "seminatural" populations maintained for 10 months, each in an area of $49m^2$. This setup simulates large populations in a "fiercely competitive environment," especially for males (Carroll et al. 2004). Food is made available *ad libitum*, but there is intense male-male competition for dominance, territories, and the chance to fertilize females, with no chance for dispersal. In this situation, both sexes suffer if they are $+/t$, compared to $+/+$, but males more so. Female heterozygotes are smaller at weaning than wildtype females, and even more so at adulthood ($\sim10\%$); they have a lower chance of surviving to adulthood, and their reproductive output is 79% of +/+ when outbred and 33% when inbred. By contrast, males show no weight effect, but +/t males are less than half as likely to secure a territory. Furthermore, if they do secure a territory, they suffer higher mortality than their +/+ competitors. +/t males produce 64% of the pups of a +/+ male when outbred and 27% when inbred. Unfortunately, there is no way of comparing exactly the relative effects of the t on males versus females, because the seminatural environment greatly accentuates male-male conflict, while probably reducing female-female conflict.

There has been considerable labwork on the effects of parental haplotype on litter size, but the results are partly conflicting. If the father is +/t and the mother +/+, litter size is invariably reduced by about 20% (Lenington et al. 1994, Ardlie 1998, Carroll et al. 2004). But while one study shows a similar 20% decline when the +/t parent is female (Ardlie 1995), another finds an insignificant 2% decline (Carroll et al. 2004). The effect of the father carrying a t is surprising, but one interpretation is that it reflects a female bias against t-bearing males, which could act to increase lifetime investment in +/+ offspring. The effect of the mother carrying a t–if it is real–presumably indicates a general effect of the t on female fitness.

A key parameter in the evolution of the t may be the expected reproductive success of t/t females. If it is very low (as would be expected from sex-antagonistic effects–a double dose of female-harming genes), then there will be little opportunity for t recombination with itself (remember that t/t males are sterile). Instead, the t will evolve as an asexual entity, displacing other copies and competing with a sexual antagonist, the rest of the genome. As noted earlier, low fitness of t/t females also increases the likelihood of selection for early-acting lethals.

In summary, the evidence provides limited support for the importance of sex-antagonistic effects, our chief obstacle being the absence of key pieces of evidence from nature.

Accounting for t Frequencies in Nature

Mathematical modelers have long tried to account for observed t frequencies on the assumption that these frequencies are at some equilibrium (e.g., Bruck 1957, Petras 1967, Durand et al. 1997). t is a simple system to model because resistant alleles are not a significant fraction of the population, and so one need consider only 2 alleles, killer and sensitive. Surveys of natural populations show that the *t* complex is patchily distributed, with an average frequency in one large survey of about 5%, ranging (in samples larger than 20 individuals) from 0–71% (Ardlie and Silver 1998). While the most extensive data are for *M. m. domesticus*, frequencies in the other 3 subspecies do not appear to be much different (Huang et al. 2001, Dod et al. 2003). Despite there being only 2 alleles, many parameters are likely to affect the equilibrium frequency of *t*, including:

- The level of drive, which averages about 0.9 with little variation around the mean, as expected from the absence of resistance alleles or suppressors (Ardlie and Silver 1996).
- The homozygous fitness effects of the *t* complex. These are often lethal and, if not, then males are invariably sterile and females probably have low fitness (Lyon 1991).
- The heterozygous fitness effects of the *t* complex. As we have seen, the *t* is associated with substantial reductions in heterozygote fitness, though the underlying mechanisms can be complex and the precise amount can vary depending upon the social environment.
- The degree of inbreeding. The *t* can only drive in heterozygotes, and anything that reduces the frequency of heterozygotes—like inbreeding— will reduce the efficacy of drive and the equilibrium frequency (Petras 1967, Durand et al. 1997).

Unfortunately, estimates of the last 2 parameters are not available for natural populations, and all we can say is that plausible combinations of parameters can be found that give equilibrium frequencies that match the observed frequencies (Fig. 2.4). The high variance in the frequency of *t* among populations suggests some (or all) of these parameters are varying as well. The one known correlate of *t* frequency is that it is higher (averaging about 10%) in smaller, more ephemeral populations, and lower in larger, more stable populations ($\sim$0%; Fig. 2.5). One possible explanation is differential inbreeding: in a detailed study of 2 allozyme loci in natural populations, a greater deficiency of heterozygotes was found in large populations (N > 50) than in small populations (N < 50; Selander 1970). Large populations may be subdivided into inbreeding demes, while small ones may be more ephemeral and composed of immigrants from different source populations.

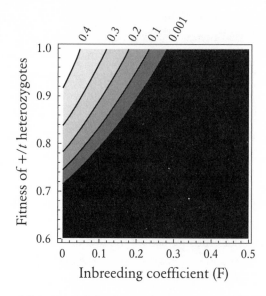

Figure 2.4 Expected frequency of *t* haplotype chromosomes as a function of the fitness of heterozygotes and the inbreeding coefficient. Contour lines connect points of equal frequency. Calculations assume that transmission in heterozygous males is 90%; the *t* is recessive lethal; and that fitness effects are equal in males and females. Note that the expected frequency increases sharply from very rare (0.001) to quite common (0.2) in a relatively small region of parameter space. The frequency of *t* in natural populations averages about 0.05, with substantial variation among populations.

Other Gamete Killers

Segregation Distorter in *Drosophila*

Segregation Distorter (SD) is a gamete killer on chromosome 2 of *D. melanogaster*. It shares many features in common with the *t* in mice, as well as some interesting differences. Heterozygous males *(SD/+)* transmit *SD* to 95–99% of their progeny (Lyttle 1991, Temin et al. 1991). Thus drive is even stronger than in the *t*. Again, the effect is destructive, but here it acts earlier, the wildtype sperm failing to develop properly and the male typically ejaculating almost exclusively *SD*-bearing sperm. Transmission is normal in females.

SD was first discovered in 1956 by a graduate student, Y. Hiraizumi, who was collecting *Drosophila* from nature and crossing them to genetically

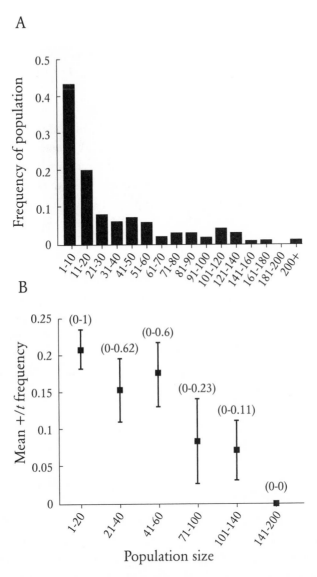

Figure 2.5 Population sizes of mice and frequency of the *t* haplotype as a function of population size. A. Frequency distribution of the sizes of populations assayed. Most commensal mouse populations are small. B. The frequency of +/*t* mice as a function of population size. Bars show standard errors; numbers in parentheses are ranges. *t*'s were present in about 40% of all populations sampled and their frequency increased as the population size decreased. Adapted from Ardlie (1998).

marked laboratory strains in order to study fitness variation on chromosome 2 (Crow 1979). Most of the time when a male was heterozygous for a natural chromosome and a laboratory one, progeny inheriting the 2 chromosomes were about equally frequent, as expected. But Hiraizumi noticed that in about 3% of matings, the male transmitted the natural chromosome to 95–100% of the progeny. It turned out that these chromosomes carried genes that destroy sperm carrying the wildtype (laboratory) chromosome. Like the *t*, *SD* has no obvious effect on the external phenotype, and the element would not have been discovered without direct genetic work.

The genetics of *SD* action is known in considerable detail. Like the *t*, *SD* is located near the centromere and it employs a set of driving elements that operate on a responder-sensitive allele to disable sperm development, while being protected itself by a responder-insensitive allele. The major driving locus and responder locus are tightly linked, separated by a map length of only 0.01 (that is, crossing-over occurs between them only 1% of the time). Most *SD*s have an inversion, and often more than 1; but unlike the *t*, these may differ between *SD* haplotypes and none is fixed. In one population an *SD* inversion increased from 1 in 6 chromosomes to 43 of 44 over a period of 35 years (Temin and Marthas 1984). Another important difference is that, when homozygous, *SD* chromosomes need not destroy one another. Flies may survive and, while some *SD* homozygotes cause male sterility, others are fully fertile: in a survey of 28 homozygous viable *SD* chromosomes, 18 were homozygous sterile and 10 fertile (Temin and Marthas 1984). Finally, *SD* appears to be a much younger element than is the *t*.

Molecular genetics of drive. The 2 most important loci in the *SD* system are *Responder (Rsp)* and *Segregation distorter (Sd;* Fig. 2.6; Kusano et al. 2003). *Rsp* is in the centromeric region of chromosome 2 and the wildtype allele is the target of the killing. It does not encode for anything, but instead consists of imperfect tandem repeats of a 240bp sequence, itself a dimer of two 120bp sequences (called *Xba*I; Wu et al. 1988; Pimpinelli and Dimitri 1989; Houtchens and Lyttle 2003). *SD* chromosomes have fewer than 20 copies of the repeat, while wildtype chromosomes have from 100 to 2400, and the sensitivity of the chromosome to the killing action is directly related to the number of copies of the repeat (those with 100–200 copies are largely insensitive, those with 700 sensitive, and those with 2400 supersensitive). In

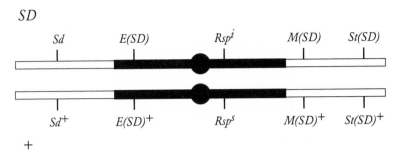

Figure 2.6 *Segregation Distorter (SD)* and wildtype (+) forms of chromosome 2 in *Drosophila melanogaster*. The order of the major genes involved in the system is shown. *Sd* is the main locus causing death of *Rsp^s*-bearing sperm, and *E(SD)*, *M(SD)*, and *St(SD)* all enhance its action and increase drive. Thick line is centric heterochromatin and solid circle is centromere; these are areas of low recombination. Adapted from Kusano et al. (2003).

short, absence of the repeats gives protection from the drivers, while an increasing repeat number gives increasing vulnerability.

The other main locus is the *Segregation distorter (Sd)* locus, which is *trans*-acting and encodes the killing action (Merrill et al. 1999). Compared to the wildtype sequence, the driving *Sd* allele has an extra 5kb, due to a tandem duplication. This resulted in the duplication of the *RanGAP* (Ran GTPase Activator Protein) gene, except the new copy encodes a protein that is missing the last 234 amino acids. It is this truncated *RanGAP* gene that causes the differential transmission of alternative *Rsp* alleles. Wildtype RanGAP protein is involved in the transport of proteins across the nuclear membrane. The truncated protein from the duplicated gene has the same enzymatic activity as the wildtype, but it goes to the wrong place in the cell. Instead of being at the outer periphery of the nuclear membrane, it is distributed diffusely in the cytoplasm and in the nucleus (Kusano et al. 2001). Having the RanGAP protein inside the nucleus is likely to disrupt nuclear transport, and thus spermatogenesis, but it is not yet clear why the dysfunction is limited to spermatids with many repeats at the *Rsp* locus. One possibility is that there is a protein that binds to (and thus is transmitted along with) the *Xba*I repeat—a *poison-tag* mechanism. This protein might be

RanGAP itself, or something that interacts negatively with this mislocalized protein. Whatever the mechanism, the activity of *Sd* leads to a failure of chromatin condensation and tail formation in spermatids carrying a sensitive *Rsp* allele as they individualize out of the syncytial state (Plate 1; Tokuyasu et al. 1977).

It is interesting that changing the promoter of the wildtype *RanGAP* gene so as to increase expression of the gene 20-fold in the male germline also causes the RanGAP protein to get into the nucleus and converts it into a killer with about 90% drive (Kusano et al. 2002). And, having a double dosage of a third locus, Enhancer of *SD* [*E(SD)*] also somehow causes wildtype RanGAP protein to get into the nucleus and causes drive of about 75%. Thus even for a single mechanism of drive, there can be multiple mutations that give the needed effect.

Enhancer of *SD* is a gene found naturally on *SD* chromosomes that, as the name suggests, enhances their drive. There are, in addition, 2 other such genes, Modifier of *SD* [*M(SD)*] and Stabilizer of *SD* [*St(SD)*], both of which are also linked to *SD* (Hartl and Hiraizumi 1976, Sharp et al. 1985; see Fig. 2.6). There are also loci on both other major chromosomes that affect the level of drive (Hiraizumi et al. 1994). These subsidiary loci have yet to be identified at the molecular level.

Thus there is little mechanistic similarity between *SD* and *t*. Both rely on interactions between at least 2 loci; but for *SD* the main killer is a new protein-coding gene and the resistant allele is a deletion, while the opposite is true for *t*–at least 1 of the killers appears to be a deletion, and the resistant allele is a new gene. Surprisingly, neither one uses a poison-antidote system. Such mechanisms are used by bacterial plasmids (Box 2.1), but why are they not more common in animals?

Resistant alleles. As we saw earlier with the *t*, it is important for the long-term persistence of a killer that resistant alleles have some cost, for otherwise they would spread to fixation and drive the killer extinct. For *SD*, resistant alleles with few *Xba*I repeats do exist in nature, but at least in the lab they appear to suffer reduced survival (Wu et al. 1989). A comparison of a wildtype chromosome with 700 copies and a deletion derivative with 20 copies showed a substantial selective advantage for the repeats of about 15% per generation. There was no effect on fertility. As expected, the relative fitness of the 2 *Rsp* alleles is reversed in the presence of *Sd*. The absence

---

BOX 2.1

Poison-Antidote Systems of Bacterial Plasmids

Many plasmids of bacteria have systems to ensure that if they are not transmitted to one of the daughter cells at cell division, that daughter cell will die (Summers 1996). For example, the R1 plasmid has 2 genes, *hok* (*h*ost cell *k*illing) and *sok* (*s*uppression *o*f *k*illing), which are encoded on opposite strands of DNA and overlap. The Hok protein is toxic: it associates with the cell membrane and leads to loss of membrane potential, arrest of respiration, and death. *sok* encodes an antisense RNA that binds to the *hok* mRNA, causing it to be cleaved. The *sok* antisense RNA is much less stable than the *hok* mRNA, and so if a plasmid-free segregant is produced, the inhibitor disappears first, leaving the *hok* mRNA free to be translated to form the toxin and kill the cell (Gerdes et al. 1988, 1992).

---

of an effect on fertility is surprising, given observations that sperm of responder-insensitive males are smaller than those of wildtype (Hauschteck-Jungen and Hartl 1978).

Though the resistant allele has a cost associated with it, this cost is not as extreme as for the *t*, and so the allele is found in appreciable frequencies in natural populations (which typically have 1–5% *SD*, 20–50% resistant, and the remainder wildtype; Lyttle 1991). Charlesworth and Hartl (1978) analyzed a model with these 3 alleles and concluded that the only way to account for the observed frequencies was if the resistant allele had a cost of 1–2% in the absence of the killer. This is smaller than the cost seen in the lab. The modeling also suggested that the *SD* chromosome needs to be associated with a substantial (e.g., 40%) reduction in male fertility, both when paired with a wildtype chromosome and with a resistant chromosome. Again, it is not clear how accurate this prediction is.

Origin. *SD* appears to have originated very recently in *D. melanogaster*. The 3 most closely related species (*D. simulans, D. mauritiana,* and *D. sechellia*) contain neither the duplicated *dRanGAP* gene nor appreciable amounts of the *Xba*I repeat sequence (Temin et al. 1991). The fact that different *SD* chromosomes have different inversions, with none having gone to fixation, also suggests a relatively recent origin. Most convincingly, DNA sequencing shows that *SD* chromosomes are very similar to each other, and there is no more divergence between *SD* and wildtype chromosomes than there is between different wildtype chromosomes (Palopoli and Wu 1996). That is, *SD* appears to have arisen more recently than the average coalescence time of wildtype sequences.

Because *SD* (like *t*) involves an interaction between at least 2 loci, neither of which is beneficial by itself, the question naturally arises, how could it have evolved? It seems to require 2 mutations, each of which alone is deleterious. As *SD* is a young complex and reasonably well understood in molecular terms, a plausible scenario can be advanced for how it originated:

- The ancestral state was presumably chromosomes without an *Sd* duplication and without many (or any) *Xba*I repeats, as found in the relatives of *D. melanogaster.*
- The first step would have been the amplification of the *Xba*I repeats. It is not clear why this occurred, but it seems to be favored by natural selection, as a deletion of the region reduces viability. Alternatively, in some genetic backgrounds the high-copy-number allele shows drive over the low-copy-number allele (Hiraizumi et al. 1994), and this may have led to its accumulation. Perhaps the repeat can sequester a functional protein—rather than (as in the presence of the *Sd* killer allele) being tagged with a toxic protein.
- Regardless of why the *Xba*I repeats accumulated, unequal crossing-over would ensure that there was substantial variation in copy number between chromosomes.
- The *Sd* duplication responsible for the killer genotype happened to occur on a chromosome that had relatively few *Xba*I repeats (or quickly recombined onto such a chromosome). The chromosome would then be relatively insensitive to the killing action, and so could show drive and increase in frequency. Consistent with this scenario, in at least some genetic backgrounds, *Sd* acts preferentially against the more sensitive *Rsp* allele (Hiraizumi, unpublished data, cited by Lyttle 1991).
- As killer chromosomes became established in the species, there would

be selection on them for ever-fewer *Xba*I repeats, to maximize drive, resulting in the low numbers of them now found on *SD* chromosomes.

In this scenario the killer hitchhikes along with a rare resistant *Rsp* allele that is initially maintained purely by mutation-selection balance. Had the resistant allele been the common type, and the susceptible allele the rare type, the killer would have had much more difficulty establishing itself–a killer can only spread by hitchhiking with a resistant allele that is initially rare (Hurst and Pomiankowski 1991a).

Spore Killers in Fungi

The fungus *Neurospora intermedia* has 2 killer complexes that in many ways are similar to *t* and *SD*, though spores are killed, not sperm cells (Turner and Perkins 1991, Raju 1994). The 2 complexes are called $Sk-2^K$ and $Sk-3^K$, and both map to the centromeric region of chromosome III. In crosses between killer (*K*) and wildtype-sensitive (*S*) strains, the spores inheriting the *S* allele are killed, and those with the *K* allele survive (Plate 2). In crosses between 2 *S* strains or 2 *K* strains of the same type, all spores are viable; and in crosses between strains carrying the 2 different killer complexes, virtually all (99.9%) spores are killed. Phenotypically similar killer complexes have been found in several other fungal species (Table 2.1). None of them has any other obvious effect on the host phenotype.

Mechanism. For *Neurospora* spore killers, the mechanism of action is unknown. Normally, meiosis is followed by a single mitotic division, and results in 8 viable spores. In $K \times S$ crosses, there is nothing obviously amiss while the 8 spores share a common cytoplasm; but when the spores form walls, 4 of them stop further development and die (Raju 1979). In some developmental mutants, sensitive and killer nuclei are enclosed in the same spore, in which case the spore will live and killer and sensitive nuclei are recovered with equal frequency (see also Raju and Perkins 1991). This indicates that resistance is dominant and based on the presence of a protein, as in the *t* complex, rather than the absence of a repeat, as in *SD*. In *Podospora anserina* there is an unusual spore killer allele that produces a prion protein, but this is not thought to be typical (Dalstra et al. 2003, Perkins 2003).

Frequency and fate. Frequencies of spore killers vary widely. $Sk\text{-}2^K$ and $Sk\text{-}3^K$ of *N. intermedia* are very rare, having been recovered just 4 times and 1 time, respectively, in 3000 isolates worldwide, all in Southeast Asia (Borneo, Java, and Papua New Guinea; Turner 2001). Resistant alleles are more common—resistance to $Sk\text{-}2^K$ is found in 43% of isolates in Southeast Asia, Australia, India, China, and Japan, but it appears to be absent from the Americas, the Congo, and the Ivory Coast. Resistance to $Sk\text{-}3^K$ was found in 26% of isolates from Southeast Asia and Australia, but not outside this region. By contrast, the frequency of $Sk\text{-}1^K$ in the congeneric species *N. sitophila* is much higher in some locations (30–40% on various Pacific islands, 100% in at least some Louisiana sites) and absent in others (Fig. 2.7). A high-frequency spore killer is also found in *Giberella fujikuroi* (~80%; Kathariou and Spieth 1982). Interestingly, only a single resistant strain was found in *N. sitophila*, well away from where the killer is common, and none was found in *G. fujikuroi*. For these 2 killers, it is difficult to see how they could possibly be at a stable equilibrium frequency (Nauta and Hoekstra 1993). This situation may have arisen because both species are associated with humans: *N. sitophila* is often found in bakeries and lumberyards (D. Perkins, personal communication), while *G. fujikuroi* is a cosmopolitan pathogen of maize, sorghum, and other crops. Perhaps human activities have increased the range of these species in recent times and have brought them into contact with isolated populations or different species from which they acquired the spore killer genes. In any case, these might be ideal model systems to document the spread of a selfish gene.

There are no other obvious effects of the killer complexes on organismal growth or reproduction. If resistant alleles are truly absent from *G. fujikuroi*, then the spore killer seems likely to spread all the way to fixation. If it does, there will no longer be any sensitive alleles to kill, and so there will no longer be any purifying selection maintaining the killing function. The complex may therefore degenerate over evolutionary time, the short-term success of fixation giving way to a longer-term failure to persist as a functional element.

Incidence of Gamete Killers

Perhaps the greatest (and most frustrating) gap in our understanding of gamete killers is some sense of how common they are. Particularly among ani-

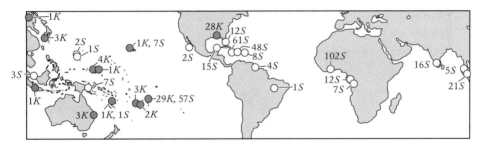

Figure 2.7 Geographical distribution of *Sk-1* killer *(K)* and sensitive *(S)* strains in *Neurospora sitophila*. Numbers are the number of strains of each type collected at that site; open circles identify sites within which all strains were sensitive. A single nonkiller resistant strain was collected from Gabon, far away from any killer. Adapted from Turner (2001).

mals, it is difficult to extrapolate from the only 2 known cases, each of which was found in a well-studied model organism by time-consuming laboratory work on the inheritance of linked phenotypic markers. In humans, several disease-causing loci have been suggested to show biased inheritance, but this has often turned out to be due to ascertainment bias—families in which many members have the disease will be better studied than those with fewer members (reviewed in Teague et al. 1998). In other cases, it is possible that the transmission ratio distortion is real (for example, retinoblastoma, Munier et al. 1992; cone-rod retinal dystrophy, Evans et al. 1994; and split hand/foot malformation, Özen et al. 1999), but the underlying mechanism is unknown, and all these are very rare in the human population, not obviously much different from that expected from recurrent mutation. It is probably safe to say that no gamete killer as effective and frequent as *t* and *SD* exists in humans, but it is not clear we can say this about any other animal. Even within humans, less effective killers may well exist. Thus, of the 3 best-characterized animal species, 2 of them have one gamete killer, and 1 has none. Are we to conclude that two-thirds of animal species have a gamete killer?

t and *SD* were both discovered initially by the non-Mendelian inheritance of linked loci. Deviations from Mendelian inheritance are occasionally reported in other species, such as a locus responsible for progressive rod-cone degeneration in miniature poodle dogs (Acland et al. 1990), a minisatellite locus in a pseudoscorpion (Zeh and Zeh 1999), and the albumin locus in rhesus monkeys (Smith and Rolfs 1984). Drive is often high (~90%) and limited to one sex, but alleles are often harmful and rare. It will take much work to determine whether these cases of biased inheritance are due to gamete killers, some other class of selfish gene, or something more innocuous (e.g., linkage to a deleterious mutation, or some genetic incompatibility between strains).

In plants, many more gamete killers have been reported (Table 2.1), each of them discovered by crop geneticists making a linkage map and then investigated to reveal the cause of the non-Mendelian inheritance. Taken at face value, the data suggest that the average number of gamete killers per plant species is about 1, give or take a factor of 10. It would be interesting to screen a natural community of plant species for individuals in which half the pollen is inviable, to see whether this estimate can be improved.

Fungal spore killers have in some ways been the easiest to detect, because the dead spores are so easily visible (Plate 2.2). Again, what data exist suggest an order-of-magnitude average of about 1 spore killer per species. But still there is much unexplained variation. A worldwide collection of *Neurospora crassa* isolates revealed not a single spore killer, whereas in *Podospora anserina,* a highly inbred species, 7 different types of spore killer have been found, 6 of them from a single population of just 99 isolates, with a combined frequency of 23% (van der Gaag et al. 2000). Most of these map to linkage group III, though at least 1 maps to another chromosome. Interestingly, crosses between different killer strains show either mutual resistance or dominant epistasis (1 strain sensitive to and killed by the other), unlike the 2 killers in *N. intermedia,* which show mutual killing. Perhaps spore killers arise more frequently in *P. anserina* than in *N. crassa* (e.g., larger global population size, or greater influx of horizontally transmitted DNA), or perhaps they remain polymorphic for longer (e.g., because they are more inbred).

Much else is equally uncertain. For example, most of the gamete killers discovered thus far are relatively effective (e.g., killing 90% or more of the wildtype gametes). Is this an accurate picture, or is there a much larger class of less effective killers that are less easily discovered? Killers might be less effective either because they kill less often, or because there is greater recombination between killer and target loci. Finally, what fraction of killers are like *t*–too harmful when homozygous to go to fixation–and what fraction go to fixation and then become invisible? Some gamete killers in plants are only detectable in species hybrids (Table 2.1). In principle, the frequency of gamete killers sweeping through populations could be estimated by inter-population crosses. Care would have to be taken, though, to ensure, first, that the populations were not so divergent that the killer would have time to degenerate and, second, that the populations were geographically isolated with no possibility of the killer transferring between them (recall that *t* has transferred recently among subspecies of mice).

Though the evidence is highly fragmentary, it seems reasonable to suppose that gamete killer complexes arise every so often within a host species, persist for a while, and then either go extinct directly or go to fixation and then go extinct. In the evolutionary lineage connecting the origin of meiosis to any extant species, this may have occurred hundreds or thousands of times, or more. These repeated attacks may have left some traces on the present-day mechanisms of meiosis and gametogenesis. One possibility is that killers have been important selective agents in the evolution of meiosis and gametogenesis—in other words, that aspects of these processes have evolved so as to suppress ancient killers. This seems especially likely when meiosis was first arising, but unfortunately those events are hidden by the mists of time. Certainly there are aspects of meiosis and gametogenesis that make the evolution of killers more difficult. Meiotic recombination is one such example, though it is less clear that breaking up killer complexes has been a quantitatively important selective agent for the evolution of recombination or affects its level today. Similar difficulties apply to some other suggestions (Box 2.2).

Maternal-Effect Killers

t and *SD* are special kinds of killer. They act in heterozygous males to kill sperm into which they have not been transmitted. Likewise, spore killers act to kill spores in which they are not found. These, in turn, are part of a larger category of genetic elements, those that gain a benefit by killing closely related, competing individuals that lack themselves. We treat first maternal-effect killers, genes that act in heterozygous females to kill progeny that do not carry copies of themselves. In this case, the victim is a diploid individual rather than a haploid gamete. A more important difference is that protection from being killed can come by inheriting a copy of the gene from the

BOX 2.2

Killers and the Evolution of Meiosis and Gametogenesis

Several aspects of normal meiosis and gametogenesis make the evolution of gamete killers more difficult, but there is little evidence as yet that any of them evolved in order to suppress gamete killers (as opposed to this being an unselected side effect).

Two-step meiosis. Diploid cells give rise to haploid cells via an indirect route, first duplicating their chromosomes and then going through 2 divisions. When combined with crossing-over, this method introduces some uncertainty over the cell division in which a gene segregates from its homolog. Haig (1993a) suggests that this 2-step meiosis evolved in order to produce this uncertainty and thereby combat a hypothetical class of selfish genes that causes a meiotic cell to kill its sister cell (e.g., by sending a toxin to the spindle pole opposite to which it moves). Such killers have not been reported. If suppressing them now is a strong selective force, one would expect to see the uncertainty maximized by having double crossovers immediately on either side of every centromere, but this is never observed (see also Hurst and Randerson 2000).

Syncytial gametogenesis. Gamete killing seems to require haploid expression of one or more genes, and such expression is made less likely because of cytoplasmic connections between developing spermatids that persist after meiosis. Hurst and Pomiankowski (1991b) suggest that these cytoplasmic connections have evolved to protect against gamete killers, but maintaining the advantages of diploidy in masking deleterious recessive alleles would seem to have larger and more general selective consequences.

Pseudohomothallism. In pseudohomothallic ascomycetes, meiospores are produced containing 2 nuclei, one of each mating type. The resultant mycelium is dikaryotic. Haploid gametes are formed that contain one or the other of these, and gametes of opposite mating type fuse, after which there is a brief diploid phase, followed by meiosis. Turner and Perkins (1991) suggest that the enclosure of 2 nuclei within a single spore may have evolved in some lineages to suppress spore killers. But the selective pressures associated with this suppression seem likely to be small compared to those associated with halving the fecundity and becoming self-fertile.

other parent (i.e., the father). This means that, like spore killers, such genes are not prevented from sweeping to fixation. A maternal-effect killer works as if it recognizes progeny that do not carry a copy of it and then kills them. A key question is, What is the benefit to the killer? The earlier the killing, the more likely a return benefit will come from reduced sibling conflict for parental investment or other shared resources. This system encourages the invasion of killer mutants when rare (Smith 1998). We begin with the four known cases and then review the expected action of natural selection.

Medea in Flour Beetles

Medea—or, maternal-effect dominant embryonic arrest—refers to a genetic factor (*M*) found in flour beetles (*Tribolium castaneum*) that acts in heterozygous (*M*/+) females to cause all +/+ offspring (i.e., those without *M*) to perish. These usually die sometime between just before hatching to as late as the second larval instar (Beeman et al. 1992). This is not some defect of the + allele, because +/+ progeny of +/+ mothers are perfectly normal. The doomed larvae appear morphologically normal, but become sluggish, uncoordinated, or paralyzed before death. *M*/+ offspring are completely protected, whether the *M* was inherited from the mother or the father. Other than the killing, there are no obvious effects of *M* factors on the rate of development, fecundity, longevity, or mating success of those possessing them (Beeman and Friesen 1999). In short, *M* appears to have no effect in a female other than to kill off progeny lacking *M*. Once again, other than its selfish effect, the selfish genetic element has no obvious phenotypic effect.

The molecular basis of *Medea* action is unknown, but evidence suggests that at least 2 linked genes may be involved. A lab mutant that fails to show the killing activity nevertheless (when transmitted through sperm) rescues zygotes from intact *M*/+ mothers (Beeman and Friesen 1999). This is reminiscent of tight linkage between responder-insensitive alleles and drivers in the *t* haplotype and in *SD*.

Worldwide collections of flour beetles indicate that there are at least 4 independent *Medea* factors found in different locations in the genome (labeled M^1-M^4), none of which rescues any of the others (Beeman and Friesen 1999). Moreover, they are both geographically widespread and patchily distributed. One *M* strain was found in half of 27 countries; most strains in Europe have it, none in India does, and the United States is sharply demarcated. Populations are typically fixed for M or for the alternative, with mixed frequencies being uncommon. All 26 populations in the United States above 33° latitude are fixed for M, but only 2 of 29 below 33°, while 21 populations lack M entirely (Beeman 2003).

That *Medea* appears to be absent from India may be due to another factor, H, which is found only in India and which segregates as a single locus and turns *Medea* factors into "suicide" genes (Thomson and Beeman 1999). For example, M^4 progeny that inherit an H allele from their father die at the fourth larval stage if raised at 25°C, and they have reduced viability at 32°C. Reciprocal crosses are more productive, though there is still some reduced viability at 25°C. The interaction between M^1 and H is even stronger, with any zygote inheriting M^1 and H being inviable regardless of which allele is maternal and which paternal. The presence of H in Indian populations may prevent *Medea* factors from spreading there, but this cannot be its function, because it is suicidal when paired with M, without any offsetting benefit.

A *Medea* factor has also been observed in *T. confusum*, despite the fact that crosses between the 2 species are entirely sterile, so its origin may have predated both species.

HSR, *scat⁺*, and *Om*[DDK] in Mice

A cytologically detectable homogeneously staining region *(HSR)* on chromosome 1 of mice *(Mus musculus)* also shows maternal-effect killing (Weichenhan et al. 1996). That is, if *HSR*/+ heterozygous females are mated to +/+ males, about 70% of the offspring recovered carry *HSR*. This drive is due to the +/+ embryos being more often resorbed by the mother *in utero*. There is no drive through males—nor if the heterozygous female is mated to an *HSR/HSR* male. That is, once again, embryos are rescued by a paternally inherited copy of *HSR*.

The molecular basis of this behavior is unknown, but the size of *HSR* is impressive. One intensively studied *HSR* consists of about 900 copies of a 100kb repeat organized into 2 bands, and some chromosomes have as many as 2000 copies, equivalent to 200Mb, or 6.7% of the genome (Weichenhan et al. 2001, Traut et al. 2001). By contrast, there are only about 60 copies of the repeat on wildtype chromosomes. The main constituent of the repeat is *Sp100-rs*, a chimeric gene formed by a fusion of the *Sp100* gene (which is interferon responsive and may be involved in transcription regulation and chromatin structure) and the *Csprs* gene (a G protein–coupled receptor). We have no idea why the repeat may have grown in size so rapidly.

HSR is patchily distributed in mice populations throughout Eurasia and, where present, its frequency varies from 4% to 76% (Agulnik et al. 1993). Even within a 5km radius, the frequency can vary among local populations from 51% to 0%. In at least 1 population, genotype frequencies do not differ from Hardy-Weinberg proportions, suggesting no gross effects on viability in nature; but it is curious that *HSR* never reaches fixation, as *Medea* so often does, even though compensation is expected to be higher in mice, with much relaxed conditions for the invasion of a rare killing gene (Smith 1998). This, along with its patchy distribution, suggests some hidden cost, or very recent evolution.

scat⁺ (severe combined anemia and thrombocytopenia) is a gene near the centromere of chromosome 8 of mice that may be another maternal-effect killer (Hurst 1993, Peters and Barker 1993). There is a mutant form of the gene (call it "–"), with the unusual property that –/– progeny of +/– mothers die early, whereas –/– progeny of –/– mothers are normal. As far as is currently known, the – allele only exists as a lab mutation and is not found in nature. It is as if the normal *scat*⁺ allele is a maternal-effect killer that arose in a population of – alleles and then swept through to fixation.

Again, little is known about the mechanistic basis of killing. Progeny suffer from severe combined anemia and intermittent episodes of severe bleeding, and in the lab about a third of affected individuals die *in utero*. Of those that are live born, 60% die in the first 2 weeks after birth. The remainder go into spontaneous remission, before a second crisis, after which 90% of live-born individuals are dead by day 50. If +/+ females are implanted with ovaries from a –/– female, and then mated to heterozygous +/– males, +/– and –/– progeny are produced, and the latter show the disease phenotype. Thus, the *scat*⁺ alleles in the mother's soma are enough to kill; killing is not (solely) through the expression of genes in the germline. Moreover, it appears that the mother is somehow turning the offspring's spleen against itself, because removal of the spleen largely cures affected individuals.

Identifying the gene responsible should allow direct testing to see whether *scat*⁺ is a maternal-effect killer (as opposed to a normal host-benefiting gene that gives rise to an unusual mutation) and, if so, how long ago it swept through to fixation, whether there is any sign of degeneration, and so on.

The *Om* locus on chromosome 11 of mice is a different kind of maternal-effect killer with a complex pattern of inheritance and we give only a simplified treatment here. Allelic differences at this locus cause matings between females of strain DDK and males of many other strains to be largely sterile, with some 90–95% of embryos dying around the time of implantation. The reciprocal crosses are normal. In crosses between DDK females and F_1 heterozygous males (DDK/DDK × DDK/+), virtually all the surviving progeny are DDK/DDK; those inheriting the paternal + allele die. That is, DDK/DDK females somehow arrange the death of progeny derived from + sperm, a result that is not dissimilar from killing the + sperm directly. Little is known of exactly how this works, except that embryos die due to an incompatibility between a maternally synthesized RNA in the oocyte and some sperm factor (Renard and Babinet 1986, Renard et al. 1994).

Further studies have shown that if the DDK allele is backcrossed for 7 generations into a wildtype background, the DDK allele appears to be dominant: if DDK/+ heterozygous females are mated to +/+ males, fertility is decreased 10- to 40-fold; and if they are mated to DDK/+ heterozygous males, there is strong drive in favor of the DDK allele (DDK/DDK, DDK/+, +/+ genotypes were recovered in the frequencies 261, 200, 16, instead of the Mendelian 1:2:1; Le Bras et al. 2000).

The Evolution of Maternal-Effect Killers

As with gamete killers, maternal-effect killers will increase in frequency only to the extent that they benefit from reduced competition. Two extreme situations can be identified. On the one hand, if siblings compete no more intensely than random members of the population, and so the benefits of killing are shared equally among the whole population, the allele will be only weakly selected when rare, with the effective selection coefficient approximately equal to the frequency of the gene (Wade and Beeman 1994). On the other hand, if all the benefits of the killing go to K/K and $K/+$ siblings of the victims, then selection will be strong, with an effective selection coefficient of 0.5. This is the same as for a gamete killer like *t* or *SD* that works only in one sex. Presumably mice are closer to this latter extreme of direct benefit than flour beetles, though with restricted larval movements in beetles, there could still be a significant release from sibling competition (McCauley and Wade 1980).

If a maternal-effect killer becomes common in a population, its associated selection coefficient will be low, because most +-bearing eggs are rescued by the paternal allele, and so there is little killing. This is the opposite of a gamete killer, in which case, without negative homozygous effects, selection coefficients steadily increase to 1. Thus, in theory, there is more scope for intermediate equilibrial frequencies in maternal-effect killers. But *Medea* factors and *scat*⁺ can go to fixation, suggesting they cause little harm to the host when homozygous.

Though the mechanism of action is not yet known for any maternal-effect killer, possibilities are easy to imagine. Perhaps the simplest is the presence, ancestrally, of a maternally supplied protein whose precise expression level is important for embryonic development. Then the killer could arise simply by a change in a control region that simultaneously reduces maternal expression and increases zygotic expression sufficient to compensate for the lost maternal supply. Or there could be a poison-antidote system. Unlike with gamete killers in animals, haploid expression is not required. In species such as mammals with extended maternal-fetal interactions, it must be all the easier for maternal-effect killers to arise. Simultaneous changes in closely linked receptors and ligands, or in proteins involved in adhesion, could give rise to a "killer" element.

Though maternal-effect killers have only recently been discovered, we suspect that in taxa like mammals with substantial postzygotic maternal investment, they may be even more common that gamete killers. We imagine they may have arisen hundreds or even thousands of times in our own ancestry, since placental mammals first arose 75mya. If so, what would we expect? If killers often work by simultaneously reducing maternal expression of an essential gene and increasing zygotic expression, then one prediction is that there would be a general shift from maternal to zygotic expression of quantitatively important genes—especially in species with maternal compensation. Could this be part of the explanation for why maternal expression is so much less developed in mammals compared to insects and amphibians, and zygotic genes begin expression so early?

Gestational Drive?

So far we have considered alleles that kill progeny in which they are absent, but consider the other side of the coin—alleles that cause mothers to invest more in those offspring in which they are present (Haig 1996a). This is especially likely in species with intimate mother/offspring connections, as in placental mammals, in which cells of the two are in direct contact and simple cellular processes can detect the presence or absence of alleles and either encourage or inhibit resource transfer between them. Haig calls this "gestational drive." By giving itself a benefit during gestation, the gene shows drive. By logic these genes are expected to arise, but they are usually expected to sweep through to fixation, after which they will not be recognized.

The general possibility that a gene could produce a signal, sense the signal in others, and then direct benefit to them was emphasized by Hamilton (1964a, 1964b) and given the colorful expression "green beard" by Dawkins (1976). That is, a gene could code for a signal (a green beard) and at the same time direct preferential action toward another individual with that signal (a green beard). At the locus involved, a relatedness of 1 is being measured, but this coefficient applies only to that and (to a diminishing degree) linked loci so that all nonlinked genes are selected to shut down the green-beard gene, insofar as conferring the benefit has a cost. But if a green-beard gene does sweep through to fixation, it would no longer be detectable as such. All offspring would be treated alike and the interaction would give the general appearance of mother-fetal cooperation (Haig 1996a). Indeed, these positive green-beard genes differ from the negative ones such as t in tending, on average, to elevate the phenotype instead of degrading it.

Especially promising for green-beard effects in mammalian pregnancies are homophilic cell-adhesion molecules (Haig 1996a). In these molecules, a part sticks outside the cell and recognizes copies of itself, while the internal part initiates intracellular action. Therefore, this single molecule (and the allele that produced it) has all the properties required for a green-beard gene. When applied to the interface of maternal and fetal tissues, one can easily imagine how greater cell adhesion would tend to encourage contact between the 2 kinds of cells and thereby the transfer of nutrients. A variety of self-adhesive molecules are known to be expressed during placental development and human pregnancy. For example, trophinin and tastin are two human cell-adhesion molecules known to be especially prominent in placental and uterine tissues (Fukuda et al. 1995).

Cell-adhesion molecules need not be homophilic to be self-benefiting. Closely linked loci can also act as green beards. This can be as simple as

CD2– and CD58–cell-surface receptors of the human immune system that are each other's principal ligand. These genes are located within 250kb of each other, perhaps as a result of a gene duplication. There are also numerous pairs of linked loci known in humans that may interact between mother and offspring in the expected way (Haig 1996a). Especially likely are ligands and their receptors and enzymes and their substrates. An important feature of this argument is that signal and detection need not be coded by 1 gene but can be coded by linked pairs. Self-benefiting cell-adhesion molecules could set up an evolutionary race with unlinked suppressors not enjoying a benefit. Consistent with this idea, 2 of 3 such placental/fetal molecules show evidence of strong, positive selection in mammals, while a cell-adhesion molecule involved in heart function does not (Summers and Crespi, 2005).

Consider another example. The genes for renin and angiotensinogen are linked together on distal 1q in humans (Gaillard-Sanchez et al. 1990). Renin cleaves angiotensinogen so as to produce angiotensis, which raises blood pressure. Other things being equal, the fetus benefits from increased maternal blood pressure, which increases nutrient flow to it. One can easily imagine a new haplotype that produces more angiotensinogen in maternal cells and more renin in placental (fetal) cells, with both products interacting in the mother to produce higher maternal investment associated with the presence of the haplotype in both parties (Haig 1996a).

A dramatic example of the role that self-adhesive molecules may play in affiliative relations comes from recent work on the amoebae of the cellular slime mold *Dictyostelium discoideum* (Queller et al. 2003). The amoebae feed as individual cells on soil bacteria, but when food runs out, they adhere in aggregations of thousands to form fruiting bodies of which 20% is nonreproductive stalk. *csA* (contact site A) is a homophilic cell-adhesion molecule that helps bring this about. By adhering to each other, these molecules help pull each other together to form a fruiting body. In competition with an equal number of *csA* mutants without cell-adhesion abilities, 82% of fruiting body cells are wildtype. This selective advantage disappears on artificial laboratory surfaces (on which aggregation is easier): the differential tendency of *csA* to form stalk tissue results in lower fitness than the free-riding mutant enjoys. In this example the interaction is symmetrical and among peers, with no obvious intragenomic conflict (Ridley and Grafen

1981). Simple green-beard genes of this type may be particularly common in microorganisms, in which individual cells can interact directly with neighbors.

Gametophyte Factors in Plants

In plants, genes have repeatedly been described that act in female tissue (the style) to kill pollen in which they are absent. These are called gametophyte factors. For example, in maize the *ga1* locus on chromosome 4 behaves as follows (Schwartz 1950, Nelson 1994). If the parent carries the *Ga1* allele, in either the heterozygous or homozygous state, and is pollinated by a heterozygous *Ga1/ga1* plant, the *ga1*-bearing pollen is a very poor competitor and fertilizes 0–4% of the ovules. This is not just some dysfunction of the *ga1* allele in pollen; if the female parent is *ga1/ga1* homozygous, the 2 types of pollen do equally well. In some crosses the *Ga1* allele suppresses but does not stop the growth of *ga1* pollen, so if only *ga1* pollen is available, there will still be a full seed set. In other crosses, plants homozygous for *Ga1* set no seed at all if fertilized only with *ga1* pollen—effectively, they kill it—and plants heterozygous for *Ga1* have a partial seed set. It is not clear whether the difference is due to alternative *Ga1* alleles or to genetic background effects. One inbred line of maize has been found to have *Ga* expressed only in pollen. That is, plants homozygous for this allele can fertilize strong *Ga1*-homozygous plants, but they do not select for *Ga1* pollen over *ga1* pollen when fertilized themselves. If we imagine this as a molecular lock-and-key system, this strain has the key, but no lock.

Little is known about how *Ga1* acts mechanistically. *ga1* pollen on *Ga1* homozygous silks does germinate and penetrate into the style, but its growth rate is progressively slower than that of *Ga1* pollen, until growth finally ceases well short of the ovules.

Allele frequencies in various races of maize are typically extreme, with many popcorns (which are considered to be primitive) and some Central American forms of dent and flint corn being homozygous for *Ga1*, while all North American flint and dent corns are homozygous for *ga1*. This is presumably the ancestral state, as *ga1* cannot invade a population fixed for *Ga1*.

While *Ga1* is the best-studied gametophyte factor locus in maize, 8 other loci with similar behavior have been described, on 6 different chromosomes. Gametophyte factors have also been reported in barley (*Hordeum vulgare;* Tabata 1961), lima beans (*Phaseolus lunatus;* Bemis 1959, Allard 1963), broad beans (*P. vulgaris;* Bassett et al. 1990) and sugar beets (*Beta vulgaris;* Konovalov 1995). In short, they are widespread in agricultural species, which are the only ones likely to have sufficient genetic work for their detection.

The breeding system is expected to have a very interesting effect on the success of gametophyte factors when they first appear and are rare. To see this, suppose that G is a gametophytic factor such that in G/g or G/G individuals, g-bearing pollen is prevented from fertilizing the ovules. If we assume random mating and no pollen limitation (and so the seed set is not affected), it is easy to show (via simulations) that G has a selective advantage of about 0.7 when rare, increasing to 1 when common. But this is the ideal case. When G is rare, very little of the pollen arriving will be G (and therefore compatible), while the G-bearing pollen that does arrive is likely to come from self or close relatives, and so may be associated with inbreeding depression. In short, random mating and full seed set are unlikely, countering the benefit in transmission. This suggests that gametophyte factors may have a difficult time getting started in self-incompatible species and spread more easily in species with a mixed mating system and little inbreeding depression. In the extreme case of pollen limitation so severe that the seed set is proportional to the fraction of arriving pollen that can fertilize the seeds, G plants suffer lower seed set and G, when rare, cannot invade a homogeneous population.

THREE

Selfish Sex Chromosomes

MOST SPECIES WITH SEPARATE males and females have 2 structurally different sex chromosomes in which either males are XY and females XX, or the opposite: females XY and males XX (sometimes called WZ, ZZ, respectively). In either case, one sex is heterozygous (or "heterogametic"), and, in principle, there is the opportunity for the sex chromosomes in that sex to show some form of drive. Because X and Y typically do not recombine with each other over much of their length (indeed, that is why they are different), and because they contain genes that determine the sex of the organism carrying them, the evolution of a driving gene on a sex chromosome raises a number of special issues. Most obviously, it means that the driver will distort the ratio of sons to daughters and, as it spreads, the sex ratio of the entire population. As a practical matter, this makes sex chromosome drive much easier to detect than autosomal drive and should ensure that our knowledge of its distribution is more reliable. Skewed sex ratios also mean that suppressors of drive will automatically be selected for, not only on the opposite chromosome but also on the autosomes, with that selection growing stronger the more the population sex ratio deviates from 1:1 (the autosomal optimal). Other adaptive responses are also possible. For example, if the sex ratio is female biased, there could be selection for a change in the sex-determining mechanism to convert what would normally be a female into a male.

The lack of recombination between X and Y and the attendant divergence between the 2 chromosomes should in some ways make it easier for a driving locus to arise (Frank 1991, Hurst and Pomiankowski 1991a, Lyttle 1991). We saw in Chapter 2 that autosomal gamete killers typically consist of at least 2 loci, a killer and a target. In order for a complex to evolve, a killer allele has to arise that can target most homologous chromosomes in the population, but not the one on which it happens to have arisen; and it must be closely linked to the resistant allele, so they are not often separated by recombination. Thus, something rather unlikely has to happen, and it is only because there are many species, all experiencing occasional mutations and chromosomal rearrangements, that killer complexes arise at all. By contrast, a killer allele that arises on, say, a Y chromosome, needs only to target any sequence on the X in order to spread, and similarly for a killer allele on an X. Recall that for *SD* of *Drosophila*, the killer allele merely targets a noncoding DNA repeat. In most species there must be many repetitive sequences found on the X but not the Y, and vice versa.

As we shall see, some of these expectations are borne out, but not others. We begin by describing in detail the logic and facts regarding the case that does best fit expectations, sex chromosome drive in the Diptera. Here, drive is common (at least in some genera), and there is good evidence that when a driving sex chromosome does arise, suppressors also evolve, both on the opposite chromosome and on the autosomes. Sex chromosome drive provides an unusual opportunity for the study of selfish genetic elements—and co-evolution by the rest of the genome—because the sex ratio itself provides a convenient measure of whose interest is being maximized. We now have extensive evidence on this for a variety of species. But driving sex chromosomes are not nearly as widespread outside the Diptera as might be expected—given both ease of detection and ease of evolution—and we suggest reasons why this may be so.

Sex chromosomes also permit some other, very interesting possibilities, such as sex change followed by drive. For example, in some mammals there is a special X* chromosome that converts X*Y males into females, in which X* drives (in part, because YY offspring are early lethals). We review these, and related cases, in lemmings and other murids. We then discuss more generally the possible effects of drive on the evolutionary stability of sex chromosome and sex-determining systems. Finally, we explore novel ways in which the X and Y could benefit themselves by aligning their body's reproduction to conditions that, for other reasons, favor the production of their sex.

Sex Chromosome Drive in the Diptera

Birds and lepidopterans (butterflies and moths) are 2 relatively well studied taxa with female heterogamety, and it is a curious fact that there are no well-established cases of sex chromosomes showing meiotic drive in either one. Among all the male heterogametic taxa, the only species known to have sex-linked gamete killers of large effect, analogous to *t* and *SD,* are from 1 taxon, the Diptera (Jiggins et al. 1999). These include both X-linked and Y-linked killers, and so we begin our survey of selfish sex chromosomes with the flies. Jaenike (2001) gives an excellent review, from which the following is largely drawn.

Killer X Chromosomes

Killer X chromosomes have been reported in 13 species of *Drosophila* fruit flies and 4 species of stalk-eyed flies, as well as the tsetse fly (Table 3.1); presumably these are only a fraction of the species in this group that have them. The 13 species of *Drosophila* are taxonomically diverse, falling into 2 subgenera and 8 species groups or subgroups, as are the 4 species of stalk-eyed flies, which fall into 3 genera. Despite this diversity, there are many similarities between the various systems, and we treat them as a single syndrome, noting species differences as appropriate. We designate the killer X chromosome as X^K, and the males carrying them as *SR* (for *sex ratio*).

Mechanism of action and genetic structure. Little is known about how killer Xs work, and nothing is known at the molecular level. In the first meiotic division, all appears normal: the X and Y first pair (in *D. melanogaster* they pair at specific sites in the ribosomal repeat clusters on X and Y; McKee et al. 2000), and then segregate normally to opposite poles. The first sign of something amiss is in meiosis II, when the sister Y chromatids fail to separate (in *D. pseudoobscura, D. simulans,* and *Cyrtodiopsis dalmanni;* Novitski et al. 1965, Cobbs et al. 1991, Montchamp-Moreau and Joly 1997, Cazemajor et al. 2000, Wilkinson and Sanchez 2001). Then, associated with this, there

Table 3.1 Taxonomic distribution of killer Xs in Diptera

Drosophilidae: *Drosophila* (fruit flies)
  subgenus Sophophora
    obscura group
      obscura subgroup: *D. obscura, D. subobscura*
      affinis subgroup: *D. affinis, D. athabasca, D. azteca*
      pseudoobscura subgroup: *D. pseudoobscura, D. persimilis*
    melanogaster group: *D. simulans*
  subgenus Drosophila
    melanica group: *D. paramelanica*
    tripunctata group: *D. mediopunctata*
    quinaria group: *D. quinaria, D. recens*
    testacea group: *D. neotestacea*
Diopsidae (stalk-eyed flies)
  Cyrtodiopsis dalmanni, C. whitei
  Diasemopsis sylvatica
  Sphyracephala beccarii
Glossinidae
  Glossina morsitans (tsetse flies)

Species in the same row are closely related and may represent a single origin. From Jaenike (2001).

is a failure of spermatogenesis. In *D. pseudoobscura, SR* males produce only half as many functional sperm as normal males (Policansky and Ellison 1970). In *D. simulans,* some *SR* males have fewer than half the number of sperm heads in normal position in the cysts, compared to normal males (Montchamp-Moreau and Joly 1997, Cazemajor et al. 2000). This last observation indicates that X^Ks can reduce sperm production by more than one-half–that some X^K-bearing sperm are affected as well–and so emphasizes the spiteful nature of the strategy.

X^K chromosomes usually have 1 or more inversions relative to the normal X chromosome–up to 5 in *D. athabasca*–and in species with metacentric Xs, these inversions can be on both arms (e.g., *D. affinis, D. paramelanica*). By analogy with the case of autosomal killers (see Chap. 2), this suggests that there are multiple interacting loci involved in producing the drive, and thus selection to reduce recombination. For some species, there is direct evidence that multiple loci are involved. In *D. persimilis,* there are at least 3 different regions of X^K that, if replaced by the homologous region from the normal chromosome of *D. pseudoobscura,* will no longer kill (Wu and

Beckenbach 1983). In *D. subobscura*, X^K has 4 inversions relative to the normal chromosome (denoted 2, 3, 5, and 7), and chromosomes with only some of these (2 alone, or 2+3, or 2+5+7) no longer kill (Hauschteck-Jungen and Maurer 1976). It appears that many loci can be involved in killing, spread right across the chromosome, and that they interact epistatically. But this association with inversions is not universal: the X^K chromosomes of *D. simulans* and *D. neotestacea* have none. In the former, the killer factor has been mapped to a 170kb region of the X chromosome (and presumably will soon be cloned and sequenced; Cazemajor et al. 1997, Derome et al. 2004). Recombinants in this region show a wide range of progeny sex ratios, perhaps suggesting again that multiple loci are involved (and in this case they interact additively).

Note that the inversions on X^K do not prevent recombination between X and Y, because that does not happen anyway (male meiosis is achiasmatic in *Drosophila*). Rather, they reduce recombination between a killer and a normal X chromosome when both occur in a female.

Suppressors and polymorphism. If a killer X chromosome arises in a population, Y chromosomes, which are the main victims, will be strongly selected to "defend" themselves and prevent the killing. In addition, killer Xs lead to skewed sex ratios and also reduce male fertility, so autosomes will be selected to side with the Ys and suppress, or even reverse, X drive. Both Y-linked and autosomal suppressors have evolved in most species; in some species, X^K can even show drag (transmission less than 50%), while normal Xs are inherited normally. As a consequence of this coevolutionary interaction between X^K and the rest of the genome, species can be polymorphic for multiple structural variants of the Y chromosome as well as multiple X^Ks.

The interaction between killers and suppressors has been well studied in *D. simulans*. In this species there appear to be 2 major categories of X chromosomes, *standard* and *sex ratio*, though with significant continuous variation within each one (Fig. 3.1; Montchamp-Moreau and Cazemajor 2002). The *sex ratio* chromosomes produce extreme sex ratios (80–100% daughters) in a susceptible background, but only slightly biased ones (54–58%) in a resistant one. In a broad geographical survey, populations were seen to fall into 3 major groups (Atlan et al. 1997; Fig. 3.2). In some areas, X^K is common (up to 60%), but resistance is also common, and so X^K is rarely expressed. In other areas, both X^K and resistance are rare or absent. Finally, in

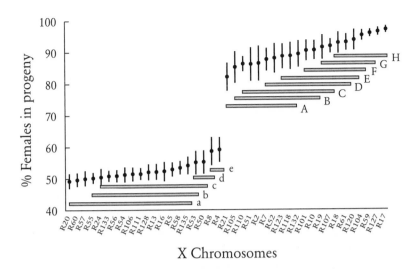

X Chromosomes

Figure 3.1 Degree of drive in 41 X chromosomes randomly chosen from nature from *Drosophila simulans*. 41 males captured from nature were bred to laboratory stocks in such a pattern as to produce an average of 28.5 male breeding replications for each X chromosome. Horizontal bars link means that are not significantly different ($p<0.05$) within each set. Note that there are 2 clusters, significant variation within each cluster, and that each cluster is significantly female-biased. Adapted from Montchamp-Moreau and Cazemajor (2002).

some areas X^K is rare, but resistance is common. Thus, in no population is X^K expressed at a high level, and the trait is usually only apparent in crosses between populations. Genetic analysis shows that suppressors are found on the Y and on both major autosomes (Cazemajor et al. 1997, Atlan et al. 2003), and that the Y effects are continuously distributed, rather than all-or-nothing (Montchamp-Moreau et al. 2001). In experimental populations fixed for X^K, both Y-linked and autosomal suppressors increase in frequency over time, as expected (Capillon and Atlan 1999).

Many of these patterns vary among species. For example, the observation that X^Ks are more likely to be effective in an alien background than in the native one is reversed in *D. subobscura*, in which X^K from North Africa produces more extreme biases in its own background (90–100%) than in a European background (~80%), where it is normally absent (Hauschteck-Jungen 1990). Also, the continuous distribution of suppressive effects by

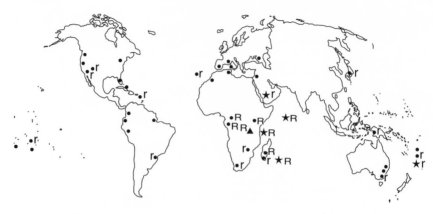

Distortion against *ST*

★ High (*d* > 0.7)

▲ Moderate (0.6 < *d* < 0.7)

• Undetected

Resistance against *SR*

R Complete

r Partial (0.6 < *d* < 0.8)

- Undetected (*d* > 0.8)

Figure 3.2 Geographical distribution of killer sex chromosomes and resistance to killer sex chromosomes in *Drosophila simulans*. Multiple individuals were collected from each locality and then crossed *en masse* with standard laboratory tester strains. Separate crosses were made to measure the extent of drive in a susceptible background (indicated by the symbol) and the degree of resistance to a strong driver (indicated by R, r, or nothing). Note that resistance to killing is more widespread than killing, and that there are no populations in which killers occur in the absence of resistance. Adapted from Atlan et al. (1997).

the Y is not found in *D. mediopunctata*, in which there appear to be 2 discrete classes of Y, susceptible and resistant (Carvalho and Klaczko 1994). Moreover, the resistant Y is only resistant against some X^K chromosomes and is susceptible to others. In natural populations, the resistant Y is in the minority (10–20%), and there are also autosomal suppressors at intermediate frequencies (Carvalho et al. 1997, 1998). In *D. paramelanica* also, there are 2 types of Y, susceptible and resistant, and again the resistant Y is resistant against some X^Ks and not others (Stalker 1961). In this case there is a geographical component to the variation, with suppressible X^Ks in the north and insuppressible ones in the south. Both species also have autosomal suppressors, which can act against both kinds of X^Ks (Stalker 1961, Varandas et al. 1997).

Things are even more complex in *D. affinis*, in which there are 2 cytologically distinguishable X^Ks (with 2 vs. 4 inversions), and Y chromosomes are highly variable in size and shape, and even presence—XO males are fertile in this species and make up 3% of the population (Miller and Stone 1962, Voelker 1972). The 4-inversion variant does well against some types of Y (mean 93% daughters in 4 strains), but poorly against one type (Y_L, 55% daughters) that the 2-inversion variant does well against (98%). Surprisingly, when the Y chromosome is absent (in X^K/O males), the 4-inversion variant shows moderate drive (average 65% daughters, with considerable variation, from 44–95%). These are the only results we know of that suggest a gamete killer need not have a defined target in order to work. Even more surprising: the 2-inversion variant self-destructs when there is no Y, giving rise only to sons. Normal Xs are inherited 50%, whether or not there is a Y.

D. athabasca has been less well studied, but still shows the correlation between X-linked killers and Y polymorphism. There are 3 behaviorally isolated semispecies in North America, and the 2 Eastern semispecies have X^K chromosomes and structurally polymorphic Y chromosomes, while the Western semispecies does not appear to have X^Ks, and Ys are monomorphic (Miller and Roy 1964, Miller and Voelker 1969). In *C. dalmanni*, X^K chromosomes produce extreme biases in a permissible background (75–100% daughters), but are transmitted *less* than 50% in a suppressive background (37%; Presgraves et al. 1997). Normal Xs are inherited 50% in both backgrounds.

Finally, some species appear to have no suppressors at all—none has been found in either *D. pseudoobscura* or *D. neotestacea* (Beckenbach et al. 1982, James and Jaenike 1990, James 1992). At least for *D. pseudoobscura*, this does not appear to be because X^K is recent. For a noncoding region within one of the inversions, X^K and normal alleles differ by 1.4%, suggesting they diverged about 1mya (Kovacevic and Schaeffer 2000).

There is clearly a great deal of variation here in the population genetics of X chromosome killers and their suppressors, the origins of which are unknown. This variation may reflect substantive differences in the biology of killing, or merely chance differences in the response of populations to a common selective agent, or indeed different time points in a common long-term evolutionary dynamic. Further progress on these evolutionary questions will, in part, come when the molecular basis of killing and suppression are uncovered.

Population biology and extinction without suppressors. The population dynamics of killer X chromosomes are expected to differ substantially depending on whether there are suppressors segregating in the population. Consider first the simplest case of no suppressors, as is seen in *D. pseudoobscura* and *D. neotestacea,* and a killer X at low frequency. If males are able to compensate fully for the lost sperm and suffer no reduction in fertility, and if the killer has no other effect on male or female fitness, it will increase in frequency, without limit, until all Xs are killers and there are no more normal Xs. At this point, if killing is complete (100% daughters), there will be no males and the population will go extinct (Gershenson 1928, Hamilton 1967).

The frequency of species going extinct due to a driving X chromosome is completely unknown. The data from *D. pseudoobscura* and *D. neotestacea* suggest that extinction is not an inevitable outcome, even if no suppressors evolve. In *D. pseudoobscura,* X^K chromosomes are rare or absent in the northern part of the species range (British Columbia), and only reach a maximum frequency of 33% further south (the range extends as far as Guatemala; Jaenike 2001). Moreover, the frequencies appear to be fairly stable, showing little change over a 50-year period in southeast Arizona (Beckenbach 1996). In *D. neotestacea* too, the frequency of X^K is 20–30%, with no reason to think that its frequency is increasing (James and Jaenike 1990). Obviously, something in the simple verbal model described here must be wrong. Two features in particular seem to be important in stabilizing X^K frequencies at an intermediate frequency: incomplete compensation in males (and diminishing compensation as the sex ratio becomes more female biased) and deleterious homozygous effects in females (Jaenike 2001).

First, consider compensation in males. Recall that X^K's *modus operandi* is to prevent Y-bearing sperm from being formed, but it does not increase the absolute number of X^K-bearing sperm, and may even slightly reduce it. Whether X^K can spread through a population depends (inversely) on the extent to which halving a male's sperm supply reduces his reproductive success. If a male typically mates infrequently, normal testicular turnover may be sufficient to replenish sperm supplies between copulations, even if sperm production is working at only 50% efficiency. But if males mate frequently, they will be unable to compensate and their fitness will suffer. For example, mating frequency probably increases with population density, and so we expect X^K to do less well in high-density populations. Crucially, a male's mating frequency will also increase as X^K spreads through a population and the sex ratio becomes more female-biased. Thus, the more frequent X^K is, the more harm it is likely to cause the male, and thus itself. This is exactly the sort of relationship that is needed to stabilize X^K frequency at some intermediate value. An additional effect in this direction comes because females may need to remate more often and *SR* males may be particularly poor at the resulting increase in sperm competition (Jaenike 1996).

Direct evidence for such effects comes from a number of sources. In population cages of *D. pseudoobscura*, X^K frequencies decline more rapidly in high-density populations than in low-density ones, and more rapidly in cages in which flies are allowed to remate than in ones in which they are not (Beckenbach 1983). In crosses between virgin males and virgin females, *SR* males sire as many offspring as normal males; but if males are made to mate repeatedly, the fertility of *SR* males is reduced to well below that of normal males (Policansky 1974, Beckenbach 1978, Wu 1983a, 1983b). Similar results have also been found in *D. neotestacea, D. quinaria, D. recens, D. simulans,* and *C. dalmanni* (James 1992, Jaenike 1996, Capillon and Atlan 1999, Wilkinson and Sanchez 2001, Atlan et al. 2004, Fry and Wilkinson 2004).

Further experiments with *D. neotestacea* show that this result is also likely to apply in nature: when males are mated immediately after capture from nature—and so would have been engaged in normal levels of mating activity—*SR* males sire significantly fewer offspring than do normal males (James 1992). Moreover, this difference is greatest when the population density in the field is greatest and is also greater when males are mated to already-mated females instead of virgin females, suggesting that *SR* males are particularly poor at sperm displacement. This is presumably a direct effect of reduced ejaculate size. In *D. melanogaster*, males that have not recently mated typically ejaculate about 5 times as much sperm as can be fitted into a female's spermatotheca, and the excess is believed to be useful in displacing preexisting sperm of others (Lefevre and Jonsson 1962).

The greater success of *SR* males when mating rates are low may account for much of the variation in X^K frequency within natural populations. In *D. pseudoobscura*, the frequency of X^K is highest in southern Arizona, where population density is low and flies are active only for a few hours early and late in the day—that is, where male mating rate is likely to be relatively low (Beckenbach 1978). In *D. neotestacea*, population densities increase from spring to late summer, with a correlated decline in X^K frequency (James

1992). Also, population density increases with increasing latitude from Virginia to Maine, and there is a corresponding (slight) decline in X^K frequency. Note that any increase in adult numbers due to the spread of X^K will tend to reduce the fitness of X^K, a density-dependent effect that may further stabilize X^K frequencies (Atlan et al. 2004).

In addition to these male fertility effects, X^K may also have deleterious effects on female fitness, particularly when homozygous, and this effect may also limit the extent to which X^K spreads. Such effects can be expected soon after the killing alleles arise, because linked loci will be made homozygous. Also, as inversions on X^K chromosomes arise, they will reduce the frequency of recombination and thus the efficacy of selection (e.g., in removing deleterious mutations). In addition, because normal chromosomes are in the majority, beneficial mutations are more likely to arise on them and may not get introduced onto X^K. Over time, X^K may fall further and further behind the normal X in terms of adaptation. Finally, to the extent that there are sex-antagonistic loci, female-benefiting mutations will be less favored on X^K than on normal Xs, because X^K is getting a larger fraction of its transmission through males (and normal Xs are expected to become more female benefiting).

So, do killer Xs ever cause species extinction? The fact is, we do not know. On the one hand, the frequency-dependent effects of killer Xs on male fertility seem almost inevitable, and the highest frequencies of X^K observed in *D. pseudoobscura* or *D. neotestacea* (both of which appear to lack suppressors) are about 30–35%, producing a population-wide sex ratio of 65–70%, which seems a long way from endangering any population. Indeed, such modest female-biased sex ratios seem more likely to increase population productivity and so reduce the likelihood of extinction. That is, killer Xs might be favored at the species level. On the other hand, looking at species that have not gone extinct is hardly a reliable guide to those that have (Carvalho and Vaz 1999, Taylor and Jaenike 2002, 2003).

Population biology with suppressors. Now consider those species in which there are Y-linked or autosomal suppressors. Though careful modeling is required, it seems that two diametrically opposed consequences may follow. First, if the frequency of X^K is primarily limited by deleterious female fitness effects, these effects will likely continue in the presence of the suppressors. If the suppressors go to fixation, X^K will have no transmission advantage in males, but it will be at a disadvantage in females, and so will be lost from the population. This may lead to populations such as are found in *D. simulans*, in which X^K is rare but resistance is common (Atlan et al. 1997; see also Hall 2004, Vaz and Carvalho 2004 for detailed modeling).

By contrast, if the frequency of X^K is primarily controlled by sperm limitation, the return to a more equitable sex ratio due to the spread of the suppressors might actually allow X^K to increase further in frequency, because a more balanced sex ratio would reduce sperm limitation. In this case, it would be possible for the spread of a suppressor to facilitate the spread of X^K, and both could go to fixation (Box 3.1).

Some (admittedly uncompelling) support for this notion comes from the relatively high frequency of X^K in some species with suppressors. In *D. simulans*, X^K can have a frequency up to 60% (Atlan et al. 1997). In *D. paramelanica*, it can be up to 40% (Stalker 1961). On the other hand, Varandas et al. (1997) argue that the presence of suppressors has reduced the frequency of X^K in *D. mediopunctata*.

Note that if X^K and its suppressors do go to fixation, killing will be detected only in between-population or between-species hybrids. Intriguingly, there is a subspecies of *D. pseudoobscura* around Bogota, Colombia, and F_1 hybrid males from a Bogota–North American cross, though only weakly fertile, produce extreme progeny sex ratios (90–95% daughters) that are not due to differential mortality (Orr and Irving 2005). One possible explanation is that a killer X has gone to fixation in the Bogota subspecies, as well as a suppressor, but the X is still active in the North American genetic background. Some of the factors affecting X suppression in Bogota are known to be autosomal. Also, *D. simulans* is thought to be fixed for a cryptic killer allele on the X (in addition to the one currently segregating in the population), which is expressed only in crosses with *D. sechellia* (Dermitzakis et al. 2000; see also Tao et al. 2001). It has been suggested that *Stellate*, an X-linked multicopy repeat in *D. melanogaster*, could be a relict killer, now suppressed by a multicopy repeat on the Y (Hurst 1992, 1996), but direct evidence fails to support this interpretation (Robbins et al. 1996, Belloni et al. 2002).

Frank (1991) and Hurst and Pomiankowski (1991a) have suggested that divergent fixation of sex-linked killers and suppressors in different populations could account for the strong tendency in hybrids for the heterogametic sex to be sterile or lethal—that is, that it could account for Haldane's Rule. This effect may have some contributory role in the Diptera; but the

BOX 3.1

Spread of a Killer X and an Autosomal Suppressor: An Example Trajectory

The figure in this box shows the spread of a killer X through a population, and the associated decline in the proportion of males. X^K does not go to fixation but instead goes to an intermediate equilibrium frequency because the fitness of X^K-bearing males is assumed to decline as the population becomes more female biased (specifically, we assume it is $Min[1,(5-4f)/3]$, where f is the proportion of females in the population). Then, in generation 200 a dominant autosomal suppressor allele is introduced that restores Mendelian inheritance and normal fertility. It spreads rapidly, increasing the frequency of males, which in turn allows X^K to go to fixation. The suppressor also, eventually, goes to fixation. It is possible, then, for killer Xs and suppressors to each go to fixation within a population, after which the only sign of biased sex ratios may be in crosses with other populations or species, when killer and suppressor are dissociated. We do not know how frequently this situation occurs in nature.

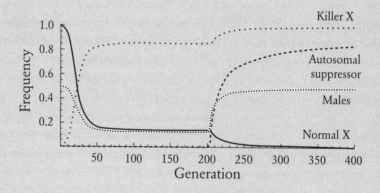

low frequency of killer or driving sex chromosomes outside the Diptera, even in taxa that do conform to Haldane's Rule (e.g., mammals, birds, Lepidoptera), suggests that it may not be of general importance. Direct tests of the theory in *Drosophila* produce mixed results. For example, 4 species pairs whose hybrids are semisterile fail to show any bias in progeny sex ratios, as might be expected if partially suppressed drive were causing the semisterility (Johnson and Wu 1992, Coyne and Orr 1993, see also Goulielmos and Zouros 1995). However, recent work has revealed several crosses in *Drosophila* in which some sterility is associated with some sex ratio bias (Tao et al. 2001; Orr and Irving 2005). For example, a small (80kb) autosomal segment of *Drosophila mauritiana*, when introgressed into the genome of *D. simulans*, generates males that produce female-biased sex ratios and also suffer from reduced fertility. More plausible general explanations for Haldane's Rule are reviewed in Coyne and Orr (2004).

Killer Y Chromosomes

Y-linked loci that disrupt the formation of X-bearing sperm have been reported in 2 mosquito species, *Aedes aegypti* and *Culex pipiens* (Wood and Newton 1991). In both species, the Y chromosome is still quite similar morphologically to the X, and presumably still shares substantial sequence similarity with it—that is, it is not the largely degenerate Y seen in many other species. In *Aedes*, both Y and X are metacentric, and the sex-determining locus (M/m males; m/m females) does not recombine with the centromere. As well as carrying the M allele, some Ys also carry a D (Distorter) allele, and MD/md males produce extreme sex ratios (>90% sons). The D locus is 1.2cM away from the M and the centromere (i.e., the rate of recombination between D and M is 1.2%; Newton et al. 1978). There also appear to be 2 other sex-linked loci, A and t, which can modify the action of D (Wood and Newton 1991).

Cytological investigations of distorter males show that the X chromosome has a high probability (~92%) of breaking during the first meiotic division (Newton et al. 1976). Almost all the breaks (~97%) are at 1 of 4 positions on the X, all of which are in a region where crossing-over with the Y is rare or absent. Because breaks are especially likely to occur in a chromosome arm if that arm has a crossover during that meiosis, the breakage may occur as part of a failed crossover (meiotic recombination is initiated by

DNA breaks; see Chap. 6). It is very interesting that Y chromosomes are also broken, though at a much reduced rate (~9%), especially because these breakpoints are the same on the Y as on the X. This is clear evidence that (once again) killing is a spiteful strategy. As a consequence, distorter males produce fewer functional sperm than normal males, and their sperm supplies are depleted more rapidly (Hickey and Craig 1966b). Curiously, distorter males are superior to other males in rate of larval development, survival through immature stages, adult longevity, and mating competitiveness (Hickey 1970). Chromosome breakage is also associated with the killer Y in *C. pipiens* (Sweeny and Barr 1978).

Geographical surveys of *A. aegypti* have shown that D is present in some populations and not in others; but wherever it is present, there is also substantial resistance, and so sex ratios are not severely biased (50–61% sons; Wood and Newton 1991). This is reminiscent of the *D. simulans* pattern. In lab crosses, X chromosomes show a more or less continuous variation in sensitivity to distortion (Wood 1976, Suguna et al. 1977). Some resistant Xs even show modest drive against MD chromosomes producing female-biased sex ratios (Wood 1976, Owusu-Daaku et al. 1997). This suggests Y self-destruction (reminiscent of X self-destruction in *D. affinis* and *C. dalmanni*).

For several reasons, Y drive is expected to more easily cause population extinction than X drive. Y drive is 3 times stronger than X because it recurs every generation, while an X drives only every third (Hamilton 1967). X drive also leads to population expansion, while the Y leads to population contraction. In addition, in a population without suppressors, a Y^K that spreads through a population would bias the sex ratio toward males and each male would have fewer chances to mate with a female. Thus, sperm limitation becomes less of a problem and there is not the same self-limitation for killer Ys as there is for killer Xs. One possible exception occurs if distorter males suffer disproportionately from sperm competition, because this becomes more common as a killer Y spreads.

Taxonomic Distribution of Killer Sex Chromosomes

This survey of killer sex chromosomes raises two obvious questions. First, why are killer sex chromosomes so much more common in dipterans than in other well-studied male heterogametic taxa like mammals? Second, within the dipterans, why does the relative frequency of killer Xs and killer Ys appear to differ between taxa (e.g., *Drosophila* vs. *Aedes/Culex*)?

Considering first the difference between dipterans and mammals, 1 explanation is suggested by the observation that, in mammals, the sex chromosomes are inactivated during male meiosis, whereas this does not occur in *Drosophila* (McKee and Handel 1993, *contra* Lifschytz and Lindsley 1972). In mammals, transcription from the sex chromosomes appears to stop sometime early in meiotic prophase; cytologically, this is associated with the sex chromosomes becoming heterochromatic, forming a "sex vesicle" or "sex body." This behavior is specific to the sex chromosomes in male meiosis: autosomes continue to be transcribed and do not become heterochromatic, and the X chromosomes behave normally during female meiosis. One consequence of this inactivation is that "backup" copies of some X-linked genes have been selected for on autosomes that are only expressed in meiotic and postmeiotic cells (Emerson et al. 2004). Another plausible consequence is that the evolution of gamete killers on the sex chromosomes is rendered more difficult.

In *Drosophila*, after chromosome pairing, the chromosomes decondense and are transcribed throughout the spermatocyte growth period, until the meiotic divisions begin, and the X does not differ much in this from the autosomes (Kremer et al. 1986). The Y chromosome is also actively transcribed in the spermatocyte, forming large "lampbrush loops" (Bonaccorsi et al. 1990). Such activity would also allow the expression of killers. Interestingly, the degree of X-chromosome activity during male meiosis varies among dipteran families (e.g., active in Drosophilidae, Trypetidae, robberflies, batflies, and Phrynidae; inactive in Tipulidae, Mycetophilidae, and Anopheline mosquitoes; McKee and Handel 1993). It would be interesting to know whether the frequency of killer sex chromosomes varies accordingly.

Why did meiotic sex chromosome inactivation evolve? One possibility is that it evolved to suppress some sex-linked killer(s) of the past (Haig and Grafen 1991). But we would then have to explain the continued activity of the sex chromosomes in male *Drosophila* as being due to some constraint, such as there being too many genes on them that are essential for spermatogenesis. Alternatively, McKee and Handel (1993) propose that inactivation is a side effect of chromatin remodeling that protects heteromorphic sex chromosomes from recombining. Such chromosome-wide protection is not possible in *Aedes* and *Culex* because X and Y do recombine, and it is not necessary in *Drosophila* because no recombination happens at male meiosis anyway. Indeed, all the dipteran taxa we listed as having sex chromosome activity have achiasmate male meiosis; among those with inactive sex chromosomes, male meiosis may be chiasmate (Tipulidae and Anopheline mosquitoes) or achiasmate (Mycetophilidae).

According to this logic, killer sex chromosomes are not found in mammals because sex chromosomes are inactivated before the killer alleles can be expressed. And not all dipterans are susceptible—only those that do not have this inactivation system, either because there is no danger of X and Y recombining (*Drosophila*), or because X and Y typically do recombine over much of their length (*Aedes* and *Culex*).

Regarding the relative frequency of X and Y drive, we suggest X drive is more common in *Drosophila* simply because their X chromosomes have many more genes than their Ys, and so there is that much more opportunity for a killer mutation to arise on the X. By contrast, in *Aedes* and *Culex* the X and Y are only slightly diverged, and presumably differ little in the number of genes. In this case, it may be that killer Ys are more likely to arise because recombination in females can break up multilocus killer complexes on the X, but not on the Y. In addition, harmful effects on female fitness may prevent a killer X from spreading, but are irrelevant to a killer Y. But with only two examples, it is not even possible to say that killer Ys are more common in mosquitoes than killers Xs.

One other factor that might be important in the origin of killers is the age of the sex chromosomes. If a sex chromosome is ancient, it seems plausible that all killer mutations that can easily arise from its genes will have long since arisen and been suppressed. But if a new sex chromosome evolves (e.g., due to movement of the sex-determining genes or a translocation of autosomal genes onto the sex chromosomes), there will be a fresh set of mutational possibilities. It is interesting in this regard that the loci responsible for X drive in *D. persimilis* and *D. pseudoobscura* are on the right arm of the chromosome, which is equivalent to an autosomal element in most other *Drosophila* species (Patterson and Stone 1952). As we have already noted, both killer Ys have arisen in species with only partially diverged sex chromosomes, which suggests recent origin. And we will see in a later section that the unusual forms of driving sex chromosomes in lemmings may be associated with recent translocations.

Evolutionary Cycles of Sex Determination

We end our review of dipteran sex chromosome killers with some speculations about how they may have been important selective agents for changes in the sex-determining mechanism. As we have seen, killer sex chromosomes lead to a skewed sex ratio, which selects for some sort of compensating change by the rest of the genome. At least in *Drosophila,* the most common type of response is for a suppressor to evolve, often on the Y, the main victim. In some cases, there may be additional responses in the sex chromosome system. The evolution of fertile XO males in *D. affinis* may be due to the presence of a killer X in that species, and in one experimental *D. melanogaster* population constructed to have a killer Y, a supernumerary sex chromosome system evolved (Lyttle 1981).

Selection may also favor changes in the sex-determining mechanism. For example, if there is a killer X and the sex ratio is female biased, a mutation that converted what would normally be a female into a male might be selected (Bull and Charnov 1977, Werren and Beukeboom 1998). This has not occurred in *Drosophila,* presumably because the Y chromosome has genes essential for male fertility, and so XX males are sterile. Similarly, in mosquitoes a new dominant feminizing mutant that turned XY individuals into females would, in the next generation, give rise to YY progeny. If the Y has degenerated at all, such individuals may be lethal or sterile, costing a female a quarter of her reproduction. Thus the evolution of heteromorphic sex chromosomes can constrain evolutionary changes in the sex-determining system (Bull and Charnov 1977). Certainly the sex-determining mechanism appears to be stable in *Drosophila,* the same hierarchy of genes being observed in *D. melanogaster* and *D. virilis,* which have been separated for some 60my (Marín and Baker 1998). But heteromorphic sex chromosomes and stasis are not universal among dipterans, and in some genera the sex-determining system is highly dynamic. In species in which sex is determined by a dominant masculinizer (*M*) gene, the gene may be found on different chromosomes in closely related species, or even in the same species (e.g., *Culex, Eusimulium, Chironomus, Megaselia, Musca;* Traut 1994, Blackman 1995). Indeed, the *M* factor in some species appears to move between chromosomes at an appreciable rate, as if it were incorporated in a transposable element (Traut and Willhoeft 1990). And in *Musca,* there are both male and female heterogametic systems.

Perhaps some of this diversity has evolved in response to the spread of driving sex chromosomes (for whose existence in *Musca* there is provisional evidence, reviewed in Jaenike 2001). After a male heterogametic system evolves, the X and Y will begin to diverge. In many dipterans there is no recombination in males (White 1973), and the 2 chromosomes will begin to diverge immediately, along their whole length; in other dipterans, reduced recombination is expected to gradually evolve between X and Y due to selection for genes with sex-specific beneficial effects ("sex antagonistic" genes; Fisher 1931, Bull 1983). In either case, as already noted, such divergence will facilitate the evolution of a killer complex. If a killer X chromosome evolves, it will produce a female-biased sex ratio, and as long as the X chromosome has not yet lost essential male-specific genes, one simple response is for the *M* allele to be copied to a new chromosome. This would rebalance the sex ratio and free the killer to go to fixation. With *M* on a new chromosome, the whole process could start again. If, the next time, a killer Y evolved, a male-biased sex ratio would result, and a simple response would be for a dominant feminizer to arise. Its spread would rebalance the sex ratio and allow the killer to go to fixation, resulting in a female-heterogametic sex-determining system, which could then reevolve to male heterogamety, and so on, until such time as the Y evolved unique and essential genes not found on the X and the sex-determining system became difficult to change. Testing these ideas will require better knowledge of the molecular genetics and evolution of dipteran sex-determining mechanisms, current areas of active research (Box 3.2). In addition to genetic conflicts, parasitic endosymbionts could also play a role in the evolutionary dynamics of sex-determining systems, as could mother-offspring conflicts (Werren and Beukeboom 1998, Werren and Hatcher 2000, Werren et al. 2002).

Feminizing X (and Y) Chromosomes in Rodents

In this section we review a class of selfish sex chromosomes that drive by first reversing the sex of their host. In all known cases, an XY male is con-

BOX 3.2

Molecular Biology of Dipteran Sex-Determining Systems

Sexual differentiation is controlled by different genes being turned on and off in males and females. In *Drosophila* this differential activity is largely controlled by a pair of transcription factors that are derived from alternative splicing of the same transcript (Schütt and Nöthiger 2000). That is, there is a gene (*doublesex, dsx*) that is transcribed into RNA and then spliced in one of two ways, making either of 2 transcription factors, DSXM or DSXF. The former activates male genes and suppresses female genes, while the latter does the opposite, activating female genes and suppressing male genes. Most somatic sexually dimorphic characters are determined by *dsx*. Homologs of *dsx* have been found in other dipterans (*Megaselia, Bactrocera, Musca,* and *Ceratitis*), and in all these species *dsx* has male- and female-specific transcripts. At least in *Drosophila,* the default is for the *dsx* transcript to be spliced to give DSXM, the male transcription factor. However, if another gene (*transformer, tra*) is active, one gets the female-specific splicing. *tra,* then, is a dominant feminizer.

In principle, these 2 genes are all one needs for a sex-determining system. There could be 2 *tra* alleles, functional and nonfunctional; the former would be a dominant feminizer, the chromosome it was on would be a Y, and the species would be female heterogametic. But female heterogamety is rare in dipterans (Blackman 1995, Marín and Baker 1998). Much more common are male heterogametic species, the most widespread system having a dominant masculinizer gene. Starting from the dominant feminizer system, the simplest change that would give a dominant masculinizer and male heterogamety would be for a dominant mutation to arise that suppressed the activity of *tra* (Schütt and Nöthiger 2000). Such a mutation (call it *M*) could be selected if, from the autosomes' perspective, the population sex ratio was female biased (due, for example, to Y drive or a cytoplasmic feminizer; Hamilton 1993, Caubet et al. 2000). The *tra*-bearing chromosome could then go to fixation, and *M* would increase in frequency to 0.25, its chromosome would be the new Y, and the species would be male heterogametic. In principle, female heterogamety could then arise via a constitutively active allele of *tra* that was no longer suppressed by *M* (Schütt and Nöthiger 2000).

In *Drosophila* itself, sex determination is more complex than it appears to be in many other dipterans, as *tra* activity is dependent upon *sxl* activity, and *sxl* activity is dependent upon the ratio of X chromosomes to autosomes (Schütt and Nöthiger 2000). We can think of two scenarios by which such a system evolved. First, suppose the ancestral state was a dominant masculinizer. Because there is no recombination in males, the X and Y quickly start to diverge, and the Y chromosome will degenerate over evolutionary time, due to the absence of recombination. As a consequence, the expression of X-linked genes (relative to autosomal genes) will differ between males and females. Now, male and female embryos will differ not only in the presence of the masculinizing protein but also in the concentration of X-linked proteins. For example, *sxl* is an X-linked gene making an RNA-regulating protein, and assuming this was also true in the ancestor, once it degenerated off of the Y chromosome, its concentration would be 2-fold greater in females than males. And if there was selection on another gene to have differential expression in the 2 sexes, the gene might evolve to be responsive to the presence of *sxl* (as opposed to responding to the presence of the masculinizer). For example, other X-linked genes, as they degenerate off of the Y, might evolve higher expression levels to compensate for being haploid in males, and then might evolve to be repressible by *sxl*, thereby reducing the impact on females. Indeed, many X-linked genes that are expressed early in *Drosophila* development are repressed in females by the presence of *sxl* (Schütt and Nöthiger 2000). In principle, the 2-fold difference in *sxl* concentrations could be amplified if *sxl* needed to bind at multiple sites to have an effect—a 2-fold difference could become 4-fold, 8-fold, and so on. Also, a transcription factor (*msl-2*) has evolved to be responsive to the presence of *sxl*, and many X-linked genes are now responsive to *msl-2*. This is how the majority of dosage compensation between the sexes is achieved. Likewise, the somatic sex determiner *tra* could itself have evolved to be responsive to *sxl*. And, at some point (perhaps early on), *sxl* evolved a more robust (and beautifully intricate) system for its own control, involving autoregulatory feedback loops, so differences in *sxl* expression between the sexes became all-or-nothing rather than 2-fold. By this scenario, it is no accident that *sxl* is sex linked or that it is active in females and not males.

Alternatively, *sxl* might have been controlling sex-antagonistic genes—genes that are beneficial in one sex but not the other—and so was itself a sex-antagonistic gene. Rather than *sxl* evolving to come under the control of *dsx* (as one usually imagines for sex-antagonistic genes), *tra* might have evolved to come under the control of *sxl*—an alternative way of achieving the same thing, namely, sex-specific expression.

verted into a female; but in some species, genes on the X cause the sex change, while in others the genes are on the Y. This turns out to be a crucial difference. Genes on an X* chromosome that cause a male to develop into a female may benefit from the early abortion of the novel YY offspring that an X*Y mother produces, but this same effect imposes an immediate cost on a novel Y* chromosome that causes sex reversal, because Y* now causes its own early abortion in half the offspring containing it. Feminizing X chromosomes have been studied intensively in several lemming species. They display a complex and interesting dynamic in which the initial spread of the feminizing X sets in motion a series of effects on maternal traits (such as ovulation rates) and on both the Y chromosome and the original X, with return effects on itself. Feminizing Ys in *Akodon* rodents have also been studied in some detail, but here the information is more in the form of an enigma. Despite some suggestive evidence, it is not yet clear how these special Ys achieve their advantage.

The Varying Lemming

In the varying lemming *Dicrostonyx torquatus*, some of the X chromosomes (designated X*) are special in that they suppress the normal male-determining function of a Y chromosome, so X*Y individuals are female (Gileva and Chebotar 1979, Gileva 1980, Gileva et al. 1982, Gileva 1987). Thus, there are 3 kinds of females: XX, XX*, and X*Y; and males are XY. When X*Y females go through meiosis, they produce both X*- and Y-bearing eggs, each of which can be fertilized by X- or Y-bearing sperm. If a Y egg is fertilized by a Y sperm, the resultant YY zygote is inviable, as it is missing all the essential genes on the X chromosome. Thus, half the Y-bearing eggs die soon after fertilization, and fully two-thirds, rather than the Mendelian 50%, of the viable progeny produced by an X*Y female carry X*. In general, rodents ovulate more eggs than they implant and implant more than they give birth to, so there is often substantial natural abortion during the gestation period. Therefore, the reproductive output of X*Y females need not be reduced by a quarter–and may not be reduced at all. In this case X* can increase in frequency when rare, but (unlike the dipteran sex chromosome killers) it is never expected to reach fixation (Bengtsson 1977). As X* increases in frequency, the population sex ratio becomes female biased, and so the benefit of changing an individual from male (the rare sex) to female (the common sex) decreases until X outreproduces X*, thereby stabilizing X*'s numbers. In the simplest situation of no fertility differences among the 3 types of females and no segregation distortion other than that due to the inviability of YY zygotes, the equilibrium frequency of XX, XX*, and X*Y females would be 60:20:20%, and together females would constitute 58% of the population (Bull and Bulmer 1981).

Several features of the *Dicrostonyx* system deviate somewhat from this simplest model. First, when X* first arose, the fertility of X*Y females was probably not identical to that of the other 2 classes of females. This is indicated by their higher ovulation rate, presumably an adaptation that arose subsequent to the origin of X* (Gileva et al. 1982). Even now, the best estimate is that X*Y females have only 87% of the reproductive output of XX or XX* females, though this is not significantly different from 100% (Gileva

1987). Reduced fertility of X*Y females may be expected both because of the death of YY zygotes and the unmasking of female-specific deleterious recessives on the single, unguarded X*. Note, though, that the fertility of X*Y females must have been at least 75% of that of the other types of females or the X* chromosome would not have spread.

The second complication is that the males, which are XY, appear to produce a disproportionate fraction of sons (~56%; Gileva 1987). Such drive is adaptive from the male's point of view. The female-biased population sex ratio produced by X* selects for males (or, more precisely, autosomal genes in males) that produce a disproportionate fraction of Y-bearing sperm, so that such males produce more sons. However, the selective dynamics here are quite complex, because Y-bearing sperm also run the risk of fertilizing a Y-bearing egg, and the maximum drive that can be stably selected for in males is 60% Y sperm, with lower values if the relative fertility of X*Y females is less than 1 (Bulmer 1988). Moreover, because the Y chromosome shows drive at the expense of the X chromosome, the frequency of the latter goes down, and therefore the frequency of X* goes up, and the eventual effect is that the population-wide sex ratio becomes more *female* biased (Fig. 3.3). It would be interesting to confirm whether males are indeed responsible for the observed segregation distortion, for that would suggest that male mammals can evolve alternative sex ratios in response to a particular selection pressure. Jarrell (1995) suggests an alternative explanation: the biased segregation may be a postzygotic effect of XX females suffering more from inbreeding depression than XY males.

If the observed female fertilities and male segregation ratios are used to calculate equilibrium frequencies of the various genotypes, the expected proportion of X*Y genotypes among females is 42% and the expected overall proportion of females is 69%. These expectations are very close to the values observed for captive populations (36% and 63%, respectively; Gileva 1987).

The Wood Lemming

The wood lemming *Myopus schisticolor* also has a dominant feminizing X* chromosome (Fredga et al. 1976, 1977). In this case, cytological studies have shown that X* has a deletion and inversion in the short arm, compared to the normal X (Fredga 1988, Liu et al. 1998, 2001). The deletion is about 7%

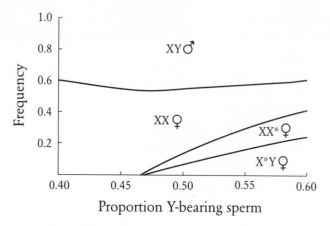

Figure 3.3 Expected frequency of different sex types in the varying lemming as a function of the segregation ratio in males. Calculated from equations in Bull and Bulmer (1981), using observed fertilities for the 3 types of females (there is no qualitative change if equal fertilities are used). Note that if males were to bias segregation in favor of X*, then X* could be driven out of the population; however, they are selected in the opposite direction, to favor Y, thereby increasing the frequency of X* and the overall sex ratio bias.

of the total length of the short arm, probably equivalent to 1Mb or more. The main difference in this species compared to the varying lemming is that X*Y females produce only X* eggs and no Y eggs (Winking et al. 1981). This is achieved by a double nondisjunction in the female germline in which, at some particular set of mitotic divisions, the X*s stick together and go to one pole and the Ys stick together and go to the other. YY daughter cells do not contain essential X-linked genes and degenerate, and so only X*X* cells proceed through meiosis and develop into X* eggs. These are fertilized by X- or Y-bearing sperm, and in either case develop into daughters. Thus, X* is passed on to 100% of the progeny of X*Y females, all of which are daughters. Consequently, one expects a higher equilibrium frequency of X* and a more extreme sex ratio bias than in the varying lemming. In the simple model of no fertility differences and no other segregation ratio distortion, the 3 females are expected to be equally frequent and fully 75% of the population are expected to be female.

The genetic basis of this double nondisjunction is not known and would be interesting to study—in particular, to determine whether it is due to X-linked or autosomal genes. While it is easy to see why an X-linked gene causing this behavior would spread through the population, it is less clear whether an autosomal gene would (Maynard Smith and Stenseth 1978, Bull and Bulmer 1981). From the autosomes' point of view, the increased fertility due to the absence of YY zygotes is a plus, but the increased frequency of daughters is a minus: at the equilibrium of 75% females, a son is 3 times more valuable than a daughter. It was once thought that the extreme population cycles of lemmings induced a sufficiently structured population in the low-lemming years that even the autosomes were selected to produce a female-biased sex ratio (Stenseth 1978, Maynard Smith and Stenseth 1978), but the data suggest otherwise (Stenseth and Ims 1993). Isozyme and chromosome frequency data fail to reveal inbreeding in low-population years, as might be expected if local mate competition were common (Gileva and Fedorov 1991). Likewise, sex ratios vary from year to year, but this variation is not associated with low- or high-lemming years (Eskelinen 1997). And preferred habitat seems to produce large, continuous populations with little clumping, even in moderate lemming years (Ims et al. 1993). Another possibility is that X* evolved a maternal-effect killer genotype, killing all male progeny *in utero,* in which case it would be in the autosomes' interest not to produce any males anyway.

Interestingly, X*Y females appear to be *more* fit than X*X and XX females (Bondrup-Nielsen et al. 1993, Fredga et al. 2000). At birth, X*Y females are about 8% heavier than XX* littermates (Fredga et al. 2000). In nature, X*Y females mature more quickly and have a higher pregnancy rate than do XX* females, which in turn have a higher rate than do XX females. X*Y females have half the oocytes of XX* females and 40% less somatic ovarian tissue but twice the frequency of growing oocytes, and there is no significant difference in mean litter size (Fredga et al. 2000). The reason for these differences is not clear—it may reflect an effect of the Y, which is expected to be more selfish *in utero* because it is normally inherited from the father (Hurst 1994; see Chap. 4), but this requires that the Y act selfishly in a female because it was selected to act that way in a male. It has no self-interest of its own to express in an X*Y female because it will not be transmitted by her (nor by any of her littermates).

Another alternative is to consider the viewpoint of X*. Because it is only found in females, it is selected to become a female-benefiting chromosome, unencumbered by selection for male benefit. In other words, all sexually antagonistic genes benefiting females are favored, not just those whose benefit in females exceeds the cost in males (Rice 1992, Chippindale et al. 2001). Set against this is the fact that recombination between the X and the X* seems to be suppressed only along the portion of the small arm of X* involved in sex reversal. The rest of the X* shows the same *G*-banding as the X (Herbst et al. 1978, Akhverdyan and Fredga 2001), suggesting that recombination has suppressed differentiation. In addition, an X*Y mother and her brood are all identically related to each other on the X*, thus reducing potential conflict. This agreement may be especially important where relative valuation of X*Y females is concerned. Because of X* drive, from the standpoint of an X* chromosome, X*Y females are twice as valuable as X*X females. This is true whether the X* expressing the preference is located in the X*Y female, her X*X littermate, or her X*Y mother. The X*Y may use the Y as a cue to appropriate action or may permit selfish Y expression because this benefits the X*. Finally, X*Y females get to enjoy a uterus in which males are absent (Bondrup-Nielsen et al. 1993). In mice and gerbils, being gestated between individuals of the opposite sex has negative reproductive consequences, apparently due to diffusion of sex hormones within the uterus (vom Saal 1981, Clark and Galef 1995).

As in the varying lemming, X*Y females are more frequent than expected on the assumption that the 3 classes of females are equally fit: data from the wild show that 45% of all females are X*Y (Fredga et al. 1993). There is no difference in survival rates between the 3 kinds of females, so the excess of X*Y females presumably reflects their greater fecundity (Bull and Bulmer 1981).

Two curious facts invite interpretation. Males appear to survive better than females because 20% of young animals are male, but 50% of old are male (Bondrup-Nielsen et al. 1993, Fredga et al. 1993). The usual pattern in microtine rodents (as in mammals more generally) is for males to show lower survival than females (Krebs and Myers 1974). Perhaps the great excess of females reduces the chance of high variance in male reproductive success, thereby putting a premium on male survival as a key factor affecting ultimate success. The second fact is that X*Y females appear to cluster with each other more tightly than do the other categories of females (Ims et al. 1993). Because the former are never gestated near males, which in mice (vom Saal and Bronson 1978) and another microtine (Ims 1987) is associated with later decreased sociality, X*Y female wood lemmings may be more sociable. In addition, from the standpoint of X*, X*Y females are uniquely valuable, perhaps leading to stronger selection for mutual support.

Finally, there appears to be some association between these feminizing X* systems and X-autosome translocations, though the significance of this is not clear. In particular, in *D. torquatus* there has been a fusion between the X and the largest autosomal arm in the karyotype, and in the sibling species *D. groenlandicus* (which also has an X* chromosome), there has been a fusion between the X and another autosomal arm (Jarrell 1995). In *M. schisticolor* the X is the largest chromosome of the complement, constituting 12.4% of the haploid genome, which is substantially larger than the 5–6% that is usual for mammals. The Y chromosome is also relatively large (5.7% of the haploid genome; Liu et al. 1998). Again, this may have come about as a result of fusions of autosomal DNA with the sex chromosomes. Perhaps these translocations brought a new set of genes to the X that are capable of mutating to feminizers.

Other Murids

There are other systems in murids that depart from the usual XX/XY, in particular, creeping voles *(Microtus oregoni)* and vole mice *(Akodon* spp.), but these systems are much more difficult to explain (Table 3.2). In each case, the system could in principle have evolved from a lemming-like one, but there is no direct evidence for this or for any lemming-like systems in closely related species.

In the creeping vole, females are somatically XO and males are somatically XY (Ohno et al. 1963). At some point in the female germline, nondisjunction of the X at mitosis produces XX and OO daughter cells. Only the former of these go on to produce eggs, and all eggs contain a single X. In the germline of males, a nondisjunction of Xs at some mitoses produces XXY and OY daughter cells. Only the latter of these go on to produce sperm cells, which are therefore either O (producing a female) or Y (producing a male). This system could have evolved from a *Myopus*-like system in which the X* chromosome shows premeiotic nondisjunction, as the X does in *M. oregoni*. Suppose a "superdominant" Y chromosome arose that could not be feminized by X*. Because it would make males, the rarer and more valuable sex, it would increase in frequency. This would reduce the population-wide sex ratio bias, which might allow X* to increase further in frequency. Careful modeling is needed here, but X* might go to fixation, fol-

Table 3.2 Taxonomic distribution of rodents with unusual sex chromosome and sex-determining systems

	Female Soma (germ)	Male Soma (germ)
Order Rodentia		
Family Muridae		
Subfamily Arvicolinae (Microtinae)		
Dicrostonyx torquatus	XX, XX*, X*Y	XY
Myopus schisticolor	XX, XX*, X*Y(X*X*)	XY
Microtus oregoni	XO(XX)	XY(YO)
Ellobius lutescens	XO	XO
E. tancrei, E. talpinus, E. alaicus	XX	XX
Subfamily Hesperomyinae		
Akodon azarae, A. mollis, A. varius	XX, XY*	XY
Subfamily Murinae		
Tokudaia osimensis	XO	XO

If the premeiotic genotype differs from the somatic genotype, it is given in parentheses afterwards. From Fredga (1994).

lowed by the superdominant Y. The sex chromosome system then would be X*X* females and X*Y males. However, the X* would still show non-disjunction in the female germline, which might lead to some reduction in female fertility. The male might then be selected to exclude X* from its gametes, instead producing Y and O sperm, as observed. This is by no means the only possible scenario, and others that do not derive from a lemming-like system are discussed by Charlesworth and Dempsey (2001).

Vole mice also reveal high frequencies of XY females (e.g., 15–40%; Bianchi et al. 1971, Bianchi et al. 1993, Hoekstra and Edwards 2000). But here the feminizing agent is a special Y* (Bianchi et al. 1971, Lizarralde et al. 1982). XY* females have been seen in 8 species of *Akodon*, and in at least 6 of them, the mutant Y* is more closely related to the normal Y of the same species than it is to a Y* of another species (Hoekstra and Edwards 2000). This suggests that the system has arisen multiple times (at least 6) and is probably evolutionarily short-lived. Curiously, XY* females of at least 1 species, *A. azarae*, appear to be *more* fit than XX females, at least in the lab, where they start reproducing earlier, have shorter intervals between litters, continue to reproduce longer, and are estimated to have about 35% higher fitness (Espinosa and Vitullo 1996, Hoekstra and Hoekstra 2001).

Again, the cause of this increase is mysterious. XY* females produce 67% daughters, and perhaps these are cheaper than sons. Also, XY* females (which necessarily have XY* mothers) have an increased chance of being gestated between 2 other females, which can increase female fitness (vom Saal 1989, Clark et al. 1986, 1998, Espinosa and Vitullo 1996). We also note that the X chromosome of XY* females is *paternally* derived, and so perhaps is more selfish, with no maternally derived X to balance it (see Chap. 4). This might explain why, even preimplantation, XY* embryos have a faster rate of development than XX embryos from the same (XY*) mother (whereas the XY and XX embryos of XX mothers show no such difference; Espinosa and Vitullo 2001). Of course, this is more directly explained by assuming it is an effect of Y*.

In addition to the fitness advantage, there is some suggestion that in XY* females the Y* chromosome has a transmission advantage, being passed on to perhaps 63% of eggs rather than the Mendelian 50%, and these 2 factors together can, in principle, account for the maintenance of Y* in natural populations (Hoekstra and Hoekstra 2001). It should also be noted that the Y* is only found in females, so (like X*) it quickly evolves to be a female-benefiting chromosome; but unlike X*, it is not constrained by recombination with the male-benefiting Y, because YY* individuals are inviable. On the other hand, the Y* is a small chromosome, smaller than the Y itself (Solari et al. 1989). One thing that needs modeling here is selection on the autosomes (and the X)–do they want to suppress (i.e., remasculinize) Y*?

Evolving the system is another matter. At its inception, the XY* system has the mirror-image disadvantage of the X*Y advantage. That is, in both systems YY individuals are lethal. With complete compensation, this means that in one system, X*Y, room is freed up for 33% more X* individuals, while in the XY* system there is a net loss of 33% of Y* individuals. There is some inconclusive evidence that the Y may drive against the X in males and, if this were ancestral and Y* drove against the X in XY* females, this would help offset the cost (H. Hoekstra, pers. comm.). But it must be noted that drive in one sex is usually completely unassociated with drive in the other and, as we have seen, a driving Y in males would be a natural evolutionary response to a feminizing and driving Y*, because each biases the sex ratio in the opposite direction.

Little is known about the genetic basis of Y*, except that, at least in *A. azarae*, it appears to be smaller than the normal Y (Solari et al. 1989), and the effect is not simply due to a deletion of the male-determining gene, *Sry* (Bianchi et al. 1993). One possibility that seems worth considering is that it evolved from a *Dicrostonyx*-like system, in which a superdominant Y evolved, which spread but did not go to fixation. Under this scenario, Y* is ancestral, and Y is derived. X chromosomes may still be cryptically polymorphic for X and X*. But there is no evidence of a lemming-like system anywhere in the genus.

It is probably not coincidental that rodents provide numerous exceptions to the general rule for mammals (both placental and marsupial) that *Sry* is found in a single copy located on the Y (Fernandez et al. 2002). These range from species in which *Sry* or the entire Y chromosome is missing, such as some mole voles and the spiny rat (Just et al. 1995, Soullier et al. 1998), to ones in which *Sry* is found in multiple copies on the Y only, as in *Microtus, Akodon,* and some murids (Bianchi et al. 1993, Nagamine and Carlisle 1994, Lundrigan and Tucker 1997, Bullejos et al. 1999), as well as *Microtus cabrerae,* in which multiple *Sry* copies are found in X chromosome heterochromatin and perhaps as few as one on the Y euchromatin (Fernandez et al. 2002). The X copies are apparently inactive, as most contain base substitutions and deletions that produce stop codons. In *M. cabrerae,* some cases of XY females have been reported (Burgos et al. 1988). *Sry* has a single exon that codes for a highly conserved high-mobility group domain (HMG box) of about 78 amino acids (Whitfield et al. 1993). HMG boxes are characteristic of a variety of proteins that bind DNA with strong affinity but varying degrees of specificity. The DNA-binding activity of the *Sry* protein resides in the HMG box and mutations therein result in XY females in house mice. It is highly conserved, but the flanking sequences are rapidly evolving in both rodents and other mammals, such as primates (Foster et al. 1992, Tucker and Lundrigan 1993, Whitfield et al. 1993).

Finally, it is worth emphasizing that all of the alternative mammalian sex chromosome systems we have discussed would appear to be susceptible to the evolution of maternal-effect killers. For example, in the *Akodon* species we have just described, the Y* chromosome is transmitted only from mothers to daughters. One can easily imagine it acquiring a gene that causes it to kill sons *in utero,* because they are guaranteed not to have inherited the Y*, thus freeing up resources for daughters. Similarly, the X* chromosome of lemmings is also only transmitted from mothers to daughters, and so it might also evolve to kill sons *in utero.* And in the creeping vole, the X chromosome is passed from mother to both daughters and sons, but the sons are an evolutionary dead end (as they are for mitochondrial genomes). There-fore, one would expect early male embryonic lethals to evolve on the X and thereby increase the total investment made in daughters. In an outcrossed population, such an X would go to fixation, driving the population extinct, unless suppressors or some other sex chromosome system evolved.

This notion that different sex chromosome systems may be differentially susceptible to the evolution of selfish elements, which in turn may affect the persistence time of the sex chromosome system, could apply more widely. One wonders, for example, how often sex-linked maternal-effect killers arise in viviparous taxa with female heterogamety, and even whether viviparity is less likely to evolve in female heterogametic taxa, and vice versa. To give one more speculative example, for males to have only a single sex chromosome (i.e., XO) is common in many invertebrate groups; but for heterogametic females, it appears to be less common (though not absent; Traut and Marec 1996). We speculate that in many cases when such a system does arise, the X will too easily evolve to show drive by preferential move-ment at female meiosis to the functional egg nucleus, thereby avoiding the polar bodies. Such behavior is common for single B chromosomes, whereas if there are two Bs, they pair and show Mendelian segregation (see Chap. 9). In species with XO females, a driving X would produce a male-biased sex ra-tio and select for a change in the sex-determining system (extinction would probably be unlikely, because female drive is not usually complete). So, if an XO female system arises, it may be short-lived, leading to a rare and spo-radic distribution on the tree of life. Other systems with unpaired sex chro-mosomes in females also appear to be rare and sporadically distributed. For example, the system of YO females, OO males appears in the frog *Leiopelma hochstetteri*, but not its congener *L. hamiltoni*, which has a more conventional system of XY females, XX males (Green 1988). A monogenous system of FO daughter–producing females and OO son–producing females (and OO males) appears in the parasitic barnacle *Peltogasterella gracilis* (Bull 1983) and we also expect it to be short-lived.

Other Conflicts: Sex Ratios and Mate Choice

The destruction of one sex chromosome by another is a particularly dra-matic action about which the 3 main parties–the X, the Y, and the autosomes–will have strong conflicts of interest and will experience quite different selection pressures. However, it is not the only issue over which there are expected to be conflicts of interest. For example, coefficients of relatedness between family members will usually differ between X, Y, and autosomes, and therefore also the selection pressures on social behavior, dispersal, and inbreeding avoidance (Hamilton 1972; see also Chap. 4 on imprinting and sex chromosomes). Here, we discuss some other sources of conflict, unconstrained by any direct facts. Indeed, for some of our ideas, it will probably be decades before the molecular genetics of birds and mammals is sufficiently well advanced to test them.

In species with chromosomal sex determination, the optimal distribution of sons and daughters from the perspective of an autosomal gene is not always the random 50% produced by default at meiosis, and there is abundant evidence of adaptive shifts in sex ratios. For example, colonial spiders very often inbreed, and so are selected to produce more daughters than sons, to reduce the competition among brothers to fertilize their sisters. In *Anelosimus eximius* and *A. domingo* (Theridiidae), for example, 90% of the eggs laid are daughters (Avilés and Maddison 1991, Avilés et al. 2000). As in most spiders, there are 2 different X chromosomes, and no Y, and males are the heterogametic sex (i.e., X_1X_2/O). The means by which the biased sex ratio is achieved is not known: male meiosis appears to be normal, with the 2 types of nuclei found in equal proportions at the end of telophase II. Presumably the male either somehow partly incapacitates O-bearing sperm or produces some detectable difference between sperm types that the female uses to control fertilization. Whatever the mechanism, sex ratio control can be quite precise: at least in *A. domingo*, the among-family variation is less than expected under a random binomial distribution. Assuming the sex ratio is partly under the control of the male genotype, there is likely to be a conflict of interest between the paternal autosomes and the X chromosomes over the extent of the sex ratio bias, with the Xs selected to have a more extreme bias (though for very high levels of inbreeding, the difference in optima is small; Hamilton 1979, Taylor and Bulmer 1980). In more outcrossed species the conflict is more intense, and at least in 2 outcrossing congeners, *A. jucundus* and *A. studiosus,* the autosomes appear to win, as family sex ratios are unbiased. Presumably this is because the sex chromosomes are minority elements in the genome (~10% of the haploid genome; Avilés and Maddison 1991).

In these species there is no Y chromosome. Had it been there, it would have been selected to sabotage or otherwise interfere with the daughter-biasing mechanism, and we wonder to what extent the efficiency of the mechanism is due to the absence of a Y fighting back. Similarly, in the nematode *Caenorhabditis briggsae,* which consists of XX self-fertile hermaphrodites and XO males, the X-bearing sperm from a male has a fertilization advantage over O-bearing sperm, thus producing more hermaphrodites than sons (~3:1 over the first 24 hours after mating; LaMunyon and Ward 1997). Again, the mechanism is not known, and it may well be encoded by genes on the autosomes. There is likely to be some conflict between X and autosomes over the optimal extent of the bias, but the fact that there is no Y chromosome under a diametrically opposed selection pressure may contribute to its smooth operation.

Some birds and mammals adjust the sex ratio of their progeny in response to a particular situation. For example, in the Seychelles warbler *(Acrocephalus sechellensis),* daughters often remain on their parents' territory and help to rear subsequent offspring. On high-quality territories with abundant food, the parents benefit from this help, whereas on low-quality territories these daughters consume more than they help. As a consequence, these birds have evolved to skew the progeny sex ratio according to the quality of the territory. In one study population, pairs without helpers on good territories produced 87% daughters and those on poor territories produced only 23% daughters (Komdeur 1996, Komdeur et al. 1997). As females are the heterogametic sex, presumably they are the ones determining the sex ratio. They might, for example, control the segregation of sex chromosomes at the first meiotic division to egg versus polar body, perhaps in response to nutritional status (Pike and Petrie 2003). Whatever the mechanism, the X and Y chromosomes will not always be willing participants. When on a poor territory, the Y will be selected to disrupt or sabotage the system, and when on a good territory to amplify it (and vice versa for the X). And these conflicts need not be manifest only at meiosis. If fat reserves are used as a proximate cue for territory quality, the Y will be selected to set down more fat and lower the threshold level of fat required to produce a female-biased sex ratio. If the central nervous system is involved, it may be selected to produce feelings of well-being in the mother-to-be, to be ever the optimist about the territory. By contrast, the X will be selected to emphasize the negative, to require an ever-higher fat content before admitting the territory is a good one, to burn off fat unnecessarily, even to find poor territories with little food more attractive. Cognitive dissonance in a bird!

In other species (e.g., the zebra finch *Taeniopygia guttata* and the blue tit

Parus caeruleus), females appear to bias the progeny sex ratio based not on the quality of the territory but that of the mate, increasing the proportion of sons when the male is attractive (apparently so the sons can inherit and express the attractive qualities of the father; Burley 1981, 1986, Svensson and Nilsson 1996, Sheldon et al. 1999). Again, we can expect similar intragenomic conflicts, and not just at meiosis, though in this case it will be the X chromosome that is optimistic, finding the male more alluring than he really is, while the Y will be dissatisfied, always ready to find fault with the male, and perhaps even finding dull males more attractive. As always, the autosomes will be somewhere in between.

In mammals it is the male that is heterogametic, but sex ratio biases are often controlled by the mother, presumably by differential fertilization, implantation, or fetal abortion (Krackow 1995). Thus, conflicts between X, Y, and autosomes will be somewhat different. For example, if there were a species in which females preferred to produce sons when mated to an attractive male, there would be conflicts in the male over how much to invest in being attractive. The Y would be selected to be particularly extravagant and vain, while the X would be selected to be more sensible and cautious about the costs of sexually selected characters. As it is, there are several species known in which dominant females produce male-biased sex ratios because maternal condition has a greater effect on the fitness of sons than of daughters (e.g., red deer, *Cervus elaphus,* and some other ungulates; Kruuk et al. 1999). To the extent that the males in these species show any kind of mate choice (e.g., differential allocation of courtship efforts and sperm supplies), genes in males on the Y will find dominant females attractive, whereas those on the X will be drawn more to subordinate females.

Conflicts of interest over mate choice can also arise if potential mates differ in the relative fitness of the sons and daughters they will produce (Hastings 1994). Thus, in female birds, the Y will be interested in choosing mates based on the quality of the daughters that will result, without regard for the quality of the sons, and the X will be largely interested in the quality of the sons (but not wholly uninterested in female effects, as the X will be passed on to the daughters of those sons). Autosomes will be more equitably interested in the quality of both sons and daughters (although still with a bias toward daughters; Trivers 1985, 2002). Thus, Y chromosomes will be selected to pay attention to costly characters that best reveal the underlying vigor of the male, knowing full well that the daughters are not going to have

Table 3.3 Typical size of the X chromosome in different vertebrate taxa

Taxon (heterogametic sex)	Size (%)
Eutherian mammals (male)	5
Marsupials (male)	3
Birds (female)	7
Snakes (female)	10

Sizes given as a percentage of the total haploid genome. From Graves and Shetty (2001).

to produce them themselves, whereas X chromosomes will be more interested in cheap gimmicks that increase a male's allure. In mammals, males tend to be less choosy than females, but still they have to optimally allocate their courtship efforts and sperm supplies (and, occasionally, paternal investment), and much the same considerations apply. In males of our own species, for example, Y chromosomes should be interested in a clear complexion and other signs of a robust immune system, which will benefit sons as well as daughters, while X chromosomes should be captivated more by breasts and hips, presumably of use only to females.

A general point is worth stressing. Classical drive requires unusual killing action or directed movement at a very precise time in development in a small part of the body. The effects we are imagining here could be found in a wide range of physiological and behavioral characters, including mate choice, reproductive effort, and reproductive timing. Selection pressures will be weaker than those associated with drive, but there are many, many ways for the genes to effect these traits. In some groups of animals, such as mammals, birds, and a few insects, these effects may be much more widespread than examples of drive. Opportunities for the Y chromosome will be limited, because it is often degenerate in size and genetic content, but the X is usually a robust chromosome with substantial coding capacity (Table 3.3).

FOUR

Genomic Imprinting

GENOMIC IMPRINTING RESULTS in parent-specific gene expression, that is, in a difference in gene expression depending on which parent contributed the gene. In the usual case, an allele is silent when inherited from one parent while the identical stretch of DNA is active if inherited from the other; but often this effect is seen in some tissues and not others, or there is only a quantitative difference in gene expression, depending on the parent of origin. To give but one example, in the mouse *Meg1/Grb10* shows maternal expression in most tissues, but paternal expression in the brain; in humans it also shows paternal expression in the brain, but biallelic expression in other tissues (Kaneko-Ishino et al. 2003). Because the difference in expression occurs when DNA is identical, the effect is said to be "epigenetic," that is, the difference is caused by some aspect of the chemistry of the proteins binding DNA or of the DNA itself, but unrelated to nucleotide sequence. Such genes are said to be imprinted.

This ability to be expressed according to parent-of-origin has striking implications for a gene's degree of relatedness to its parents—and, thus, to all other relatives differentially related through them. Consider an autosomal gene's relatedness to its mother. An unimprinted gene has no information where it came from and computes only the average chance of one-half. An imprinted gene calculates exact relatedness, maternal = 1, paternal = 0. These sharply divergent kinship coefficients create 2 kinds of conflict. One occurs over evolutionary time in which the spread of selfish paternally active genes naturally selects for opposing maternally active genes (and vice versa), while both select (albeit less strongly) for opposing unimprinted genes. The second kind of conflict is imagined to occur within an individual whenever the opposing genes act against each other. Unlike most forms of conflict described in this book, the conflict here is not over which genes to transmit at what rates but rather over how much an individual should help or hinder a given relative.

The major focus of conflict between maternally and paternally active genes is parental investment. Degrees of relatedness are maximally divergent and a critical resource is at issue. Consider a fetus developing and growing within its mother's womb. Because they are not genetically identical, there will inevitably be some conflict between the fetus and the mother over how much investment should be transferred from the mother during the pregnancy, with the fetus selected to acquire more than the mother is selected to provide (Trivers 1974). A similar conflict also exists between maternally and paternally derived genes in the fetus: selection on maternally derived alleles is influenced by the fact that a copy of the gene is also found in the mother and will be passed on to half of her other offspring (Haig 1993b). This will tend to dampen selection to be ever more demanding. However, the paternally derived allele is (barring inbreeding) not found in the mother, and it will only be selected to reduce demand insofar as the mother has (or will have) other offspring with the same father. In the extreme case, in which each of a female's offspring is fathered by a different male, paternally derived genes in each offspring are selected to take as much as possible from the mother, without regard for her future survival or reproduction. At the opposite extreme, if females only mate with a single male in their lifetime and vice versa, there is no conflict because the father suffers just as much as the mother from any decrement in maternal fitness. Conflict–though less intense–is also expected over actions affecting other relatives, whenever these are differentially related through mother or father (see Fig. 4.1).

This general approach to the meaning of genomic imprinting may be called the kinship theory of imprinting and is due primarily to the work of David Haig, who first introduced the approach (Haig and Westoby 1989, Haig and Graham 1991, Moore and Haig 1991) and who has been most active in developing it further (reviewed in Haig 2000a, 2002). Other theories have been advanced to explain the adaptive significance of imprinting (reviewed in Hurst 1997), but those to date seem to lack both supporting logic and evidence (Haig and Trivers 1995, Wilkins and Haig 2003a), while Haig's

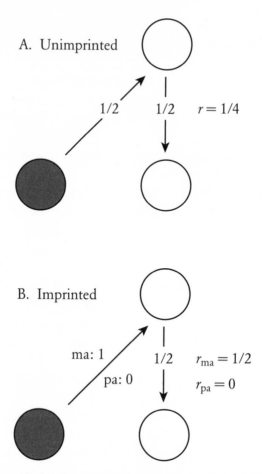

Figure 4.1 Degrees of relatedness (r's) under imprinting. A. r's to mother and mother's offspring for an unimprinted gene (1/2 to each, r to maternal half-sib = 1/4). B. r's to mother (1 and 0) and mother's offspring (1/2) for a maternally (ma) and paternally (pa) active gene, respectively (r to maternal half-sib is 1/2 and 0).

theory explains a broad pattern of evidence, with no facts clearly in contradiction. One need assume no more than Mendelian genetics and the action of natural selection. Once parent-specific gene expression appears, it must by logic evolve the biases Haig describes, with paternally active genes acting to further paternal relatives–in other words, patrilines–and maternally active genes, matrilines. If there are additional selection pressures we need to take into account to understand imprinting, we do not yet know what they are. (We exclude here, and throughout this chapter, "imprinting" in the sense of paternal genome loss in scale insects and others, which we explain according to a different logic in Chapter 10.)

Because our own genes are imprinted and a number of developmental disorders are associated with failures of imprinting, including cancers (reviewed in Hall 1999, Tycko 1999, Walter and Paulsen 2003), this category of selfish gene has been studied with unusual intensity. In addition, because the mouse, with most of the same imprinted genes, is a model genetic organism, there is a rich experimental and genetic literature regarding the mechanisms (and effects) of imprinting, especially early in development.

In this chapter, we begin our account with a review of the role of imprinting in mother-offspring interactions over parental investment, especially fetal growth. We then review the molecular mechanisms underlying imprinting, focusing in particular on the idea that conflict is expected not only between maternally and paternally active genes but also between the various components of the imprinting machinery itself—and that this may help explain much of the molecular complexity. We then expand the discussion to other sorts of social interactions and to the special factors that apply to imprinted genes on sex chromosomes. Finally, we review what is known, and what might be predicted, about the distribution of imprinting outside of mammals.

Imprinting and Parental Investment in Mammals

At least 100 genes are thought to be imprinted in the mammalian genome (out of a total of about 30,000; Reik and Walter 2001b, www.mgu.har .mrc.ac.uk/imprinting/imprinting.html, www.otago.ac.nz/IGC). Those that have been described are disproportionately expressed in the placenta and involved in fetal growth. We start with a detailed look at 2 well-studied loci in mice, oppositely imprinted and with strong opposite effects on early growth.

Igf2 and *Igf2r:* Oppositely Imprinted, Oppositely Acting Growth Factors in Mice

Igf2 and *Igf2r* are oppositely imprinted genes with strong contrary effects on fetal growth in mice. They were the first 2 imprinted genes discovered in mammals (Barlow et al. 1991, DeChiara et al. 1991), and they provided a paradigm for the underlying kinship logic of imprinting (Haig and Graham 1991). *Igf2* is imprinted in both mice and humans. It acts as if it has a complex, conditional strategy (reviewed in Stewart and Rotwein 1996). It is inactive in almost all adult tissue, save the choroid plexus and leptomeninges, in which both alleles (paternal and maternal) are active. By contrast, it is very active in almost all tissues of the fetus, including extraembryonic; but only the paternal allele is active, with the maternal being transcriptionally silent. Inactive copies can be generated by homologous recombination and when such a copy is inherited from the father—so no *Igf2* is expressed in the fetus—the individual mouse is 40% smaller at birth and throughout life but is otherwise properly proportioned (DeChiara et al. 1991). *Igf2* produces insulin-like growth factor 2, which is a growth promoter. It stimulates cells to divide. When found in 2 active copies in humans, it is often associated with Beckwith-Wiedemann syndrome, a fetal overgrowth disorder with frequent tumors (Rainier et al. 1993, Ogawa et al. 1993).

In mice and humans, the action of *Igf2* is opposed by the action of *Igf2r*, or mannose-6-phosphase-receptor, except that this gene is maternally active in mice but biallelic (active from both alleles) in humans, in which there is also no good evidence that *IGF2R* inhibits prenatal growth (for a recent review, see Dahms and Hancock 2002). In vertebrates it targets proteins tagged with mannose 6-phosphates (as well as the digestive enzymes) to the lysosomes, where they are degraded. In mammals, it has evolved a secondary binding site for *Igf2*, which it then degrades. Its deletion in mice causes offspring to be born that are 125–130% of normal birth weight (Ludwig et al. 1996). Its evolution, along with that of *Igf2*, is summarized in Table 4.1. Neither is imprinted in monotremes and both are in marsupials, the secondary binding site in *Igf2r* having also evolved in the interval. It is reasonable to suppose that *Igf2* became imprinted first, in response to the appearance of intrauterine life, and that *Igf2r* evolved its secondary binding site next and became imprinted, but there is no evidence on this (Wilkins and Haig 2001, 2003a).

Igf2's imprinting status does not change in placentals, but for unknown reasons *Igf2r* becomes biallelic in tree shrews, flying lemurs, and other primates, including ourselves. It is not at all clear why imprinting of these genes is not also found in monotremes. They do lay eggs, but these develop within the mother for up to 4 weeks, with maternal nourishment, and they increase in size from 4mm to 20mm in diameter (John and Surani 2000).

Table 4.1 *Igf2* and *Igf2r* in mammals

Gene	Imprinting Status
Monotremes	
Igf2	Not imprinted
Igf2r	Igf2-binding site absent
Marsupials	
Igf2	Imprinted
Igf2r	Binding site present and imprinted
Placentals	
Igf2	Imprinted in rodents, artiodactyls, primates
Igf2r	Imprinted in rodents, artiodactyls, not in primates

References in Wilkins and Haig (2003a).

Perhaps lacking a placenta, the egg has few ways to induce greater investment. After birth, offspring lick milk provided by the mother; so all during this time, increased growth rate associated with paternal genes would seem to give an advantage, assuming multiple paternity of a female's lifetime reproductive output. As expected, neither *Igf2* nor *Igf2r* is imprinted in chickens (Nolan et al. 2001; Yokomine et al. 2001).

Here are 2 oppositely imprinted genes in mice with strong counteracting effects on growth that cancel out to give little or no net effect—precisely what is expected of internal genetic conflict based on evenly matched paternal and maternal alleles. There are additional complexities we shall consider shortly. For example, *Igf2* is located on mouse and human chromosomes in a cluster of at least 10 imprinted genes and there are complex interactions between them. One of these, the oppositely imprinted and nearby *H19*, shares many control elements in common with *Igf2*, its mRNA interacts with that of *Igf2*, and some knockouts cause *Igf2* to become biallelic while others do not (reviewed in Runge et al. 2000, Arney 2003).

Growth Effects of Imprinted Genes in Mice and Humans

Detailed physiological and genetic evidence is now available for at least 27 imprinted genes in mice (most of which are also imprinted in humans), and this evidence shows a striking pattern (comprehensively reviewed by Tycko and Morison 2002; Table 4.2). Most of the genes have early effects on

Table 4.2 Imprinted genes in mice, their effects, and whether these effects support the kinship theory of imprinting

Imprinted Gene	Effect	Support kinship theory?
PATERNALLY ACTIVE		
Igf2	Increases size at birth (40%).	+
Ins1, Ins2	Imprinted expression in yolk sac, biallelic in other tissues (including pancreas). *Ins* known to affect early growth positively, but parent-of-offspring effect not yet demonstrated.	0
*Kcnq1ot1	Antagonizes expression of growth inhibitor $p57^{KIP2}$.	+
Dlk1	Same gene causes skeletal muscle overgrowth in sheep but time of action and cost to mother uncertain.	(+)
Sgce	Absence of gene in humans associated with obsessive-compulsive behavior and panic attacks.	0
Rasgrf1	Affects postnatal growth positively, with maximum effect at weaning (Itier et al. 1998).	+
Peg1/Mest	Increases placental and embryonic growth, which is associated with greater postnatal growth and survival. Positive effect on maternal behavior, such as pup retrieval (see text).	+
Peg3	Increases placental size and weight at birth (20%) and nursing by pups. Positive effect on nursing by mothers and on oxytocin-positive neurons in the hypothalamus (see text).	+
Nnat	Appears to have positive effect on embryonic growth, but data only from uniparental disomy. Overexpressed in embryonic cancers. Produces a neuronal protein with strongest expression during perinatal period.	(+)
Ndn	Absence associated with early postnatal lethality in mice and failure to thrive in humans (Prader-Willi syndrome) including hypotonia and failure to nurse; obsessive food-seeking behavior later in life (for this and other references to Prader-Willi, see Haig and Wharton 2003).	(+)
Snrpn	Active in many fetal and adult tissues, encodes a splicing factor component expressed at highest levels in neurons. Knockout has no obvious phenotype.	0
*Pwcr1	Absence associated with failure to thrive, early postnatal lethality, and Prader-Willi syndrome in humans. Expressed predominantly in brain, encodes a small nucleolar RNA that may affect processing of serotonin.	(+)

Table 4.2 (continued)

*Ipw	No obvious effect on phenotype at birth. Encodes an abundant spliced RNA that builds up in cytoplasm and shows high expression in brain and other tissues.	0
*Xist	Causes paternal X-chromosome inactivation in the placentae of mice, either by direct action or via its oppositely imprinted antisense transcript *Tsix*.	0

MATERNALLY ACTIVE

Igf2r	Decreases placental size and size at birth (25%) by binding to and degrading *Igf2*. In humans, not imprinted (or only in some individuals) and affects cognitive ability in children (Chorney et al. 1998).	+
*Meg3	Encodes an abundant, nontranscribed, spliced RNA, found in many fetal and adult tissues, esp. adult skeletal muscle; closely linked to *Dlk1* and appears to down-regulate overgrowth of skeletal muscle associated with *Dlk1*, but cost of overgrowth to mother unknown.	(+)
Gnas	Encodes subunit of signaling protein that decreases insulin sensitivity and fat cell growth.	+
Ube3a	Imprinted in hippocampal and cerebeller neurons, increases learning and long-term potentiation of synaptic transmission. Absence contributes to Angelman syndrome in humans, which includes neonatal hyperactivity, opposite of the oppositely imprinted Prader-Willi syndrome. Imprinted in fibroblasts, lymphoblasts, and neural-precursor cells (Herzing et al. 2002).	0
*H19	Closely linked to the oppositely imprinted *Igf2*, it encodes an abundant, spliced RNA and acts in *cis* to down-regulate *Igf2* expression via shared insulator and enhancer sequences. Silenced in most Wilms' and other embryonic tumors in humans.	+
Ascl2/Mash2	Expression limited to placenta; increases spongiotrophoblast growth at the expense of giant cells, which secrete placental lactogens (which actively solicit maternal investment; Haig 1993b). Giant cells penetrate farthest into maternal decidua; limitation on giant cells suggests limitation on maternal investment. Absence causes death in midgestation.	(+)
Cd81	Weakly imprinted only in extraembryonic tissues, biallelic in others, codes for a membrane protein, expressed in many tissues; no evidence on imprinted phenotype.	0
Kcnq1	Imprinted only in fetal tissues of mice and humans, biallelic in adults, encodes a voltage-sensitive potassium channel. Absence increases stomach size 3-fold but no evidence of an imprinted phenotype.	0

Table 4.2   (continued)

p57^(Kip2)/Cdkn1c	Imprinted in multiple fetal tissues, encodes a kinase inhibitor, evidence from *in vitro* and *in vivo* suggests that it inhibits cell proliferation in various tissues.	+
Tssc3	Strongly imprinted in placenta (and liver) of mice and humans and weakly imprinted in other fetal and adult tissues; encodes a small cytoplasmic protein, restrains placental growth but without reducing fetal growth.	(+)
Esx1	Imprinted in mouse placenta, X-linked so probably imprinted as a result of general paternal X-inactivation (see *Xist*). Absence of gene increases size of early placenta but, at later stages, depresses fetal growth, probably due to placental pathology.	(+)
Gatm	Catalyzes the rate-limiting step in the synthesis of creatine, an important molecule in energy metabolism. Imprinted in placenta and yolk sac but not embryo; effect on offspring and maternal resources unknown (Sandell et al. 2003).	0
Grb10	Potent growth inhibitor: deleting maternal copy leads to placental and fetal overgrowth, especially liver but not brain, and offspring are 30–40% larger at birth, independent of *Igf2* pathway (Charalambous et al. 2003); associated with Silver-Russell syndrome in which doubled maternal copy gives growth retardation.	+

* codes for untranslated RNA.
Except where otherwise noted, all information (including almost every evaluation regarding kinship theory) is from Tycko and Morison (2002).
  + strongly supports kinship theory; N=10
  (+) weakly supports kinship theory; N=8
  0 neutral; N=9
No cases contradict theory.
Total: N=27

growth (18 of 27). When organized according to whether the genes are paternally active or maternally active and whether they affect growth positively or negatively, 10 strongly support the kinship theory, 8 weakly (at least 1 piece of evidence absent), and none is clearly opposed. In one case *(Rasgrf1)*, the expected growth effect occurs after birth but before the end of weaning (Itier et al. 1998). In addition, some of the neutral cases are at least suggestive—for example, early growth effects not yet shown to be imprinted. In any case, the pattern is so strong as to be impervious to minor decisions regarding evaluation of individual cases (for detailed analysis of selected cases, see also Haig 2004b).

Additional supporting evidence regarding growth effects comes from mice chimeras (individuals produced in the lab that consist of a mixture of wildtype cells and others). Mice chimeras in which the added cells are androgenetic (double dose of paternal) are, as expected, usually larger than pure wildtype individuals, which in turn are larger than chimeras in which the added cells are parthenogenetic (Fundele et al. 1997). These differences tend to disappear after weaning, again as expected because there is little scope for internal genetic conflict over growth when the resources to support growth are supplied from outside the family.

Humans are very unusual in the length of the period of parental investment—well past weaning and usually into young adulthood. Thus, important imprinted effects concerning parental investment may concern postweaning life. With this in mind, Haig and Wharton (2003) have recently reinterpreted the symptoms of Prader-Willi syndrome, which results from the absence of the paternal copy of a segment of chromosome 15. For the first few years of life, children suffer lassitude and low appetite, as expected from loss of paternally active genes during nursing; but then appetite becomes voracious and undifferentiated, that is, the children eat anything and even forage afield for new foods, traits that would tend to reduce cost to the mother. For a detailed analysis of imprinted effects on calcium metabolism in pregnancy, see Haig (2004a).

Though the data summarized in Table 4.2 provide compelling evidence that kinship conflicts have played an important role in the evolution of genomic imprinting, this evidence does not mean that every imprinted gene must have been selected for its kin effects. Some genes may be imprinted as a pleiotropic effect of imprinting at a neighboring locus and some because in the past there was selection for a change in expression level, and the first appropriate mutation happened to work in a parent-specific manner. Once the imprinting machinery has evolved, we should expect non-kin effects to occur at least some fraction of the time.

Evolution of the Imprinting Apparatus

The Mechanisms of Imprinting Involve Methylation and Are Complex

In order for maternal and paternal alleles to have different expression levels despite having the same nucleotide sequence, there must be some "epigenetic" difference between them. Considerable progress has been made in the past 10 years in understanding the molecular basis of this difference in mice and humans (Reik and Walter 2001a, 2001b, Kaneko-Ishino et al. 2003, Murphy and Jirtle 2003, Verona et al. 2003). Chief in importance is the role of DNA methylation.

In the DNA of mammals (and many other taxa), when a cytosine (C) is followed by a guanine (G), the C may have a methyl group attached (Alberts et al. 2002). Methyl groups can be added or removed from particular C residues, and methylation states can be maintained or not through cell divisions, according to which enzymes are active. Methylation is therefore a kind of reversible mark that can be put upon DNA. The presence of a methyl group changes the proteins that bind to the DNA, and thus the expression level of the gene. Among unimprinted genes, methylation is usually associated with gene silencing—so, for example, a gene required only in 1 tissue may be methylated and therefore silent in other tissues. Among imprinted genes, maternal and paternal alleles usually have different patterns of methylation; however, there is no consistent association between methylation and silencing. *H19* and *Snrpn* show biallelic expression in methylation-deficient mice (indicating methylation is associated with silencing), but *Igf2* and *Igf2r* show no expression (indicating methylation is associated with activation; Reik and Walter 2001a). Overall, it seems that for about half of imprinted genes, the methylated allele is the active one (Kaneko-Ishino et al. 2003).

Because imprints must be reversed from one generation to the next, it is not surprising that in both male and female germlines almost all DNA methylation is first removed, and then the DNA is methylated *de novo* in a sex-specific manner (Murphy and Jirtle 2003). The removal occurs in the primordial germ cells of male and female fetuses and occurs as these cells are entering the developing gonad. Virtually all methyl groups are removed, except those associated with repetitive DNA. It is not yet clear if the methyl groups are lost by active demethylation or passively (by failure to maintain methylation states through cell divisions). After the removal, during gametogenesis, the DNA is then methylated *de novo*. In males, methylation occurs in germline cells before meiosis (Reik and Walter 2001b). In females, *de novo* methylation occurs in oocytes, while they are arrested in meiosis I but still growing. Some genes are modified earlier than others—they are not all modified simultaneously—suggesting that different mechanisms may be used for different target genes (Obata and Kono 2002).

Imprinted genes typically occur in clusters: in mice, fully 80% of imprinted genes are near another imprinted gene (Reik and Walter 2001a; Fig. 4.2). Indeed, the association is so strong that the best way to find an imprinted gene is often to look near other imprinted ones. A single cluster can contain both maternally and paternally expressed genes. At least in the clusters that have been most extensively studied, there is a differentially methylated region (DMR) in which the pattern of methylation differs between sperm and eggs (Reik and Walter 2001a, Kaneko-Ishino et al. 2003, Spahn and Barlow 2003). Though initially confined to a small region, the differential methylation between maternal and paternal chromosomes can spread out from the DMR along the chromosome, in *cis*, thereby affecting the expression of multiple genes in the vicinity. For example, there is a DMR on mouse chromosome 7 that is methylated in sperm but not in eggs (Rand and Cedar 2003, Verona et al. 2003). After implantation there is a complex series of secondary methylation imprints mediated by the control element, in both directions (Srivastava et al. 2003). As a consequence the promoter of the paternal *H19* gene becomes methylated and is silenced, and a region upstream of the paternal *Igf2* is methylated, which activates the gene.

Before these secondary methylation events can occur, the primary methylation of the DMR must survive a genome-wide demethylation that occurs soon after fertilization (Mayer et al. 2000, Reik and Walter 2001b). First, the paternal genome is actively demethylated just hours after fertilization, while the 2 parental genomes are still separate in the pronuclei. Then the maternal genome is gradually demethylated by the failure to maintain methylation states through DNA replication. After these demethylations, there is a genome-wide wave of *de novo* methylation, which again the differential methylation of the DMRs must survive. We do not yet know how this occurs.

It might seem tempting to equate the germline methylation of a gene with the imprinting of that gene, and so *Igf2r*, for example, which is methylated in the female germline, would become a "maternally imprinted" gene (e.g., Kaneko-Ishino et al. 2003). But there is a danger here if we conclude that the evolutionary innovation that led to parent-specific expression was some change in the female germline—in other words, that the ancestral state was no methylation in either germline. In the case of *Igf2r*, it turns out that the DMR contains an 8bp sequence that directs methylation in both male and female germlines, and a separate 6bp sequence that prevents methylation only in the paternal germline (Birger et al. 1999). It is highly plausible,

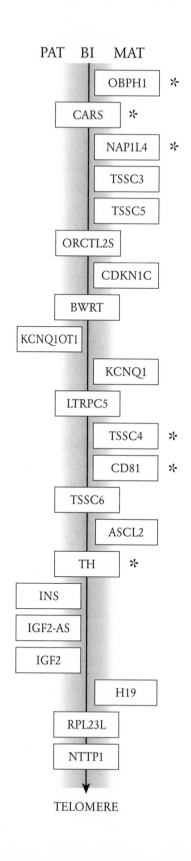

then, that the ancestral state was biallelic methylation, and the evolutionary innovation was the gain of this 6bp sequence that prevented methylation in the male germline. In this case, we might say that imprinting is the *failure* to methylate.

Once the maternal and paternal alleles have acquired their differential methylation, the final step in genomic imprinting is the use of those differences to affect gene expression. That is, the imprints must be read. Some of the mechanisms by which this occurs are summarized in Figure 4.3. The first is the simplest and the most common: methylating a promoter in order to inactivate the gene. The second involves methylation of promoter regions of antisense transcripts that interact in *cis* to prevent expression of the corresponding sense transcripts. The third mechanism is regulation of genes by differential methylation of boundary elements within a CpG island. The methylated allele binds factors that block access of promoters to downstream enhancers. Finally, differential methylation may result in differential binding of silencing factors, which then repress the promoter in *cis*. With this mechanism methylation is associated with the active allele because it inactivates the silencers.

Much about the molecular biology of imprinting remains unknown. Because CpG islands and clustered direct repeats attract methylation, it is perhaps not surprising that they are found close to the promoters of most imprinted genes (Reik and Walter 2001b). But how do they attract methylation differently by parent of origin? Likewise, how does imprinting work at *Mash2,* which continues to show maternal-only expression even in methylation-deficient mice (Kaneko-Ishino et al. 2003)? And why do imprinted genes have few and small introns (Haig 1996b, Hurst et al. 1996b, McVean et al. 1996). Highly transcribed genes have smaller introns, at least in humans and *C. elegans* (Castillo-Davis et al. 2002), so perhaps imprinted

Figure 4.2 Clustering of imprinted genes in humans. A section of human chromosome 11p15.5 containing a large cluster of imprinted genes. PAT = paternally active, MAT = maternally active, BI = biallelic. For 6 genes (*) imprinting status is uncertain in humans, so the mouse status is given instead. Note there are only 8 unimprinted genes (central column) compared to 14 imprinted in the cluster, even though unimprinted genes are 100 times more frequent in the genome at large. Adapted from Reik and Walter (2001a).

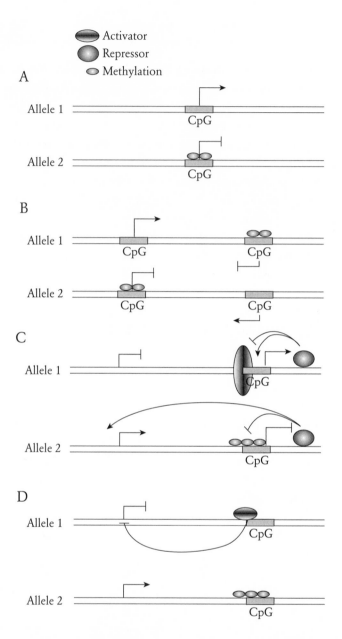

Figure 4.3 Alternative mechanisms for reading imprints. In each case allele 1 is active and allele 2, inactive. A. Silencing by methylation of promoter or nearby CpG island. B. Similar regulation by antisense transcripts. C. Differential methylation of boundary elements within a CpG island: methylation prevents the binding of proteins (vertical disk) that block the access of upstream promoters to downstream enhancers. D. Methylation leads to differential binding of silencing factors that repress the promoter in *cis*. Adapted from Reik and Walter (2001a).

genes are highly transcribed in order to counteract opposing genes and because only 1 allele is active. For many other traits, we lack even hypotheses. Why are maternal and paternal copies in imprinted regions often replicated asynchronously (Rand and Cedar 2003, Gribnau et al. 2003)? And why do imprinted regions show higher rates of recombination in the male but not female germline (opposite the usual pattern; Paldi et al. 1995)?

Despite all these uncertainties, one thing that does seem clear is that the molecular mechanisms responsible for genomic imprinting are very complex. For the *Igf2-H19* regions, some 30 different deletions, insertions, and transgenes have been characterized that reveal a complex regulatory system that we still do not understand in full (Arney 2003). The next section shows how this complexity is an expected feature of the internal tensions within the system, especially between the imprinting machinery and the imprinted genes.

Conflict Between Different Components of the Imprinting Machinery

In thinking about the evolution of imprinting in mammals, we can reasonably imagine that the ancestral state (before intimate mother-fetal interactions evolved) was equal expression of the 2 alleles, at a level that was optimal for the offspring. Then there are two possibilities for the origin of imprinting (Sleutels and Barlow 2002). First, there may have been preexisting differences in the methylation activities of the male and female germlines, and a gene evolved a *cis*-acting regulatory sequence that attracted methylation differentially in the 2 germlines, with the consequence that it was differentially expressed according to whether it was maternally or paternally inherited. In principle, a growth promoter like *Igf2* might have evolved both "down-regulate-when-maternal" and "up-regulate-when-paternal" mutations (and vice versa for a growth suppressor like *Igf2r*). Under this scenario, the gene has evolved a facultative expression pattern, with no change in the methylation machinery itself. After a succession of such mutations, the final state is 1 silent allele and the opposite set at its optimal level of expression. This is the scenario modeled by Haig (1997).

Alternatively, there might initially have been no evolution of the gene itself or its *cis*-acting regulatory sequences, but rather evolution of the *trans*-acting imprinting machinery, starting initially with one that marks the target gene equally in the 2 germlines and evolves to mark it differentially. Again, we can imagine evolutionary changes in the imprinting machinery of both germlines, in opposite directions. Rather than a gene evolving a facultative expression pattern, we imagine parents evolving to manipulate expression levels in their offspring.

As an aside, each of these scenarios could include evolutionary changes in both male and female germlines. While we could say that a gene is maternally or paternally *expressed,* or *methylated,* it does not make sense to say that a gene is maternally or paternally *imprinted*–it is both.

If both *cis*-acting targets and *trans*-acting Imprinters can evolve, they can coevolve. And this brings up a subtlety in the kinship theory of imprinting that may not at first be apparent. The conflict over kinship that underlies genomic imprinting does not just occur between maternal and paternal alleles of an individual, but also occurs within the imprinting process itself, that is, between *trans*-acting Imprinter loci and their *cis*-acting target: the two may "disagree" over whether an imprint should be applied (Burt and Trivers 1998a). This conflict arises because the target gene is always transmitted along with its own imprint, but an unlinked Imprinter gene is transmitted along with its imprinted target only half the time. For a maternally derived allele in a fetus, the fetus is as important as the mother (both contain a single copy of the gene); but for an Imprinter allele active in the female germline, the fetus is worth only half as much as the mother (because it has only a one-half chance of being in the fetus).

Similar considerations apply in the male germline, and the general rule is that the target gene will be selected to make the fetus more demanding than will the imprinting apparatus (Box 4.1). Therefore, if the imprint is to silence a growth promoter or activate a growth suppressor, the imprinting machinery may be selected to apply the imprint, but the target sequence selected to resist it. In principle, this could lead to a coevolutionary chase through sequence space in which the target keeps changing to prevent being recognized, and the Imprinters keep evolving to recognize it. The conflict can also go the other way. If the effect of the imprint is to silence a growth suppressor or activate a growth promoter, the target may be selected to acquire an imprint, but the imprinting machinery selected not to apply it. In this case the target might evolve to mimic other imprinted genes, and the imprinting apparatus therefore selected to make ever-finer discriminations. These conflicts between Imprinters and their targets must follow as surely as those between maternal and paternal alleles.

And that is not all. There is another class of genes involved in imprinting, with yet another opinion on the optimal level of expression of a gene: the unimprinted offspring genes that maintain the imprint through development and read it at the time of expression. Even if the imprinting machinery is selected to apply an imprint and the target selected to acquire it, unimprinted offspring genes may be selected to erase the imprint, or ignore it. Thus, for, say, *Igf2*, there are 5 classes of genes that may affect fetal expression levels, each with its own opinion about the optimal level of expression.

History of Conflict Reflected in the Imprinting Apparatus

Thus the various classes of genes involved in imprinting are expected, from first principles, to evolve according to complex and contrarian selection pressures. We suggest that the unusual complexity of the imprinting apparatus is the result of a long history of constant adjustments and counter-adjustments among the various components, as such complexity is unlikely to arise in a few mutational steps and seems unnecessary otherwise. More specifically, the fact that a particular C in the maternally derived *Igf2r* is methylated in the oocyte and the 2-cell embryo, unmethylated at the 4-cell stage, and then remethylated at the 8-cell stage is suggestive of a history of conflict (Moore and Reik 1996). The genome-wide demethylations that occur around this time may also represent attempts to erase unwanted imprints. The rapid active demethylation of the paternal genome soon after fertilization is presumably the result of maternally supplied RNAs and proteins, whereas the more gradual demethylation of the maternal genome is presumably due to the action (or inaction) of unimprinted offspring genes. This suggests that demethylation will not be seen in the absence of extended mother-fetal interactions—which is true of zebrafish (Macleod et al. 1999).

Unwanted imprints may be erased, or they may simply be ignored. In mice and humans, *Igf2r* and *Grb10* show maternal-specific methylation, and in mice this is associated with maternal-specific expression, but in humans expression is typically biallelic, as if the methylation differences have no effect (Killian et al. 2001, Arnaud et al. 2003). Indeed there appear to be many regions of the genome in which maternal and paternal copies are differentially methylated, but there is no detectable difference in gene expression (de la Casa-Esperón and Sapienza 2003). We wonder whether some fraction of these represent past attempts by parents or the target loci to manipulate

BOX 4.1

Genetic Conflicts in Genomic Imprinting

Evolutionary changes in genes affecting fetal growth often involve a benefit to one party (mother or fetus) and a cost to the other. In order for a new mutation to be selected, the ratio of benefits to costs (each measured as changes in the organism's expected number of future offspring or reproductive value) must be greater than some threshold. These thresholds differ for the different classes of genes involved in genomic imprinting, as shown in the figures in this box. The top figure covers mutations that benefit the fetus at a cost to the mother, while the bottom covers the reverse situation (labeling of lines is in inverse order to that above). For maternal Imprinter loci and maternally imprinted loci, these thresholds are constant, but for the paternal equivalents, the thresholds depend on k, the change in the father's reproductive value caused by a unit change in the mother's reproductive value. k may be close to 1 for monogamous species, and close to 0 for promiscuous or polyandrous species. Finally, for unimprinted progeny genes, the thresholds are the average of those for maternally and paternally imprinted genes.

The top figure also gives the stopping rule for maternal investment for the different types of genes (continue to invest until the benefit:cost ratio falls below the specified level), whereas the bottom figure can be read as indicating how much more important the fetus is than the mother for the various genes. Note that for $k = 1$, all conflict is parent versus offspring, and that as k gets smaller, the conflict gradually becomes more maternal versus paternal. From Burt and Trivers (1998a).

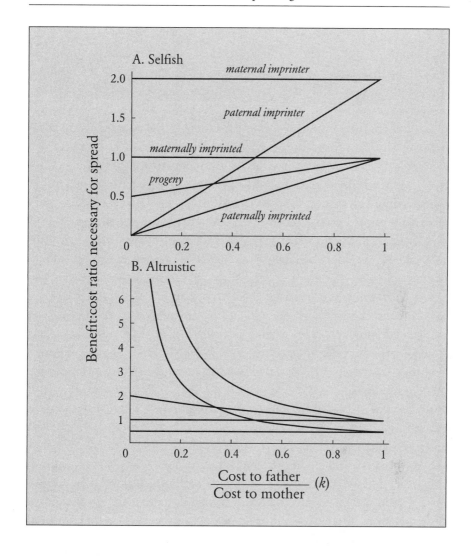

offspring expression levels that the offspring has evolved to ignore. We also wonder whether these genes are especially likely to be expressed in the placenta and affect fetal growth.

Even the clustering of imprinted genes may result from a history of conflict between the different components of the imprinting machinery. If a gene is selected to acquire an imprint but Imprinter genes are selected not to apply it, the former may take advantage of imprinting mechanisms operating at other nearby loci where Imprinters and targets are both positively se-

lected. Consistent with this idea are interconnected causal pathways acting both in *cis* and in *trans* within these clusters, such as between *H19* and
Igf2. Alternative explanations for clustering posit a general difficulty in becoming imprinted (Haig 1997) or an imprinting process that is costly
(Mochizuki et al. 1996).

Kinship conflicts may also account for some of the apparent regularities
in the mechanisms for reading methylation imprints. Take, for example, the
evolution of *cis*-acting antisense RNAs, all 5 known examples of which are
associated with the repression of a paternal allele (Reik and Walter 2001a).
One possible explanation for this pattern is that these first evolved to work
in an unimprinted way—recent work on mutations in the globin gene show
how *cis*-acting antisense RNAs may arise by a single mutational step (in this
case, an 18kb deletion; Tufarelli et al. 2003). If so, their evolution implies
past selection on the embryo for reduced expression, which leads us to
expect they will more often be associated with demand inhibitors than
with demand enhancers. It follows that they are more likely to evolve to be
methylated and suppressed when paternally inherited (Arney 2003). This
would explain why the 5 imprinted genes in which the silent allele is associated with an antisense RNA are all maternally expressed (i.e., probably demand inhibitors).

Reik and Walter (2001a) list 8 protein-coding genes that are maternally repressed and maternally methylated, but none that is paternally repressed
and paternally methylated. This difference could be due to the constraint of
active demethylation of the paternal genome immediately after fertilization—it might be that it is easier for the male germline to evolve to remove
methyl groups than to add them and have them stick through the global
demethylation. Alternatively, it could be because maternal methylation and
silencing are more stable over evolutionary time than paternal methylation
and silencing (Wilkins and Haig 2002). If a growth enhancer like *Igf2* is maternally silenced and the paternal allele is expressed in the offspring at the
optimal level for either itself or the unimprinted progeny genes (or between
the two), the maternal Imprinter genes, the maternal copy of *Igf2*, and the
unimprinted progeny genes will not be selected to turn on expression of the
maternal allele. By contrast, if a growth suppressor like *Igf2r* is paternally silenced and the maternal allele is expressed in the offspring at the optimum
level either for itself or the unimprinted progeny genes (or between the
two), and if the species is tending toward monandry (i.e., $k>0.5$ in Box 4.1),

the paternal copy of the gene will be selected to remain silent. However, the *trans*-acting paternal Imprinter loci may be selected to reactivate the paternal gene—in other words, not to methylate it. Put more simply, if expression levels are mostly controlled by unimprinted progeny genes, maternal Imprinter genes and their targets will usually be selected in the same direction, whereas paternal Imprinter genes and their targets may be selected in opposite directions (particularly for high k), and so imprinting in the paternal germline may be less stable than that in the maternal germline. Either of these explanations could also account for the pattern with the antisense RNAs. In addition, subtler effects are being discovered almost daily and this literature will provide rich material for understanding conflict during the production of imprints in gametogenesis and subsequent maintenance in the offspring.

Evolutionary Turnover of the Imprinting Apparatus

One possible outcome of these conflicts between different components of the imprinting machinery is a constantly dynamic pattern of natural selection, as evolutionary changes in one component of the imprinting machinery selects for an evolutionary response by another. At first sight, the imprinting machinery appears to be relatively stable, because the vast majority of genes that are imprinted in mice are also imprinted in humans. Exceptions are *Igf2r*, *U2af-rs1*, and *Impact*, which are imprinted in mice but not in humans (Kaneko-Ishino et al. 2003). Likewise, *Mash2* is imprinted in mice but not in deer mice (*Peromyscus* spp.) and the reverse is true for *pPll-v* (Vrana et al. 2001). But there may still be a substantial amount of small-scale shifting, adjusting, and balancing, especially insofar as such variables as degree of polyandry or matrilocality change over time. As with many other classes of selfish genes, a good way to test for this is to look at genomic imprinting in species hybrids. If the machinery has been stable since the common ancestor of the 2 species, the expression of imprinted genes in hybrids should be the same as in the parental species, whereas if the machinery has been evolving, it may not work properly in the hybrids. Thus far, the only extensive data on genomic imprinting in species hybrids is for 2 species of deer mice thought to have diverged some 100,000 years ago (Vrana et al. 1998, 2000, 2001). Interestingly, the expression of most imprinted genes tested differs between reciprocal F_1 hybrids of *P. maniculatis* and *P. polio-*

Table 4.3 Expression of imprinted genes in *Peromyscus* hybrids

Gene	Wildtype	*P. polionotus* × *P. maniculatus**	*P. maniculatus* × *P. polionotus**
Igf2	Pat	Pat	Pat
Peg3	Pat	Bi	Pat
Snrpn	Pat	Bi	Pat
Mest	Pat	Bi	Pat/Bi
pPl1-v	Pat	Pat	Pat
H19	Mat	Bi	Mat
Igf2r	Mat	Mat/Bi	Mat/Bi
Grb10	Mat	Bi	Bi

*crosses are female × male
Pat: paternal; Mat: maternal; Bi: biallelic
From Vrana et al. (1998, 2001).

notus, indicating that there has been recent evolution of these genes and/or the imprint recognition machinery since the species diverged (Table 4.3).

The evolution of the imprinting machinery in these 2 lineages may have been accelerated by a change in the mating system. Like most mammals, *P. maniculatus* is polyandrous (the progeny of a single female often have different fathers, even within a single litter) whereas *P. polionotus* is mostly monogamous. This change is expected to have resulted in fetuses being less demanding, and consequently in mothers having less need to dampen fetal demand. Interestingly, while there was no consistent trend among the 3 maternally expressed genes investigated, among the paternally expressed genes there were 3 genes for which one of the hybrids showed biallelic expression, and for all 3 genes the hybrid was of *P. polionotus* females × *P. maniculatus* males. One interpretation of these results is that *P. maniculatus* females silence these genes more strongly than do *P. polionotus* females, or the imprints "stick" better. This difference is in the expected direction of imprinting being somewhat relaxed in the monogamous species. It would be interesting to follow this up—to see how general the result is across both genes and species pairs, and to get at the underlying genetics of the difference.

Another possible manifestation of perpetual evolution would be rapid sequence evolution of the *cis*-acting targets of imprinting and the proteins that interact with them. It would be interesting to compare rates of evolution of these regions with those of unimprinted genes. An early study suggested that imprinted genes do not have an obviously elevated rate of sequence evolution (McVean and Hurst 1997), but we are not aware of a similar study on the *cis*-acting control regions.

Intralocus Interactions, Polar Overdominance, and Paramutation

It is well accepted that sequences near a gene can evolve to affect the expression of that gene, in *cis*. If it were mutationally possible, they would also be selected to affect the expression of the allele on the homologous chromosome. Consider, for example, a sequence near the growth suppressor *Igf2r*. According to kinship theory, it will be selected to reduce the expression of *Igf2r* when paternally inherited. It might be thought that the furthest it can evolve in this direction is to silence the paternal allele. However, it is very likely that *Igf2r* expression from the maternal allele will still be higher than is optimal for the paternal allele. Were it possible, the paternal allele would be selected to reduce expression from the maternal allele. Such selection pressures might account for some of the complex interactions between maternal and paternal chromosomes at many imprinted regions:

- *Igf2* (paternally active) and *H19* (maternally active) are closely linked, oppositely expressed genes, and deletion of the maternal *H19* not only increases methylation (and expression) of the maternal *Igf2* allele, but also reduces methylation of the paternal *Igf2* allele (Forné et al. 1997).
- The callipyge ("beautiful buttocks") mutation in sheep is a single base pair change in the *DLK1-GTL2* imprinted domain that causes muscular hypertrophy in $+^{Mat}/CLP^{Pat}$ heterozygotes, but not any other genotype, including *CLPG/CLPG* homozygotes (Georges et al. 2003). This unusual form of inheritance has been called "polar overdominance" and appears to result from *trans*-repression of a paternally expressed growth promoter (*DLK1*) by a maternally expressed RNA-encoding gene (*GTL2* and/or *MEG8*).
- *Rasgrf1* is an imprinted gene in which the paternal allele is normally methylated and activated, and the maternal allele is unmethylated and silent. Herman et al. (2003) made a mutant form of *Rasgrf1* that, when paternally inherited, not only becomes methylated and activated itself but also somehow causes the wildtype maternal allele to become methylated and activated. This "epimutation" of the maternal allele was stable, in that it was not silenced when passed through the female

germline. Moreover, in some instances after passing through the female germline, it somehow managed to silence the wildtype paternal allele in the next generation! The mutant *Rasgrf1* was created by replacing an internal region controlling its own methylation by a similar region from *Igf2r* (which is maternally methylated and activated). Similar *trans*-effects have been seen at the *Ins2* gene of mice (Duvillié et al. 1998) and at some unimprinted loci in plants (so-called paramutation; Chandler et al. 2002).

• There is a region of repeated sequences upstream of the human *INS* gene and the number of repeats affects the expression of both *INS* and *IGF2*. In Caucasians, alleles separate into 2 main size classes: Class I (26–63 repeats) and Class III (140 to >200 repeats). Class I genes are associated with higher expression of *INS* and *IGF2* in placenta and pancreas, and they have a slight transmission ratio distortion (54%), which is independent of the sex of the parent and the sex of the child (Eaves et al. 1999). Presumably this reflects higher survival of fetuses with a Class I allele. Class I alleles are also associated with a lower frequency of polycystic ovary syndrome, but a higher frequency of type I diabetes. However, the latter association is highly unusual (Bennett et al. 1997). The most common Class I allele (called 814, with 42 repeats, frequency in the United Kindom is 0.18, and 0.25 of Class I alleles) appears to predispose to disease when inherited from the mother or when inherited from the father if in the father it was paired with another Class I allele. But if it was paired with a Class III allele, it does not predispose toward disease, as if it carried with it the protective function of the Class III allele. Thus, disease susceptibility appears to depend on the identity of the untransmitted paternal allele. Other Class I alleles do not behave in this way.

The molecular basis of these *trans*-effects is unknown. One possibility is suggested by the observation that, for at least some imprinted domains, the maternal and paternal regions are paired in mitotic cells at late S phase (LaSalle and Lalande 1996). In *Drosophila* it has been shown that a control sequence on one chromosome can affect the expression of a gene on the homologous chromosome, and that this effect is dependent on the somatic pairing of homologous chromosomes ("transvection" effects; Kennison and Southworth 2002). Another possible mechanism for *trans*-effects is antisense RNA production.

Transmission Ratio Distortion at Imprinted Loci

Some imprinted regions show not only parent-specific expression but also parent-specific drive. One such example is in the distal region on mouse chromosome 12 (which contains the imprinted loci *Dlk1* and *Meg3/Gtl2*). Croteau et al. (2002) crossed 2 strains of mice, in both directions, backcrossed the hybrid females to one of the parental strains, and then monitored the genotypes of the resulting offspring. Genotypes were monitored for both implanted embryos (at 7.5 days postcoitum (dpc)) and for pups (at 3 weeks postpartum). Individuals were scored according to whether they were male or female and whether they had inherited the maternal or paternal chromosome from their mother. There were thus four genotypes, and under Mendelian inheritance we would expect 1:1:1:1 ratios. But at 7.5 dpc there was a significant overabundance of males inheriting their mother's paternal allele (Table 4.4). The cause of this is uncertain; if it was due to differential postfertilization mortality, it implies some strong positive interaction between the Y chromosome and the mother's paternal allele in, say, the probability of implantation. Then in the period from 7.5 dpc to 3 weeks postpartum, males are more likely to die than females and progeny inheriting the grandpaternal allele are more likely to die than those inheriting the grandmaternal allele, and so the class with the highest initial survival had the lowest subsequent survival (Table 4.4).

The causes of these mortality differences are unknown. Regarding the mortality from embryos to pups, the higher mortality of males is also seen in humans and brings the sex ratio at birth closer to 50:50. Croteau et al. (2002) suggest that the increased mortality of individuals with the grandpaternal allele is because of a failure to erase and reimprint in the maternal germline. If so, this seems like a high frequency of "mistakes," which itself requires explanation. By comparison, the frequency of spontaneous imprinting errors leading to Prader-Willi syndrome is thought to be 10^{-5} to 10^{-6} (Naumova et al. 2001). Clearly there will be strong selection on target genes to acquire the correct imprint, and on zygotic genes to maintain it correctly, if the alternative is death. Might the maternal allele somehow be sabotaging the correct imprinting of the paternal allele, in order to drive? Alternatively, perhaps there is an imprinted maternal-effect killer locus in the region—a parent-specific version of *HSR* or *scat*[+] (see Chap. 2).

Parent-specific drive has also been reported for some imprinted regions of human chromosomes, though the pattern differs from locus to locus, with

Table 4.4 Transmission by female mice of their paternal and maternal alleles at distal chromosome 12 to male and female offspring

	Mendelian Expectations		Numbers of 7.5dpc Embryos		Numbers of 3-Week Pups		Relative Survival*	
	Pat	Mat	Pat	Mat	Pat	Mat	Pat	Mat
Male	1	1	153	106	285	294	0.62	0.93
Female	1	1	101	109	238	326	0.79	1.00

*Relative survival from embryos at 7.5 days postcoitum (dpc) to 3-week pups.
From Croteau et al. (2002).

no obvious logic (Naumova et al. 2001). For example, in a survey of 31 three-generation families at the chromosomal region including *Igf2*, females have an increased probability of dying if they inherit the paternal allele from their mother, but it does not appear to matter which they inherit from their father. By contrast, males have an increased probability of dying if they inherit the paternal allele from their father, but it does not matter which they inherit from their mother. However, at the region including *Igf2r*, both males and females are more likely to die if they inherit the paternal allele from their mother, but it does not matter which allele they inherit from their father. (Curiously enough, this region is homologous to proximal chromosome 17 in mice—which includes the *t* haplotype.) And, at the region including *Grb10*, the only effect seems to be that females are more likely to die if they inherit the maternal allele from their father. If these results are corroborated and found to be general, they represent perhaps the most intriguing mystery in the evolution of genomic imprinting.

Biparental Imprinting and Other Possibilities

We noted earlier that 1 possible route to imprinting is for *trans*-acting Imprinter genes to evolve to apply imprints onto the control regions of a particular target gene, and that this could be viewed as an attempt by the parent to manipulate expression levels in the progeny. Under monogamy (k~1 in Box 5.1), there is little or no conflict between the maternal and paternal alleles in an individual, but there is still the opportunity for substantial conflicts between parents and offspring, as each individual fetus or neonate will want more maternal investment than is optimal for either the mother or the father. This raises the possibility of selection for *trans*-acting Imprinter loci that act identically in male and female germlines to repress a growth promoter or activate a growth suppressor. Currently, genomic imprinting is defined as differences in gene expression according to the sex of the parent, but it may be useful to think more broadly about parental attempts to influence gene expression in the next generation. Sometimes the parents will disagree, leading to maternal/paternal differences in gene expression, but sometimes they will agree. If mothers and fathers are manipulating fetal expression levels in the same direction, we could call this biparental imprinting.

Haig (2000a) raises another interesting possibility: might genetic memory extend back more than one generation? It is easy to imagine both social situations that select for such a pattern (e.g., sex-biased dispersal patterns, as seen in most mammals) and molecular mechanisms by which imprinting might be cumulative when genes are passed down the same-sex lineage. But we know of no direct observations on the matter. A related issue is whether components of the imprinting apparatus may themselves be imprinted. If so, one would have to consider kinship coefficients over 3 generations. Suggestive evidence comes from observations that genes affecting the methylation of artificial transgenes can themselves work in a parent-specific manner (Allen and Mooslehner 1992). On the other hand, the expression of imprinted genes is maintained in androgenetic and parthenogenetic embryos, suggesting that genes involved in the maintenance and reading of imprints are not themselves imprinted (Sleutels and Barlow 2002).

More generally, Haig (1994, 1996c) wonders whether imprinting might sometimes be facultative—for example, paternal alleles, or the paternal Imprinting apparatus, adjusting in response to the likelihood of the male mating again in the future with the same female. And more speculatively, Haig (2003) wonders whether oppositely imprinted genes might, over time, evolve complex strategies of cooperation, defection, and punishment that are conditional on the expression of oppositely imprinted genes. It would be very interesting if such genes could create stable paternal and maternal personalities that interacted as individuals. If so, interactions with parental molding are expected to be important, because one set of genes in the offspring, associated with one personality, has interests exactly aligned with one parent but not the other, and vice versa. Does the degree of reciprocity between a couple affect the degree of reciprocity within their children?

Other Traits: Social Interactions after the Period of Parental Investment

Of course, kinship interactions extend beyond the parent-offspring relationship in many species. In these, imprinting is expected whenever effects of genes on patrilines differ from those on matrilines. For example, in many mammals females live their lives in close proximity to relatives and have many important interactions with them. These may include warning others of danger, shared use of space (and defense of it from conspecifics), social grooming, shared aspects of reproduction, cooperative food gathering (in some species, hunting), and so on. Insofar as these actions affect matrilines and patrilines differently—as often they must—there is potential for imprinting.

The chief difference between these effects and conventional parent-offspring effects is that conventional effects are expected to act early in development, on starkly different degrees of relatedness, and to have profound ramifying consequences for juvenile growth and survival. By contrast, later kinship interactions may select for effects that have little to do with size but a lot to do with behavior, brain physiology, interactions with others, and internal psychological conflict. In our own species, the latter may interact in a complex way with systems of deceit and self-deception, which may also promote mental subdivision. A second difference is that there is now a very large body of evidence on early effects, while the evidence on later effects is often fragmentary or indirect. At the same time, the theoretical possibilities for the later effects are often more complex and interesting than those for early effects (leaving aside conflict over the mechanisms of machinery, which applies to both). In short, in the discussion that follows, there is a much higher ratio of theory and speculation to fact than in what has gone before.

Consider alarm calls. In many ground squirrels adult females warn other closely related females of the approach of predators. In Belding's ground squirrels, these females are more related to each other on their maternal side than on their paternal, and they adjust frequency of calling to the number of closely related females living nearby (Sherman 1977, 1980). For unimprinted genes, alarm calling is a unitary event with a single goal: warning genetically related females of danger. For imprinted genes there may be psychological bifurcation and conflict. We can easily imagine maternally active genes taking the lead in generating a sentry-like mentality, with the approach of a predator eliciting a loud cry warning that "a predator is coming, a predator is coming!" while a fearful, inner paternal self urges the female to keep silent, to run and to hide. Degree of sociality between neighbors also correlates positively with kinship, and Holmes and Sherman (1982) have made the striking discovery that 1-year-old females in nature prefer their littermate full sisters over their littermate half-sisters. Nothing is known of imprinting in this group, but paternally active genes would gain the most from discriminating correctly between siblings based on their paternity.

In mice, 8 of the 27 imprinted genes in Tycko and Morison's (2002) survey have behavioral and/or neuronal effects. And there is growing evidence in humans that imprinted genes have effects on adult behavior. Paternally active genes on the X chromosome appear to affect sociability and social intelligence in women, while paternally active genes located on chromosome 2 are associated with schizophrenia and handedness in one human population (Francks et al. 2003). Also, a series of studies have shown bipolar illness in people with Prader-Willi syndrome (Vogels et al. 2003 and references therein). The condition results from an imprinting failure and has many other effects on early development. The severity of autism, epilepsy, and Tourette syndrome may also depend on which parent donated the susceptibility gene (Isles and Wilkinson 2000). That adult behavior should show imprinting effects is expected, but the interpretation of these effects is obscure, largely because detection of the traits is based on gross dysfunction.

Maternal Behavior in Mice

The mother is usually equally related to her offspring through her maternal and through her paternal genes. Other things being equal, we expect no bias in her maternal behavior to evolve via imprinting. But at least 2 imprinted genes are known to have striking effects on maternal behavior in mice (Lefebvre et al. 1998, Li et al. 1999, Curley et al. 2004). Both are paternally active (*Peg1/Mest* and *Peg3*) and together these copies have positive effects on such diverse maternal traits as nest-building, retrieval of pups, eating of the placenta, oxytocin levels in the brain, and thermoregulatory behavior toward pups.

There are at least two reasons why selection may have favored paternal genes for these maternal traits. Inevitably, maternal investment will, to some degree, subtract from investment available for other relatives. Insofar as a female is more related to these relatives on her maternal chromosomes, her paternal ones will show a stronger interest in maternal behavior and investment (Haig 1999b). Or, consider possible relatedness between mates (Wilkins and Haig 2003b). This will tend to be higher early in a female's reproductive life on the paternal side (e.g., father-daughter) and it will tend to decline with time. Thus, paternal genes will be more closely related to current progeny than to later, and are expected to favor more parental investment now.

But there is a coincidence. Both of these genes also enhance prenatal growth (as expected for paternally active genes). Why should these 2 different kinds of effects be associated in 2 different imprinted genes? To us the most plausible scenario is that 1 imprinted effect came first—most likely the effect in offspring favoring early growth (because this would usually be more strongly selected)—and that this imprinted gene then attracted additional imprinted functions (in this case, affecting maternal behavior in adult females). Strong selection in one context may sometimes also be associated with strong selection in the other. With many maternal half-siblings, for example, selection favors both selfish paternal growth effects in offspring, as well as greater paternal control of maternal investment in adult females.

Whether or not this scenario turns out to be correct, the general point remains that once imprinting evolves at a locus, that locus experiences a new set of selection pressures (due to its new set of kinship coefficients). This may cause it to evolve new functions or expression patterns that it did not previously have. Not only does the function of a gene influence whether it evolves to be imprinted, but the imprinting status of a gene should also influence the evolution of new function. Effects in both directions could contribute to the correlation between imprinting status and function.

Inbreeding and Dispersal

When an individual inbreeds, he or she is related to the partner through the mother or the father or both. When related through mother and father (e.g., toward a full sibling), maternal and paternal genes usually calculate the same degree of relatedness to the partner (in this case, one-half). In the more usual cases, r is measured only through one parent (e.g., half-sibling or cousin). Consider a female about to engage in an extrapair copulation with her maternal half-brother (that is, they will not raise the child together). Maternal genes in her see the half-sibling as related by one-half, increasing the relatedness to their own offspring by one-fourth. Against this must be set the costs of inbreeding depression. If the gain in relatedness outweighs the costs of inbreeding, maternally active genes will be selected to promote inbreeding. By contrast, paternally active genes are expected to look askance at this kind of inbreeding. They enjoy no gain in relatedness but surely suffer the costs of inbreeding, so paternally active genes will be selected that discourage inbreeding on the maternal side. We can easily imagine maternally active genes in females that think kissing cousins are cute, while paternally active genes look on in a sullen and moralistic way, emphasizing the biological defects so generated!

Exactly the opposite attitudes to inbreeding are expected in the male about to mate with his maternal half-sister: his maternal genes will worry about saddling the half-sister with low-fitness offspring, while his paternal genes will take a devil-may-care attitude and urge him on—go for it! Again, age effects are possible to imagine, with paternal genes more likely to favor inbreeding between age-mates (paternal half-siblings) than maternal genes. Haig (1999a) has undertaken a general analysis of incest along these lines, considering the various categories in turn—full sibling, father-daughter, and so on.

Dispersal is also a social act whose evolution is influenced by kinship, insofar as it reduces the probability of mating and competing with relatives. If we imagine an ancestral state with no (sex bias in) dispersal, the average relatedness of an individual to the entire social group (as opposed to just littermates) would usually be greater through paternal genes than maternal, due to higher variance in male reproductive success. This will be particularly true if we consider relatedness to individuals of the same age, who are the more likely mates and competitors. Thus, dispersal away from the social group will initially be more advantageous to paternal genes than to maternal genes, and we might expect paternally active genes to have taken the lead in its evolution.

Low male parental investment and high variance in male reproductive success should also select for male dispersal. First, high variance in male reproductive success will select for dispersal away from male relatives, such as brothers, to avoid cutting into each other's reproductive success. New, successful male genotypes—absent special opportunities for cooperation—will benefit from migrating away from each other. Also, opportunities for cooperation may be more likely for members of the investing sex, who share many activities in common toward offspring. Where there is substantial inbreeding depression, each sex may be selected to avoid inbreeding. For casual encounters, the investor loses more, so females will be more strongly selected to avoid inbreeding than males. Such choice will decrease the number of suitable mates for a female, and will directly decrease male reproductive success, thus further selecting for dispersal.

For all of these reasons, we imagine that selection has rapidly favored a sex-biased pattern of dispersal, such that in many species with negligible male parental investment, males migrate further from place of birth and places of mating than do females. In any case, this is the common pattern and it has immediate implications for how paternal and maternal relatedness develop within groups over time. In a simple model, Haig (2000b) showed that male dispersal creates a set of female relatives who tend to be more related to others on the maternal than paternal side, although this differential of relatedness varies with the relative age of the 2 individuals. For example, if a single male simultaneously fertilizes many females, paternal relatedness will reach maximal value for age-mates. In baboons in nature, pairs born within 2 years of each other are less likely to engage in sexual consortships than pairs born at greater intervals (Alberts 1999). In semi-free-ranging rhesus macaques, the closest female affiliative relationships are between maternal half-siblings, but paternal half-siblings are favored by females over non-kin, especially when they are age-mates (Widdig et al. 2001).

When reversals in sex-biased dispersal take place (e.g., humans and chimps compared to other primates) we might expect even the *direction* of imprinting of some genes to vary from species to species.

Kin Recognition

Recent studies show striking evidence of imprinting effects in the kin recognition abilities of humans and mice. In our own species, women have been shown to prefer odors from men who have HLA alleles similar to their own paternally derived alleles, while the degree of similarity to their maternally derived alleles seems not to matter (Jacob et al. 2002). This is not simply a matter of preferring the smell of father, because the degree of similarity to the untransmitted paternal alleles was also unimportant. It is as if, somewhere in the olfactory system, only the paternal HLA alleles are expressed and odors are compared to it.

Imprinting is also involved in the kin recognition abilities of mice. Males and females from reciprocal crosses of 2 inbred lines that were transferred as embryos into females of a third line avoid the smell of female urine from the same strain as their genetic mother (Isles et al. 2001, 2002). They have no such differential response to male urine. Apparently there is something in the urine of females that individuals are comparing to their maternally derived genes, and if there is a match, they find it aversive. The fact that both males and females are responding suggests it has something to do with spacing rather than inbreeding avoidance.

In another study on mice, males and females from reciprocal crosses of inbred lines were cross-fostered at birth to new mothers of either of the 2 parental strains (Hager and Johnstone 2003). The amount of milk transferred from mother to pups was greater when pups were raised by a female from the same strain as their genetic mother than when they were raised by a female from the same strain as their genetic father. One interpretation of this result is that females have evolved to recognize and favor their own offspring by cueing in to pups' maternally derived alleles. Such recognition abilities would be useful in mice, whose young are nursed in communal nests with females sometimes caring for the young of others.

Functional Interpretation of Tissue Effects in Chimeric Mice

Some of the most intriguing facts regarding imprinting are indirect, coming from the study of chimeric mice, which are mice whose cells are a mixture of types. Wildtype cells are mixed with cells whose double set of genes came from a single parent, either the father (androgenetic) or the mother (gynogenetic or parthenogenetic). These chimeras present less radical phenotypes than might have been expected. At all loci, there is some degree of normal gene expression (from the wildtype cells, which typically make up more than one-half of the chimera), while the strength of the uniparental effect is expected to reflect relative cell frequency in the chimeras. By contrast, in knockout mice or uniparental disomies, some uniparental effects are imposed on all cells.

As mentioned earlier, data from numerous chimeras show a general effect

Table 4.5 Mouse tissues showing a bias in chimeras toward paternal (androgenetic) or maternal (parthenogenetic or gynogenetic) cell contribution

Paternal	Maternal
Extraembryonic	Embryo
Hypothalamus	Cortex
Chondrocytes	Dentin
Enamel	Olfactory mucosa

Data from Fundele and Surani (1994), Fundele et al. (1994, 1995a, 1995b), Allen et al. (1995), and Keverne et al. (1996).

in which the size of the chimera increases as the frequency of gynogenetic cells decreases and (separately) as the frequency of androgenetic cells increases (Keverne et al. 1996, Fundele et al. 1997; see Fig. 4.4). More strikingly, differently derived cells proliferate and survive in some kinds of tissues and disappear from others. Although the creation of chimeras is time-consuming and sample sizes small, consistent patterns emerge (Table 4.5). Androgenetic cells do well in the skeleton (chondrocytes), the hypothalamus, and the cell layer that produces enamel of teeth, but not in the brain as a whole (Fig. 4.4), while maternal ones do well in the brain itself, especially the neocortex, and in the cell layer that produces the dentin of teeth. Maternal cells also survive preferentially in the olfactory mucosa and vomeronasal tissue. In very early development, androgenetic cells proliferate in extraembryonic tissues and gynogenetic cells in the embryo proper. Androgenetic cells also show up in brown adipose fat (though there is wide variation in the original data). The assumption is that these differences in relative survival of cells reflect differences in the activity of imprinted genes within them. For some tissues we already know this to be true. Paternally active genes (but not maternally active) are found in the mouse hypothalamus and in brown adipose fat tissue (Curley et al. 2004, Plagge et al. 2004).

The meaning of these facts is mostly obscure. If neocortex is especially oriented toward social interactions involving kin and associated conscious mentation, a maternal dominance may make sense, while the hypothalamus, involved directly in appetites, might represent a more narrowly selfish orientation, hence paternal dominance. Psychologically it is easy to imagine one inner voice saying, "I like family, family is important," while the other counters, "I'm hungry!" Enamel is believed to be tissue expensive in rare

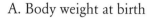

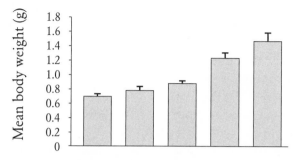

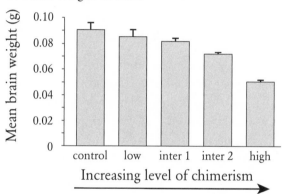

Figure 4.4 Total size at birth and brain size at birth as a function of the percentage of cells in mouse chimeras expressing only paternally active genes. Body and brain size at birth are plotted as a function of different categories of chimeras according to percentage of androgenetic (double paternal) cells. Control (none), low, intermediate 1, intermediate 2, and high percentage of androgenetic cells. Note that brain size goes down while body size goes up, so relative brain size declines even more sharply with increasing frequency of paternally active genes. Courtesy of E. B. Keverne, data in Keverne et al. 1996.

minerals, while brown adipose fat is rich in energy for immediate consumption; hence both may show a selfish (paternal) bias. It is also possible that paternal genes may favor slower teeth eruption as this is associated with longer nursing and greater maternal interbirth intervals (Lahn et al. 2001).

It is interesting that adult male chimeric mice with a large complement of parthenogenetic cells in the brain are quick to aggress against other males (Allen et al. 1995). The interpretation of this finding is not obvious, but we prefer the following. If dispersal is sufficiently male-biased that maternally active genes are selected to take the lead in causing that dispersal, they should also cause the dispersing males to avoid each other. Quick aggression as maternal relatedness is detected would seem a natural device to space out maternally related males. The intention is not to injure the opponent, only to drive him away quickly. Our argument rests on the assumption that males with parthenogenetic cells are detecting each other as close maternal relatives, perhaps because of their shared parthenogenetic cells, especially in the olfactory mucosa and olfactory bulb.

Deceit and Selves-Deception

It is generally recognized that natural selection may favor intraspecific deception in a wide range of contexts, including sexual misrepresentation, false warning cries, parent-offspring relations, and a host of others (Trivers 1985). It is less widely recognized that selection to deceive and to avoid being detected may select for self-deception in the deceiver, the better to hide the ongoing deception from others. If conscious knowledge of ongoing deception is accompanied by stress in the deceiver, the deception may be rendered unconscious so as not to reveal the stress. In short, mental fragmentation—with some internal conflict—may serve the individual by hiding useful information from others (reviewed in Trivers 2000).

Genomic imprinting opens up new vistas in self-deception, which might almost be called "selves deception." While not the only source of internal genetic conflict over actions toward relatives, genomic imprinting pits half of the nuclear genotype against the other half, easily selecting for biased information flow within the organism that may resemble classical self-deception. If the neocortex helps allocate resources to others by giving signals from the hypothalamus a given weight in its calculations, selection for paternal effects in the hypothalamus may favor a heightened signal so as to compensate for maternal efforts toward down-regulation from the neocortex, and vice versa.

Haig (1996d) has modeled the analogous situation of mother-fetal conflict in which fetal hormones have replaced maternal ones as control agents for various maternal states pertaining to blood flow, blood sugar levels, and so on. In mother-fetal conflict the two are separate organisms related by one-half in which the fetus has preferential access to the mother's bloodstream, where it can augment her hormone levels for desired downstream effects. In conflicts due to genomic imprinting, the two are unrelated genome halves of the same individual that may be differentially expressed in different parts of the body, favoring biased forms of communication between these parts. These could also involve hormones and be subject to the same escalation of hormone production by some tissues, countered by increasing inertness to the hormones in others. This, of course, is on top of any *within*-tissue conflict between maternal and paternal genes (over levels of hormone production, for example, or sensitivity to its effects).

Insofar as a paternal orientation may often be the more selfish one and we wish to suppress evidence of such selfishness to others, there may be a tendency for the paternal viewpoint to be more often unconscious. Whether stable alternative personalities could be associated with paternal and maternal genes is, of course, completely unknown.

There is growing evidence that the 2 brain hemispheres are asymmetrical regarding consciousness and self-deception (reviewed in Trivers 2000). For example, the right side of the amygdala is activated during unconscious perception, while the left is activated during conscious (Morris et al. 1998). Injuries to the left side of the brain causing right-side paralysis are never associated with denial and rationalization, while a small minority of right-hemisphere injuries show bizarre denial of the paralysis and then rationalization when the individuals are shown films proving otherwise ("my arthritis is acting up"; Ramachandran 1997). This is interpreted as reflecting the natural tendencies toward self-deception of the left side of the brain, unchecked by the right. Given these facts, we would be surprised if some imprinted genes affecting brain function were not found to do so asymmetrically.

Imprinting and the Sex Chromosomes

So far we have restricted our discussion to imprinting of autosomal genes, and the imprinting of sex-linked genes has a number of additional considerations. For example, the Y chromosome in mammals is always paternally derived. Therefore, we can expect it to evolve the selfishness of a paternally derived gene, but without the need for reversible imprints (Hurst 1994).

Similarly, the X chromosome in males is always maternally derived, and so genes on it that are expressed only in males are expected to evolve the self-restraint of a maternally derived gene, again without the need for reversible imprints.

Female mammals are typically XX, and the evolution of imprinted genes expressed in them depends on the system of dosage compensation. In female marsupials, the paternal X chromosome is inactivated or ejected from somatic cells, and so only the maternal X is expressed (VandeBerg et al. 1987, Johnston et al. 2002). Therefore, it will evolve as in males, to have the same self-restraint as any maternally derived gene. Again, reversible imprints are unnecessary (beyond whatever is needed for the silencing or ejection of the paternal chromosome). But in eutherian mammals, females are mosaics, with the maternal X silenced in some cells and the paternal X silenced in others (Lyon 1961). Thus, both chromosomes are expressed, each in about half of the organism. Here we expect the usual pattern of the paternally derived X being more selfish than the maternal X in promiscuous species.

There are additional possibilities for imprinting of sex-linked genes under monandry. If all the progeny of a mother have the same father, the paternal X chromosome in a female will be found in all her sisters, but (obviously) none of her brothers. Thus, a paternally active X-linked gene that consumed brothers to benefit self or sisters would spread. A maternally active X gene would value brothers and sisters equally and half as much as self (thus more nearly approximating mother's best wishes). The opposite argument applies to the Y chromosome in a male, which would value self and brothers to the exclusion of sisters.

The kind of effect that might be produced is suggested by evidence from gerbils and mice that the sex of one's intrauterine neighbor has a strong effect on development (Clark and Galef 1995). A female developing between 2 males is masculinized to her own disadvantage, development is slowed, and—compared to a female raised between 2 females—lifetime reproductive success (in the lab) is cut by one-half (Clark et al. 1986, Clark and Galef 1988). The same pattern is found for males: benefits from male neighbors, costs from female (Clark et al. 1992, Clark and Galef 1995). But mice released into nature show little effects of prior intrauterine position on survival and reproduction (Zielinski et al. 1992). Is this an example of imprinted sex chromosomes favoring their own kind *in utero*? Paternal Xs are expected to try to create a female-benefiting uterine environment and Ys a male-benefiting one, regardless of cost to the opposite sex, while maternal Xs and autosomes want a more gender-neutral environment. Unfortunately, little is known of the degree of polyandry in nature for either gerbils or mice, or whether degree of monandry across species correlates with degree of intrauterine effects. The few observations from gerbils suggest that a female in the wild often mates with more than 1 male per season (Ågren et al. 1989).

Note that a mammal species switching from polyandry to monandry would simultaneously reduce selection for imprinting on autosomes while selecting for sex-specific benefits from the paternal X and Y. An interaction between autosomes and the paternal X could occur in the monandrous species. Paternal Xs could activate paternal autosomal genes otherwise newly silenced or might oppose movement to biallelic expression of maternal genes. A Y in males might have similar effects.

Another factor in the evolution of sex-chromosome imprinting is that it could be used as a way to encode sexual dimorphism, at least in eutherians (Iwasa and Pomiankowski 2001). For example, if the paternally derived allele is silenced, the gene is expressed in all the cells of a male but in only half the cells of a female. And, if the maternally derived allele is silenced, the gene is expressed in half the cells of a female but not at all in a male. Such effects may be particularly important in the evolution of sexually dimorphic expression patterns for early-expressed genes, before circulating sex hormones are available as a signal of sex.

Skuse et al. (1997) have made a striking discovery concerning the human paternal X chromosome. It appears to have an imprinted gene or genes that affect positively the degree of sociability and social intelligence. The evidence comes from women with only 1 X chromosome (Turner's syndrome): their single X may be maternal (less sociable and socially aware) or paternal (more so). By inference, these paternal X genes improve a female's ability to detect the feelings and emotions of others, to interpret their body language, to understand her effects on them, to reason with them when upset, and so on. There are also subtle effects on memory, including one test on which females with only the paternal X do better (Bishop et al. 2000). It is also noteworthy that some of these tests imply that imprinting affects hemispheric specialization (for a paternal X effect on brain morphology, see Kesler et al. 2003). It is as yet unclear whether imprinting in this case evolved due to kinship conflicts or sexually dimorphic selection pressures.

As we have seen, the mammalian system of dosage compensation itself depends on imprinting, though slightly differently in marsupials and eutherians. Even within eutherians, there are differences: in mice, while most tissues show random X inactivation, in the extraembryonic trophoblast it is always the paternal X that is inactivated. Interestingly, this appears to be due to a paternally active gene on the X that causes the rest of the X to become inactive (Lee and Jaenisch 1997). In humans, all tissues appear to show random X inactivation.

Why might the paternal X be selected to shut itself down in mouse trophoblast (and perhaps also in marsupials, though it is not clear yet whether this is due to an X-linked gene)? Haig (2000a and pers. comm.) suggests an answer based on kinship conflicts. He argues that on the X chromosome allelic substitutions at unimprinted loci are expected to be biased in favor of matrilines. This is true as long as there is no differential imprinting of sons versus daughters because then 2 Xs arrive from matrilines (mother) each generation and only 1 from a patriline (father). If these effects are sometimes dosage-sensitive, a paternally active gene may be favored that reduces dosage of the matrilineally biased genes by shutting down the entire chromosome.

This, we imagine, would usually be associated with a substantial cost, and we prefer the simpler idea that X inactivation evolved in parallel with the degeneration of the Y chromosome, so the X- and Y-bearing gametes from a male were largely equivalent in gene expression. Random X-inactivation might then have replaced paternal X-inactivation, in order to mask the effect of deleterious recessive mutations (Chandra and Brown 1975). Interestingly, a very similar system of dosage compensation has been reported from a mole cricket (Gryllotalpa fossor), an insect with no obvious scope for kinship conflicts (Rao and Padmaja 1992).

The evolution of random X-inactivation in placental mammals may have had its own interesting consequences. Haig (pers. comm.) points out that, if Xs are differentially imprinted, there will be selection to resist silencing and instead induce it in the other chromosome. In this way, a higher proportion of the organism would show the preferred expression pattern. Again, this is a situation that might lead to rapid evolution, and it is interesting in this respect that within mice there is significant allelic variation at the X-linked Xce locus that controls X-inactivation, with the result that, in heterozygotes, the proportion of cells with one X active versus the other can be very different from 50:50 (Johnston and Cattanach 1981, Plenge et al. 2000). And, in *Peromyscus maniculatis* × *P. polionotus* hybrids, *P. maniculatus* Xs are expressed about 85% of the time in female embryos (Vrana et al. 2000).

The fact that females are mosaics also introduces the possibility of cell lineage selection. In mice there appears to be a tendency for cells with a paternally active X chromosome to outnumber those with a maternally active chromosome (Cattanach and Beechey 1990). One explanation is that, as expected, cells with a paternally active X have a slightly higher optimal growth rate. It would be interesting to look for consistent differences among tissues (as in chimeric mice). In addition, clones of cells with the same active X can be seen in morphology (e.g., coat color patterns in cats) and this is also true of the brain: alternating radial bands of cells in the cortex express either predominantly the paternal or the maternal X chromosome (Tan et al. 1995). Are these bands ever in conflict, especially in monogamous species—and, if so, what kinds of mental states are thereby generated?

Genomic Imprinting in Other Taxa

Within-individual kinship conflicts can be expected in many taxa outside of mammals. In a striking case of evolutionary convergence, genomic imprinting is also found in flowering plants, and, specifically, in their placenta-like structure, the endosperm. There are other groups with intimate parent-offspring connections, and we shall speculate on the possibility that imprinting has evolved in them as well.

Flowering Plants

The first imprinted gene was described in a plant—the maternally active endosperm pigment gene in maize (Kermicle 1970)—but relatively few have been described since then. Baroux and colleagues' (2002) review lists only 4 imprinted genes in maize endosperm and 1 major imprinted gene, *MEDEA*, in *Arabidopsis*—with several associated *FIE* genes, also known now to be imprinted (Danilevskaya et al. 2003). *MEDEA* conforms to kinship expectations because knockouts of maternally inherited copies result in embryo overgrowth. The genes in maize are uninformative because they all show the unusual feature that imprinting is specific to some inbred strains and not others. We know little or nothing of the selection pressures that have gone into forming different inbred lines; depending on selection for kernel size, selection on imprinting could differ markedly. For example, selection for large kernels may favor biallelic expression at paternally active growth-promoting loci and nonallelic expression at maternally active, growth-inhibiting loci—in either case, removing the gene from imprinted status. *Arabidopsis* is self-fertilizing but presumably only partially or recently so; otherwise, there would be no selection for imprinting. Further reviews of imprinting in plants can be found in Grossniklaus et al. (2000) and Messing and Grossniklaus (1999).

Most of the evidence for imprinting in plants comes from the effects of altering the normal 2:1 balance of maternal and paternal chromosomes in the endosperm. This can be done by crossing parents of different ploidy. Crossing a tetraploid female (4x) with a diploid male (2x) produces an endosperm with a 4:1 ratio (female-biased) while the reciprocal cross generates a 2:2 ratio (male-biased). Evidence from many such crosses produces a consistent picture: maternal excess is associated with restrained growth and small size of the endosperm, while paternal excess is associated with greater growth, at least initially (Haig and Westoby 1991). There are, of course, exceptions. In maize endosperm with an extra set of paternal genes (2:2), there is premature appearance of starch but the critical nutrient-transporting transfer cell layer fails to develop and seeds become shrunken (Charlton et al. 1995).

Recent work on *Zea mays* helps to explain the common observation that in the endosperm of flowering plants, paternal excess promotes cell proliferation—and maternal excess prevents cell division (Leblanc et al. 2002). Maternal genomic excess (4x × 2x crosses) forces endosperm cells to enter early into endoreduplication (replication of DNA without subsequent cell division), resulting in multiple copies of the genome per nucleus. Paternal genome excess (2x × 4x) prevents its establishment. As expected, the evidence suggests that there are maternally active gene(s) involved in mitotic arrest and paternally active ones that initiate reentry into S phase. In *Arabidopsis thaliana* a double dose of maternal chromosomes inhibits development of the endosperm, resulting in a smaller embryo, while a double dose of paternal promotes growth of the endosperm and embryo (Scott et al. 1998). Triple doses lead to more extreme initial phenotypes followed by abortion. Using a transgenic line with reduced methylation in crosses with normal plants, Adams et al. (2000) were able to reproduce the effects of different ploidies. Hypomethylation prevents inactivation of genes, in effect, doubling expression of oppositely active genes. It is striking that development is less severely affected if both parental genomes are hypomethylated, suggesting that activation of oppositely imprinted pairs of genes is occurring.

But why a triploid endosperm in the first place? Haig and Westoby (1991) argue that imprinting itself may be involved. Assume an original diploid endosperm that evolves paternally active and maternally active genes. Assuming some of these are dosage-dependent, doubling of the maternal genome may give maternally active genes an additional advantage.

There is enormous variation in the relative size of the endosperm and the embryo of flowering plants. In maize 90% of the seed is endosperm; in peas it is 10%. We would expect correlated effects on imprinting (and asexual reproduction). That is, relatively large endosperm should be associated with imprinting of the endosperm and little or none in the embryo, while the reverse is expected for relatively small endosperm. In the former, imprinting does not act as a barrier to asexual reproduction, especially with pseudogamous apomixis, in which the endosperm is fertilized but not the embryo (Haig and Westoby 1991). By contrast, groups of plants with large embryos, compared to endosperm, are expected to see asexual reproduction evolve less frequently.

Other Taxa Predicted to Have Imprinting

Within-individual kinship conflicts are not limited to mammals and plants, but arise wherever there is postconception maternal investment, and also when later kin interactions are asymmetrical for paternal and maternal alleles. Birds, for example, have postconception parental investment and so may be expected to show imprinting, but the circumstances are complicated by monogamy, extrapair copulations, and the chance of exploiting the opposite-sexed parent in biparental families (Trivers and Burt 1999). Here we review other likely candidate taxa in which imprinting has not yet been observed but might be expected.

Viviparous vertebrates. In mammals and flowering plants, offspring are in intimate contact with their mothers (or maternal tissue) during critical stages of early development. They are in a position to actively extract additional resources. The various mechanisms that the human fetus appears to employ to gain additional resources from the mother have been beautifully described by Haig (1993b). All of these mechanisms are subject to selection for imprinting. It is at present unknown whether related kinds of imprinting have evolved outside of mammals and flowering plants, but they are certainly expected in viviparous species such as some lizards, snakes, frogs, fish, and arthropods, all of which have evolved intimate parent-offspring associations during early development. In pipefish and seahorses, intimate connections are established between the offspring and their father—and broods may consist of offspring contributed by more than 1 mother, so that maternally active genes would be expected to play the same selfish role normally played by paternally active genes in species such as ourselves.

Haplodiploid social insects. The haplodiploid social insects—such as ants, bees, and wasps—provide special opportunities for imprinting to evolve throughout the lifecycle (Haig 1992, Queller 2003). Because individuals spend their lives investing in relatives to whom they are usually differentially related by sex, paternal and maternal chromosomes in females are often thrown into conflict about a series of reproductive decisions, including investment in self, investment in reproductives, ratio of investment in male and female reproductives, survival of the queen, encouragement of laying workers, and so on.

In their inheritance, haplodiploid species are exactly like X chromosomes—2 sets of chromosomes in a female, 1 in a male. Under complete monogamy (single insemination of the queen), a worker ant's paternal genes are identically related to those of other females but unrelated to those of males. Regarding the ratio of investment, paternal genes prefer all-female, while maternal, at their equilibrium, 1:1 (Haig 1992, Queller 2003). In hostile interactions between 2 sister workers (or incipient queens), paternal genes would be expected to value the other female as much as self while maternal genes would see the other as related by only one-half. These are just a few of the many critical variables over which imprinting is expected to evolve (Queller 2003).

Counterintuitive effects are also expected. Because the mother is equally related to her daughters, a selection pressure exists for her to silence her own genes affecting interactions among her daughters—the better to express her mates' because the latter will express her own interests! (Of course, any tendency toward multiple mating by females reduces these effects.) The direction of imprints may easily change in response not just to degree of polyandry but also to variables such as queen number. But unlike diploid systems, there is not much conflict expected between Imprinter genes and imprinted (Queller 2003). For example, males, being haploid, pass on all their genes to daughters so there can be no such conflict.

Although suggestive evidence for imprinting in sex determination of a haplodiploid species has been uncovered for the hymenopteran parasitoid *Nasonia vitripennis* (Dobson and Tanouye 1998), there is no evidence whatsoever on the kinds of effects predicted by kinship theory. In that sense, eventual work on imprinting in the Hymenoptera should provide a relatively pure test of the kinship theory of imprinting, because striking effects are predicted on many aspects of social insect biology. Indeed, if these arguments are correct, social insects may serve eventually as much as an example of internal genetic conflict as they now do for between-individual cooperation (Queller 2003).

Diploids on haploids. Postzygotic investment and the potential for maternal-paternal conflicts of interest also occur in those species in which a diploid phase grows on a haploid. For example, in *Neurospora* and other hyphal ascomycetes, a single fertilization event produces a dikaryon and eventually leads to the production of many haploid spores, and paternally derived genes will be selected to increase the number of spores, more or less regardless of the consequences for the supporting maternal mycelium. Will the filamentous ascomycetes be the next group in which genomic imprinting is discovered? They do have methylation (Foss et al. 1993), and, interestingly, apomixis derived from sex is rare among fungi (Carlile 1987), consistent with both maternal and paternal genomes being required for development.

Similar diploid-on-haploid life cycles occur in bryophytes (e.g., mosses) and red algae. The traditional explanation for this postfertilization amplification of the diploid phase is that it represents a way for the mother to increase the number of progeny from a single rare fertilization event. Alternatively, perhaps it is a way for the offspring to parasitize the mother. If so we would expect paternally imprinted genes to have taken the lead in this. Indeed, much of the evolution of these diploid tissues and individuals may have been pushed along by paternally active genes attempting to better exploit the haploid mother.

FIVE

Selfish Mitochondrial DNA

IN ADDITION TO THE GENES in their nuclei, the vast majority of eukaryotic organisms also have genes in their mitochondria (so-called mtDNA). Mitochondria are cellular organelles found in the cytoplasm and responsible for oxidative phosphorylation and much of the cell's ATP production. That is, they convert food and oxygen into units of cellular energy. This alternative location for genes derives from the fact that mitochondria are descendants of bacterial endosymbionts and still contain their own DNA. Most importantly for the topic of this book, the replication and transmission of mitochondrial genes is fundamentally different from that of nuclear genes, which translates into differences in the pattern of selection. In the typical diploid eukaryotic cell, there are 2 copies of each nuclear gene, each one is replicated exactly once during the S phase of the cell cycle, and at mitosis each daughter cell receives 1 copy of each chromosome. The consequence is that, barring mutation or gene conversion, a heterozygous cell gives rise to heterozygous daughter cells—there is no segregation of nuclear alleles at mitosis, and no change in allele frequency. Birky (2001) describes the nuclear genome as "stringent." By contrast, mitochondria have "relaxed" genomes: the number of genomes per cell varies widely (e.g., 20 to 10,000), different mitochondrial chromosomes can be replicated different numbers of times during the cell cycle (Clayton 1982), and daughter cells typically inherit a more-or-less random partition of the chromosomes. Thus, at mitosis there can be both segregation and changes in gene frequency. Even in nondividing cells, mitochondria are continually replicating and dying, and allele frequencies can change over time. These differences greatly increase the scope for within-individual selection, both among cells and within cells, and therefore for the evolution of selfish mitochondrial variants that are harmful to the host organism, but are still able to spread and persist because of a within-individual transmission advantage in the germline (Eberhard 1980).

Another major difference between mitochondrial and nuclear genes is that, in most eukaryotes, mitochondria are usually transmitted only by one sex, usually the female. This difference has the opposite effect of the first, as uniparental inheritance greatly reduces within-organism variation and thus the opportunity for within-organism selection. But it also means that mitochondrial genes are selected only to increase female fitness, whereas nuclear genes are selected to increase male and female fitness equally. Thus, the different patterns of transmission (uni- versus biparental) lead to conflicts over investment in male versus female function.

Mitochondria are not the only ancient, vertically inherited, mutually obligate endosymbionts—for example, all plants have chloroplasts, and many insects have bacteria to aid their digestion. These too have "relaxed" modes of replication and are expected to show within-individual selection, and thus the opportunity for selfish variants to arise. But we focus in this chapter on mitochondria, as the oldest, most widespread, and best-studied example. We begin with a brief survey of mitochondrial genomics, because the size and content of their genomes should in some sense define their evolutionary potential. We then review (1) examples of within-individual mitochondrial selection; (2) the idea that uniparental inheritance evolved specifically to prevent such selection; and (3) the resulting conflict between nuclear and mitochondrial genes over allocation to male and female function, particularly as manifested in widespread male sterility in flowering plants. As we will see, though nucleus and mitochondria cooperate intimately in making a functioning organism and typically cannot live without each other, all is not peace and harmony.

Mitochondrial Genomics: A Primer

In the time since mitochondria were free-living bacteria, the most striking change in their genome has been the extreme loss of genes, either absolutely or by transfer to the nucleus (Lang et al. 1999). The total number of proteins encoded by mitochondria ranges from 97 (in the free-living flagellate *Reclinomonas*) down to 3 (in malaria-causing *Plasmodium*). In some protists the process is thought to have gone to completion: there are no mitochondria, but there are other organelles called hydrogenosomes that appear to be derived from mitochondria but do not contain any DNA at all (Embley et al. 2003). In all species, the vast majority of proteins in mitochondria are encoded in the nucleus.

Among animals, mitochondrial genomes are relatively uniform, typically encoding 12–13 proteins, all involved in electron transport and oxidative phosphorylation, as well as rRNAs and tRNAs necessary for their synthesis. Most animal mitochondrial genomes are compact: human mtDNA, for example, is only 17kb in size, of which only 7% is noncoding. Somewhat larger mtDNAs (20–42kb) have been found in some insects, molluscs, and nematodes, but they do not encode more proteins.

Choanoflagellates are the sister group to animals, and the mitochondrion of at least 1 species is somewhat less degenerate, encoding 26 proteins. It is also less compact, at 77kb, of which 53% is noncoding. The main difference in gene content involves ribosomal proteins: the choanoflagellate mtDNA has 11 and animal mtDNA has none. This suggests that some streamlining of mtDNA occurred concomitantly with the evolution of multicellularity and tissue differentiation (Burger et al. 2003).

Fungi are like animals in that their mitochondria have also lost all, or nearly all, their ribosomal protein genes. However, while some are compact (e.g., 19kb, 11% noncoding in the fission yeast *Schizosaccharomyces pombe*), others are not (e.g., 86kb, 59% noncoding in the budding yeast *Saccharomyces cerevisiae*). Among plants, the mitochondria encode 40 proteins in some species, down to 24 in others (Adams et al. 2002). Again, most of the variation is in the ribosomal proteins, with some mitochondria encoding 14 and others none. At least 2 of these ribosomal protein genes have transferred repeatedly between distantly related plant species over evolutionary time, by unknown means (Bergthorsson et al. 2003). What evidence exists suggests that most of the losses are transfers to the nucleus. In general, plant mtDNAs are large and vary greatly in size, ranging from 200–2500kb, with the vast majority of this variation thought to be in intergenic regions (Palmer 1990).

All recent movements of functioning genes from the mitochondrion to the nucleus have taken place in plants or protists and, where the mechanism is known, have usually used reverse transcription from the mitochondrial mRNA (e.g., Covello and Gray 1992, Grohmann et al. 1992, Wischmann and Schuster 1995). The intermediary step in all transfers is, presumably, active copies in both nucleus and mitochondria. Active mitochondrial and nuclear copies of *cox2* have actually been observed in some legumes (Adams et al. 1999). In this situation, why might the nuclear gene often persist and the mitochondrial one not? We can imagine many possibilities. For example, if the gene was lost by deletion, selection for the deletion might be stronger in the mitochondria than in the nucleus, either because of selection within the individual for shorter mitochondrial genomes or because of selection among organisms to reduce DNA requirements by having 2 nuclear copies rather than thousands of mitochondrial copies per cell. Or the nuclear gene might be better for the organism because of a lower mutation rate or more precise transcriptional or translational control, or because it was not clonally inherited or would be selected to work well in both sexes. It would be interesting to know whether mitochondrial genes are typically silenced by nuclear genes before they disappear, or whether they disappear "of their own volition."

Mitochondrial Selection within the Individual

"Petite" Mutations in Yeast

Selfish mitochondrial mutations have long been known in baker's yeast, *Saccharomyces cerevisiae* (Ephrussi et al. 1955, Dujon 1981, Clark-Walker 1992, Chen and Clark-Walker 2000). In this species, mtDNA is biparentally inherited: the 2 gamete types (**a** and α) are of equal size, and they contribute mitochondrial DNA equally to the resulting zygote. Haploid cells contain about 25–50 copies of the 80kb mitochondrial genome, organized into 10–20 DNA-protein clusters called nucleoids (MacAlpine et al. 2000). If the mitochondrial genome is defective, the strain is unable to respire, but it can still grow and replicate by fermentation. Nevertheless, the cells replicate more slowly than normal, and so the resultant colonies are "petite" (designated ρ^-). Such strains arise spontaneously at a surprisingly high frequency—about 1% per cell division. At the molecular level, many types of mitochondrial mutations can disrupt respiration and give a petite phenotype, but we especially expect to observe mutants that have a within-cell replication advantage, because most of the genomes in a cell must be mutant before respiration is knocked out. A within-cell replication advantage is most easily observed by crossing petite and normal strains, propagating the progeny until they have only one kind of mtDNA and then scoring them. Many mutants have a within-cell advantage: with so-called hypersuppressive petites, more than 95% of the progeny end up petite (Fig. 5.1). Hypersuppressive petites typically are deleted for more than half of their mitochondrial genome, with the remaining fragments tandemly repeated to make up approximately the same size genome, but enriched for so-called *ori* sequences. These are thought to be *cis*-acting sequences that promote DNA replication (possibly by acting as origins of replication, or something analogous), and this is thought to give the hypersuppressive petite genomes their replication advantage (MacAlpine et al. 2001). Many of the deletions are thought to occur by recombination between specific 20–50bp GC-rich sequences, of which there are about 200 copies dispersed around the mitochondrial genome (Clark-Walker 1992).

Despite their transmission advantage, petite mutations are not able to spread in natural yeast populations because inbreeding is frequent (Johnson et al. 2004) and the mutations impose a large cost on the organism. Nevertheless, such observations raise the question of what determines the evolutionarily stable number of *ori* sequences in normal mitochondrial genomes. In *S. cerevisiae*, there are 7 or 8 such sequences, all similar in organization and primary sequence (Faugeron-Fonty et al. 1984). Curiously, only 4 of them are active, the others being inactivated by short insertions. As *S. cerevisiae* is largely inbred, the mitochondria inherited from the 2 parents will usually be identical, and selection for supernumerary *ori* sequences may have been relaxed. Other yeasts are different: *S. castellii*, for example, does not have recognizable *ori* sequences; and while it spontaneously generates petite mutations at about the same rate as *S. cerevisiae*, they are much less likely to be hypersuppressive (Petersen et al. 2002). In the more distantly related *Candida glabrata*, an asexual species, there are no GC-rich sequences dispersed around the mitochondrial genome, and petites are formed much less commonly, at a frequency of about 10^{-6} per cell division (Clark-Walker 1992, Chen and Clark-Walker 2000).

Within-Individual Selection and the Evolution of Uniparental Inheritance

Selfish mitochondrial genomes like the petite mutations of yeast have a replication advantage over normal mitochondrial genomes in within-organism

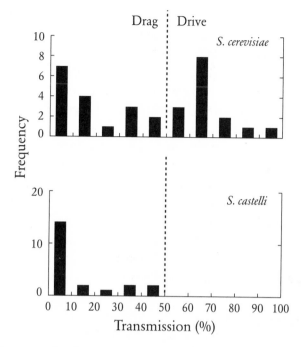

Figure 5.1 Frequency distribution of transmission rates for spontaneous petite mutations in 2 species of yeasts. Mutant mitochondria showing drive (transmission greater than 50%) are more common in *Saccharomyces cerevisiae* than in *S. castellii*, but the reason for this is not known. Data from Petersen et al. (2002).

selection, but they lose out in conventional among-organism selection. Therefore, if a population is polymorphic for 2 mitochondrial types, one more selfish than the other, a nuclear gene that somehow reduces the efficacy of the within-individual selection will tend to become associated with the less selfish type, and so can increase in frequency due to among-organism selection. That is, there will be selection for nuclear genes that modify mitochondrial behavior so as to reduce the efficacy of within-organism selection (just as there is selection on nuclear genes to suppress drive at unlinked loci). Even in yeast, the opportunity for within-individual selection is limited by the rapid segregation of parental types, which is much faster than if the genomes were well mixed and independent. Mechanistically, this is due to the clustering of mitochondrial genomes into nucleoids, and due to slow mixing in the zygote (Jensen et al. 2000). The evolution of these characters may have occurred partly to prevent selection for selfish mitochondrial mutants (Hurst and Hamilton 1992).

An even better way to limit within-organism mitochondrial selection is to ensure that only one parent transmits mitochondria to the next generation—that is, to impose uniparental inheritance (Grun 1976, Hoekstra 1990, Hastings 1992, Randerson and Hurst 1999). This is the normal pattern in eukaryotes, though the underlying mechanisms responsible are quite diverse (reviewed in Birky 1976, 1995). In animals, the egg is usually much larger than the sperm and contains many more mitochondria. For example, in *Xenopus* frogs, the ratio is about a million to 1 (10^8 versus 10^2). Crayfish sperm are reported not to have any mitochondria at all. Thus, purely as a matter of dilution, one would expect an overwhelming bias toward maternal inheritance. But in many species, even this is not left purely to chance. In mammalian spermatogenesis, proteins in the mitochondrial membrane are tagged with ubiquitin, a small protein that acts as a marker for protein degradation and recycling (Sutovsky et al. 2000). All components of this system are nuclear-encoded: the ubiquitin tag, the membrane proteins that are tagged, and the proteins applying the tag. The ubiquitin is apparently masked as the sperm travels through the male reproductive tract, but then is unmasked and amplified after fertilization. This is thought to lead to the destruction of paternal mitochondria at or before the 8-cell stage of preimplantation development. The notion that males specifically tag their mitochondria for destruction is further supported by the observation that mitochondria from spermatids are destroyed if artificially injected into embryos, but not mitochondria from liver cells (which would have no need for a tag; Shitara et al. 2000).

The existence of such a mechanism is particularly noteworthy given that paternal mitochondria are in any case outnumbered by maternal mitochondria by a factor of 10^3 to 10^5 (Ankel-Simons and Cummins 1996). One possible contributing factor is that the mitochondrial mutation rate may be higher in the male germline than in the female germline, and so male mitochondria may, on average, be of lower quality. But even if paternal mitochondria were mostly defective, their low frequency in the fertilized zygote should mean that the selection pressure for destroying them is relatively weak, were it not for the possibility of within-individual selection. Examples of such selection in mammals are reviewed later; here we simply note that the one known instance of paternal inheritance in humans was of a defective mitochondrion. A 28-year-old man with severe lifelong "exercise intolerance" was found to have maternally derived mitochondria in most of his body, but paternal mtDNA in his muscles; and the paternal mtDNA had a novel 2bp frameshift deletion in one of the genes that obliterated downstream function (Schwartz and Vissing 2002). (There was also recombination between maternal and paternal mtDNA in his muscles; Kraytsberg et al. 2004.)

Other animals also have mechanisms for the destruction of paternally derived mitochondria (reviewed in Birky 1976, 1995, Eberhard 1980). In some sea urchins, the paternal mitochondria appear to be destroyed in the egg. In honeybees, many sperm enter a single egg and about a quarter of all mitochondrial DNA in newly laid eggs is paternally derived; but the paternal mtDNA is degraded or replicates more slowly than the maternal mtDNA and, by the time the organism hatches out as a larva, the paternal mtDNA is undetectable. In the tunicate *Ascidea*, paternal mitochondria do not even make it into the egg, being shed before the sperm passes through the chorion membrane. Analogous mechanisms are also seen outside the animals. In the single-celled alga *Chlamydomonas rheinhardtii*, the fusing gametes are of equal size, but only mitochondria from the "−" mating type gamete are inherited; those from the "+" parent are destroyed after gamete fusion. In another alga, *Derbesia tenuissima*, male gametes are smaller than female ones, and while the male gamete has a large and apparently functional mitochondrion, the DNA inside it is destroyed during gametogenesis, prior to fertilization (Lee et al. 2002). The slime mold *Physarum polycephalum* has a particularly baroque system in which mitochondrial transmission is determined by a linear hierarchy of alleles at the nuclear *matA* locus, with the "losing" mtDNA being actively degraded soon after gamete fusion (Kawano and Kuroiwa 1989, Meland et al. 1991). And, in many plant species mtDNA is either stripped from mitochondria or severely degraded during pollen maturation, prior to fertilization (reviewed in Mogensen 1996). For example, in 9 of 16 species from a range of families, mtDNA disappeared from the generative cells during anther development (Miyamura et al. 1987). Finally, in redwood trees and banana trees, mitochondria are paternally inherited (Neale et al. 1989, Fauré et al. 1994), which presumably goes against the numerical trend, suggesting that there is some active mechanism for destroying maternal mitochondria.

In short, uniparental inheritance in many species is not merely a function of numerical dilution. Mechanisms have evolved for the active elimination of the mtDNA from one parent. The only explanation plausible to us is that these mechanisms have evolved to suppress selfish mtDNA–that is, uniparental mitochondrial inheritance is a system evolved by the nucleus to ensure mitochondrial quality. Especially striking are mechanisms of uniparental inheritance in which one parent or gamete type somehow cripples its own mitochondria to keep them from being transmitted. The possibility that nuclear genes might evolve to suppress selfish mtDNA by sabotaging their own mitochondria (as opposed to those being transmitted by the other gamete) has been dismissed on theoretical grounds (Hastings 1992, Randerson and Hurst 1999), but the facts suggest this dismissal may be premature. The theoretical difficulty is that the population must be polymorphic for mitochondria varying in degree of selfishness in order to select for nuclear modifiers of mitochondrial inheritance, and in simple models the spread of selfish mitochondria does not easily evolve to a stable polymorphism–either they are lost or go to fixation. But there are two counterarguments. First, uniparental inheritance can protect against selfish organelles that are too harmful to go to fixation, but nevertheless arise repeatedly. The petite mutations of yeast are like this–recall they arise at a 1% frequency. In this case a stable intermediate frequency of the selfish mutant type is reached, analogous to conventional mutation-selection balance, and uniparental inheritance can be effective in reducing the probability of inheriting a mutant organelle (and reducing the equilibrium frequency of the mutant type). Second, if the organelle is only mildly selfish, such that heteroplasmy is often maintained from one meiotic generation to the next, with only a shift in frequencies, and if fitness declines more than linearly with the frequency of the selfish type, a polymorphism can be maintained and uniparental inheritance selected for. Formal modeling of both possibilities is clearly desirable.

What role do the mitochondria play in all this? They are hardly expected to be indifferent to their own destruction in the germline and will be selected to avoid this, if possible. But there does not appear to be any good example of such avoidance behavior. At first glance, the fact that paternal leakage is more common in some interspecific mouse matings than in intraspecific ones (Kaneda et al. 1995) seems suggestive of a coevolutionary arms race between nuclei and mitochondria. But this paternal leakage appears to come about because the ubiquitin tagging and destruction system is disrupted in some hybrid matings (Sutovsky et al. 2000) and, as we have seen, this is a wholly nuclear-encoded system. Perhaps mutations allowing mitochondria to escape destruction just do not arise. Similarly, we are not aware of mtDNA from one parent being involved in the destruction of mtDNA from the other. In a sea urchin *(Paracentrotus lividus),* just minutes after the first cell division of a newly fertilized egg, large paternal mitochondria are seen surrounded by small maternal ones; their membranes touch and the paternal ones begin to disintegrate (Anderson 1968). But even if the maternal mitochondria are involved in the destruction of the paternal mitochondria, most of their proteins are encoded in the nucleus, and so this gives no compelling evidence of mtDNA involvement. Nevertheless, evidence for these sorts of direct action may yet be found—an obvious place to look would be in species that have recently changed from maternal to paternal inheritance. Presumably ancient mitochondria had more influence on their own transmission, before they became so degenerate.

The idea that uniparental inheritance evolved to prevent the spread of selfish mitochondria has been extended to binary mating systems (e.g., male/female; $+/-$; a/α). Perhaps these also exist in order to allow uniparental transmission (i.e., for 1 mating type to be a transmitter and the other not; Hurst and Hamilton 1992, Hurst 1995). Support for this idea comes from ciliates and mushrooms, 2 groups in which it is common for mating cells not to fuse, but instead to trade nuclei. There is thus no opportunity for mitochondria from the 2 parents to mix and compete. Consistent with the hypothesis, it is common in both groups not to have a binary mating system, but instead a multipolar system with many mating types (e.g., 30 or more).

But this explanation for binary mating types cannot be universal (Hurst 1995). Both cellular and acellular slime molds also have multipolar mating systems, and each has full gamete fusion. And ascomycete fungi typically have a binary mating system, but this has nothing to do with mitochondrial transmission. For example, in *Saccharomyces* yeasts, there are 2 mating types, but mitochondrial inheritance is biparental. In *Neurospora crassa,* there are 2 mating types, each of which can act both as a male parent and as a female parent, but mitochondria are inherited from the maternal parent, regardless of mating type. Instead of organizing mitochondrial inheritance, the binary mating system of ascomycete fungi is used to organize pheromone production and reception: the **a** mating type produces the **a** pheromone and has a receptor for the α pheromone, and the α mating type produces the α pheromone and has a receptor for the **a** pheromone. Such a division of labor is presumably needed to prevent one's receptors from being clogged up with one's own pheromone. It may be relevant in this context that many ciliates and mushrooms do not use dispersible pheromones to attract a mate. In the case of ciliates, the pheromones of some species remain attached to the cell membrane (Görtz et al. 1999). In mushrooms, pheromones are only activated after mating and are used to distinguish self from non-self (Casselton 2002). Perhaps the abandoning of dispersible pheromones is also important in the evolution of multipolar mating systems.

It has also been suggested that selection to kill off one's own organelles could account for the small size of sperm and, more generally, the evolution of anisogamy (reviewed in Randerson and Hurst 2001). As a mechanism for ensuring uniparental inheritance, this would be terribly crude. While it might give some slight advantage to those making smaller-than-normal gametes, surely this is trivial compared to the more general advantage of reducing the cost of a gamete, and thus being able to make more of them (Randerson and Hurst 2001, Bulmer and Parker 2002).

Within-Individual Selection under Uniparental Inheritance

Even if uniparental inheritance is absolute and there is no paternal leakage, this does not wholly eliminate the possibility of selfish mtDNA increasing in frequency due to within-individual selection. Due to the "relaxed" nature of mitochondrial replication, variant mtDNA that arise by mutation during the lifetime of an organism may spread due to within-individual selection. At least in the simplest case—in which there is a mitochondrial bottleneck in the germline and females produce mostly homoplasmic eggs—the effect of within-individual selection is to change the "effective" mutation rate: mutants with a within-individual selective advantage will behave as if they have an elevated mutation rate, while mutants with a within-individual disadvantage will behave as if they have a depressed mutation rate. For recurrent mutations that are harmful to the organism, the classic equations of mutation-selection balance will apply (e.g., Crow and Kimura 1970), and the equilibrium frequency of the mutant is $\hat{q} = u_e/s$ where u_e is the effective mutation rate (taking into account within-individual selection) and s is the (among-individual) selection coefficient against the mutant. This formula applies if

$u_e < s$. When the within-individual selection is strong, or among-individual selection is weak, it may be that $u_e > s$ and the mutant will go to fixation in the population.

In principle, then, within-individual selection of harmful mitochondrial variants could increase the mutational load on the population, and nuclear genes will be selected to reduce this when possible. There are two ways they may do so. First, they could reduce the actual mutation rate. And there is some evidence this has happened: in mammals, DNA polymerase γ, encoded in the nucleus, is the enzyme responsible for replicating the mitochondrial genome, and it is among the least error prone of all DNA polymerases (ca. 2 errors per million base pairs; Gillham 1994). *In vitro*, it is 2 to 4 times less likely to make a base substitution than pol-α, the enzyme responsible for replicating nuclear DNA, and more than 10-fold less likely to make a frameshift mutation (Kunkel 1985). Perhaps because of this high accuracy, pol-γ is also among the slowest DNA polymerases, incorporating 270 nucleotides per minute per strand, 200 times slower than the polymerase of *E. coli* (Clayton 1982). Despite the greater accuracy *in vitro*, rates of silent substitution over evolutionary time are higher in mtDNA than nuclear, perhaps because of the unavoidable proximity to oxygen radicals produced in the mitochondria, or perhaps because some repair mechanisms are absent (Gillham 1994). In plants, the silent substitution rate in mitochondria is actually an order of magnitude lower than in the nucleus, which suggests a very accurate polymerase (or low damage from oxygen radicals).

Second, nuclear genes will be selected to create a cellular environment such that within-individual selection on mitochondria is aligned as closely as possible with among-organism selection. In modern day eukaryotes, nuclear genes are largely responsible for regulating mitochondrial replication and destruction, and the manner in which they do so will determine the mitochondria's selective environment. The results can be counterintuitive (Chinnery and Samuels 1999, Birky 2001). For example, consider the apparently sensible possibility that nuclear genes regulate mitochondrial copy number according to the local needs for oxidative phosphorylation: the more ATP required, the more mitochondria are produced. Though perhaps satisfactory in the short term, the consequence of this strategy is the selection of ever less efficient mitochondria. To understand this, note that under this scenario an abundance of ATP will inhibit mitochondrial replication. Suppose there are two types of mitochondria in the cell, one more efficient than the other. Then an extra copy of an efficient mitochondrion will in-hibit further mitochondrial replication more than an extra copy of the inef-ficient mitochondrion. Consequently, the cell will gradually come to be populated by the inefficient mitochondria—not necessarily what the organism wants, either for itself or to pass on to its offspring!

There is compelling evidence for within-organism selection in mice that have been artificially made to carry 2 types of mitochondria. In one set of experiments, mice were made heteroplasmic for mitochondria from 2 lab strains, NZB and BALB (Jenuth et al. 1997, Battersby and Shoubridge 2001). In hundreds of mice, in a variety of different nuclear backgrounds, the NZB mitochondria increase in frequency in the liver as the mouse ages, with a se-lection coefficient of about 14% per mitochondrial generation (Fig. 5.2). NZB mitochondria also increase in frequency in the kidneys. Surprisingly, selection is reversed in blood and spleen, where the BALB mitochondria come to predominate. Finally, in other tissues, there is no apparent selec-tion. The molecular basis of the selection is unknown: there are only 15 amino acid differences in all the proteins coded by the 2 mitochondria, and there is no detectable difference in respiratory efficiency. Battersby and Shoubridge (2001) suggest that the differences are in death rates of the 2 types, but this needs confirmation.

People with mitochondrial diseases are typically heteroplasmic for defec-tive mitochondria (homoplasmy would be lethal), and these too can show tissue-specific patterns of selection (Chinnery and Turnbull 2000, Chinnery et al. 2002). In particular, the defective mutants are typically lost from rap-idly dividing tissues such as bone marrow, but may accumulate in nondivid-ing cells such as skeletal muscles and the central nervous system (recall that, unlike nuclear DNA, mtDNA is continually degraded and replaced even in these nondividing cells). Consequently, mitochondrial diseases tend to be progressive and to most affect metabolically active, nondividing tissues. The differences observed between proliferative and nonproliferative tissues may simply be due to differences in the relative importance of within-cell selec-tion favoring the defective mitochondria (e.g., because the cell regulates mi-tochondrial copy number based on ATP requirements) and among-cell se-lection favoring wildtype mitochondria (Taylor et al. 2002).

Studies of cultured human cells suggest that, at least under some circum-stances, there can also be within-cell selection for shorter mitochondrial genomes. First, if mitochondrial copy numbers are artificially repressed 90%

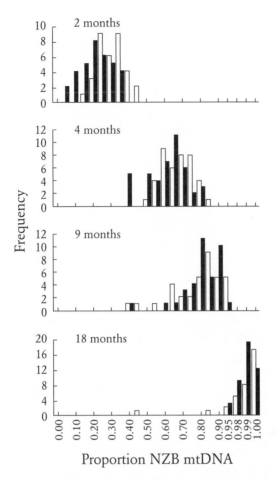

Figure 5.2 Changes in the frequency of a mitochondrial genotype with age. Mice were made heteroplasmic for mtDNA from 2 lab strains, NZB and BALB, and then the frequency of the 2 types monitored by single-cell PCR of liver cells. Data are from 2 representative mice at each age. The increase in the frequency of NZB mtDNA with age indicates a selective advantage within the organism. Adapted from Battersby and Shoubridge (2001).

by ethidium bromide treatment and then allowed to recover, mitochondrial genomes with a 7.5kb deletion (= 45% of the total length) recover in number more rapidly than do wildtype genomes in the same cells (Diaz et al. 2002). This difference may occur simply because in this environment mitochondria are being replicated by the cell as fast as possible, and shorter genomes are more rapidly replicated. Second, the equilibrium or steady-state amount of mitochondrial DNA per cell was found to be identical in cultures homoplasmic for wildtype genomes, those with a 7.8kb deletion (= 47%), and those with an 8.8kb duplication (= 53%; Tang et al. 2000). These results suggest that nuclear genes sometimes regulate mitochondrial copy number based on the total amount of mtDNA (by, for example, limiting the pool of nucleotides available for mitochondrial replication). This will select for shorter, more compact genomes.

Of all tissues, selection in the female germline will be particularly important for mitochondrial evolution. In mammals, the studies published to date do not find as dramatic within-individual selection in these cells as is found in some somatic tissues (Jenuth et al. 1996, Chinnery et al. 2000, Inoue et al. 2000). It would be interesting to compare rates of mitochondrial turnover in oocytes with those in, say, skeletal muscles. One possible sign of substantial selection in the female germline is the curious pattern that in humans, mitochondrial diseases due to a point mutation may be passed on from one generation to the next (heteroplasmically), but diseases due to a deletion are not (Chinnery and Turnbull 2000). In both cases, the ATP-producing function of the mitochondria can be completely knocked out. The mechanistic basis of this difference is not known, but one possibility is some sort of differential selection in the female germline.

Another possible indication of selection in the female germline is the extreme compactness of the mitochondrial genome in mammals, and animals more generally. This is consistent with the possibility that at some point in the female germline, mitochondrial copy numbers are regulated by the total mass of mtDNA (Birky 2001). Plant mitochondria are anything but compact, and the same logic would predict that copy numbers in plants are regulated by some other means.

Finally, selection in the female germline has been experimentally demonstrated for naturally occurring mitochondrial variants of *Drosophila simulans* that differ in sequence by 2–3% (de Stordeur 1997). In these studies, females were made heteroplasmic by microinjection, but in some populations of *D. simulans* heteroplasmic females occur naturally at an appreciable frequency (1–12%; James and Ballard 2003). This suggests at least occasional paternal leakage of mtDNA, which increases the importance of germline selection in mitochondrial evolution. In other species (e.g., rabbits, *Oryctolagus cunicu-*

lus), most or all females are heteroplasmic for length variation, but the relative importance of high mutation rates and different forms of selection in maintaining this polymorphism has yet to be clearly distinguished (Casane et al. 1994).

DUI: Mother-to-Daughter and Father-to-Son mtDNA Inheritance in Mussels

We conclude our review of within-individual mitochondrial selection with a description of the peculiar situation in mussels and their relatives, in which mitochondria are typically passed father-to-son and mother-to-daughter. This so-called doubly uniparental inheritance (DUI) was first discovered in species of the marine mussel *Mytilus* (Zouros et al. 1992, 1994a, 1994b, Skibinski et al. 1994). DUI was soon shown in the closely related *Gukensia demissa* (Hoeh et al. 1996), as well as freshwater mussels of the family Unionidae, separated by more than 400 million years (Hoeh et al. 1996, Liu et al. 1996). Doubly uniparental inheritance has also recently been discovered in a clam (Passamonti and Scali 2001, Passamonti et al. 2003) so the trait may be ancestral in the evolution of the bivalves. (By contrast, in *Cepea nemoralis,* a related land snail with a high degree of mtDNA variation, there is no evidence of deviation from pure maternal inheritance; Davison 2000.)

Females display the mtDNA from their mothers only, while males are heteroplasmic. In somatic tissue males show both forms, but in the testes the paternal mtDNA predominates and most males pass on exclusively (or almost exclusively) the paternal set. Thus, 2 forms of conspecific mtDNA, F (female) and M (male) mitotypes, evolve side by side in almost complete isolation and they may show extensive sequence divergence—for example, 10–20% in *Mytilus,* depending on locality and species (Rawson and Hilbish 1995, Stewart et al. 1995).

The 2 sexes are believed to start life with the same relative numbers of maternal and paternal mitochondria—in other words, 5 very large paternal mitochondria and tens of thousands of maternal ones (Cao et al. 2004). Subsequent adult distributions are known only for *M. edulis* (Garrido-Ramos et al. 1998; see also Dalziel and Stewart 2002). Most females (~70%) contain no M mitotypes and, where these are found, they are found in small amounts in some tissues, varying between individuals. A few females (~5%) appear to show small traces of M in the gonads, but it is almost impossible to study germ cells uncontaminated by nurse cells and there exists no clear evidence of M mtDNA in unfertilized eggs. By contrast, in males M mitotypes predominate in testes, while Fs do so in all other tissues (Stewart et al. 1995, Garrido-Ramos et al. 1998). The most sensitive tests show M mitotypes in all tissues of all individuals and suggest a regular pattern of M mitotype abundance as follows: testes>>adductor muscle, digestive gland>foot, gill, and mantle. By maturity, M mitotypes so predominate in testes that Fs are virtually undetectable. On the other hand, in nature and the lab there are rare males whose maternal and paternal mitotypes both resemble the typical F form (reviewed in Cao et al. 2004).

How are these patterns achieved? And how do the 5 paternal mitochondria in males manage to overwhelm tens of thousands of maternal ones, at least in testicular tissue? Recent evidence suggests how this may happen (Cao et al. 2004). In females, the 5 paternal mitochondria are randomly dispersed in the cytoplasm and appear to go randomly to descendent cells (in early cells, without replication). By contrast, in males paternal mitochondria aggregate themselves immediately in the fertilized egg and (without replication) they pass together into the first cleavage cell leading to the germplasm. In the first 5 cell divisions, these mitochondria remain aggregated and pass preferentially to 1 daughter cell at each cleavage (Plate 3). Presumably, this is always the germinal cell. Because both germinal and adductor muscle cells are mesodermal, these findings also provide an explanation for elevated frequency of paternal mitochondria in adductor cells.

Male mitotypes evolve more rapidly than do female. That is, nonsynonymous substitutions are significantly higher in M mitotypes than in Fs, and the ratio of nonsynonymous to synonymous substitutions is more than twice as high in Ms (Stewart et al. 1996). This suggests that faster M evolution results neither from more frequent replication during male gametogenesis nor from greater damage to sperm mitochondria by free radicals, because these effects should act on synonymous and nonsynonymous sites equally (Stewart et al. 1996). A more plausible explanation is that F mitotypes are fully exposed to selection every generation in females, while M mitotypes are shielded from the direct action of selection by the overwhelming presence of F throughout the male body (Saavedra et al. 1997). That less functionally constrained organelles will show more rapid evolution is suggested by the rapid evolution of chloroplast DNA in nonphotosynthetic plant species (Wolfe et al. 1992). Male mitotypes may also be subject to greater directional selection through sperm competition, which is known in bivalves to impose strong, directional selection on nuclear DNA, primarily on sperm proteins but also on egg receptors (Yang and Bielawski 2000, Galindo et al. 2003).

F mitotypes sometimes colonize the male lineage and displace the existing Ms—a striking fact that can be seen in *Mytilus*, in which some Ms are more related to Fs than to other conspecific Ms (Hoeh et al. 1997). In one population of *M. galloprovincialis* from the Baltic Sea, the majority of males lack an M type, suggesting a very recent invasion of M types (Ladoukakis et al. 2002). (In this species, the F genome is more variable within populations and the M between populations.) Takeover of Ms by Fs may also occur between species. In European *M. trossolus,* most of the M types come from Fs of *M. edulis.* Apparently, recently introgressed F molecules first displaced the *M. trossolus* F and then were able to colonize males and displace the local M (Quesada et al. 1999, Quesada et al. 2003). Many males are heteroplasmic for 2 paternal mtDNA types and 1 maternal form, so that cotransmission by males of 2 forms must be a common occurrence. mtDNA variants often differ by length, with the short-length variants preferably transmitted by males (Zbawicka et al. 2003).

A very similar system is found in a freshwater mussel *Anodonta grandis* (Liu et al. 1996). Mitochondrial inheritance is doubly uniparental and in different locales female mitotypes hardly differ (~0.4% sequence divergence) while male mitotypes differ by 11.5%. But unlike the Mytelidae, the Unionidae show no evidence that F mitotypes ever replaced M mitotypes and the split between the 2 seems to be very ancient, at least 200mya (Curole and Kocher 2002, Hoeh et al. 2002).

Why has DUI evolved—and why in the bivalves? One possibility is provided by considering sex-antagonistic effects of mtDNA, effects that are negative in one sex and positive in the other. If such effects are large in bivalves, especially in the gonads, DUI permits mtDNA to specialize in 2 types, each adapted to 1 of the sexes. Sexual dimorphism in clams and mussels is known only from gonadal tissue. Suppose in an early bivalve there was some small amount of mitochondrial transmission from fathers to offspring ("paternal leakage") and a mutation arose that was strongly beneficial in testes or sperm, but weakly deleterious in other male tissues and in females. The mutation could increase in frequency to some intermediate level. At that point the population would be polymorphic for 2 types of mitochondria, and nuclear genes would be selected to arrange the internal cellular environments so the new mitochondria would be selected for in testes and the original mitochondria would be selected for in all other tissues. Note that morphology today is suggestive of extreme sexual dimorphism: 5 large paternal mitochondria versus thousands of small, maternal ones (Cao et al. 2004). From the mussel's standpoint, deriving its paternal mitochondria from previous paternal ones may give a better version than having to differentiate them anew each generation from maternal mitochondria. If this argument has any merit, perhaps other species with highly dimorphic (egg/sperm) mitochondria also show DUI.

Thus we imagine that an ancestral bivalve evolved DUI in response to strong selection on male mtDNA for novel functions connected to intense sperm competition. Because there is no sexual dimorphism outside the gonads, selection on the newly emergent M mitotype may favor little or no representation within somatic tissue.

The mtDNA of *Mytilus* has a number of unique features. It is unique among animals (along with nematodes) in lacking ATPase-8, in having an extra tRNA for methionine (one using an anticodon, TAT) and in having a unique gene order that shows no homology with other coelomate metazoans (Hoffmann et al. 1992). Could these facts be related to DUI? One possibility is that DUI leads to more recombination, which increases the rate of evolution. Recombination in males has been confirmed in both *M. galloprovincialis* and *M. trossulus,* and evidence suggests a high rate in both (Ladoukakis and Zouros 2001, Burzynski et al. 2003). But for this to affect rate of F evolution, there must be at least occasional paternal leakage in females. Another possibility is that heteroplasmy in males also encourages conflict between the maternal and paternal lineages, which has the potential to greatly accelerate rates of genomic change.

How is sex determined? In principle, mtDNA itself could determine sex—for example, when male mitotypes predominate in the gonads, these become testes (Zouros 2000). But if this is how sex determination is achieved, how is the 50:50 sex ratio that the autosomes would prefer enforced? Mitotype frequencies, and thus sex, should evolve to be under autosomal control, even if they begin under mitochondrial. In any case, recent genetic evidence demonstrates that the sex ratio is under maternal, nuclear control in blue mussels (Saavedra et al. 1997, Kenchington et al. 2002). In 2 species of

Mytilus, sex ratios are highly variable from one pair to another, and the variation is associated with the mother and not the father. This variation is heritable in a manner that excludes mtDNA as the controlling element. Instead, the nuclear genome of the mother determines sex of progeny, perhaps through alleles at a single locus. A noteworthy feature of sex ratio variation is that all-female families are common but all-male families are uncommon and majority-male families are found instead. It will be most interesting to learn exactly how sex determination operates in these species, both genetically and developmentally.

Cytoplasmic Male Sterility

Uniparental Inheritance Implies Unisexual Selection

Though uniparental inheritance has the advantage (from the nuclear point of view) of suppressing within-individual mitochondrial selection, it introduces a new danger. If mitochondria are transmitted only by one sex, all that will matter for their evolution is how well they perform in that sex, with performance in the other sex being irrelevant. For example, in a species with maternal transmission, if there are recurrent deleterious mutations that cause only mild harm to the female, the mutant class can reach relatively high frequencies, even if the harm to the male is substantial (Frank and Hurst 1996). Something like this may be happening in humans (Ruiz-Pesini et al. 2000). In a survey of mitochondrial genotypes of couples at infertility clinics, one relatively common mitochondrial haplotype ("T") was found to be associated with moderately reduced sperm motility (asthenozoospermia; Fig. 5.3), apparently because this haplotype is slightly less efficient at oxidative phosphorylation and ATP production. If, as seems plausible, sperm motility is among the most sensitive physiological parameters to ATP production, this slightly reduced efficiency may have much less negative effect on female fitness (or could even be beneficial), and as a result the haplotype may have reached higher frequencies than it otherwise would have. Among Caucasian populations, the frequency of haplotype T ranges from 4% in the Druze, to 12% in Germans, to 22% in Swedes. These are the kinds of effects that would select for nuclear genes to become ever more involved in mitochondrial functions.

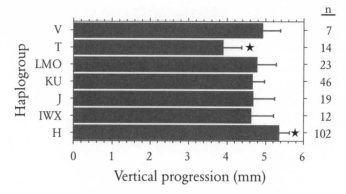

Figure 5.3 Swimming ability of human spermatozoa carrying different mitochondrial genotypes. The assay measures the distance swum from the semen into a capillary tube in 30min. n is sample size; * indicates groups that are significantly different. Differences in swimming speed could affect male fertility but would not normally be subject to selection because mitochondria are not transmitted by males. Adapted from Ruiz-Pesini et al. (2000).

Disproportionate Role of mtDNA in Plant Male Sterility

The dangers of uniparental inheritance are even more clear in flowering plants. Most plants are hermaphrodites—in other words, produce both pollen and ovules—but transmit their mitochondria only through ovules, and so mitochondrial mutants that completely abolish pollen production, causing cytoplasmic male sterility (CMS) will be positively selected if, in so doing, there is even a slight increase in ovule production (Lewis 1941). Two ways mtDNA might do this are (1) allowing some scarce resource(s) to be reallocated from pollen to ovule production, or (2) preventing self-fertilization and thus inbreeding depression. By contrast, nuclear genes are more willing to tolerate inbreeding depression because self-fertilization is associated with increased gene transmission. Because nuclear genes are transmitted equally through pollen and ovules, a nuclear gene causing male sterility will spread only if it more than doubles female fertility. Therefore, the spread of a CMS gene usually selects for nuclear suppressor genes that counteract the CMS gene and restore male fertility. Thus, the evolution of uniparental inheritance sets up a powerful conflict of interest between mito-

Table 5.1 Number of species, genera, or families showing nuclear and cytoplasmic inheritance of male sterility in hybrid crosses

Type of Cross	Nuclear	Cytoplasmic
Within species	0	23
Within genus	13	41
Within family	1	3

From Kaul (1988, Tables 2.2, 3.2–3.4).

chondrial and nuclear genes over both the optimal allocation to male and female function and the optimal levels of outcrossing.

The result of this conflict is widespread male sterility. Even as of 1972 (the last tally of which we are aware), CMS had been described in 140 species, within 47 genera, from 20 families (Laser and Lersten 1972). It occurs in 2 different situations. First, some plant populations have 2 kinds of individuals, hermaphrodites and male-steriles (i.e., females). Such populations are said to be gynodioecious. Gynodioecy itself has been described in 350 species from 39 families (Table 10.1 in Kaul 1988) and, among European flowering plants, is the second most common gender system, after hermaphroditism (7.5% versus 72%; Richards 1986). In most gynodioecious species that have been investigated, male sterility is at least partially maternally inherited (Sun 1987 and references therein). And, even within a single species, there are often multiple CMS genotypes (Kaul 1988, Frank 1989). In maize, for example, there are 3 distinct CMS genotypes (T, S, and C), as there are in rapeseed (*pol, nap,* and *ogu*). The inheritance of male sterility usually has a nuclear component as well, indicating that nuclear restorers are also common.

The second situation in which male sterility is common is in the descendants of crosses between hermaphrodites, either from different species or from different populations of the same species. The male-sterile progeny are sometimes seen in the first generation and sometimes in subsequent generations of hybrid crosses or backcrosses. And, in hybrids too, cytoplasmic inheritance of male sterility predominates (Table 5.1). Presumably this occurs because, in at least 1 of the species, a CMS gene has arisen and its restorer has swept to fixation, reestablishing hermaphroditism, and male sterility is only uncovered when the 2 genes are dissociated (Box 5.1). That is, hermaphrodites are restored male-steriles. The data in Table 5.1 suggest that

---

BOX 5.1

Coevolution of CMS Genes and Nuclear Restorers: An Example Trajectory

The figure in this box shows the fate of a CMS gene that increases ovule production by 20%, introduced at a frequency of 10^{-6} into a population with a selfing rate of 20%. Initially it increases in frequency. But as it becomes more common, pollen becomes less available, and male-steriles suffer more from pollen limitation than do hermaphrodites because they are unable to self. Therefore the CMS gene does not go to fixation, but instead reaches some intermediate equilibrium frequency (in this example, pollen limitation is modelled by assuming that the probability of a non-selfed ovule getting fertilized is proportional to the frequency of hermaphrodites). Then, in generation 100 a nuclear restorer gene is introduced, which spreads rapidly and goes to fixation. As it does so, the frequency of hermaphrodites, and thus the amount of pollen, increases, thus allowing the CMS gene also to increase in frequency. In the example shown here, the nuclear restorer goes to fixation before the CMS gene, with the result that the population is entirely hermaphroditic and there is a hidden mitochondrial polymorphism. The fate of the CMS gene then depends on its relative female fitness in the presence of the restorer, selection at linked loci, and drift.

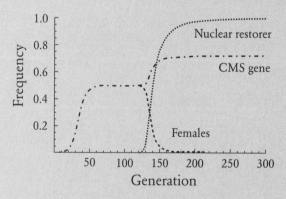

If the CMS gene reaches an appreciable frequency, selection on nuclear restorers will be intense, meaning that they can spread even if they have negative side effects. But a restorer with negative side effects may not go to fixation, in which case the population will remain gynodioecious. If, at the same time, the CMS gene goes to fixation, the inheritance of male sterility will appear to be nuclear rather than maternal. Or, if both the CMS gene and the restorer have negative side effects, the population may remain polymorphic at both loci. In computer simulations, gene frequencies can show complicated dynamics over time. The coevolutionary dynamics of CMS genes and nuclear restorers has been modelled extensively (Lewis 1941, Lloyd 1976, Charlesworth and Charlesworth 1978, Charlesworth and Ganders 1979, Delannay et al. 1981, Frank 1989, Gouyon et al. 1991, McCauley and Taylor 1997, Bailey et al. 2003, Jacobs and Wade 2003).

this is not a rare event, though exactly what fraction of species pairs shows CMS on hybridization is not known.

Note that these findings do not occur because male fertility is somehow especially sensitive to mitochondrial dysfunction: induced mutations causing male sterility have turned out to be nuclear in 37 species and cytoplasmic in only 4 species (Tables 2.3 and 3.5 in Kaul 1988). As of 1994, over 20 nuclear genes that can mutate to male sterility were known in maize, compared to the 3 CMS genotypes (Gabay-Laughnan and Laughnan 1994). Note too that, as expected, male sterility is more common than female sterility. There are only 7 well-documented cases of species with mixed hermaphrodite and male individuals ("androdioecy"; Vassiliadis et al. 2002), and we are not aware that hybridization of hermaphroditic plants often leads to female sterility. (That said, it would be interesting to look for this in the few taxa with paternal transmission of mitochondria, and test for paternal transmission of mitochondria in the few species with androdioecy.)

CMS may be the tip of the iceberg in terms of cytoplasmic attempts to influence sex allocation or the mating system. A reduction in pollen pro-

BOX 5.2

The Uses of CMS in Hybrid Seed Production

Many agricultural crops are hybrids: the seeds planted in the field are derived from parents of 2 different, carefully selected lines. Most crops are self-compatible, and so it is not easy to produce hybrid seed without some selfed seed mixed in. CMS genes have been very useful in solving this problem (Wise and Pring 2002). One of the parental lines is bred to carry a CMS gene and so does not produce any pollen. It is then grown adjacent to the other line, which acts as the pollen donor, and hybrid seed is collected from the CMS line. For crops like sugar beet that are grown for their vegetative parts, it does not matter if the F_1 hybrid is also male-sterile. But for crops like corn from which the seed is harvested, the F_1 hybrid must produce pollen, and so the pollen donor line must be homozygous for a nuclear restorer. CMS is known from such crops as beet, carrot, onion, maize, sorghum, rice, rye, sunflower and wheat (Schnable and Wise 1998).

The most famous CMS genotype is the T-cytoplasm of maize, which was discovered in the 1940s. At the time, corn breeders could prevent self-fertilization only by laboriously removing tassels from the plants by hand. By the 1960s the T-cytotype came to dominate the worldwide corn seed industry. But it was susceptible to a new strain of the southern leaf blight fungus *(Cochliobolus heterostrophus)*, and in 1970 an epidemic destroyed more than 700,000,000 bushels of corn and wiped out 17% of the plants in the United States alone, thus ending the use of T-cytoplasm in seed production (Tatum 1971, Ullstrup 1972). T-cytoplasm maize is also vulnerable to attack by a second fungus, *Mycosphaerella zeae-maydis*, for much the same reason. The CMS gene, T-*urf13*, encodes a protein that assembles in the inner mitochondrial membrane, and the fungi produce toxins that bind to the protein in a way that causes the mitochondria to leak. Levings (1990) speculates that there is some anther-specific substance that also binds to the URF13 protein, causing the mitochondria to leak, and this is how the CMS dysfunction is limited to anthers.

duction of 10% would be difficult to observe. One would like to see careful comparisons of hybrids and allotypes (individuals with organelles from one species and nuclei from another) with the parental species in terms of pollen and seed production, flower size, timing of anthesis, and so on.

Mechanisms of Mitochondrial Action and Nuclear Reaction

The mechanistic basis of CMS has been relatively well studied, particularly in crop species, in which CMS genes are especially useful in producing hybrid seed (Box 5.2). But we still do not have a clear picture for any species of exactly how it works. Morphologically, CMS manifests itself in different species in almost every conceivable way, including the complete absence of male organs, meiotic failure, the abortion of pollen at any step in its development, failure of dehiscence, and failure of pollen germination (Laser and Lersten 1972, Budar and Pelletier 2001). Even within a species, different CMS systems can produce completely different morphologies. In carrots, for example, there is a "brown anther" type of CMS, in which pollen growth ceases prior to maturation, and a "petaloid" type in which petals or petal-like structures form instead of stamens and anthers (Nakajima et al. 1999). Homeotic-like transformation of stamens into (sterile) pistils (female structures) is known in wheat, and floral malformations also occur in tobacco (Bereterbide et al. 2002, Murai et al. 2002; Plate 4; see also Fig. 5.4).

In many species the dysfunction is thought to originate in the tapetum, the specialized tissue surrounding the developing pollen grain (Schnable and Wise 1998). All nutrients used by the developing pollen grain must pass through the tapetal layer, which acts as a storage organ and also synthesizes many parts of the pollen wall (Conley and Hanson 1995). The tapetal cells, in turn, are relatively rich in mitochondria (Lee and Warmke 1979, Balk and Leaver 2001). In sunflowers, the death of tapetal cells and meiocytes associated with CMS has recently been shown to be due to the initiation of a programmed cell death pathway (Balk and Leaver 2001). This is interesting because mitochondria are well known to be involved in programmed cell death pathways (at least in mammals and yeasts), via the release of cytochrome c into the cell cytoplasm. In addition, cell death is essential for the normal development of fertile anthers: tapetal cells eventually have to lyse and release lipid compounds that coat the pollen; other cells die and shear to allow pollen release; and still others die to provide the spring mechanism

H   MS1   MS2

Figure 5.4 Hermaphrodite (H) and male-sterile (MS) flowers in *Plantago lanceolata*. The 2 different male-sterile types are caused by different mutations. Development of MS1 flowers is normal up to the stage of anther differentiation, resulting in a flower in which only the stamens are affected. Filaments are short and anthers, protruding just outside the corolla tube, rapidly turn brown. Pollen sacs can be discerned, but they do not contain any pollen. Development of MS2 flowers is even more aberrant, and in the extreme form stamens are completely absent. Sometimes the style cannot leave the narrow corolla and gets coiled up inside, possibly reducing the chance of pollination. On the other hand, pistils of these flowers occasionally have 3-seeded capsules, compared to the normal 2. Adapted from Van Damme and Van Delden (1982).

for dehiscence. CMS in sunflowers may simply result from the premature activation of programmed cell death pathways necessary for normal male fertility. In some other species, CMS may result from the mitochondria failing to induce normal cell death.

The genetic basis of CMS has been uncovered in about a dozen cases. In each one, CMS is due to a novel protein-coding gene in the mitochondria that has been formed by genomic rearrangements bringing together fragments of other genes (reviewed in Schnable and Wise 1998, Budar and Pelletier 2001, Budar et al. 2003; Fig. 5.5). For example, male sterility in the T-cytoplasm of maize is due to the T-*urf13* gene, which is a chimera of part of the *atp6* gene, the 3' flanking region of the 26S rRNA gene, an unidentified sequence, and part of the coding region of the 26S rRNA gene. Male sterility in *Petunia* is caused by the *pcg* gene, which is a chimera of *atp9*, *coxII*, and an unidentified sequence. CMS genes in rapeseed, sunflowers, sugar beets, and carrots all contain fragments of *atp8* (= *orfB;* Nakajima et al. 2001). Typically, the CMS-causing gene is close to and cotranscribed with an essential mitochondrial gene, and it encodes a protein that binds to the mitochondrial membrane. In most species the gene is expressed in all tissues, but we do not know how the phenotypic effect is limited to pollen development (Budar et al. 2003). One possibility is that the CMS protein interacts with some unknown "factor X" that exists only in the male reproductive tissues to produce the harmful effects. Alternatively, the CMS genes may simply cause a slight decrease in mitochondrial efficiency, to which the male reproductive tissue is uniquely susceptible. Given that CMS mutants can only spread if they increase female fitness and many CMS types are used in crops selected for high productivity, any overall decrease in mitochondrial efficiency is probably slight.

The genetics and mode of action of nuclear restorer genes have also been investigated (reviewed in Schnable and Wise 1998, Budar et al. 2003). Most of these appear to interfere with the processing of the CMS mRNA after transcription (recall that most CMS genes are cotranscribed with a nearby essential gene). In species with multiple CMS types, each of the different types is suppressed by a different restorer. For example, in maize, *cms*-T is restored by *Rf1* and *Rf2; cms*-S is restored by *Rf3; cms*-C is restored by *Rf4;* and each of these restorers is found at a different locus. Interestingly, in rapeseed *(Brassica napus)*, the *nap* and *pol* CMS types have different restorers, but they map to the same place in the genome and may be allelic (Li et al. 1998).

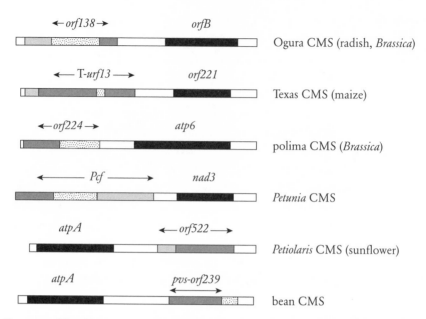

Figure 5.5 The chimeric structure of CMS genes. Shown are 6 different CMS genes, each of which is derived from fragments of 1 or more normal mitochondrial genes and regions of unknown origin (indicated by the differential shading). These genes presumably arose by recombination. Also shown are the normal mitochondrial genes with which the CMS genes are cotranscribed. Adapted from Budar and Pelletier (2001).

Such specificity of CMS and restorer genes can help maintain polymorphism in natural populations (Frank 2000). On the other hand, there is also redundancy: more than 1 nuclear gene can suppress a particular CMS gene. For example, *Rf8* and *Rf** are genes in maize that can, at least in part, substitute for *Rf1* to restore fertility (Dill et al. 1997). All 3 genes mediate T-*urf13* transcript accumulation. Nuclear restorers are typically dominant, but can act either sporophytically or gametophytically—that is, if the plant is heterozygous for the restorer, either all the pollen will be functional or only those pollen grains that have inherited the restorer gene. Restorer genes for the S-cytoplasm of maize are gametophytic (Kamps et al. 1996). Restorer genes have been cloned from *Petunia* and radish plants, and in each case they contain multiple copies of a 35–amino acid pentatricopeptide repeat (PPR), a motif thought to bind RNA (Bentolila et al. 2002, Koizuka et al. 2003).

(There are over 200 PPR-containing genes in the *Arabidopsis* genome, two-thirds of which are predicted to be targeted to organelles; Small and Peeters 2000.)

CMS and Restorers in Natural Populations

The high frequency of male sterility in hybrids between populations or species (Table 5.1) suggests that CMS is often a transient phenomenon: a CMS mutant arises, spreads through a population, and selects for a nuclear restorer, which goes to fixation, leaving the population back where it began, fully hermaphroditic. The underlying tension is then only revealed when mismatched CMS and nuclear elements are thrown together in crosses between populations. Unfortunately, the phenomenon of transient sweeps has not been studied by evolutionary geneticists using modern molecular techniques, and so our information about it is largely anecdotal. Such successive sweeps happen–sometimes. This is a research opportunity crying out for study. One promising system has recently been described by Barr (2004): *Nemophila menziessi* (Hydrophyllaceae) populations are typically hermaphrodite, but females are found in a hybrid zone between white- and blue-flowered morphs, and experimental crosses between colors produce a higher frequency of females than those within morphs.

What has attracted more research are species with apparently stable gynodioecy, that is, coexistence of hermaphrodites and male-steriles. Three genera in particular have been well studied: *Plantago* (Plantaginaceae), *Thymus* (Labiatae), and *Silene* (Caryophyllaceae). In each genus, gynodioecy is widespread: in *Silene*, for example, there are several hundred species and gynodioecy is likely to be the ancestral gender system (Desfeux et al. 1996). *Thymus* also has many gynodioecious species and no fully hermaphroditic ones (Manicacci et al. 1998). Gynodioecy in these species may therefore be relatively ancient.

In each of these genera, the genes responsible have yet to be identified, but experimental crosses suggest that the underlying genetics of male sterility is often complex. In *P. lanceolata*, 3 different CMS types are known (de Haan et al. 1997a, 1997c) and, for the most common one, there appear to be 5 different loci segregating in natural populations for restorer alleles, 2 dominant and 3 recessive (Van Damme 1983). For a less common CMS type, there are 3 dominant restorer alleles, each at a different locus. In *S. vulgaris*, there are 3 different CMS types: for one there were no restorers detected in the population; for another there were 2 dominant epistatic restorers (i.e., restorer alleles needed to be present at both loci in order to produce pollen); and for the third there were 3 dominant restorers, 2 epistatic and 1 independent (Charlesworth and Laporte 1998). In both species, all 3 CMS types can be found in a single population. In the only *T. vulgaris* CMS genotype investigated, there were 3 different restorer loci, with 1 of them (recessive) having to interact with either of the other 2 (1 dominant, 1 recessive) to produce pollen (Charlesworth and Laporte 1998). All this complexity is reminiscent of the various suppressors of *SD* and *SR* that have been found in *Drosophila* populations and species (see Chaps. 2 and 3). CMS systems have probably been highly dynamic over evolutionary time, with both mtDNA and nuclear DNA continually evolving new ploys and counterploys.

Gynodioecy is relatively stable in these genera—most populations contain both females and hermaphrodites—but it is not yet clear how stable the underlying genetics is, nor whether there is a perpetual flow of new mitochondrial and nuclear mutations. The same CMS type has been found in different populations of *Plantago*, and similarly for *Silene* (de Haan et al. 1997b, Charlesworth and Laporte 1998; see also Laporte et al. 2001 for *Beta vulgaris*). In *Thymus*, mitochondrial genotypes are highly variable (reviewed in Tarayre et al. 1997), and it is not clear whether CMS genotypes in different populations have common or independent origins. Whether or not the genes themselves are changing, their frequencies certainly are. This is most obvious in the frequency of females, which varies widely among populations: 0–34% in *P. lanceolata*, 0–75% in *S. vulgaris*, and 5–95% in *T. vulgaris* (Van Damme and Van Delden 1982, Manicacci et al. 1998; McCauley et al. 2000). Variation in the underlying gene frequencies can be even greater: for example, in *Plantago lanceolata*, 2 populations (BM and HT) with similar frequencies (7% and 9%) of a particular male-sterile morphology had almost 2-fold differences in the frequency of the underlying CMS genotype (57% and 31%)—but this difference was hidden by a corresponding difference in the frequency of restorers (de Haan et al. 1997a, 1997b, 1997c). Populations of this species also show changes in the frequency of females over time. In *T. vulgaris*, older populations (measured as time since last disturbance) show a lower frequency of females (Couvet et al. 1990).

Further progress in studying the dynamics of CMS and restorers in natural populations must await identification of the genes themselves, or markers linked to them.

CMS, Masculinization, and the Evolution of Separate Sexes

In cases like the ones we have been considering, in which a CMS mutation arises in a population and a nuclear restorer does not sweep through to fixation, there is selection for other nuclear responses as well. In particular, because reproductive investment in the population is female biased, there is selection for mutations that masculinize the hermaphrodites—in other words, cause them to allocate more to pollen production, even at the expense of reducing ovule production (Cosmides and Tooby 1981, Ashman 1999, Frank and Barr 2001, Jacobs and Wade 2003). The consequence is to increase the difference in ovule production between male-steriles and hermaphrodites, with reduced selection for restorers as one possible result. Indeed, if the difference in ovule production becomes more than 2-fold, even nuclear male sterility genes will be selected.

The extent of masculinization has been studied in 3 species of *Thymus* and it varies substantially (Manicacci et al. 1998). Masculinization is greatest in *T. mastichina:* hermaphrodites are estimated to invest some 10 times more energy into pollen production than seed and consequently produce only about a third the number of ovules as females do. This is also the species with the highest frequency of females (72%). In other gynodioecious species, the process has gone even further: in *Hebe subalpina,* for example, hermaphrodites produce only one-fortieth the number of ovules as do females; and in *Fragaria virginiana,* more than half the "hermaphrodites" never set fruit (Lloyd 1976, Ashman 1999).

The limit of this process is complete masculinization with the species consisting of separate males and females and no hermaphrodites—that is, dioecy (Maurice et al. 1993, 1994, Schultz 1994). Separate sexes have evolved many times among the flowering plants, with gynodioecy thought to be the most common ancestral state (Barrett 2002). According to the tabulations of Maurice et al. (1993), of 308 taxonomic families of flowering plants, 55% have 1 or more dioecious species. But of the 35 families in which gynodioecy is known, fully 89% of them also have dioecious species. Thus there is some evidence of comparative association (though these data need to be corrected for differences in family size). Genera with both gynodioecious and dioecious species are known from 17 families.

What happens after dioecy? The evolution of separate sexes does not remove the conflict between nuclear and mtDNA over optimal sex allocation. It merely alters the way in which this conflict manifests itself. In particular, in plants with separate sexes, mtDNA will be selected to kill male embryos, so as to free extra resources for their sisters. Killing could usefully occur at any early stage—at germination, or while the seed is still developing on the mother, or even prior to fertilization if Y-bearing pollen can be targeted. We are not aware of any such mitochondrial killers being described; but if they can evolve to target pollen, they should be able to evolve to target males. In principle, such conflicts may even contribute to the poor evolutionary prospects of plants with separate sexes. Although dioecy is widespread in flowering plants, it is also rare, occurring in only 6% of species, and dioecious clades are far less species-rich than their sister taxa (Heilbuth 2000). Another contributing factor could be increased investment in unproductive male function. Outcrossed dioecious species typically have a 50:50 sex ratio, whereas hermaphroditic ones may enjoy all the benefits of outbreeding and sexual selection, while investing as little as 5% of their resources each generation in male function (Charnov 1982). Thus, in causing a hermaphroditic species to become dioecious, CMS genes could put their species at a competitive disadvantage with overlapping hermaphroditic ones, possibly forcing early extinction.

Pollen Limitation, Frequency Dependence, and Local Extinction

Before masculinization evolves, it is an inevitable consequence of the spread of a CMS gene (at least in the short term) that less pollen will be produced. This could result in seed production being limited by the availability of pollen. If the hermaphrodites are self-compatible, they will be less affected by pollen limitation than are the females, who rely completely on pollen from other plants. Thus, the relative fitness of females may decline as they become more common. This would cause the CMS gene to come to some intermediate equilibrium frequency rather than going to fixation (Lewis 1941; see also Box 5.1). Note that this is similar to the advantage of a driving X chromosome in *Drosophila* populations, which declines as it increases in frequency due to the reduction in the frequency of males (Chap. 3). Exactly this sort of frequency dependence has been observed in *Silene vulgaris*, in both experimentally manipulated and natural populations (McCauley and Brock 1998, McCauley et al. 2000; Fig. 5.6).

In principle, such frequency dependence may not arise in self-incompatible species that show random mating, because both hermaphrodites and fe-

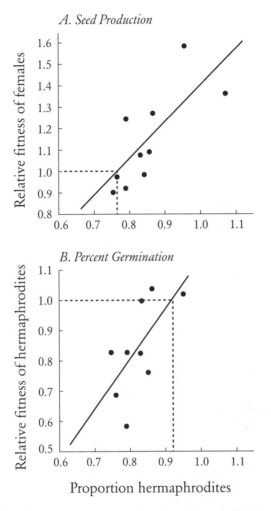

Figure 5.6 Frequency-dependent effects on the relative fitness of females and hermaphrodites in *Silene vulgaris*. Data are from different natural populations that vary in the relative frequency of hermaphrodites vs. females. A. Y axis is the ratio of the square root of the number of seeds per capsule for females and hermaphrodites. B. Y axis is the ratio of the arcsine square root percentage germination for hermaphrodites and females. X axis in both cases is the arcsine square root of the proportion of hermaphrodites in the population. The dashed lines indicate the proportion of hermaphrodites at which the fitness measures of the 2 sex types are equal. Note that the effect on seed production will tend to stabilize the polymorphism, whereas the effect on germination will have the opposite effect. Adapted from McCauley et al. (2000).

males rely equally on foreign pollen. In the absence of nuclear restorers, the CMS gene could go to fixation and the population would go extinct (Charlesworth and Ganders 1979). But any real population is likely to show some spatial structure and local mating, so hermaphrodites will tend to be surrounded by other hermaphrodites, which could allow them to persist (McCauley and Taylor 1997). But if a selfed seed is better than none, there may then be strong selection for self-compatible mutations.

Resource Reallocation Versus Inbreeding Avoidance

Nuclear and mitochondrial genes are in conflict over pollen production for reasons of both sex allocation and inbreeding. On the one hand, optimal allocation to female function will be higher under mitochondrial control than under nuclear control; this difference will be greatest in outcrossing species and tend to disappear with increasing inbreeding. On the other hand, mtDNA—which is transmitted through only one sex—gains no relatedness to progeny through inbreeding. It wants to avoid selfing (assuming fertilization is achieved by outcrossing) if there is *any* inbreeding depression; a nuclear gene wants to avoid selfing only if fitness is reduced by more than 50% (because gene transmission to selfed offspring is doubled). More generally, a cytoplasmic gene will be selected to incur a greater cost, and risk a higher probability of not being fertilized, to avoid inbreeding than will a nuclear gene.

It is not clear which of these—resource reallocation or inbreeding avoidance—is more important in the evolution of CMS. Male-sterile individuals do typically produce more seeds than do hermaphrodites; but, as we have just seen, these differences can be inflated by the masculinization of hermaphrodites subsequent to the spread of a CMS mutation. Consider *Plantago lanceolata*, in which compensation has been studied with unusual care (Poot 1997). The species is self-incompatible. Thus, the primary benefit from CMS is expected to come through an increase in female function in male-sterile individuals. As expected in a species with male-sterile individuals, investment in flowers in hermaphrodites is biased toward male function—as much as 65% of the total. Progeny from between-population crosses raised under nitrogen-limited conditions in a growth chamber reveal that hermaphrodites, females, and intermediate individuals consume the same amount of nitrogen, despite the fact that stamens and pollen are relatively expensive in nitrogen. Females and intermediate individuals convert the nitrogen saved into additional reproductive and vegetative (= future reproductive) tissue, and females set about 50% more seeds than hermaphrodites. Incidentally, there is considerable variation in form and percentage of compensation depending on the particular population cross and cytotype used. This is reminiscent of the high variability found between cytotypes and their nuclear restorers within single species.

In the gynodioecious *Sidalcea oregana,* no differences are found between the sex types in total allocation to reproduction, but females allocate more biomass, nitrogen, phosphorous, and potassium to seeds (Ashman 1999). In *Hebe subalpina* and *Cucurbita foetidissima,* more biomass is allocated to seeds in females than in hermaphrodites (Delph 1990, Kohn 1989). It would be interesting to compare these (possibly) evolved differences to changes in fertility associated with new mutations, experimental emasculations, and organelle introgressions, which would give information on the advantage to a newly arisen CMS gene. Interpretation is also made difficult because "hermaphrodites" include wildtype cytoplasms without restorers, wildtype cytoplasms with restorers, and male-sterile cytoplasms with restorers, as these genotypes cannot yet be distinguished, except by laborious crosses (e.g., Van Damme 1984).

For some species—for example, those in which anthers and pollen are produced but the pollen fail to develop after reaching the stigma—inbreeding avoidance would seem to be the only possible function. The advantage to the CMS gene of not selfing, assuming no pollen limitation, is $s\delta/1-s\delta$, where s is the selfing rate of the hermaphrodites and δ is the inbreeding depression suffered by selfed seeds. At least in some mixed-mating populations, this advantage can be substantial (e.g., 2-fold; Kohn and Biardi 1995, Schultz and Ganders 1996). In self-incompatible species, $s = 0$, and so the benefit to the mitochondria of avoiding inbreeding depression should be much reduced (though perhaps not wholly absent because self-incompatible hermaphrodites can be pollinated by maternal half- or full siblings, but not if they carry a CMS gene). It is interesting in this context that gynodioecy appears to be severely underrepresented in self-incompatible species, though we are not aware of any formal comparative analysis (Charlesworth and Ganders 1979). This suggests that the avoidance of selfing is—from the mtDNA point of view—an important function of CMS. It would be interesting to know whether CMS is more common in hybrids of related self-compatible populations (species) than of related self-incompatible populations (species). (Note that there is an alternative explanation.

If CMS does arise in a self-incompatible population, it may lead to extinction or to self-compatibility.)

Importance of Mutational Variation

It is a truism that phenotypes can evolve only if they first arise by mutation. In every case that has been studied so far, CMS is due to mutations in mtDNA, not in chloroplast DNA. Yet chloroplasts are also usually maternally inherited, and the selective advantage of a chloroplast gene causing male sterility should be about the same as that of a mitochondrial gene. This suggests that such mutations arise less frequently in chloroplast DNA than in mtDNA, perhaps because the chloroplast genome is less recombinationally dynamic (Palmer 1990; recall that mitochondrial CMS genes typically originate from genome rearrangements). It could also be because mitochondria are more intimately involved in anther and pollen production than are chloroplasts and so may more frequently mutate to disrupt these processes. A null mutation causing mtDNA malfunction in anthers may be entirely sufficient. We noted earlier that plant species differ significantly in their mitochondrial gene content, and it would be interesting to know whether this has any effect on their propensity to generate CMS mutations.

Mitochondria in animals are like chloroplasts in plants: in principle, the selection pressures are there for CMS or its equivalent, but the appropriate mutations apparently occur rarely, if at all. Many animal taxa are hermaphroditic, but mitochondrial mutations causing testicular or spermatic failure have yet to be reported. Many arthropods have maternally inherited bacterial parasites that have evolved to feminize or kill males (Box 5.3), but none has been reported with mitochondria that do this. And in mammals, a mitochondrial variant that destroyed male embryos, thereby creating space for reproductive sisters, would be strongly selected, but has yet to be described. We attribute these failures of evolution to the relatively small size and recombinational inactivity of mitochondrial genomes in animals.

Given that plant mitochondria have somewhat sophisticated capabilities, such as the ability to convert male plant parts into (crude) female parts, it is interesting to speculate what else they may do. We have already suggested that they may increase seed production quite independently of abolishing pollen production, and that they may kill male embryos or otherwise skew

BOX 5.3

Maternally Inherited Reproductive Parasites of Arthropods

Many arthropods are host to intracellular bacteria or protozoa that, like mitochondria, are inherited through the egg, from mother to offspring (O'Neill et al. 1997). Like the mitochondria inhabiting the same cells, these maternally inherited endosymbionts are selected to skew host reproduction toward females; unlike mitochondria, many of them have successfully evolved to exploit this opportunity, and in a variety of ways (O'Neill et al. 1997, Charlat et al. 2003):

Male killing. In several insect and mite species, there are maternally inherited bacteria that have evolved to kill male embryos. At least for *Spiroplasma* in *Drosophila,* they do this by inducing embryo-wide apoptosis (G. Hurst, pers. comm.). Killing a male can increase a bacterium's probability of being transmitted to the next generation if, for example, juvenile females eat their dead brothers or have more resources available in their absence, thus increasing their probability of surviving to adulthood.

Diploid feminization. In amphipod crustaceans such as *Gammarus,* there are microsporidian protozoans that interfere with normal sexual differentiation and cause genetic males to develop as females. In isopod crustaceans such as *Armadillidium* and some lepidopterans, there are bacteria doing the same thing. Infected females produce twice as many daughters as uninfected ones, allowing the endosymbiont to be transmitted to twice as many granddaughters.

Haplodiploid feminization. In some haplodiploid wasps and mites, there are bacteria that induce unfertilized eggs, which would normally develop into males, to develop instead into parthenogenetic females. In wasps the bacteria cause the chromosomes to double, and so the females are diploid, whereas in *Brevipalpus* mites the females remain haploid (Weeks et al. 2001). Again, infected females produce twice as many daughters as uninfected ones, thereby allowing the endosymbiont to be transmitted to twice as many granddaughters.

Cytoplasmic incompatibility. In many insect species there are bacteria that in males somehow modify the sperm such that eggs from uninfected females that are fertilized by those sperm die; eggs from infected females are viable regardless of whether the father was infected. In this way the bacteria decrease the fitness of uninfected females, thereby increasing the (relative) fitness of infected females. Recently, one of these bacteria has swept through the California population of *D. simulans* over a 5-year period (Hoffmann and Turelli 1997).

Strikingly, there is a single genus of bacteria, *Wolbachia*, in which all 4 of these phenotypes have been observed. This rich diversity of reproductive parasitism contrasts with the apparent absence of any similar adaptations in animal mitochondria, and perhaps suggests what nuclear-mitochondrial relations may have been like in early eukaryotes, when the symbiosis was young.

the sex ratio in dioecious species toward females. In some species, plants adjust their allocation to male versus female function in response to the existing population sex ratio, as judged by the amount of pollen arriving on the stigma (more pollen indicates a more male-biased population). For example, in *Begonia gracilis* (Begoneaceae), plants produce separate male and female flowers, and plants receiving a lot of pollen subsequently produce relatively more female flowers than those receiving little pollen (López and Domínguez 2003). And in some dioecious *Rumex* species (Polygonaceae), females produce relatively more daughters when they receive a lot of pollen compared to when they receive little (Rychlewski and Zarzycki 1975, Conn and Blum 1981). Mitochondria may be expected to interfere with these mechanisms, to bias sex allocation toward female function even under low pollination conditions. Mitochondrial genes will also be more strongly selected to maintain self-incompatibility than will nuclear genes (especially those linked to the self-incompatibility locus), and may in some species have evolved a critical role, acting either in the style or in the pollen. But even for mitochondria in plants, constraints on the mutational spectrum should be critical. Nuclear genes have a much larger mutational spectrum than does mtDNA, because they have many more genes (and noncoding DNA), and presumably this is a major reason why the nucleus usually wins the conflict with the mitochondria—that is, why most plants are hermaphrodites, not gynodioecious, with male sterility often only uncovered in population or species hybrids.

CMS and Paternal Transmission

We have already noted that CMS can select for a diverse array of nuclear responses, including restoration of male fertility, masculinization of hermaphrodites, and self-compatibility. We end our review of CMS by noting that, in principle, it also selects for a reversal of the very thing from which it springs: maternal inheritance. In a population polymorphic for a CMS gene, mitochondria in pollen are less likely to carry the gene than mitochondria in ovules. Thus, nuclear genes may be selected to promote paternally derived mitochondria over maternally derived ones. This will be true of nuclear genes acting in the mother, the father, or the offspring. The result could be a degree of paternal "leakage" or even outright paternal inheritance. The comparative biology of mitochondrial inheritance in plants is not well understood—all we know is that occasionally there is paternal leakage, and occasionally there are species with predominantly paternal inheritance. Perhaps occasionally such reversals evolve in species troubled by CMS genes, as a way of reducing the chance of transmitting so selfish a gene.

Other Traces of Mito-Nuclear Conflict

This chapter reviews the two main forms of conflict between mitochondria and nuclei. First, mitochondria may replicate faster than is optimal for the organism, in order to swamp a competitor. This is particularly likely with biparental inheritance, because of the greater within-organism genetic variation. Second, with uniparental inheritance, mtDNA evolves to skew reproduction toward the transmitting sex. But there have undoubtedly been other forms of conflict during the more than a billion years that mitochondria have been living inside eukaryotic cells. Presumably conflicts were particularly prevalent before mitochondria became so degenerate (much as more recent endosymbionts of arthropods are more often in conflict with their hosts; see Box 5.3). Such ancient conflicts may have left traces in the modern workings of mitochondria and nuclei. In this section we review several possibilities; all of them are speculative.

Mitochondria and Apoptosis

It is a striking fact, already alluded to, that mitochondria play an active role in apoptosis (programmed cell death; Zamzami et al. 1996, Green and Reed 1998, Thornberry and Lazebnik 1998, Blackstone and Kirkwood 2003). Although it is not clear what exactly precedes what, an early sign of death is the opening of pores or ion channels in the mitochondrial membrane, followed by osmotic expansion of the mitochondrion until the outer wall bursts, releasing caspases and other chemicals into the cytoplasm (Green and Reed 1998). At the same time, disruption of electron transport and therefore energy metabolism is another early sign, but ATP production is affected much later. Indeed ATP is itself necessary for some downstream apoptotic events, presumably because destruction is active and requires energy. Mitochondria contain an apoptosis-inducing factor, AIF[5], which is normally confined to mitochondria but which translocates to the nucleus when apoptosis is induced (Susin et al. 1999). There it causes chromatin condensation and large-scale fragmentation of DNA. AIF[5] also induces mitochondria to release the apoptogenic proteins cytochrome c and caspase-9, which itself activates further downstream executor caspases. (The caspases are a family of cysteine-dependent aspartate-specific proteases whose members act as both initiators and executors of apoptosis.)

Cells lacking mtDNA are capable of apoptosis, and both AIF[5] and cytochrome c are coded by nuclear genes, but several mitochondrial genes appear to be involved in apoptosis. In human hematopoietic myeloid cell lines, inhibition of the mitochondrial gene NADH dehydrogenase subunit 4 decreases cell viability and generates markers of apoptosis (Mills et al. 1999). In the same kind of cell lines, a mitochondrial antisense RNA for cytochrome c oxidase can induce large-scale DNA fragmentation within 36 hours and cell death within 3 days (Shirafuji et al. 1997). In human colonic epithelial cells, elevation of mitochondrial gene expression appears to be important in inducing apoptosis of these cells in culture (Heerdt 1996).

How might mitochondria have come to be central players in apoptosis? Perhaps in some ancient eukaryote mitochondria were sometimes selected to kill their host cell, much as they now are in the tapetum of sunflowers. Kobayashi (1998) speculates that early mitochondria were selected to kill cells to which they had not segregated, analogous to systems now observed for some bacterial plasmids (Box 2.1). Interestingly, the one mitochondrial gene that is variably present in animals, called A8, has no detectable homology to any prokaryotic ATP synthase, but its hydropathy profile shows a significant similarity to that of *hok*, a gene involved in bacterial plasmid-maintenance systems (Jacobs 1991). Nuclear genes may then have evolved to co-opt the phenotype and express it when it was in their own interests to do so. The genes originally responsible for the action may no longer be in the mitochondria but instead have moved to the nucleus.

Mitochondria and Germ Cell Determination

Another organismal function in which mitochondria seem inexplicably to be involved is germ cell determination (Amikura et al. 2001). In many animal species, there is a histologically distinct region of the egg cytoplasm, called the germplasm, which is the part that will go on to form the germline, eventually producing eggs or sperm. This germplasm consists largely of germinal granules (also called polar granules) and mitochondria. The germinal granules are electron-dense structures that act as repositories for the factors required for germ cell formation. In *Drosophila*, they are rich in mitochondrial-type ribosomes, which are ribosomes consisting of mitochondrially encoded RNAs and some of the same nuclear-encoded proteins as used in mitochondrial ribosomes. Such ribosomes are not typically found outside of mitochondria in other tissues. Similar observations have been made for planarian, sea urchin, ascidian, and frog embryos. How might mitochondria have become involved in germ cell determination? We speculate that in the early evolution of tissue differentiation in animals (when, based on choanoflagellates, mitochondria probably had more genes than they do now, including many for ribosomal proteins; Burger et al. 2003), mitochondria would have been selected to ensure that the cell or embryonic region in which they were found ended up as part of the germline, and evolved mechanisms to ensure this happened. These mechanisms may then have been co-opted by the nucleus.

Mitochondria and RNA Editing

In many organisms there is a minority of genes in which the DNA is transcribed into RNA, and then enzymes act on the RNA to change the coding sequence—for example, inserting or deleting a single base or changing one base to another (Gray 2001). That is, the RNA is edited. Why might such a system have evolved—if the alteration is beneficial, why would it not simply occur at the DNA level directly? Gray (2001) suggests that perhaps the editing machinery originally evolved neutrally, by genetic drift, but now is maintained by selection because the genome has acclimatized to it. But this seems like a lot to ask of random drift unaided by selection. An alternative scenario is suggested by the observation that editing is disproportionately common for mtRNA, while the enzymes responsible are encoded in the nucleus. We speculate that in some cases RNA editing originally evolved as a way for nuclear genes to rewrite or override the instructions of mitochondrial genes that were antagonistic to nuclear interests. In other cases, partial editing may have evolved so as to make 2 proteins from a single gene. In either case, once the machinery has evolved, the genome may then "degenerate" to the point at which editing is essential for life (Gray 2001).

SIX

Gene Conversion and Homing

IF AN RNA OR PROTEIN MOLECULE is damaged, the cell will typically degrade the entire molecule and replace it with a new one. But cells cannot do this with DNA. A chromosome is too big and the copy number is too low, so it is far more efficient to repair the chromosome than to replace it. If only one DNA strand is damaged, the other may be used as a template for repair; alternatively, if both strands are damaged, the homologous chromosome may be used. DNA repair involves DNA synthesis, sometimes of just 1 nucleotide, and sometimes of thousands or more, depending on the type of damage and the method of repair (Alberts et al. 2002). This repair-associated DNA synthesis has given rise to multiple forms of drive.

On the one hand, there is biased gene conversion (BGC). Like other forms of drive, it is a process by which a heterozygote transmits one allele more frequently than the other. This drive does not arise because the gene encodes a protein that acts to bias inheritance; rather, drive arises as a by-product of biases in the mechanisms of DNA repair. This is therefore a more passive form of drive. It is also relatively weak, with effective selection coefficients of 1% or less. But BGC has the potential at least to be pervasive, with effects across the entire genome, and several lines of evidence suggest it can play a major role in shaping the composition of eukaryotic genomes. It is also likely to have played an important role in the evolution of the meiotic machinery.

At the other extreme are homing endonuclease genes (HEGs), a class of selfish genes that are anything but passive. These are optional genes, with no known host function, that spread through populations by exploiting their host's DNA repair systems. They encode an enzyme that specifically recognizes and cuts chromosomes that do not contain a copy of the gene; the cell then repairs the damage and, in so doing, copies the HEG onto the cut chromosome. HEGs cause minimal damage to their hosts, even when homozygous, and can sweep through a population with minimal hitchhiking effects on flanking DNA.

The enzyme encoded by a HEG has a relatively sophisticated and unusual capability—cleaving DNA at a specific 15–30bp recognition sequence—and this can be used for more than just promoting the spread of a HEG. So, for example, yeast cells have evolved a complex mechanism for mating-type switching, the central enzyme in which is clearly derived from a homing endonuclease. Humans too are finding uses for HEGs, with several available in molecular biology catalogs. They have a role in genetic engineering: already HEGs are used for one method of gene targeting in *Drosophila* (Rong and Golic 2003), and there is the potential for their use in human gene therapy (Chandrasegaran and Smith 1999). As we shall see, they may even be used to genetically engineer entire populations.

In this chapter we review the biology of BGC and HEGs. We also include a brief review of group II introns, a class of genes that are in many ways similar to HEGs but that use a completely different mechanism of drive based on their own DNA polymerase. We end with a discussion of how artificial HEGs could be used for the control and genetic engineering of pest populations.

Biased Gene Conversion

In cells that are heterozygous for 2 alleles, A and B, one allele is sometimes converted into another and the cell becomes homozygous (AA or BB). This is not due to conventional mutation: though rare, conversion occurs many orders of magnitude more frequently than *de novo* mutations (e.g., AA or BB homozygous cells giving rise to AB heterozygotes). Rather, genetic information is transferred from one chromosome to another, without the donor chromosome itself being affected. Gene conversion does not appear to be an end in itself, but rather an incidental by-product of other, more fundamental cellular processes. And it is widespread: conversion probably occurs with some frequency at most loci in most sexual eukaryotes. As well as occurring between 2 alleles of a single locus, it can also occur "ectopically," between alleles at different loci, if they are homologous (e.g., members of a gene family), and so it can be a mechanism for spreading sequences around the genome. Importantly, gene conversion is often biased, so AB heterozygotes give rise more often to one type of homozygote (say, AA) than the other (BB). As a result, the A allele shows drive and will increase in frequency in the population unless opposed by some other force (Gutz and Leslie 1976). Thus, genes can increase in frequency and persist in populations simply because they are good converters. And in a large multigene family with hundreds of loci, all producing the identical product, the effect of a new mutation on the organismal phenotype may be very small, and its fate determined much more by its conversion properties than by its phenotypic effects (Hillis et al. 1991).

Molecular Mechanisms

Detailed studies of gene conversion suggest that it is an incidental by-product of more fundamental processes like DNA repair, meiotic crossing-over, and (perhaps) chromosome synapsis.

Gene conversion and DNA repair. DNA consists of 2 complementary strands, and so if damage occurs to 1 strand, the other can be used as a template for repair. But if there is damage to both strands, the only possible template for repair is another piece of DNA. Suppose, for example, a piece of DNA suffers a break across both strands. Sometimes the cell simply joins the 2 ends back together again ("nonhomologous end-joining"), but the broken ends are susceptible to degradation and this type of repair often leads to the loss of 1 or more base pairs at the join. A more precise alternative is to use the homologous chromosome to align the broken ends and act as a template to replace whatever may have been lost by degradation (Fig. 6.1). In this way genetic information can be copied from one chromosome to another; and if the 2 molecules are different, the break-and-repair event would be detectable as a gene conversion event. This recombinational repair pathway is probably the main cause of gene conversion in mitotic cells (Orr-Weaver and Szostak 1985). Rates of mitotic gene conversion are greatly

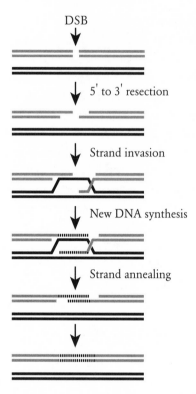

Figure 6.1 Gene conversion associated with DNA repair. Shown here is the simplest form of double-strand break repair, via the synthesis-dependent strand annealing (SDSA) pathway. After resection of the two 5′ ends of the broken chromosome, each of the 3′ ends invades the intact chromosome and uses it as a template for DNA synthesis. Once this has proceeded sufficiently, the two 3′ strands anneal, and the chromosome is then made whole. The repair process therefore involves the copying of sequence information from the intact chromosome to the broken one. Adapted from Pâques and Haber (1999).

increased by X-rays, UV irradiation, and some chemical mutagens (Kunz and Haynes 1981).

Gene conversion and meiotic crossing-over. Rates of gene conversion are 2 to 3 orders of magnitude more frequent in meiotic cells than in mitotic cells, and much of this increase is because gene conversion occurs as a by-product of meiotic crossing-over (Orr-Weaver and Szostak 1985, Pâques and

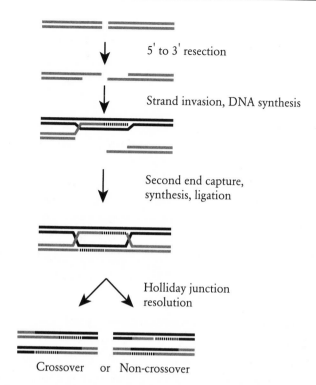

5' to 3' resection

Strand invasion, DNA synthesis

Second end capture, synthesis, ligation

Holliday junction resolution

Crossover   or   Non-crossover

Figure 6.2 Gene conversion associated with meiotic recombination. Crossing-over is initiated by an enzyme cleaving one chromosome, followed by resection of the two 5' ends. The double-strand break repair (DSBR) pathway is more complex than the SDSA pathway of Fig. 6.1 and can lead to both crossover and noncrossover products (only 2 of the 4 possible resolutions are shown). In either case, sequence information is copied from the intact chromosome to the cut one. Adapted from Allers and Lichten (2001).

Haber 1999). Crossing-over is initiated by a double-strand cut made in 1 of the chromatids (Sun et al. 1989, Wu and Lichten 1994; Fig. 6.2). On either side of the cut, the 5' strand of DNA is digested away, leaving a 3' single strand overhang of several hundred nucleotides (Sun et al. 1991). This chromatid is then "repaired"; and, though the cut occurs after DNA replication, the preferred template is the homolog, not the sister chromatid (Schwacha and Kleckner 1994). If the individual is heterozygous in this region, there will be some mismatch in the heteroduplex DNA; and if no other changes are made, these sites will segregate 5:3 in favor of the uncut allele. However, the mismatches may themselves become targets of the cell's mismatch repair system, in which case they will segregate either 6:2 or 4:4, depending on the direction of repair. The region of heteroduplex DNA formation is much larger than the double-strand gap (which is a few nucleotides at most), and so most gene conversion events will occur within it (Sun et al. 1991).

Gene conversion and chromosome synapsis. Only about one-third of gene conversions in *Saccharomyces* and *Neurospora* are associated with crossing-over of flanking genes, leaving the other two-thirds unaccounted for (Perkins et al. 1993, Stahl and Lande 1995). One idea is that these are byproducts of an earlier stage in meiosis, in which DNA strands interact with one another, looking for complementary regions, so as to find and properly align with the homologous chromosome (Carpenter 1984, 1987, Smithies and Powers 1986). Recent molecular evidence is consistent with this idea: crossover and noncrossover gene conversion events occur by different pathways, and the latter occur earlier in meiosis (Allers and Lichten 2001).

Biases. The importance of gene conversion as a source of drive derives from the fact that it is often biased. This bias can arise in two ways. First, if 1 allele is somehow more susceptible to double-strand damage or is more likely to be the allele at which meiotic recombination is initiated by a double-strand break, it tends to lose out in gene conversion. Second, during all these molecular processes, there is base-pairing between strands of DNA from the 2 homologous chromosomes (Figs. 6.1 and 6.2). If there are any mismatches between them, they are likely to be acted on by the cell's mismatch repair system, which removes 1 of the bases and replaces it with 1 complementary to the remaining base. Biases have been well documented in other contexts of mismatch repair (Brown and Jiricny 1987, 1988). In one experiment, DNA sequences constructed to have a G/T mismatch and introduced into cultured monkey cells were repaired 92% of the time to G/C and 4% to A/T, while 4% were left unrepaired; C/T and A/G mismatches also show a repair bias to G/C (Table 6.1). The extreme bias associated with G/T mismatches is thought to be an adaptation to deal with the high frequency of C→T mutations due to methylation (Brown and Jiricny 1987). Consistent with this idea, biases in yeast (which does not have methylation) are much less pronounced (Table 6.1). But whether the biases associated with gene conversion are selected themselves to compensate for mutational

Table 6.1 Biases in mismatch repair in monkey and yeast cells (percentages)

| | Repaired to: | | |
Mismatch	G:C	A:T	Unrepaired
Monkey			
G:T	92	4	4
A:C	41	37	22
C:T	60	12	28
A:G	27	12	61
Yeast			
G:T	53	37	10
A:C	44	34	21
C:T	48	33	18
A:G	48	36	16

Plasmids with mismatches of the indicated type were introduced into cells and then recovered to test for biases in the direction of repair. Numbers are percentages. From Brown and Jiricny (1988) and Birdsell (2002).

biases or are unselected side effects of biases in DNA repair is not yet clear (Bengtsson and Uyenoyama 1990). For homogeneous mismatches (A/A, C/C, G/G, T/T), there cannot be any overall bias to G/C, but biases have been observed in which the nucleotide is replaced according to the identity of the neighboring nucleotides (Brown and Jiricny 1988). In principle, this could lead to the accumulation of particular combinations of nucleotides.

Effective Selection Coefficients Due to BGC in Fungi

Quantifying the frequency of gene conversion and the extent of bias is most easily done in ascomycete fungi, in which the 4 products of a single meiosis can be observed together. As with other forms of drive, we can quantify the strength of gene conversion by the effective selection coefficient, which is the selection coefficient that would give an equivalent change in gene frequency. Table 6.2 shows an example dataset and outlines the relevant calculations, and Table 6.3 summarizes data on segregation ratios for various loci in 3 ascomycete fungi (see also Lamb 1984, 1998). For these loci, segregation ratios that deviate from 4:4 typically occur in 1–5% of meioses, most mutants show a statistically significant bias in which allele is preferentially

Table 6.2 Segregation ratios at the *g3* spore color locus in *Sordaria brevicollis*

Mutant	Segregation Ratio (mutant:wildtype)									Total	%gc	%mb	s_e
	4:4	6:2	2:6	5:3	3:5	7:1	1:7	8:0	0:8				
YS94	2919	178	3	13	0	1	0	2	0	3116	0.6	98	-0.03
RW9	2608	63	9	17	18	0	0	0	0	2715	1.6	75	-0.01
YS51	2706	31	91	36	54	3	2	0	1	2924	6.4	32	0.01
YS52	2868	23	80	46	45	2	0	0	0	3064	5.6	36	0.01
YS53	2928	20	79	23	65	1	0	0	1	3117	5.4	23	0.01
YS50	2630	274	2	12	0	3	0	11	0	2932	1.0	99	-0.05
YS46	2988	235	2	9	11	1	0	4	0	3250	0.8	95	-0.04
YS48	3333	272	0	7	5	1	0	4	0	3622	0.5	98	-0.04
YS47	3092	222	11	3	0	0	0	9	0	3337	0.7	96	-0.04
YS45	2521	181	1	10	0	1	0	0	0	2714	0.4	99	-0.03
YS49	2821	137	11	7	1	0	0	0	0	2977	0.6	92	-0.02

%gc: the percentage of asci showing gene conversion (i.e., deviating from 4:4 segregation).

%mb: the percentage of non-4:4 asci that are biased in favor of the mutant. All are significantly different from 50% (p<0.001).

s_e: the effective selection coefficient for the mutant when rare, calculated as $(2d-1)$, where d is the average proportion of ascospores carrying the mutant.

Data from Yu-Sun et al. (1977).

Table 6.3 Frequencies of gene conversion and effective selection coefficients in 3 fungi

Species (reference)	Locus	N	% gc	s_e	% sig
Ascobolus immersus (1)	*b1*	31	2.9	0.007	77
	b2	31	18.0	0.062	90
Sordaria brevicollis (2)	*g3*	11	2.2	0.026	100
	g4	10	1.4	0.008	80
	g5	18	1.5	0.007	89
Saccharomyces cerevisiae (3)	*	30	4.8	0.003	33

*30 different loci.

N: number of different mutations studied.

% gc: average percentage of asci showing gene conversion (i.e., deviating from 4:4 segregation).

s_e: average effective selection coefficient of the mutant allele when rare for all mutations, calculated as $|2d-1|$.

% sig: percentage of mutations showing significantly biased gene conversion (p<0.05).

References: (1) Leblon (1972a); (2) Yu-Sun et al. (1977); (3) Fogel et al. (1979).

recovered, and effective selection coefficients in the range of 0.1–1% are common. Such selection coefficients are small compared to most others discussed in this book, but even under inbreeding they are probably larger than needed to overpower drift. Indeed, for large fungal populations one wonders whether there are any nucleotide sites where random drift is more important than biased gene conversion. Unfortunately, it is not clear whether these loci are typical of others in the same species, nor do we have similar data for plants and animals.

BGC and Genome Evolution

Because BGC is such a pervasive form of drive, it has the potential to mold the structure and composition of entire genomes. The most tantalizing evidence in this respect concerns the GC content of mammalian genomes. Because of the rules of base pairing, the amount of G in DNA is always equal to the amount of C, and similarly for A and T, but the proportion of GC pairs versus AT pairs is allowed to vary and does so dramatically, even within a single genome. In the human genome, for example, the GC content of third-codon position sites (GC3) varies from 40% to 80%, with variation on the scale of about 300kb (Galtier et al. 2001). As we have already noted, the mismatch repair system is biased toward producing GC pairs, and so if BGC is important, one expects there to be higher GC content in regions of the genome with higher recombination rates. Several observations are consistent with this expectation (see also Hickey et al. 1994, Galtier 2003):

- There is a genome-wide correlation between the GC content of a region and the local rate of recombination—not only in mammals, but also in *Drosophila, Caenorhabditis,* and yeast (Birdsell 2002, Marais 2003). The correlations tend to be weak, but are much stronger if only recent nucleotide changes are included, as expected if local rates of recombination are not stable over long evolutionary timespans (Meunier and Duret 2004).
- X-linked genes of humans (which recombine) have a higher GC3 content than their Y-linked paralogs (which do not; 51% versus 45%; Birdsell 2002).
- In the last 3 million years, in the lineage leading to present-day mice, a gene called *Fxy,* which in other mammals is X-linked, has been duplicated with the new copy having its last 7 exons in the "pseudoautosomal" region where X and Y chromosomes recombine at male meiosis. This region has perhaps the highest rate of meiotic recombination in the whole genome because it is small, but nevertheless it must have 1 crossover with every male meiosis in order to ensure proper segregation. As expected, there has been a dramatic increase in the GC3 content of these exons, from about 55% to 86% (Perry and Ashworth 1999, Montoya-Burgos et al. 2003), while exons 1–3 have remained with the ancestral GC3 content.

Surprisingly, intergenic regions and introns have lower GC contents than the GC3 contents of the ORFs on either side, and pseudogenes have lower GC contents than their functional counterparts (Birdsell 2002). It is not clear at this time how to interpret these observations. Nor can we explain the apparent quantitative discrepancy between studies of BGC in mammals and fungi. The latter suggest effective selection coefficients due to BGC of 0.1–1%; if values were as large as this in mammals, essentially all otherwise-neutral sites would be 100% GC.

Homologous chromosomes can differ not only by nucleotide substitutions but also by insertions and deletions (indels), from 1bp up to many thousands (e.g., a transposable element insertion on one chromosome). If indels also show a consistent pattern of BGC, this will, over time, affect the size of the genome. Again, there are several suggestive observations:

- In the filamentous fungi *Ascobolus* and *Sordaria,* gene conversion biases are particularly strong for alleles induced by mutagens that insert and delete nucleotides, and are less strong for alleles differing by base substitutions (Leblon 1972a, 1972b, 1979, Yu-Sun et al. 1977).
- If heteroduplex DNA molecules are constructed in which 1 strand has a 12–246bp insert, and these are introduced into mammalian cells, the mismatch is corrected with a 2:1 bias toward the deletion (Weiss and Wilson 1987, Bill et al. 2001). Short (12bp) palindromic inserts can show the opposite bias. If the same biases operate at meiosis, they would lead to genomes shrinking, but the spread of short palindromic repeats.
- For the *Fxy* gene that recently duplicated to the pseudoautosomal region of mice, 3 introns have been compared between ancestral and derived copies, and all 3 shrank dramatically after the duplication (from 5.2, 5.0, and 10.0kb to 0.9, 1.5, and 1.0kb, respectively; Montoya-Burgos et al. 2003). The molecular mechanisms underlying these facts are not yet clear. It could be that deletions show BGC relative to insertions. Alternatively, an elevated GC content implies a simpler DNA sequence (at the limit there are only 2 bases instead of 4), which is more likely to form stable secondary structures (due to pairing between Gs and Cs on the same strand), and this may affect the relative frequency of insertion and deletion mutations during DNA synthesis.
- There is a genome-wide trend in mammals for introns to be shorter in GC-rich regions of the genome, and gene density to be higher (Montoya-Burgos et al. 2003).

Whatever the molecular basis, the results are consistent with recombination leading over evolutionary time to local contractions of the genome. It will be very interesting to see whether this is generally true, and also whether there are any exceptions—in other words, whether there is ever a bias in favor of an insert, due either to differences in genomic location or to differences in the sequence of the insert. An occasional bias in favor of inserts could, in principle, account for the vast number of pseudogenes and transposable elements that have gone to fixation at particular sites in many genomes, including our own (see Chap. 7).

In addition to GC contents and genome size, part of the repetitive nature of genomes may also derive from BGC. The pseudoautosomal introns of the *Fxy* gene, while they have shrunk, have also acquired minisatellite repeats (tandem repeats of 7–121bp motifs; Montoya-Burgos et al. 2003). And minisatellites are disproportionately common in GC-rich regions of the genome. Moreover, the direct analysis of human minisatellites in meiosis indicates they frequently undergo conversion-like events involving the transfer of repeats between chromosomes, which result almost exclusively in the gain of repeats (Jeffreys et al. 1999, Buard et al. 2000). Indeed, Jeffreys et al. (1999: 1675) suggest that "a deep and fundamental connection exists between minisatellite instability and meiotic crossover such that minisatellites may represent crossover parasites spawned by the recombination machinery." Though the data are as yet fragmentary, it seems at least possible that BGC has played a pervasive role in shaping genome size and structure.

BGC and Evolution of the Meiotic Machinery

The molecular investigations of gene conversion and meiotic crossing-over described earlier reveal a very curious feature: the sites at which recombination is initiated by a double-strand break are subject to gene conversion and are lost as a result of the event itself (Nicolas et al. 1989). This curious property of recombination has some interesting evolutionary implications (Burt 2000).

First, consider a species in which there is selection for increased recombination. In principle, one might think that such selection would lead to the accumulation of *cis*-acting sequences at which recombination is initiated, due to their being tightly linked to the site of action. But because these sequences are lost as a result of initiating recombination, it is difficult to see how they could increase in frequency, particularly when they are rare at a locus and therefore usually heterozygous. Our understanding of the molecular mechanics of recombination suggests that this expectation is unrealistic and that the *trans*-acting genes encoding the recombinational machinery will be selected instead. Put another way, recombination is something *imposed* on the recognition sites rather than something they have evolved to attract, and recombination is thus an example of intragenomic conflict. Conflicts over rates of recombination have previously been discussed in the context of autosomal killers (see Chap. 2).

Second, existing *cis*-acting recombinogenic sequences are susceptible to being lost because mutations that are less recombinogenic have an inherent advantage from BGC (Boulton et al. 1997). To take a particular example, a 14bp polyA sequence in the promoter region of *ARG4* of *Saccharomyces* is a *cis*-acting recombination enhancer, and an allele with this sequence deleted would have an effective selection coefficient due to BGC of about 0.008% (calculated from Table 2, lines 1 and 3, in Schultes and Szostak 1991, and assuming 1% outcrossing). The obvious question is, What maintains this 14bp sequence in the population? There seem to be 3 possible answers: (1) the advantage of recombination is sufficient to pay this cost and the deletion, should it arise, is selected out; (2) the 14bp sequence has some other function that maintains it in the population; or (3) the use of the sequence as a recognition site for recombination is recent and transient: when the appropriate mutation does arise, it will spread to fixation. If the advantage of crossing-over is not sufficient to pay the cost of gene conversion, the recombination machinery may continually have to evolve to recognize new sequences, forcing each in turn to extinction. Perhaps this sort of dynamic contributes to changes in fine-scale rates of recombination (Meunier and Duret 2004, McVean et al. 2004).

Recent work on a recombinational "hotspot" in the human MHC locus is very similar (Jeffreys and Neumann 2002). This is 1 of 6 hotspots in the MHC class II region that together account for essentially all meiotic recombination activity in the region. The hotspot, called *DNA2,* is in the noncoding intergenic region downstream of *HLA-DOA.* In this region there are a number of single nucleotide polymorphisms (SNPs) segregating within human populations. For one of them, a G/A polymorphism, the G is associated with a reduced frequency of initiation of recombination, with the result that in G/A heterozygotes, when there is a recombination event in the region, it is transmitted to 76% of sperm rather than the Mendelian 50%. The frequency of recombination in the region in heterozygotes is 3.7×10^{-5} (it is a hotspot relative to neighboring regions of the genome), and so the effective selection coefficient is 0.00002. This is a small value, but may nonetheless have had a modest role in increasing the frequency of G to its current level (52% in northern Europeans; gorilla and chimpanzee sequences show that G is the derived state, having arisen from A by mutation).

Homing and Retrohoming

The practice of repairing a broken chromosome by using the homologous chromosome as a template for repair has obvious advantages in helping to ensure that no base pairs are lost. But it is also a process that is exploited by 2 major classes of selfish genetic elements, homing endonuclease genes (HEGs, covered in this chapter) and DNA transposons (covered in the next). The study of HEGs began in 1970, with reports of strongly biased transmission ratios in a specific region of the yeast mitochondrial genome (Coen et al. 1970), and the logic of how they operate was worked out over the next 16 years (Colleaux et al. 1986). For unknown reasons, HEGs are disproportionately common in organelle genomes. They are also mostly found in eukaryotes of simpler organization—fungi, algae, and other protists—and this fact, as we shall see, probably has to do with the ease of horizontal transmission in these taxa. Sequences homologous to HEGs are found in the mitochondria of some plants and some sea anemones, but whether they act in the same way is unknown. HEGs are also found in some bacteria with unusual chromosomal conjugation systems (e.g., some myco-bacteria and archaea), and in some bacteriophages (Burt and Koufopanou 2004). The enzymes made by HEGs fall into 3 or 4 independent protein families, indicating multiple origins (Gimble 2000, Chevalier and Stoddard 2001). Each of the families is diverse, and so likely to be old, but otherwise nothing is known about where or when they originated.

How HEGs Home

HEGs are among the smallest selfish genetic elements (1–2kb) and drive by an elegant mechanism with 2 principal ingredients (Fig. 6.3). First, they encode a sequence-specific endonuclease—in other words, an enzyme that recognizes a specific sequence of DNA and cuts it at that site (Plate 5). The recognition sites are relatively long (e.g., 15–30bp) and typically occur only once in the host genome. Second, HEGs are inserted in the middle of their own recognition sequence, disrupting it, and thus protecting their own chromosome from being cut. If the HEG is nuclear, the only time the en-

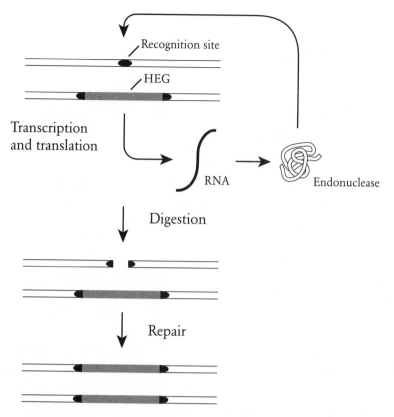

Figure 6.3 How HEGs home. HEGs encode a protein that in heterozygous cells recognizes a sequence on the homologous chromosome and cleaves it; the HEG is located in the middle of the recognition sequence, and so the chromosome carrying the HEG is protected from being cut. After cleavage, the cell's broken chromosome repair system is stimulated, and that system uses the HEG-containing chromosome as a template for repair. As a consequence, the heterozygote is converted to a homozygote.

zyme comes into contact with an intact recognition sequence is in $+/-$ heterozygotes, where 1 chromosome contains the HEG ($+$) and the other does not ($-$). In these heterozygotes, the enzyme simply cuts the empty chromosome, the one with the intact recognition site. The presence of a broken chromosome turns on the cell's broken-chromosome repair system. As we have seen, this typically uses the homologous chromosome as a template to create a patch between the 2 ends. In the case of breaks caused by HEGs, the intact chromosome that is used as a template for repair contains the HEG itself. Therefore, the HEG is copied across to the other chromosome during the repair process, as part of the patch. In this way heterozygous +/− cells are converted into homozygous +/+ cells. This process is called "homing."

For HEGs in yeast mitochondria, homing is much the same as in the nucleus because mitochondria are biparentally inherited and mitochondria from the 2 parents fuse in the diploid cell, allowing direct interaction of the 2 parental mitochondrial genomes (see Chap. 5). If haploid strains with and without a mitochondrial HEG are crossed, the resultant diploid cell initially has 2 types of mitochondrial genomes. If these diploid cells are allowed to divide mitotically, eventually the mitochondrial variants segregate, with most cells carrying only the HEG$^+$ genomes and a small minority carrying only HEG$^-$ genomes.

HEGs are also found in the organelles of filamentous fungi, plants, and protists. Organelles in these taxa are usually transmitted by one parent only, in contrast to the biparental inheritance of mitochondria in yeast (see Chap. 5). Nevertheless, at least some of these HEGs can still home. For example, in *Chlamydomonas* algae, soon after 2 haploid cells fuse, the chloroplast DNA from the mating type "−" cell is actively digested, resulting in the inheritance of cpDNA from only the mating type "+" cell (Nishimura et al. 2002). However, if the former contains a HEG not found in the latter, the HEG can still be transmitted to 90–100% of the progeny (while neutral markers that are not close to the HEG are transmitted to none; Bussières et al. 1996). This suggests that homing occurs before the mating type "−" cpDNA is completely digested. Perhaps such HEG leakage also occurs in other taxa with uniparental organelle inheritance.

HEGs Usually Associated with Self-Splicing Introns or Inteins

Although some HEGs occur in the DNA between host genes, most occur in the middle of host genes, but still cause minimal disruption to the function of these genes because they are associated with elements that splice the HEG out of the host RNA (i.e., self-splicing introns), or out of the host protein (i.e., self-splicing inteins). For example, the first HEG discovered, ω, is contained within a self-splicing group I intron, which in turn is inserted within an rRNA gene of yeast mitochondria (Fig. 6.4A). The HEG and

A

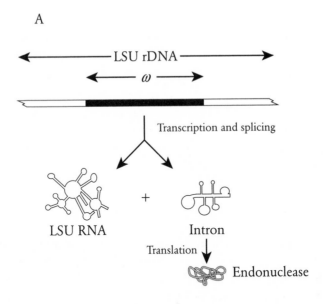

LSU rDNA

ω

Transcription and splicing

LSU RNA   +   Intron

Translation

Endonuclease

B

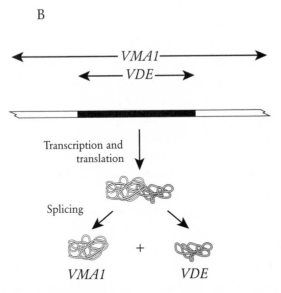

VMA1

VDE

Transcription and translation

Splicing

VMA1   +   VDE

Figure 6.4 Association of HEGs with self-splicing introns or inteins. A. The ω element of yeast mitochondria consists of a HEG inserted into a self-splicing group I intron, which in turn is inserted in the host LSU ribosomal gene. The endonuclease recognizes a sequence in the host gene, and intron and HEG home as a unit. B. The *VDE* element of yeast nuclei consists of an endonuclease domain inserted into a self-splicing intein domain, which in turn is inserted into the host *VMA1* gene. Again, the 2 domains home as a unit. In both cases, self-splicing preserves the function of the host gene.

intron function and drive as a unit, with the enzyme recognizing a sequence in empty rRNA genes that do not contain either HEG or intron, and both the HEG and intron are copied across during repair. By contrast, the only nuclear HEG of yeast occurs not in an intron, but in an intein (Fig. 6.4B). The host gene (called *VMA1*) is transcribed into RNA, which is translated into protein, and the HEG insert then splices itself out, giving rise to a functional host protein and a functional endonuclease. By associating with self-splicing introns or inteins, HEGs help themselves by greatly reducing the harm done to the host.

HEGs and Host Mating System

Homing can occur with relatively high efficiency. For example, with the nuclear *VDE* endonuclease of yeast, heterozygotes segregate 4:0, 3:1, and 2:2 with frequencies of 73%, 15%, and 12%, respectively (Gimble and Thorner 1992), giving a transmission rate of $d = 0.9$, instead of the Mendelian 0.5. For the mitochondrial ω and *ENS2* HEGs of yeast, $d = 0.99$ and 0.85, respectively (Jacquier and Dujon 1985, Nakagawa et al. 1992).

Despite the high levels of drive, these genes are expected to spread slowly through their host populations because yeasts usually inbreed. Homing occurs only in heterozygous individuals, and anything that affects the frequency of heterozygotes also affects the frequency of drive, and thus the rate at which a HEG spreads. The most important factor affecting the frequency of heterozygotes is the mating system and in particular the propensity to outcross versus inbreed. In an outcrossed, random mating population, a HEG with $d = 0.99$ and no effect on host fitness can increase in frequency from 1% to 99% of the population in 9 generations. But in a wholly inbred population there are no heterozygotes and the HEG does not increase in frequency at all. These expectations have been confirmed experimentally for *VDE*, which increases in frequency in outcrossed populations but not in inbred ones (Fig. 6.5). This dependency of population spread on the mating system confirms that the spread is due to drive, and not due to some beneficial effect on host fitness. Indeed, effects of the HEG on host fitness, either positive or negative, were undetectable (<1%) in the laboratory.

As we have said, yeasts like *S. cerevisiae* typically inbreed. In one population of the close relative *S. paradoxus,* the frequency of heterozygotes was

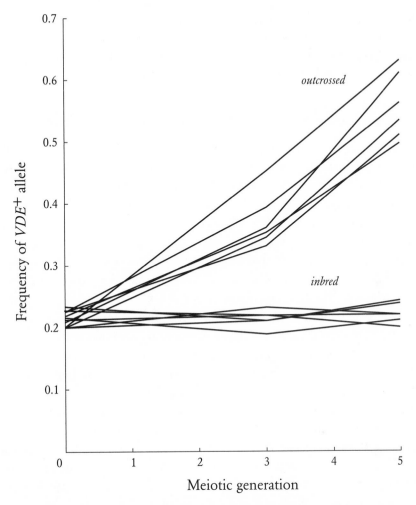

Figure 6.5 Spread of a HEG as a function of host breeding system. Experimental populations of yeast were initiated with a starting frequency of *VDE* of 20–25% and then propagated either by outcrossing or by inbreeding. Each line is a different replicate population (n = 6 for each treatment). The ability of *VDE* to spread in outcrossed populations but not inbred ones is expected for a gene that relies on drive. Adapted from Goddard et al. (2001).

only 1% of that expected under random mating (Johnson et al. 2004). Thus, the effective selection coefficient associated with a HEG would be only a hundredth of what it otherwise would have been, and it would take 100 times as long for the gene to spread through a population (Box 6.1).

Evolutionary Cycle of Horizontal Transmission, Degeneration, and Loss

Strong drive combined with minimal harm to the host should lead to a HEG spreading through a population and going to fixation. But what then? Once fixed in the population, there are no more empty targets to cut and so no selection against mutations that destroy the enzyme's ability to recognize and cleave the target site. Moreover, if there is even a small cost to the host of producing a functional endonuclease (e.g., because it occasionally cleaves DNA at other sites), the frequency of these nonfunctional mutants will increase due to natural selection. How, then, can functional HEGs persist over long evolutionary timespans?

The answer appears to be that HEGs occasionally move from one species to another, and the purifying selection necessary for long-term persistence derives from the fact that only functional elements can successfully spread through the recipient species. As long as, on average, a HEG in a species transfers to at least 1 new species before degenerating in the original species, it can persist over long evolutionary timespans.

Evidence for this model comes from surveys of ω and *VDE* in a series of closely related yeast species (Goddard and Burt 1999, Koufopanou et al. 2002). For ω, out of 20 species, 3 had functional elements, 11 had nonfunctional elements, and 6 had no element at all. The fact that these different states were not clustered on the host phylogeny indicates relatively frequent transitions among the various states (Fig. 6.6A). Moreover, the phylogeny of the element is substantially different from the host phylogeny, indicating frequent horizontal transmission (Fig. 6.6B). These results support a cyclical model of ω acquisition, degeneration, and loss, followed by reacquisition, within the host species (Fig. 6.6C). Note that loss requires precise excision of the element to reconstitute the recognition sequence. Such excisions have been observed in the laboratory and are thought to arise by reverse transcription of spliced RNAs (i.e., a spliced transcript of the host gene occasionally gets reverse transcribed into DNA, which then recombines with the genomic copy, producing what appears as a precise deletion of the

BOX 6.1

Selection Coefficient of a Driving Gene under Inbreeding

There is a pleasingly simple expression for the effective selection coefficient of a driving gene in inbred host populations, derived as follows. Haploid cells are of 2 types, $+$ and $-$, and diploids can be formed by the 3 possible types of fusion: $+/+$, $+/-$, and $-/-$. Diploid frequencies in one generation are $x = p^2 + Fpq$; $y = 2pq(1 - F)$; and $z = q^2 + Fpq$, respectively, where p and q are the frequencies of $+$ and $-$ ($p + q = 1$), and F is Wright's inbreeding coefficient. $+/-$ individuals produce $+$ and $-$ meiotic products with frequencies d and $(1-d)$, respectively. The frequencies of $+$ and $-$ haploids in the next generation will then be $p' = x + dy$ and $q' = z + (1-d)y$. The frequencies in subsequent generations can be calculated by iteration, as can the number of generations required to increase from one frequency to another. The effective selection coefficient can be calculated as:

$$s_e = \left(\frac{p'}{q'} \middle/ \frac{p}{q} \right) - 1$$

$$= \frac{cH}{1 - cHp}$$

where $H = 1 - F = y/(2pq)$ is the frequency of heterozygotes relative to Hardy-Weinberg expectations and $c = 2d - 1$ is a measure of the strength of drive ($c = 0$ for Mendelian inheritance). s_e is the selection coefficient a Mendelian allele would need to have in a random mating population to have the same change in frequency as the driving gene. When the driving allele is rare ($p \sim 0$), this becomes $s_e = cH$; when it is common ($p \sim 1$), this becomes $s_e = cH/(1 - cH)$.

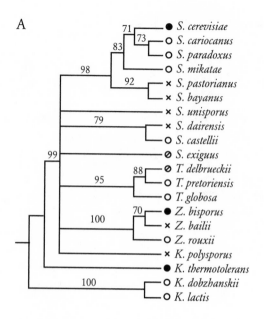

A

● Intron with
    functional HEG

◑ Intron with
    non-functional HEG

○ Intron with
    no HEG

× No Intron

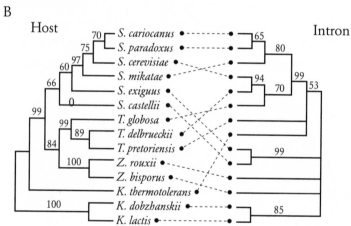

B

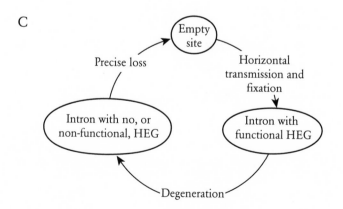

C

intron; Dujon 1989). A best estimate gives about 75 complete iterations of the cycle in the history of the 20 species, though confidence limits are wide (10 to infinity!). A similarly dynamic picture is apparent for *VDE* elements (Koufopanou et al. 2002, Posey et al. 2004). As expected, degeneration is associated with a relaxation of purifying selection (Koufopanou and Burt 2005).

These results demonstrate that, on an evolutionary timescale, HEGs regularly transfer between species. If, as we are suggesting, this has been *necessary* for their long-term persistence, HEGs that are better able to transfer successfully should persist for longer than those that are less able. As a consequence, they should show adaptation for horizontal transmission. A prerequisite for successful transfer is that the recognition site exists in the recipient species, for only then can the HEG drive through the population. Thus, we expect there to have been selection against HEGs with poorly conserved recognition sequences, and extant HEGs, which have persisted, should have unusually well conserved recognition sequences. The very fact that most HEGs target a host gene rather than an intergenic region is consistent with this idea, as genes are usually better conserved than intergenic regions. For *VDE* there are 3 additional lines of evidence (Fig. 6.7). First, the specific host gene targeted is unusually well conserved compared to random genes. Second, the 31bp recognition site is in an unusually well conserved region of the host gene. Third, not all 31bp are equally important for recognition, and the 9 most important (as identified by site-directed mutagenesis) are better conserved than the rest. These important sites are all nonsynonymous positions, at which changes in the DNA change the amino acid. That is, *VDE* has apparently evolved to "pay attention" to the conserved nonsynonymous positions and ignore the more variable silent sites. These various features appear to indicate adaptation by *VDE* for horizontal transmission and thus are evidence that successful horizontal transmission is necessary for HEG persistence.

Figure 6.6 Evolutionary history of ω in yeasts. A. Phylogenetic distribution of different forms of ω. Numbers on branches are bootstrap support values. The intermingling of the different states on the phylogenetic tree indicates repeated evolutionary change. B. The phylogeny of ω and its host are significantly different, indicating horizontal transmission. C. Cyclical model of ω gain, degeneration, and loss within host lineages. Adapted from Goddard and Burt (1999).

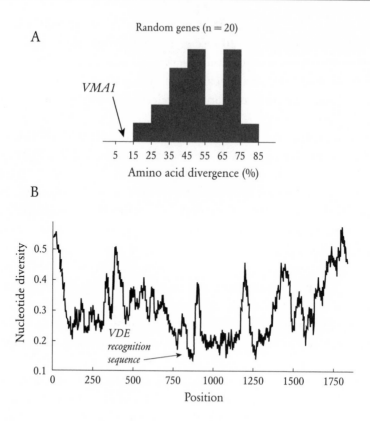

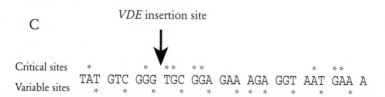

Figure 6.7 Conservation of the VDE homing site. A. Frequency distribution of amino acid divergence between *Saccharomyces cerevisiae* and *Candida albicans* for 20 random genes. Arrow indicates divergence of *VMA1* (the host gene for *VDE*). B. Sliding window analysis of nucleotide diversity in *VMA1* from 6 ascomycete fungi. The position of the *VDE* recognition sequence is indicated. C. The *VDE* recognition sequence. Stars mark nucleotide sites critical for effective homing (as found by site-directed mutagenesis) and those variable among the 24 species of yeast. Arrow indicates *VDE* insertion site. *VDE* recognizes nonvariable nucleotide sites in a highly conserved region of a well-conserved gene. Adapted from Koufopanou et al. (2002).

These results raise the question of how HEGs that target intergenic regions can persist over evolutionary time, given that these regions are usually far less well conserved than are genes. The best-studied such HEG of yeast is *ENS2* (Nakagawa et al. 1991, 1992, Mizumura et al. 1999, 2002). The corresponding endonuclease is unusual in being a heterodimer between the product of this gene and a nuclear-encoded heat shock protein (HSP70). The latter normally functions for host benefit, as a chaperone for proteins that need to be transported into the mitochondria. Interestingly, the effect of dimerizing with HSP70 is to greatly reduce the specificity of the core endonuclease, with the result that it cuts at more than 30 locations in the mitochondrial genome, rather than just once. Perhaps the association with HSP70 evolved to expand the diversity of sequences that can be recognized and cut, and this allows it to persist in a less well conserved part of the genome. A side-effect is an increase in the rate of recombination throughout the mitochondrial genome. This HEG is also unusual in that the target recognition site needed for homing is to one side of *ENS2*, not interrupted by it, and the gene spreads because it is closely linked to a less susceptible site. This is reminiscent of the linkage between driver and responder loci in meiotic drive (see Chap. 2).

A requirement for regular horizontal transmission may also explain why HEGs are mostly found in simpler organisms (fungi, algae, and so on) in which access to the germline is easier. In yeasts, possible mechanisms of horizontal transmission include interspecific hybridization, vectoring by predacious yeasts, and uptake of naked DNA from the environment (Koufopanou et al. 2002). Horizontal transmission has also been observed for a mitochondrial HEG in plants (Cho et al. 1998), as it has for other plant mitochondrial genes (Bergthorsson et al. 2003), but it is not known how this happens. This HEG appears to have recently invaded plants, rather than stably cycling among them.

Finally, the problem of weak or nonexistent purifying selection may be even more acute for HEGs that target nuclear rRNA genes. These host genes typically occur in tandem arrays of tens or hundreds of copies, and so it is possible for functional, nonfunctional, and empty alleles all to occur in the same cell. Thus a functional HEG might cause an empty rRNA gene to be cut, and the latter repaired using a nonfunctional allele as template. That is, nonfunctional elements could parasitize the functional elements, making it more difficult for functional HEGs to persist. Such parasitism may account for the great preponderance of HEG-less group I introns compared to those containing HEGs in nuclear rDNA (a ratio of about 60:1; Haugen et al. 2004).

HEG Domestication and Mating-Type Switching in Yeast

Haploid yeast cells are 1 of 2 mating types, **a** or α, and only cells of opposite mating type can fuse to form a diploid cell. Mating type is genetically determined: cells have either an **a** allele or an α allele at the appropriate locus. But although genetically determined, mating type is also plastic: when dividing mitotically, **a** cells regularly give rise to α cells, and vice versa, because yeast has a sophisticated mechanism for switching mating types. In addition to the **a** or α allele at the active mating-type locus, they also have silent copies of both genes elsewhere on the same chromosome, and they regularly replace the active allele with a copy of the other allele. The process of gene replacement begins with an enzyme cleaving the DNA at the active mating-type locus; the break is then repaired using the appropriate silent gene as a template, a copy of which is inserted into the active site as a result. This mechanism has obvious similarities to homing, and the enzyme responsible for making the initial cut, HO, is clearly derived from *VDE* (Fig. 6.8). It is a domesticated HEG, which no longer drives but works for the benefit of the host.

It is interesting to speculate what exactly that benefit may be. The most obvious result of mating-type switching is that it allows mating between mitotic clonemates that otherwise would have the same mating type—that is, it allows an extreme form of selfing. One possible advantage for a yeast cell of mating with a close relative is that it is then less likely to acquire a new array of selfish genes—for example, selfish plasmids (Box 6.2) or RNA viruses (Wickner 1996). Certainly one consequence of inbreeding is to reduce the spread of selfish genetic elements. As we shall see in Chapter 7, yeast are missing 2 widespread classes of transposable elements, and perhaps this is because they are so highly inbred. Thus the domestication of 1 selfish genetic element may have made it more difficult for others to spread, and this effect may even have been the reason for its domestication.

Group II Introns

Group II introns are in many ways similar to HEGs. They are optional genetic elements in the organelles of relatively simple eukaryotes

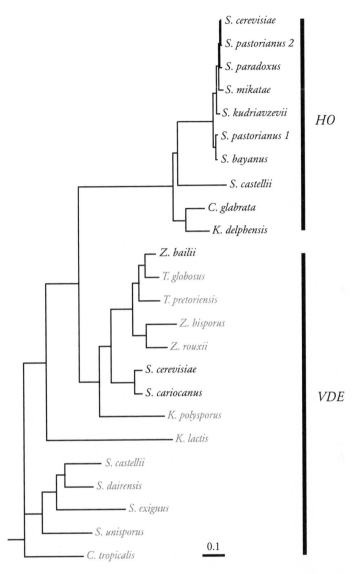

Figure 6.8 Domestication of a selfish gene. *VDE* is a HEG of yeasts and *HO* is a host gene that encodes a sequence-specific endonuclease involved in mating-type switching. The 2 genes are homologous, and phylogenetic analysis shows that the *HO* sequences form a clade (with 99% bootstrap support) that is nested within the *VDE* sequences (91% support), indicating that *HO* genes arose once from a *VDE* ancestor. Note also that the branch leading to the *HO* clade is the longest internal branch, as expected if domestication was associated with an increase in the rate of protein evolution. *VDE*s in gray are defective and have slightly longer terminal branches than the functional *VDE*s (black), consistent with a relaxation of purifying selection. Adapted from Koufopanou and Burt (2005).

BOX 6.2

2μm Plasmid of Yeast

Many strains of the yeast *Saccharomyces cerevisiae* carry a 6.3kb circular plasmid in their nucleus that, unlike many bacterial plasmids, appears to be wholly parasitic, contributing nothing to the host (Broach and Volkert 1991). It encodes 4 proteins, all of which are involved in plasmid maintenance, and it imposes a demonstrable cost on the cell, lengthening the generation time by 1.5–3% (Futcher and Cox 1983, Mead et al. 1986). Typically, there are 60–100 copies per haploid cell, for a total of about 4% of nuclear DNA. Plasmid-bearing cells give rise to plasmid-free cells with a frequency of 10^{-4} to 10^{-5} per cell generation; and in asexual or wholly inbred populations, the plasmid-free cells would be expected to take over. However, the plasmid is maintained in populations because of drive: if a carrier and a noncarrier mate, essentially all the subsequent meiotic products will carry the plasmid. Laboratory experiments have shown, as expected, that the plasmid can spread in outcrossed populations, but not inbred ones (Futcher et al. 1988).

The 2μm plasmid drives because it can replicate more than once per cell cycle, and so can accumulate within cells. Even if a cell is transformed with only a single copy of the plasmid, the plasmid will amplify until the normal copy number is attained. The mechanism by which it does this is ingenious. Normally, a small circular DNA molecule replicates by having a single origin of replication, from which replication forks proceed in both directions until they meet at the other side, and the plasmid is replicated once. However, the 2μm plasmid has a pair of inverted repeats of about 600bp on opposite sides of the plasmid, one of which is very close to the origin of replication. The plasmid also encodes a protein (FLP1) that causes the inverted repeats to recombine. Now, if this protein is active during plasmid replication and causes recombination between a replicated repeat and an unreplicated repeat, instead of traveling in opposite directions, the replication forks would be traveling in parallel, chasing each other around the plasmid. This would copy the plasmid many times in a single cell cycle. Subsequent recombinations would reorient the replication forks, allowing replication to terminate, and would resolve multimeric plasmids into monomers. In this way the plasmid could increase in copy number relative to host DNA, even if its origin of replication fired only once per cell cycle.

The most puzzling aspect of the 2μm plasmid is how rare it appears to be. Structurally similar plasmids have been found in *Kluyveromyces* and *Zygosaccharomyces* yeasts, but apparently they have yet to be found in other fungi, or in any plant or animal. We do not know why this should be. The fact that it is nuclear should mean that even extreme anisogamy should not be a problem for its biparental inheritance. Moreover, the key gene, *FLP1*, is fully functional when introduced into mammals, insects, and plants (Jayaram et al. 2002). The very high selfing rate of baker's yeast may be relevant here, both in preventing the accumulation of nonautonomous defective plasmids and in preventing the plasmid from evolving ever greater copy number, and thus virulence.

The top figure shows the structure of the 2μm plasmid. Also shown are the steps in overreplication. A bidirectional replication event (b) is converted by FLP-mediated recombination into 2 unidirectional replication forks (c), leading to plasmid amplification (d, e). A second recombination event (f) terminates amplification (g). Products h and i are multimeric and monomeric plasmids, respectively, and resolution of h yields plasmid monomers (j, k, etc.). Adapted from Broach and Volkert (1991).

that have no known host function and that contrive to copy themselves from intron[+] chromosomes to intron[−] chromosomes. The means by which they accomplish this, though, is completely different from that used by HEGs.

Three introns have been well studied (*aI1* and *aI2* of yeast and *L1.LtrB* of the bacterium *Lactococcus lactis*), and the precise details of how they get cop-

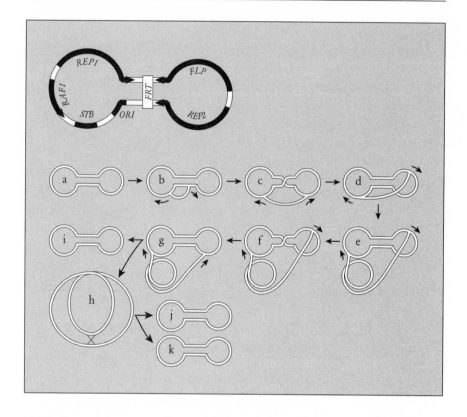

ied from intron⁺ to intron⁻ chromosomes vary among them (Fig. 6.9). In the simplest pathway, the introns encode a multifunctional protein that helps in splicing the intron out of the RNA transcript of the intron⁺ host gene (i.e., it acts as a "maturase"). The intron and protein remain associated, and together they recognize intron⁻ copies of the host gene and reverse splice the intron into the sense strand of DNA. The protein then nicks the antisense strand and reverse transcribes the intron into DNA. Final ligation of the intron DNA to the flanking host gene is done by host repair mechanisms.

As for HEGs, the recognition sequences used by group II introns are long (e.g., 30–35bp) and not all positions within these sequences are equally important (Belfort et al. 2002). Both the intron and the protein are involved in recognizing the target site, with the former based on RNA-DNA base pairing. The fact that recognition depends in part on RNA-DNA base pairing means that it is relatively easy to engineer group II introns with novel recognition sequences (Guo et al. 2000).

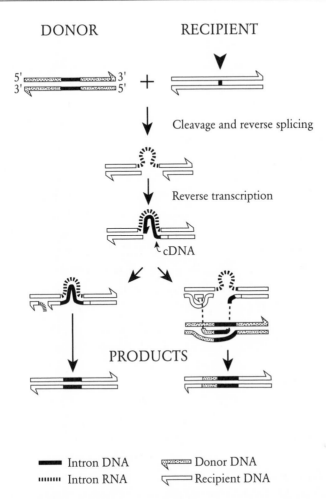

DONOR   RECIPIENT

Intron DNA   Donor DNA
Intron RNA   Recipient DNA

Figure 6.9 Retrohoming of group II introns. Shown here are two variants of the retrohoming pathway. First the intron and associated protein cleave the target DNA and the intron is reverse spliced into 1 strand. It is then reverse transcribed into cDNA using the opposite strand as a primer (target primed reverse transcription). The simplest pathway (left) is then for the RNA to be displaced or degraded, and then the second DNA strand is synthesized and loose ends are ligated to the target DNA. Alternatively (right), the cDNA may recombine with the donor before second strand synthesis, leading to transfer from donor to recipient chromosomes of sequences flanking the intron (marked by stippled DNA in the product). Adapted from Belfort et al. (2002).

For the yeast elements, this "retrohoming" pathway–so named because it involves reverse transcription of RNA into DNA–can be complemented by purely DNA-based homing: the intron and protein can together cause a double-strand break in the target site that is repaired using the intron[+] gene as a template. In 2 different crosses, this alternative DNA-based pathway accounted for 10–40% of homing events (Eskes et al. 1997).

As with HEGs, once a retrohoming group II element goes to fixation, one might expect it to slowly degenerate, particularly in the reverse splicing, nicking, and reverse transcription activities, as there would be no target left to attack. Movement, either to a new site in the genome (transposition) or to a new gene pool (horizontal transmission) would then be required for long-term persistence. Transposition to a new location has been observed in the lab for both yeast and bacterial elements (Mueller et al. 1993, Sellem et al. 1993, Cousineau et al. 2000), though the frequency with which this occurs successfully in nature is not known. Phylogenetic information that might reveal evidence of transposition, horizontal transmission, degeneration, or loss is scarce. For algae, a survey of a particular site in the *cox1* gene of 19 species (including haptophytes, diatoms, yellow-green algae, and brown algae) showed that 2 species had an intron with an ORF, and 17 species had no intron (Ehara et al. 2000). Both species are thought to have recently acquired the intron by horizontal transmission from separate unknown sources, though the sampling is still sufficiently sparse that other interpretations are possible. Interestingly, for the diatom species *(Thalassiosira nordenskioeldii)*, individuals collected from the Arctic and Atlantic Oceans had substantially different ORFs, suggesting multiple independent acquisitions by a single species. One of the ORFs had a frameshift mutation that would make it nonfunctional.

Other group II introns are degenerate and show long-term persistence without movement. In higher plants, introns in the chloroplast *trnK* gene and mitochondrial *nad1* gene encode proteins (matK and matR, respectively) that function as maturases, splicing the intron out of the host gene transcript, but they are missing key domains that are important in reverse transcriptase activity and are thought to be incapable of retrohoming (Zimmerly et al. 2001). Nevertheless, each is widespread in higher plants and appears to have been maintained for long evolutionary timespans purely by vertical inheritance. Presumably, they have taken on some role beneficial to the hosts (e.g., trans-splicing of other introns in each organelle?).

It might seem surprising that group II introns are also found in bacteria, given the requirement for some sort of biparental inheritance to show a transmission advantage. In this respect, it is most interesting that they appear to be associated with the most sexual parts of the bacterial genome, the plasmids and insertion sequences. Fully 18 of 20 bacterial group II introns found in the databases by Zimmerly et al. (2001) were found in such mobile DNA. By contrast, group II introns of eukaryotic organelles tend to be found in housekeeping genes.

The evolutionary origins of retrohoming group II introns are a matter of speculation (Curcio and Belfort 1996). A genetic element with so many components (RNA plus multifunctional protein with domains for splicing and reverse splicing, DNA nicking, and reverse transcription) presumably arose by fusion, but of what? Some speculate that they are holdovers from a previous "RNA" world. Subsequently, they are thought to have given rise to both LINE-like transposable elements (see Chap. 7) and to the conventional spliceosomal introns found in the nucleus of most eukaryotes.

Artificial HEGs As Tools for Population Genetic Engineering

One of the many pleasures of working on selfish genetic elements is the possibility that they may one day be used to ameliorate human suffering. There are a small number of species that cause substantial harm to the human condition—species that cause disease, transmit disease, or reduce agricultural output—and there has long been the hope that selfish genetic elements could be used to reduce population numbers or otherwise render these species less noxious. In the 1970s the killer Y chromosome of *Aedes* mosquitoes came close to being used as part of a control program in India (Curtis et al. 1976). HEGs and group II introns are among the simplest of selfish genetic elements and, despite their relatively recent discovery, among the best understood in terms of molecular mechanisms. Molecular biologists are already using HEGs and the enzymes they encode as reagents in the laboratory for mapping genes and for engineering organisms (Belfort et al. 2002). Moreover, there is much work on how one might artificially engineer a HEG to recognize and cut any specified DNA sequence (e.g., Bibikova et al. 2001, Chevalier et al. 2002, Seligman et al. 2002); for group II introns, this problem has already largely been solved (Guo et al. 2000). Such retargeted selfish genes may be useful in genetic engineering, functional genomics, and gene therapy. In this section we explore the possible uses of such designer selfish genes for population control and population genetic engineering (Burt 2003). We phrase the discussion in terms of HEGs, but many of the comments apply equally to group II introns if they can be engineered to target nuclear genes. In Chapter 7 we meet another class of site-specific selfish genes (see Box 7.1), and these may also prove useful. We shall begin with strategies for reducing the size of a pest population.

The Basic Construct

Consider an artificial HEG with the following properties (Fig. 6.10): (1) It is engineered to recognize and cut a sequence in the middle of an essential gene, and the HEG is inserted into the middle of its own recognition sequence, simultaneously disrupting the gene and protecting the chromosome from being cut. There would be no self-splicing introns or inteins, and so the insertion of the HEG would disrupt the host gene. (2) The target gene is chosen such that the knockout mutation has little phenotypic effect in the heterozygous state but is severely deleterious when homozygous (i.e., the knockout is recessive). (3) Finally, the HEG is under the control of a meiosis-specific promoter, so heterozygous zygotes develop normally but transmit the HEG to a disproportionate fraction of their gametes. This last condition can be relaxed, depending on when the target gene is expressed. If the target gene is expressed only in larvae or in somatic tissues, the promoter can be adult-specific or germline-specific.

If such a construct is introduced at low frequency into a population, it will initially appear mostly in the heterozygous state, and so it will show drive but few harmful effects. It will therefore increase in frequency, until reaching some equilibrium at which the harmful effects balance the drive. If we assume the population is large and mates at random, the knockout is a recessive lethal, and drive occurs equally in males and females, the equilibrium frequency of the HEG ($\hat{q}$) can be shown to be $\hat{q} = c$, where c is the probability that the HEG$^-$ allele in a heterozygote is converted to a HEG$^+$ allele ($c = 0$ for Mendelian inheritance). The load imposed on the population (i.e., the fraction of the reproductive effort that is rendered unproductive) is then equal to the frequency of homozygotes, $L = \hat{q}^2$, and the mean fitness of the population is 1 minus this, or $1 - c^2$. As we have seen, some HEGs of yeasts can show extreme drive, with $c = 0.99$ (Jacquier and Dujon 1985, Wenzlau et al. 1989). If one considers, conservatively, an engineered

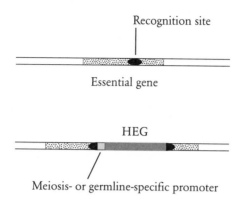

Recognition site

Essential gene

HEG

Meiosis- or germline-specific promoter

Figure 6.10 A construct for biological control. A HEG is engineered to recognize a sequence in an essential gene for which the knockout phenotype is recessive. The HEG is inserted into the middle of its own recognition sequence, disrupting the host gene but protecting the chromosome from being cut. The HEG is also put under the control of a promoter chosen so that heterozygotes develop normally but transmit the HEG to a disproportionate fraction of the progeny. Such a construct will spread through a population and impose a genetic load. Adapted from Burt (2003).

HEG with $c = 0.9$ (we make this same assumption in all numerical examples in this section), the equilibrium mean fitness of the population is 0.19. That is, four-fifths of zygotes produced are killed by the HEG. Moreover, this load arises relatively quickly. If the HEG is introduced into 1% of the population, it will take only 12 generations for the load to equal 90% of its equilibrium value (Fig. 6.11). If one can manage to release only an initial frequency of 0.01%, it will take 19 generations.

A key feature of this construct is that it is evolutionarily stable, in the sense that the mutant forms most likely to arise as the construct spreads through a population will be selected against and lost. Mutant HEGs that lose the ability to recognize or cut the target DNA, or that are active in somatic as well as germline tissues, or that show less sequence specificity, will all be selected against and disappear from the population with little effect. A further attractive feature is that the manipulation is fully reversible. If we target a gene that, when knocked out, is strongly deleterious, there will be strong selection in favor of resistant alleles—sequences that are functional but are not recognized and cut by the HEG. One could engineer resistant alleles by, for example, using the degenerate property of the genetic code to

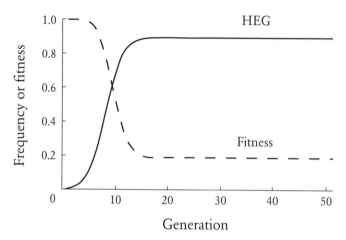

Figure 6.11 Spread of an engineered HEG and resulting decline in mean fitness of the population. Calculations assume an initial release frequency of 1%; the target gene is a recessive lethal; in heterozygotes the HEG⁻ allele has a probability of $c = 0.9$ of being converted to HEG⁺; and the population is random mating. Note that equilibrium is reached in fewer than 20 generations. Adapted from Burt (2003).

create a DNA sequence that coded for the same amino acid sequence but differed in nucleotide sequence from the target. Releasing such resistant alleles could be used to effectively "recall" a HEG, as the resistant allele would spread through the population, driving the HEG extinct (Fig. 6.12).

Increasing the Load

For many pest species, killing four-fifths of the zygotes may have little effect on the population dynamics, merely relaxing density-dependent pressures on survival and reproduction. It is therefore worthwhile investigating how we could increase the load further. First, we might be able to increase the rate of drive. If one could engineer a highly effective HEG with $c = 0.999$, targeting a single recessive lethal gene would give an equilibrium mean fitness of 0.002 (i.e., 99.8% of reproductive output would be wasted and only 0.2% would be viable), probably sufficient to eradicate most pest populations. However, such high levels of drive may not be achievable.

Second, we could choose other genes as targets. For many species, killing

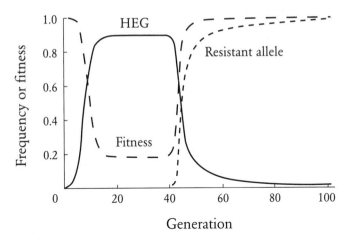

Figure 6.12 Recalling an engineered HEG. A resistant allele is introduced at 1% in generation 40; all other parameters are as in Fig. 6.11. Note the rapid decline in the frequency of the HEG and the recovery of population mean fitness. Adapted from Burt (2003).

males is worse than useless, because it reduces the frequency of the HEG but does little to reduce population growth rates or equilibrium density, these being largely determined by female productivity. One way to avoid the waste of killing males would be to target a gene that, when knocked out, kills only females. For the drive assumed here ($c = 0.9$), targeting such a gene would give an equilibrium mean fitness of 0.055, a 3-fold reduction compared to targeting a recessive lethal. Knockouts that cause females to be sterile would be equally effective. Indeed, knockouts causing male sterility would also have the same effect, if they had no effect on the male's fertilization success (e.g., eggs were fertilized but development was then aborted).

Finally, we could target many loci simultaneously. In the simplest case, in which the drive and phenotypic effects at 1 locus are independent of genotype at the other locus, the equilibrium mean fitness is the product of the fitnesses of the 2 loci separately. If we engineered HEGs to target n different loci essential for female fertility and $c = 0.9$ at each, mean fitness would be 0.055^n. If we targeted 5 loci, mean fitness would be 5×10^{-7}, enough to drive any population extinct. In *Drosophila melanogaster*, at least, there are thought to be some 3000 essential genes and more than 100 required for fertility (Ashburner 1989, Miklos and Rubin 1996, Ashburner et al. 1999). Most prospective target species are likely to have an abundance of suitable target loci.

Preventing Natural Resistance and Horizontal Transmission

The simulation shown in Figure 6.12 demonstrates that, if a functional host gene exists that is resistant to the HEG, it will increase rapidly in frequency and drive the HEG extinct. Care must therefore be taken to minimize the likelihood that such sequences exist, or arise before the population is eradicated. For example, mutagenesis experiments and structural studies could be used to choose target genes, and sites within genes, that are unlikely to be able to change (at the amino acid level) without seriously compromising function. And we would want to target sequences that are not too variable.

"Combination therapy"–the use of multiple drugs simultaneously–slows the evolution of resistance in human pathogens (White et al. 1999, Palumbi 2001), and the same approach can be used with engineered HEGs, virtually without limit. First, we could release, say, 10 different HEGs, attacking 10 different sites along the length of a single gene. The more HEGs that are released, the lower the likelihood that resistant sequences will exist for all of them and, even if they do exist, the longer it will take for a multiply resistant sequence to be stitched together by recombination. The genetic load imposed in the meantime could be enough to drive the population extinct. The second form of combination therapy would be to target multiple loci simultaneously. Even if resistance could evolve at each locus separately, by combined attack the genetic load could be sufficient to drive the population extinct.

Probably the most serious challenge to this program comes from the possible suppression of drive by the larger genome. If a mutation is present or arises in the host that reduces or eliminates homing activity, but is otherwise neutral, it will spread rapidly. Of critical importance to a possible counterattack is whether the resistance is specific to a particular HEG or whether it protects against all HEGs. If resistance is specific, one may be able to swamp the population with multiple HEGs; but if resistance is general, alternative approaches will need to be explored. Depending on the nature of resistance, possibilities include targeting the resistance alleles directly, targeting genes to increase the cost of resistance, or (if resistance already has a cost) targeting a series of nonessential genes whose knockouts are less costly than is resistance. One possibility is that introduced HEGs spread in a population until they encounter 1 or more genes in the host that suppress drive. Then decimated areas may be recolonized by these newly emerging resistant genotypes, with coevolutionary responses in nature requiring a continuing series of responses in the lab. Certainly it would be essential to study resistance in the laboratory (using as large and diverse populations as possible) before releasing HEGs out of doors.

Another issue worth thinking about beforehand is how to reduce the likelihood of horizontal transmission from one species to another. As we have seen, this occurs for HEGs in yeasts and in plant mitochondria, and probably in other taxa too. Nevertheless, these results should not be exaggerated: evidence that horizontal transmission occurs regularly on an evolutionary timescale of millions of years does not mean it is a substantial risk during a 10-year population control program. For horizontal transmission to occur, the DNA containing the HEG would somehow have to get into a germline nucleus of another species and be sufficiently intact to be transcribed and then used as a template for repair. This is unlikely to occur in many prospective target species and, indeed, may be the main reason why HEGs appear to be absent from animals with segregated germlines. One way to reduce further the probability of horizontal transfer is to engineer the HEG not to recognize the homologous sequence in related species. As we saw in yeasts, this seems likely to be an effective way to limit horizontal transfer (Koufopanou et al. 2002). In addition, we could target a region with low overall nucleotide similarity among species, to reduce the likelihood of homologous recombination.

Population Genetic Engineering

So far we have focused on how to impose a genetic load sufficient to control or eradicate a population. More subtle approaches may often be desirable. In particular, we may not want to eradicate a population but rather to genetically transform it such that it is less noxious. That is, we might want to do population genetic engineering. In conventional lab-based genetic engineering of individual organisms, there are 3 basic types of manipulation: removing a gene, replacing a gene, and introducing a novel gene. In principle, artificial HEGs could be used for population-wide equivalents of all 3. For purposes of illustration, we consider the example of engineering a mosquito population so it can no longer transmit malaria.

Gene knockouts. Suppose, for example, there is a gene in mosquitoes that is not essential for it to live and reproduce but is essential for it to transmit malaria. We could design a HEG against it, and if the knockout was not otherwise too harmful, the HEG would spread to fixation with little or no effect on mosquito numbers. Many genes could be knocked out in this way.

Gene replacements. Suppose that there is a gene essential to the mosquito that can be changed in a way that is lethal to the malarial parasite but has minimal effect on the mosquito. We could engineer a HEG that attacked the unwanted sequence but not the desired one, and then release individuals carrying the HEG and the desired resistant allele. As shown in Figure 6.13, the resistant allele would come to predominate relatively quickly after introduction, with only a relatively small and temporary reduction in mean fitness.

Gene knockins. Finally, suppose we have discovered or designed a new gene that, when introduced into a mosquito, would render it unable to transmit malaria. Promising candidates for such a gene have already been described (Ito et al. 2002). We may then wish to drive this gene into a population. To achieve this, the novel gene could be linked to a HEG targeting a neutral region of the genome and the whole construct inserted into the recognition site. As the HEG spread through the population, it would bring the novel gene with it. Other strategies are also possible. For example, a HEG could be engineered to target an essential gene. As it spread, it would select for a resistant gene. We could then introduce a resistant allele with the novel gene linked to it, possibly using an inversion. This strategy would greatly expand the size of the gene we could introduce and even allow multiple, linked genes to be introduced simultaneously. This strategy would also reduce the rate at which nonfunctional mutant genes arise, if DNA replication associated with cell division has a lower error rate than that associated with DNA repair and gene conversion.

Regardless of how exactly the novel gene is driven into the population, a key limitation of this approach is that, if it is harmful, the construct will not be evolutionarily stable. Mutant constructs in which the novel gene is deleted or otherwise defective will be selected, because they cause less harm to

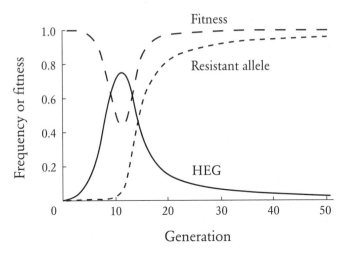

Figure 6.13 Population-wide gene replacement. The HEG and the resistant allele are introduced simultaneously at a 1% frequency; all other parameters are as in Fig. 6.11. Note the rapid spread of the resistant allele with only a transient decline in mean fitness of the population. Adapted from Burt (2003).

the host but still have the transmission advantage. But if the novel gene is not too harmful, and the mutation rate not too high, it can persist for a considerable length of time before going extinct. Suppose, for example, the resistant allele in Figure 6.13 has linked to it a novel gene that is fully dominant, reduces fitness by 10%, and mutates to a nonfunctional form with a frequency of 10^{-6}. This novel gene would reach a frequency above 50% within 15 generations and above 95% within 40 generations, and it would remain above 95% for about 4000 generations. Full dominance is critical here, for then heterozygous mutants have no selective advantage. Introducing multiple copies of the gene, linked to 1 or more resistant alleles, should also help to reduce selection in favor of nonfunctional mutants and prolong the population transformation.

Some authors have suggested using transposable elements or cytoplasmic incompatibility agents as vectors to drive novel genes through a population (Ribeiro and Kidwell 1994, Turelli and Hoffmann 1999), but the use of HEGs is likely to have a number of advantages. Constructs using transposable elements are expected to be less stable, due to high mutation rates during transposition, and they give little control over genomic location and copy number. Constructs using cytoplasmic incompatibility agents are expected to spread more slowly, require larger introduction frequencies, and do not allow the gene to be in the nucleus. Both alternatives are probably more likely to transfer the novel gene to another species than is a HEG construct, because at no time is the latter separate from the host chromosome. Introducing the novel gene linked to a resistance locus should be even safer. Finally, population-wide transformations using a HEG should be fully reversible, by releasing a HEG engineered to target the novel gene.

Other Uses

Two other potential uses for engineered HEGs are worth mentioning. First, it has long been recognized that if a Y chromosome were to show drive and spread to fixation in a population, the sex ratio would become male biased. If the drive was extreme, the population could be driven extinct for want of females (Hickey and Craig 1966a, Hamilton 1967; see Chap. 3). Lyttle (1977) engineered such a driving Y chromosome in *D. melanogaster* by creating translocations between the Y and a driving *SD* chromosome 2 and found it could rapidly drive laboratory populations extinct (Lyttle 1977). And in *Aedes* and *Culex* mosquitoes there are Y chromosomes that cause the X chromosome to break during the first meiotic division and thereby show drive (see Chap. 3). These naturally occurring Ys suggest the following strategy: insert onto a Y chromosome 1 or more endonuclease genes that recognize and cut sequences specific to the X chromosome, and put the endonuclease genes under the control of premeiotic-specific promoters. Then, during spermatogenesis, the X chromosome would be cut and, as there would not be an appropriate template for repair, the Y chromosome would show drive. It would spread through the population and, if the drive was sufficiently extreme, the population could be eradicated. Such an approach would not rely on recombinational repair or homing. In some species the sex chromosomes are inactivated prior to meiosis, which could complicate the design of the construct; but this is not so in all species, including many dipterans (see Chap. 3).

Second, all the manipulations discussed so far are "inoculative," in that the release of relatively few engineered individuals will drive the manipulation. Often this will be an advantage, but not always. If one wanted to eradicate only one population, for example, and leave others in the species range undisturbed, inoculative methods may not be appropriate. "Inundative" strategies such as the release of sterile males (Knipling 1979) are inherently self-limiting, and so more appropriate for such population-specific targeting. Engineered HEGs could be used in an inundative strategy if they were to cause dominant female lethality or sterility. Knockouts causing dominant female-specific effects are rare; but if the HEG was engineered to be active in all tissues, then even if a zygote started heterozygous, the organism would be converted to a homozygote. Thus, we could still target a recessive female-specific locus. Females inheriting the HEG would be dead or sterile, and males would pass on the HEG to the next generation. As long as the HEG was not perfectly efficient ($c<1$), it would slowly disappear from the population but could cause a substantial load before doing so. Thus, simply by changing the promoter, the threat of rare emigrants to neighboring populations could be avoided. The use of such engineered HEGs would be more efficient than the release of sterile males, allowing either fewer individuals to be released or larger populations to be targeted (Thomas et al. 2000).

SEVEN

Transposable Elements

TRANSPOSABLE ELEMENTS ARE, at the same time, the most widespread selfish element, the best studied, the most complex to master, and the element showing the most unusual form of drive. The last of these is the key. While other elements compete for representation at a given locus, transposable elements accumulate by copying themselves to new locations in the genome. As a consequence, there may be tens or hundreds of active copies of a single transposable element dotted around the genome of a single individual, and different individuals may have their insertions in different places in the genome. This abundance and variety come despite the fact that transposition rates per element are low relative to the activity of other selfish genes (jumping every fourth generation is a very high rate, and more commonly it occurs every hundred, or thousand, or more generations).

Being small and adapted to integrating themselves into novel places in the genome, transposable elements are also capable of moving between species. Because of this expansive form of drive—both within and between species—transposable elements appear to have colonized all eukaryotic species. They have radiated into a bewildering array of subtypes, a veritable zoo of elements, with many different strategies and substrategies of drive, and new ones described almost daily. Most species have multiple different types or families of transposable elements, each present in multiple copies per genome.

Due to their abundance, transposable elements are almost guaranteed to have profound effects on their hosts. About half of our own genome is derived from transposable elements. And, in addition to transposition itself, these elements can cause a bewildering array of chromosomal rearrangements. As with other types of mutations, some fraction of these insertions and rearrangements will be beneficial to the host and positively selected. Transposable elements may thus be important sources of mutation that would not occur by other means. And transposable elements can also be domesticated (rarely) by the host—as we shall see, just such an event was critical to the evolution of the vertebrate immune system. These beneficial effects notwithstanding, transposable elements are still best considered as parasites, not as host adaptations or mutualists.

Transposable elements were first discovered by Barbara McClintock, in 1952. The key observation was that the linkage arrangement of some genes in maize could change from one generation to the next. For decades they were little studied, but with the advent of molecular biology an enormous amount of work has been done on them. This is partly because they are so widespread, present in all the model genetic organisms (including prokaryotes) and humans. Moreover, since 1982 they have been used as tools for genetic engineering, first in *Drosophila* and now in a growing number of other insects. In addition, some transposable elements are closely related to retroviruses, including HIV, the cause of AIDS. A recent and valuable collection of review chapters on "mobile DNA," much of it on transposable elements, runs to 50 chapters, more than 1150 pages of text and thousands of references (Craig et al. 2002).

It is not possible here to review or even tabulate every well-known transposable element in every species (as, for example, we have done for gamete killers). Instead, we review what seems to us the most important and interesting, focusing on their evolutionary logic. As for any selfish gene, knowing how transposable elements actually work is an essential guide to thinking about how they are likely to evolve. We therefore begin by describing in some detail what transposable elements look like and how they jump.

Molecular Structure and Mechanisms

There are 3 main types of transposable elements, which have little in common in structure or mechanism except that they are relatively short (typically 1–10kb) and encode 1 or more proteins, all of which appear to be involved in copying the element to a new location in the genome. Otherwise, their mechanisms are quite different. Briefly, DNA transposons typically encode 1 protein, a transposase, which recognizes the ends of the element, cuts it out, and reinserts it elsewhere in the genome. These cut-and-paste mechanisms lead to an increase in copy number, because they take advantage of DNA repair mechanisms (analogous to HEGs; see Chap. 6) or of differences in the replication times of genes in different locations. The other 2 classes transpose via an RNA intermediate, through the action of reverse transcriptase, which copies RNA transcripts into DNA. The simpler class (LINEs) typically encode 1 or 2 proteins, whereas the more complex one (LTR retrotransposons) encodes 5 or 6 and is thought to be an amalgam of the other 2 types. All 3 classes of element are susceptible to parasitism by shorter elements (typically 100–1000bp) that are copied by the transposition machinery of the intact elements but do not themselves encode any proteins. Finally, all 3 classes of element are widespread and found in animals, fungi, plants, and protists, though there is a tendency for the retroelements to be even more abundant than the DNA transposons. More detailed reviews of transposable element structure and function can be found in the volume edited by Craig et al. (2002).

DNA Transposons

The simplest class of transposable elements are the DNA transposons, which include the *Ac* and *Ds* elements of maize, the first transposable elements to be discovered, the well-studied *P* elements of *Drosophila melanogaster*, and many others. DNA transposons are typically 1–10kb depending on the type, encode 1 (or rarely 2) proteins, and have inverted repeats at the ends, usually 10–500bp long (Fig. 7.1). The protein is a transposase that recognizes specific DNA sequences at or near the ends of the element, cuts the element out of the genome, and inserts it somewhere else (Plate 6). There is a tendency for the new insert to be near the old one. For example, in one study of *Ac*, about 60% of transposition events were to the same chromosome and, of these, 40% were within 4 centiMorgans of the donor site (Greenblatt 1984). *P* elements also tend to jump locally (Tower et al. 1993, Zhang and Spradling 1993). Though transposition is essentially cut and paste, the copy number of the transposon may nonetheless increase in at least two ways.

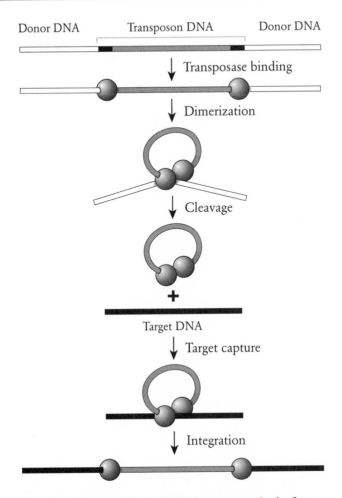

Figure 7.1 Cut-and-paste transposition of DNA transposons. In the first step, transposase (shaded spheres) binds to specific recognition sequences (black bars) in the inverted repeats at the ends of the transposable element. The transposase then dimerizes, cleaves the transposable element out of the host DNA, and inserts it into another site in the host genome. Adapted from Davies et al. (2000).

(1) *P* elements take advantage of the cell's broken chromosome repair system, like HEGs, but in a slightly different way (Engels et al. 1990, Nassif et al. 1994; Fig. 7.2). Elements are excised at a point in the cell cycle between DNA replication and mitosis, creating a double-strand gap. The element goes on to insert elsewhere in the genome. Meanwhile, the double-strand

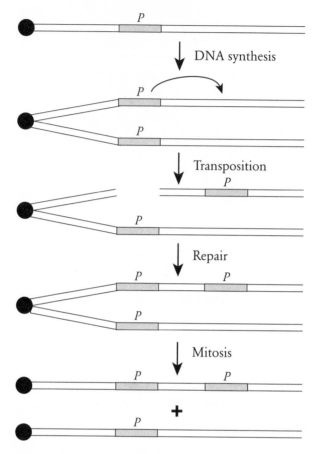

Figure 7.2 Increase in copy number of *P* elements by recombinational repair. Transposition occurs after DNA replication and before cell division, and the gap left by the excised element is repaired using the sister chromatid as a template (see Fig. 6.1 for details). After cell division, both daughter cells have the old insert, and 1 of them also has the new insert. Here the new insert is on the sister chromatid of the donor, but it can be anywhere in the genome. Solid circle represents the centromere.

gap turns on the cell's repair system, which uses a homologous template to fill in the missing DNA. Usually the sister chromatid (with a copy of the *P* element) is used, in which case both chromatids once again have the *P* element. Occasionally, the repair process is interrupted, and a *P* element with an internal deletion is formed. At other times (~15% on autosomes), the homologous chromosome (which will usually not have a *P* element at the same site) is used as a template, in which case the excised *P* is not replaced—this event appears instead as a precise excision event.

The same mechanism is used by *Minos* elements of *Drosophila hydei* (Arcà et al. 1997), *Tc1* elements of *Caenorhabditis elegans* (Plasterk 1991, Plasterk and Groenen 1992), the bacterial transposon Tn7 (Hagemann and Craig 1993), and presumably many other DNA transposons.

(2) For *Activator* (*Ac*) elements of maize, movement is usually associated with loss of the element at the donor site; this and other observations have suggested an alternative mechanism for increase in copy number (Fig. 7.3). Again, elements are excised at some point between DNA replication and mitosis. The resulting gap is then usually closed by joining the broken ends back together without using the sister chromatid as a template for repair. Analysis of empty donor sites indicates that excision is precise only 10% of the time (Schwarz-Sommer et al. 1985). The element then inserts elsewhere in the genome. Most of the time (62% in one study) it transposes to an unreplicated segment of the genome, and so the DNA replication fork passes through it a second time and there is an increase in copy number (Fedoroff 1983). Most of the other times it inserts into the sister chromatid of the donor, and there is no increase in copy number.

It is not yet clear why the gap left by an excised *Ac* element is repaired simply by ligating the 2 ends rather than by recombinational repair, as in *P* elements. Repair of *Mu* element excision events in maize appears to occur by end-joining in somatic tissues but by recombinational repair in germline tissues (Walbot and Rudenko 2002). It is not known if this reflects an adaptation by the element or some tissue-specific difference in how maize repairs its DNA. Yamashita et al. (1999) note that some elements with little recombinational repair, such as *Ac* of maize and *Tam3* of *Antirrhinum majus*, have complex hairpin structures at their ends that may prevent them from participating in recombinational repair. The function of these hairpin structures is not known, though perhaps they are structures recognized by the transposase. Presumably their function is important, for, at least according to the hypothesis, their presence substantially reduces the increase in copy number associated with transposition. Another possible factor is the type of ends left after excision: *Ac* elements are thought to leave covalently closed hairpin loops in the flanking DNA (Kunze and Weil 2002), which may be less prone to recombinational repair than the free ends left by *P* element excision.

Recently an entirely new class of DNA transposons, called helitrons, has

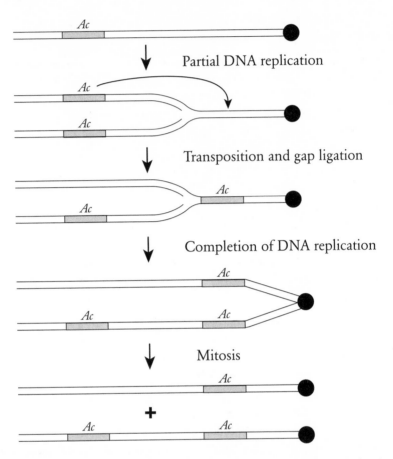

Figure 7.3 Increase in copy number of *Ac* elements by jumping across the replication fork. *Ac* elements preferentially move from replicated DNA to unreplicated DNA, and so a replication fork passes through an element twice in a single cell cycle. After cell division, both daughter cells have the new insert, and one of them also has the original insert. Here the new insert is on the same chromosome as the donor, but it can be anywhere in the genome. Solid circle represents the centromere.

been discovered from sequence analysis of the *Arabidopsis*, rice, and *C. elegans* genomes (Kapitonov and Jurka 2001). These elements encode 2 or 3 proteins and do not have inverted repeats at their ends. Based on their similarity to certain bacterial elements, the authors hypothesize that helitrons transpose by "rolling-circle replication" in which replication is an inherent part of transposition (essentially, 1 strand of the transposon is transferred to the target site, the other strand remains at the donor site, and both single strands are replicated to give double-stranded DNA; for further details, see Garcillán-Barcia et al. 2002). Homologs have also been found in selected vertebrates, insects, and fungi (Poulter et al. 2003). Some other bacterial transposons move by yet another mechanism in which replication is an inherent by-product of transposition (replicative transposition, e.g., *Mu*, *Tn3*, and, sometimes, *Tn7;* Craig et al. 2002). However, similar transposons have not yet been found in eukaryotes, perhaps because transpositions by such elements often result in chromosomal rearrangements that are strongly selected against in sexually reproducing populations.

LINEs and SINEs

The second main class of transposable elements are the LINEs (long interspersed nuclear elements, also known as retroposons or non-LTR retrotransposons). LINEs transpose via an RNA intermediate. LINEs encode a multifunctional enzyme with domains for DNA binding, DNA cleavage, and reverse transcription of RNA into DNA. Many LINEs also encode a second protein that binds RNA, but its function is not yet clear (Seleme et al. 1999). Retrotransposition is thought to occur by the following steps (Luan et al. 1993, Boeke 1997; Fig. 7.4). First, an element is transcribed into RNA. Unlike most host genes, which have their promoter(s) upstream of the transcription start site, many LINEs have an internal promoter. By carrying its own promoter, the element increases the probability that it will be transcribed regardless of where it happens to land in the genome. The RNA then moves to the cytoplasm, where it is translated to make its 1 or 2 proteins. As they are being made, or shortly thereafter, the protein(s) bind to the very mRNA molecule from which they are being translated. After translation, the protein-RNA complex moves back to the nucleus, and the protein cuts a single strand of the host genome at the point of insertion.

Most LINEs can integrate at many different places in the genome (with a preference for A-T rich regions), but a substantial minority of LINEs are site-specific and insert themselves only into particular places in multicopy host genes (Box 7.1). In either case, the exposed 3′ end of host DNA is used to prime reverse transcription of the RNA into DNA. (Like normal host DNA polymerases, reverse transcriptase cannot synthesize a complemen-

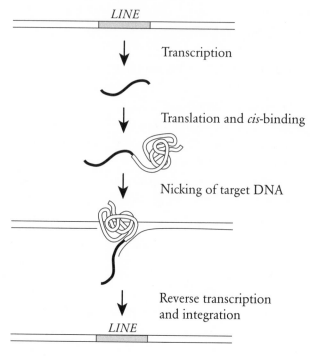

Figure 7.4 Transposition of LINEs via an RNA intermediate. A LINE is transcribed into RNA and then translated into 1 or 2 multifunctional proteins that bind to the RNA template from which they were synthesized, carry it to the nucleus, nick the host DNA, and then use the exposed 3′ end to initiate reverse transcription (target-primed reverse transcription). After reverse transcription and synthesis of the second strand of DNA, all the loose ends are ligated to the host genome.

tary strand *de novo* but must instead have something preexisting that it then extends.) Reverse transcription starts at the 3′ end of the mRNA and works up to the beginning of the transcript. Very often it does not get all the way to the front, and a 5′ truncated element (with the front part missing) ends up being inserted. After reverse transcription (making a hybrid RNA-DNA molecule with the DNA strand attached at one end to the host genome), the other strand of host DNA is cut and somehow the second strand of the LINE element is synthesized, replacing the RNA with DNA, and all loose ends get ligated to the host genome to create an integrated LINE element.

BOX 7.1

Site-Specific LINEs

Some LINEs insert only at particular sites in the host genome. For example, *R1* is a LINE in *D. melanogaster* that inserts specifically at a particular nucleotide position in the 28S rRNA genes; *R2* is another LINE that inserts at another position 74bp upstream of *R1* (Eickbush 2002). Either insert renders the rDNA gene nonfunctional, and the hosts (and LINEs) can persist only because there are typically hundreds of copies of the rDNA gene per individual, only a fraction of which are disrupted (in most species, 5–20%, in some species up to 50%; Eickbush 2002). These elements are widespread in all lineages of arthropods. Site-specific LINEs have also been found in spliced leader RNA (SL RNA) genes of trypanosomes and nematodes, in tandem TAA repeats of some insects and vertebrates, and in the pentanucleotide repeats at the telomeres of *Bombyx mori* (Eickbush and Malik 2002). Site specificity is achieved by the protein recognizing and nicking only a particular DNA sequence. Amusingly, the site-specific LINEs also include *Tx1* elements of *Xenopus* frogs, which specifically target a DNA transposon *Tx1D* (Christensen et al. 2000). And, somewhat less specifically, *Zepp* elements of *Chlorella* algae have a tendency to insert into preexisting copies of themselves (Higashiyama et al. 1997). Transposable element wars!

How does selection maintain site specificity? Presumably mutations arise that relax this sequence specificity, allowing the element to integrate at many more places in the genome. Why have such mutants not spread and driven the site-specific elements extinct? Site-specific elements may have several advantages (Eickbush and Eickbush 1995, Eickbush 2002). First, inactivating one of hundreds of copies of a host gene may cause less harm than inserting more or less randomly in the genome, particularly in small genomes. Second, randomly placed inserts run the risk of getting into sites where they cannot be expressed, or where their expression is not appropriately regulated. Third, accidental ectopic recombination between nonallelic copies is less deleterious for elements in a tandem array like the rDNA cluster than for copies randomly placed in the genome. It would be interesting to investigate more formally the circumstances under which such advantages could compensate for the release from competition for sites that would accrue to a randomly inserting mutant. It is interesting that the common ancestor of all extant LINEs was itself probably site-specific, and that site specificity has been lost at least once and probably also regained at least once (Malik et al. 1999, Eickbush and Malik 2002).

Several important features distinguish LINE transposition from that of DNA transposons. First, mutations that occur during the copying process happen to the element at the new site, not the one at the ancestral site. (In a public lecture, M. Singer compared LINEs to faxes, in which the original stays at home and a potentially degraded copy goes elsewhere.) Second, excision is no part of the transposition process, which means that once acquired, LINEs are unlikely to be lost from a site. Finally, and most importantly, the *cis* activity of LINEs—the fact that the reverse transcriptase protein predominantly uses as template the very same mRNA molecule from which it was translated—greatly simplifies the evolutionary dynamics of these elements, as we shall see throughout this chapter. For example, defective LINEs are created with great frequency, due to truncations and point mutations (transcription and reverse transcription being significantly more error-prone than DNA replication); but once created, they are much less likely to replicate again than are functional elements. The same need not be true of DNA transposons.

That said, the *cis* preference of LINE proteins for their own mRNAs is not absolute, and SINEs (short interspersed nuclear elements) are a subclass of transposable elements that do not encode any proteins themselves but instead have evolved to parasitize the LINE retrotranspositional machinery. Many SINEs are chimeric elements, with a 5′ region derived from a host tRNA gene, a middle region of unknown origin, and a 3′ tail derived from the 3′ tail of a LINE, the part thought to be recognized by the protein (Okada et al. 1997, Ogiwara et al. 1999). *Alu* elements of humans are an exception, being derived from the genes for 7SL RNA.

LTR Retroelements

The final class of transposable elements are the LTR (long terminal repeat) retroelements. These are more complex than the other 2 classes of transposable elements, typically encoding 2 structural proteins (capsid and nucleocapsid) and 3 enzymes (protease, reverse transcriptase [including RNaseH], and integrase). The reverse transcriptase is homologous to that of LINEs and the integrase is homologous to some transposases, suggesting that LTR retroelements may have originated as a chimera between the other 2 classes of transposable elements (Eickbush and Malik 2002). The 2 structural proteins are cleavage products of a single polyprotein encoded by a single ORF (*gag*), and the 3 enzymes are cleavage products of a polyprotein encoded by a second ORF (*pol*). While there is some variability among specific elements (Levin 2002, Eickbush and Malik 2002), the basic outline of the replication cycle is as follows (Fig. 7.5; for more details see Coffin et al. 1997):

First, the element is transcribed into RNA. This begins at a site near the middle of the upstream LTR and ends downstream of the corresponding point in the downstream LTR. Transcripts can then be translated into protein (a "somatic" fate) or encapsidated, reverse transcribed, and inserted back into the genome (a "germline" fate). A particular RNA molecule cannot act simultaneously as a template for protein synthesis and as a template for DNA synthesis. We examine the selection pressures optimizing the decision to be one or the other in Box 7.2. After translation, the gag proteins assemble into a capsule or particle containing the genomic RNA, the enzymes, and a tRNA. There is a specific "packaging signal" recognition sequence in the genomic RNA that interacts with the gag proteins, and only RNAs with this sequence typically get packaged. Once the particle has assembled, the reverse transcriptase becomes active and copies the RNA into a double-stranded cDNA, using the tRNA as a primer to initiate synthesis. There are 2 RNA genomes packaged per particle, and (at least in retroviruses) the polymerase can jump between them during replication, thus

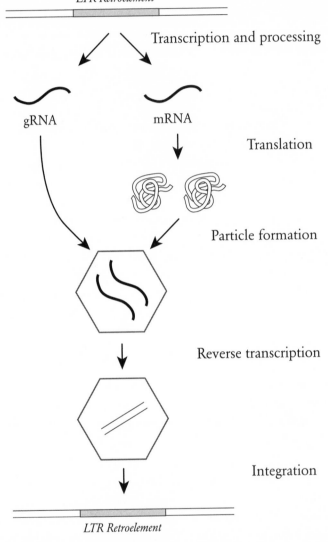

Figure 7.5 Transposition of LTR retroelements via an RNA intermediate. RNA transcripts can have 2 different fates. First, they can act as normal mRNA and therefore as templates for protein synthesis (a somatic fate). Typically 5 or 6 different proteins are synthesized, including those for capsid formation, reverse transcription, and integration. Alternatively, RNA transcripts can act as gRNA, which is encapsidated and reverse transcribed into cDNA, which in turn is integrated into the host genome (a germline fate). Note that gRNA from one element may get copied by proteins made from the mRNA of a different element.

BOX 7.2

The Two Fates of LTR Retroelement Transcripts

A particular RNA transcript of an LTR retroelement cannot simultaneously act as a template for translation (i.e., as an mRNA) and be encapsidated for reverse transcription (i.e., as a genomic or gRNA). Sometimes these RNAs even have different structures: for example, the gag protein of *copia* (and all retroviral env proteins) is translated from spliced mRNAs that do not have the full complement of genes, nor the recognition sequence for encapsidation (the "packaging" signal; Boeke and Stoye 1997). Some members of the DIRS1, Ngaro, and Penelope retroelement families have introns, which presumably are spliced out of mRNAs but not out of gRNAs. In some retroviruses the two fates appear to be mutually exclusive even without differential splicing, as if the passage of a ribosome over the RNA made it incapable of being encapsidated (Levin and Rosenak 1976, Sonstegard and Hackett 1996). And even if a particular RNA molecule can act as a template and be encapsidated, both cannot happen simultaneously.

The proportion of transcripts that are mRNA versus gRNA is likely to be under the influence of *cis*-acting control regions in the RNA (that, for example, affect their affinity for spliceosomes, or ribosomes, or the gag protein). What, then, is the optimal ratio of the two types? Nee and Maynard Smith (1990) propose a simple model and derive the optimal (or evolutionarily stable) proportion of mRNA as $p = 1/(n + 1)$, where n is the average copy number of the element in the genome. The precise quantitative result depends on the simplifying assumptions used, but the qualitative result is likely to be general: the higher the copy number, the lower the proportion of transcripts that should be templates for protein synthesis. This is because elements will be increasingly selected to rely on proteins made by other elements. Paired comparisons of endogenous and exogenous retroviruses in vertebrates would be an ideal system in which to test this prediction: exogenous viruses should behave as if $n=1$ and so should have a higher ratio of mRNA:gRNA than endogenous viruses. Nee and Maynard Smith also note that competition among retroelements to be genomic RNAs may in part explain why transcripts are often abundant but transposition rates low. For example, *copia* transcripts can comprise as much as 3% of the polyadenylated RNA in some *Drosophila* cell lines, and yet the transposition rate is less than 10^{-3} per element (Rubin 1983). *Ty1* RNA is 5–10% of total polyadenylated RNA in haploid yeast cells (Roeder and Fink 1983), but again transposition rates are low.

synthesizing recombinant cDNA (Temin 1991). The LTRs are reconstituted at this step, in a way that they are always identical at the time of insertion (unless the 2 RNAs have different LTRs and there has been recombination; Telesnitsky and Goff 1997). Particle assembly and reverse transcription may occur in the cytoplasm or in the nucleus (Boeke and Stoye 1997). After reverse transcription the particle falls apart and the integrase recognizes the 2 ends of the cDNA and inserts them into the host genome.

At least in yeast, there is an alternative pathway for the cDNA to get into the genome: the double-stranded cDNA can recombine with a preexisting copy of the element, either converting the old sequence to the new one or (if the LTRs interact) generating a tandem pair of elements. In 1 experiment with *Ty5* elements of baker's yeast, about 7% of insertions were by homologous recombination with a single preexisting copy, of which 70% resulted in gene conversion and 30% in tandem elements (Ke and Voytas 1997; see also Sharon et al. 1994). It is not clear whether this alternative pathway is an accident or an evolved adaptation.

Though LTR retroelements are like LINEs in that they use reverse transcriptase, there are some important differences between the 2 classes. First, the existence of the LTRs means that there can be recombination between the 2 ends of an element, which will delete the element, leaving only a single copy of the LTR (Fig 7.6). Thus, though excision is not part of the trans-

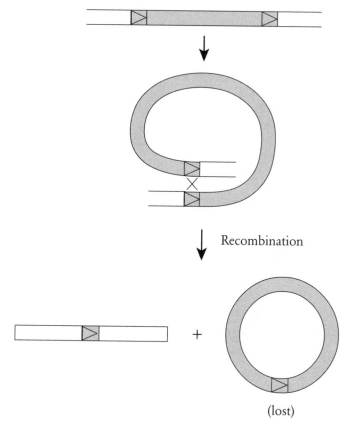

Figure 7.6 Looping out of LTR retroelements. Recombination between the 2 long terminal repeats of an element leads to deletion of the element, with only a single copy of the repeat left behind. Excision rates are usually less than insertion rates.

position pathway, it is more common than for LINEs. Second, elements need not encode any functional proteins to transpose, as these can be supplied by another element, in *trans* (Xu and Boeke 1990). But they do require a number of *cis*-acting sequences in order to be transcribed, packaged, reverse transcribed, and integrated, most of which are near the ends (in or near the LTRs; Vogt 1997). In all these respects, LTR retroelements are more like DNA transposons than LINEs.

There are 2 main subclasses of LTR retrotransposons, which differ slightly in the arrangement of their coding regions: the *Ty1-copia* elements and the

Ty3-gypsy elements. Both are widely distributed among fungi, plants, invertebrates, and "lower" vertebrates, but appear to be largely absent from birds and mammals. Also related to these are the vertebrate retroviruses, including HIV. Retroviruses encode, at minimum, 1 extra structural protein, *env*, needed to construct the viral envelope that allows transfer between cells; more complex retroviruses such as HIV may encode up to 6 more small proteins as well (Vogt 1997). In vertebrates of all sorts, there is a third class of LTR retrotransposons, the endogenous retroviruses (ERVs), which are retroviruses that have integrated into the germline and resumed their ancestral habit of being transposable elements. Active ERVs have been found in baboons, mice, and chickens (Boeke and Stoye 1997). At least 6 other lineages of LTR retroelements have acquired *env*-like genes, though it is not yet clear whether any of them are regularly infectious between individuals (Eickbush and Malik 2002). The *env* gene of *gypsy* elements of *Drosophila melanogaster* allows them to transfer from follicle cells to oocytes in females (Song et al. 1997). In the laboratory, *gypsy* elements can also be transferred between individuals: flies lacking *gypsy* can acquire it if they grow up in a medium of homogenized pupae that contain the element (Kim et al. 1994).

In addition to these well-known families, there are a number of other less well studied families, including the BEL, VIPER, DIRS, and Ngaro families (Vázquez et al. 2000, Goodwin and Poulter 2001, 2004). Members of the latter 2 families encode tyrosine recombinases (homologous to bacterial genes and the FLP recombinase of yeast) instead of an integrase, and they are thought to transpose via a circular intermediate rather than a linear one. Some also have introns, as do members of the even more distantly related *Penelope*-like elements (Arkhipova et al. 2003).

Population Biology and Natural Selection

Transposable elements accumulate by copying themselves around the genome rather than by sticking to a single locus, and there can be tens or hundreds of active copies of an element in a single individual. This complicates our thinking about their spread and evolution. Considering just a single insert, there can be a trade-off in how active it will evolve to be. The more active it is, the more copies it will make per generation, but also the more harm it will cause the host, and so the faster it will disappear from the population. Moreover, the evolutionarily stable level of activity may depend upon the number of elements in the genome: chromosomal repair systems may not be able to cope with 50 DNA transposons excising simultaneously and, were that to happen, the host cell would die. Having so many copies of an element in a single individual also means that the proteins encoded by one element may well end up copying another element, and so there is the possibility of social interactions between elements, particularly parasitism. We will build up to these complications slowly, first considering the case of a single homogeneous population of elements and their spread through a host gene pool.

Transposition Rates Low But Greater Than Excision Rates

Transposable elements have evolved sophisticated mechanisms for increasing in copy number in the genome, but how often does transposition actually occur? There is no simple answer to this question, except to say that transposition rates are low (compared to per capita replication rates for other selfish genetic elements) and highly variable. The best data come from lab experiments with *D. melanogaster.* In one study, for *P* elements in "dysgenic" crosses, in which the male has many elements and the female has none, the transposition rate was about 0.25 (i.e., 1/4 of the elements transposed each generation); while in nondysgenic crosses in which both males and females have many elements, the rate drops to about 0.005 per element per generation (Eggleston et al. 1988). If there are 50 elements in the genome, this works out to about 1 transposition in the whole genome every 4 generations. In a study of *P* elements in inbred lines, there was only 1 net transposition in about 350,000 opportunities, for a rate of 2.9×10^{-6} (Domínguez and Albornoz 1996). Other transposable elements in these same inbred lines also had low transposition rates (*hobo:* 0/305,760; *FB:* 2/ 291,200 = 6.9×10^{-6}). In another study of inbred lines, no transpositions were observed for either *hobo* or *FB4* out of 28,000 and 45,500 opportunities, respectively (95% upper bound on transposition rates of 1.1×10^{-4} and 6.6×10^{-5}; Nuzhdin and Mackay 1995). In yet another study, of outbred lines, *hobo* moved at an appreciable rate, though the data did not allow quantification (Harada et al. 1990).

Transposition rates of LINEs and LTR retroelements (per element per generation) are not dissimilar: in two *Drosophila* experiments, the rates were typically about 10^{-4}, with some variation (at least an order of magnitude) among families (Nuzhdin and Mackay 1995, Maside et al. 2000; see also Vieira and Biémont 1997, Suh et al. 1995). Though small, this rate is substantially greater than the rate of excision, which averaged about 10^{-6}. The few excisions that were observed were of LTR retroelements and presumably arose by LTR-LTR recombination; there is no similar mechanism known by which LINEs can be deleted. Excision rates of ERVs in mice are similar, at about 4×10^{-6} (Seperack et al. 1988, Frankel et al. 1990). For *Ty1* elements of baker's yeast, excision rates are about 2×10^{-5} (Winston et al. 1984). Surprisingly, estimated transposition rates are lower than this (3×10^{-7} to 1×10^{-5}; Curcio and Garfinkel 1991). These are laboratory measurements, and we suspect that transposition rates in nature must be greater than those in the lab.

Though small compared to those of other selfish genes, these rates of accumulation are still sufficient to have a dramatic effect on evolutionarily trivial timescales. For example, with a transposition rate of 10^{-4}, the abundance of elements is expected to double in about 7000 generations ($\approx$600 years for *D. melanogaster*), or increase 10-fold in 23,000 generations ($\approx$2000 years), if unopposed by selection (assuming all new inserts are functional, calculated using $y = (1+t)^g$, where y is the proportional increase in copy number, t is the transposition rate, and g is the number of generations).

Natural Selection on the Host Slows the Spread of Transposable Elements

In reality, the spread of a transposable element through a host population will be somewhat slower than these calculations suggest, as effects on the host are usually deleterious, and so natural selection on the host population will work against them. Again, the best data are for *Drosophila*. High rates of *P* element transposition (*P* element "dysgenesis") cause chromosome breakage and rearrangements, which in turn cause reduced fertility (Fig. 7.7) and increased embryonic lethality. Moreover, some 5–10% of insertions cause recessive lethal mutations (Cooley et al. 1988). The nonlethal inserts also appear to reduce fitness: in one study of *P* element insertions on the third chromosome, viability in the laboratory was reduced about 5.5% per insert when heterozygous and 12.2% when homozygous (Mackay et al. 1992). In another study, viability was reduced about 1.4% per insert on the X chromosome when hemizygous (i.e., in males; Fig. 7.8).

Effects on host fitness have also been investigated for *mariner* elements of

MALE   FEMALE

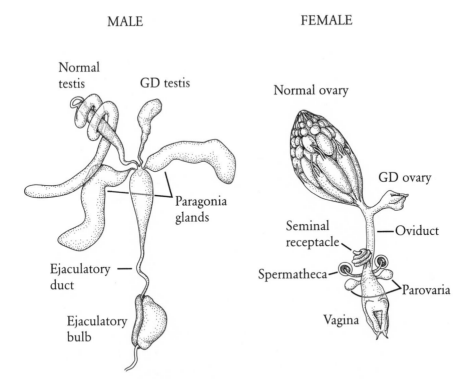

Figure 7.7 Gonadal dysgenic (GD) sterility resulting from high levels of *P* element activity in *D. melanogaster*. For both the male and female reproductive tracts, the left gonad is drawn from a normal adult and the right one from a GD sterile adult. Note that the gonad is greatly reduced in both sexes. Adapted from Engels (1989).

D. simulans. Flies with just 2 mobile elements die about 4 days earlier in the lab than those with none (57.6 vs. 61.4 days; Nikitin and Woodruff 1995).

Again, LINEs and LTR retroelements are not dissimilar. High frequencies of *I* element transposition (a LINE of *Drosophila melanogaster*) are associated with elevated rates of dominant lethality and other mutations (Finnegan 1989). High activity of *Penelope* elements (LTR retroelements of *D. virilis*) are thought to cause gonadal dysfunction, chromosomal rearrangements and nondisjunction, male recombination, and increased mutation rates (Evgen'ev et al. 1997). Some of these effects appear to occur because *Penelope* activity mobilizes other transposable elements, including *Ulysses* (another LTR retroelement), *Helena* (a LINE), and *Paris* (a DNA transposon).

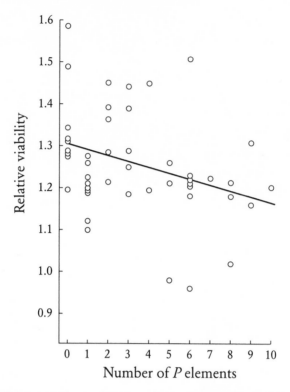

Figure 7.8 Decline in viability as a function of increasing numbers of *P* elements in *D. melanogaster*. *P* elements were mobilized in the lab to generate X chromosomes with differing numbers of inserts. Each additional insert reduces male survival by about 1.4%. Due to hemizygous expression, inserts on the X are expected to be more harmful than those on the autosomes. Adapted from Eanes et al. (1988).

The deleterious nature of transposable elements is also apparent in other ways. About 50% of visible mutations in *D. melanogaster* are due to transposable element insertions (Finnegan 1992). In mice, of 160 spontaneous mutations at 86 different loci, 4 were due to *L1* insertions and 13 to various LTR retroelements (*IAPs*, *ETns*, and *MaLRs*), for a total of about 10% of mutations (Kazazian and Moran 1998). In humans, the number is much lower: about 0.1% of new mutations that cause genetic disease are *L1* insertions and another 0.1% are *Alu* insertions (Kazazian 1999). Transposable elements can also be harmful in causing ectopic recombination between elements in nonhomologous locations. Recombination between ectopic copies has been observed in the lab for many families of transposable elements in *D. melanogaster* (Lim and Simmons 1994). In humans, 0.2% of new mutations that cause disease are due to ectopic recombination between *Alu* elements (Deininger and Batzer 1999). In addition, ectopic recombination between *Alu*s in somatic cells has been implicated in a number of cases of cancer. Finally, formation of transpositional intermediates such as RNA transcripts and retroelement particle formation is also likely to put a burden upon the physiology of the host cell. There have been debates about which of these various sources of harm is the more important, but with no clear resolution (Hoogland and Biémont 1996, Charlesworth et al. 1997, Biémont et al. 1997, Nuzhdin 1999).

Rapid Spread of *P* Elements in *D. melanogaster*

We saw in Chapter 1 that the ability of transposable elements to spread through a population has been dramatically—and fortuitously—demonstrated for *P* elements in *D. melanogaster*. In that species, the elements invaded and established themselves in the gene pool in just a few decades, having been introduced somehow from *D. willistoni*, a distant relative (Engels 1992). Most strains of *D. melanogaster* isolated from the wild before 1960 do not contain *P* elements, while strains collected since that time are increasingly likely to carry *P*s, and now they appear to be present in all natural populations (Kidwell 1983, Anxolabéhère et al. 1988; see Table 1.1). This ability to invade and establish when rare has also been observed in lab populations (Daniels et al. 1987, Kidwell et al. 1988, Good et al. 1989; Fig. 7.9). As expected, the invasion of the *D. melanogaster* gene pool was accompanied by an increased mutation rate (rev. in Quesneville and Anxolabéhère 1998). In Russian populations, the wave of *P* element invasion was associated with a phase of high mutability, and in Japanese and Korean populations the arrival of *P* elements was associated with a transient increase in the detrimental load per chromosome.

Further insight into the invasion comes from analyzing the frequencies of *P* elements at particular sites in the genome. Surveys of natural populations show that, at any one site, insertions tend to be rare (Ronsseray and Anxolabéhère 1986, Ajioka and Eanes 1989, Biémont et al. 1994; Fig. 7.10). This indicates that *P* elements have not spread because some inserts are beneficial and increased in frequency by natural selection. Rather, most inserts appear to be deleterious (though neutral or very small beneficial effects can-

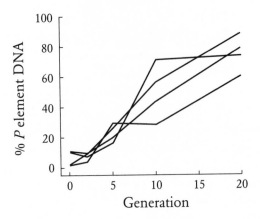

Figure 7.9 Spread of *P* elements in experimental populations of *D. melanogaster*. Populations were started by introducing flies with *P* elements into a population without. Two different donor strains were used, and 2 different starting frequencies (2% and 20%), for a total of 4 treatment combinations. Y axis is the amount of *P* element DNA relative to a control strain. The mean of 2 replicate populations is shown for each treatment. Note the speed of the invasion. Adapted from Good et al. (1989).

not be ruled out because there has not been sufficient time for such mutations to increase in frequency). There is one notable exception to the low site occupancy, namely, a high frequency of insertions at sites near 1 tip of the X chromosome. This exception appears to prove the rule, for these insertions have apparently been selected because they repress *P* element activity (Ronsseray et al. 1991, Stuart et al. 2002)! The fact that repressors are positively selected is further evidence that *P* elements are parasitic. It would be interesting to know whether this element has retained the ability to transpose, or whether it has lost that ability and completely gone over to "the other side." Because transposition is costly, repressors that are unable to transpose should be even more favored by natural selection on the host population.

Net Reproductive Rate a Function of Transposition Rate and Effect on Host Fitness

Transposable elements can invade and persist in a host gene pool even if every single insert is unconditionally harmful to the host organism and re-

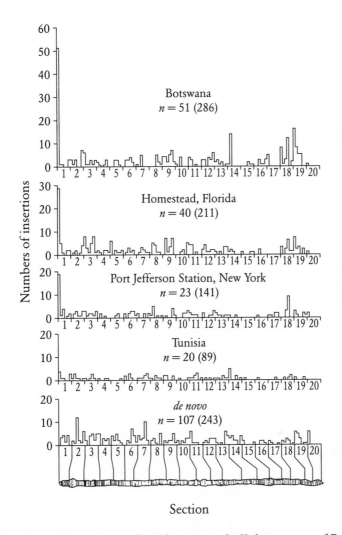

Figure 7.10 Site-occupancy profiles for *P* elements on the X chromosome of *D. melanogaster*. The X axis represents different cytologically detectable sections of the X chromosome, from one end to the other, and the Y axis shows the number of chromosomes carrying a *P* element in the different sections. Data from 4 different populations are shown, as well as for inserts generated in the lab *de novo*. *n* is the total number of X chromosomes studied in that population, with the total number of inserts in parentheses. Note that for all sections of the X except the tip, *P* elements are present in only a minority of chromosomes. The high frequency of inserts at the tip of the X is not due to a high rate of insertion, indicating it is due to positive selection. Adapted from Ajioka and Eanes (1989).

duces its fitness. Deleterious inserts are usually driven extinct by natural selection on the host, but as long as an insert transposes to a new location on
average more than once before this happens, the average number of copies
per genome will increase. Consider the following calculation. A new insertion occurring in a single individual that reduces host fitness by an amount s
is, in the next generation, expected to be in $1-s$ individuals, in the following generation, $(1-s)^2$ individuals, and so on. The total expected number of
individuals carrying a new insertion before it goes extinct is then $1 + (1-s)$
$+ (1-s)^2 + (1-s)^3 + \ldots = 1/s$ (Crow 1957, Nei 1971). For example, an insertion that reduces fitness by 10% will on average be found in 10 individuals before it goes extinct; an insertion that reduces fitness by 0.01% will be
found in 10,000 individuals. Each of these individuals will on average have
fitness $1-s$, and so the total number of opportunities for the element to
transpose will be $(1/s)(1-s)$. If the transposition rate is t, the net reproductive rate of the element will be:

$$R = t\,(1/s)\,(1-s)$$
$$= t\,(1/s-1)$$
$$\sim t/s \text{ (assuming } s<<1).$$
Eqn. 7.1

This is the average number of daughter insertions that a new insertion will
give rise to before going extinct. If R is less than 1, the element will decline
in frequency in the gene pool; and if R is greater than 1, it will increase in
frequency. At equilibrium, R=1. Note that for the equilibrium to be stable,
R must decrease as copy number increases. That is, either the transposition
rate per element must decrease or the harm caused per element must increase with increasing copy number. Charlesworth and Charlesworth (1983)
present a more detailed model of transposable element population dynamics, and come to the same conclusion. Note also that different insertions
will have different effects on host fitness (i.e., different values of s), and it is
the harmonic mean of these that is important. An element can persist even
if it produces mostly lethal insertions ($s = 1$), as long as it also produces
some of very small effect.

We noted earlier that the transposition rate for P elements in nondysgenic
laboratory crosses is about 0.005. If this rate is typical of natural populations, at equilibrium inserts should reduce host fitness by about the same
amount (i.e., 0.5%). This estimates either the harmonic mean effect of new

inserts, or the arithmetic mean effect of a random sample of inserts segregating in natural populations. This value is less than the 1.5–5.5% reduction observed for new inserts in the lab (see earlier). Presumably at least part of this discrepancy is due to the difference between harmonic and arithmetic means.

If there are 50 P elements in the genome transposing at a rate of 0.005 per generation, this works out to about 1 transposition per genome every 4 generations, and fitness is reduced on average by about 3.5%/4 = 0.875% every generation due to P element transposition. This value is not too different from the estimated total reduction in fitness due to all new mutations (≈ 1 to 2%; Burt 1995, Charlesworth et al. 2004). Finally, if s for extant insertions is 0.005 and the inserts interact multiplicatively, the fitness of flies with 50 inserts will be $(1-0.005)^{50} = 0.78$ relative to those with none. That is, P elements alone would be reducing population mean fitness by about 20%. Allowing for epistatic interactions in the deleterious effects, this load may be reduced by a factor of 2, to 10% (Charlesworth and Barton 1996). But these are unusually active elements. Taking all other families of elements together, if we assume there are 500 inserts, each transposing at a rate of 10^{-4}, and if we allow for epistatic interactions, the mean fitness of the population will be reduced by 2–3% (Charlesworth and Langley 1989).

The net rate of reproduction R is also a measure of the fitness of a particular transposable element variant. Variants with smaller than average R will decline in frequency, those with larger R will increase in frequency, and the population of transposable elements will evolve to maximize R.

Reducing Harm to the Host

Equation 7.1 captures the intuitively obvious notion that the fate of a transposable element depends not only on the transposition rate but also on the fitness effects on the host. We should therefore expect transposable elements to have evolved adaptations for reducing the harm done to the host. One such adaptation is germline specificity: transposition in the somatic tissue of an organism is harmful to the host without benefiting the transposable element, and so is selected against (Charlesworth and Langley 1986). Many elements in *D. melanogaster* have evolved such tissue specificity. For P elements, transposition is suppressed in somatic cells because the third intron is not spliced out of the P transcript, and so there is no functional transposase. Artificially created *P* elements that are missing the intron produce functional transposase and transpose in somatic cells, thereby significantly reducing host lifespan (Woodruff and Nikitin 1995). Mutations in the region 12–31bp upstream of the intron can abolish the germline specificity of the splicing, suggesting that a protein normally binds to this region in somatic cells and prevents splicing (Chain et al. 1991, Siebel et al. 1995). Another DNA transposon, *hobo,* also transposes only in the germline, but the mechanism is completely different, being based on tissue-specific transcription (Calvi and Gelbart 1994). Similarly, *I* elements, which are LINEs, are transcribed only in ovaries and transpose only during female meiosis (McLean et al. 1993, Udomkit et al. 1996). Among LTR retrotransposons, *gypsy* elements are expressed only in the follicle cells at the anterior end of the developing oocyte, and *ZAM* elements only in those at the posterior end; both then invade the oocyte before the vitelline membrane forms (Song et al. 1997, Leblanc et al. 2000). Thus, within a single species, representatives of all 3 main classes of transposable element have independently evolved mechanisms for being inactive in the somatic cells, and so of reducing the harm done to the host. The *P* element example is interesting because the intron adds about 200bp (7%) to the length of the element, which may impose some small cost in reduced rate of successful transposition in the germline (though there are also 2 other introns with no known function). Nor is this behavior restricted to fruit flies: for example, *L1* LINEs of humans are also active predominantly in the germline (Moran and Gilbert 2002).

Another way to limit the damage caused by transposition is to insert preferentially into "safe" sites in the genome. We have already noted that some LINEs target a particular DNA sequence, and many other transposable elements show regional specificity. Such targeting is well developed in the LTR retroelements of the baker's yeast *S. cerevisiae.* This species has a dense genome (ca. 70% coding, compared to 15% in *D. melanogaster* and 1.5% in humans), and so random insertions are particularly likely to disrupt gene function (Boeke and Devine 1998). *Ty1, Ty2, Ty3,* and *Ty4* elements all target the regions upstream of tRNA genes and other host genes transcribed by RNA polymerase III (Kim et al. 1998). In baker's yeast there is usually an intergenic region of 500–2000bp upstream of such genes, larger than most intergenic regions, and so these regions are likely to be safe havens for the elements. At least 2 of these elements, *Ty1* and *Ty3*, are distantly related and use very different mechanisms to target this region, indicating convergent evolution. Interestingly, there may be a cost to this site specificity in reduced rate of transposition, because one explanation for the large intergenic region is that RNA polymerase II transcription is down-regulated there, and *Ty* elements are transcribed by pol II (Bolton and Boeke 2003). *Ty5* elements also show site specificity, in this case for "silent chromatin," including telomeres. This site specificity is not based on a particular DNA sequence, but rather on the host proteins bound to these particular regions of chromosomal DNA, as is also the case for *Ty1* and *Ty3*.

A third adaptation for mitigating harmful effects is the ability to be spliced out of host transcripts, shown to varying degrees for a number of elements (Gierl 1990, Rushforth and Anderson 1996, Lal and Hannah 1999, Kunze and Weil 2002). Surprisingly, there is no record of a transposable element associated with a self-splicing intron or intein. We do not know why this is. Our best hope may be to construct a self-splicing transposable element and see how it behaves.

Transposition Rate and Copy Number "Regulation"

Transposition rates of 10^{-4} are low compared to rates of accumulation for other selfish genetic elements and compared to nucleic acid manipulations that are beneficial to the host (e.g., splicing, recombination). The low rates may reflect relatively successful suppression by the host. Alternatively, they may reflect self-suppression by the transposable elements themselves. That is, perhaps selection on the elements does not act to maximize transposition rates, but to bring them to some intermediate optimum. Equation 7.1 shows that, for such prudence to evolve, an increase in transposable element activity must increase the harm done to the host more than it increases the rate of successful transposition. The number of hosts a particular insertion appears in before going extinct depends on the harm done to the host, which in turn may depend in part on how active the element is. We liken this to being in quicksand: try to move either too slowly or too fast, and you sink; the optimal activity is intermediate. But selection for an intermediate optimum is not inevitable. If most of the harm caused by transposable elements is due to the presence of the DNA, regardless of how active it is (i.e., insertional mutations and ectopic exchange), a reduction in activity may not lead to any substantial reduction in harm, and selection will tend to maximize activity (Charlesworth and Langley 1986).

It is interesting in this regard that mutant elements sometimes have higher transposition rates than wildtype. For example, the transposition rate of *Ty1* elements of yeast, which is normally 10^{-5} to 10^{-7} in the lab, can be increased more than 1000-fold to 0.05 by giving it a stronger promoter (Voytas and Boeke 2002). For *Ty5*, transposition rates can be increased 40-fold by changing just 2 amino acids in gag (Gao et al. 2002). Similarly, changing 1 or 2 amino acids in the transposase from the *Himar1* element of a hornfly *(Haematobia irritans)* can make it several-fold more active (Lampe et al. 1999, Fischer et al. 2001). And changing the codons of mouse LINE-1 elements to those used by highly expressed genes (without changing the encoded amino acid sequence) increases transposition rates more than 200-fold (Han and Boeke 2004). But it is not yet certain whether these various mutations are disrupting some element adaptation for prudent transposition, or whether they are preventing host detection and suppression. It would be interesting to put wildtype and engineered elements into direct competition in a laboratory population.

A closely related issue is what determines transposable element copy number. We saw earlier that in order for there to be a stable equilibrium copy number of transposable elements per genome, either the transposition rate per element (*t*) has to decline with increasing copy number or the deleterious effect of each insert (*s*) has to increase. Presumably at least 1 of these 2 relationships holds for all transposable elements. *P, hobo,* and *I* elements of *D. melanogaster* all show the phenomenon of hybrid dysgenesis, in which progeny from crosses between males who carry the element and females who do not show greatly elevated rates of transposition. This suggests that at low copy numbers the transposition rate (per element) will be higher than at high copy numbers, for at very low copy numbers, when only 1 of the 2 parents will typically have an element, half the time the transposition rate should be high. Indeed, if *I* elements are introduced into a strain, either by crossing or by transgenics, copy number increases over several generations until there are 10–15 per haploid genome, after which there is no further change (Chaboissier et al. 1998, Bucheton et al. 2002). The plateau is hit because their activity is inversely related to copy number. *I* elements from *D. teissieri* behave similarly when introduced into *D. melanogaster,* reaching an equilibrium copy number of 10–15 per haploid genome, despite the fact that in its native species there are typically only 2–4 copies (Vaury et al. 1993). Inverse relationships between transposable element activity and copy number have also been suggested for *mariner* elements of *Drosophila* and *Ac* elements of maize (Lohe and Hartl 1996, Kunze and Weil 2002).

Why do transposition rates decrease as copy number increases? Is this self-regulation by the elements themselves or suppression by the larger host genome, and how is the effect achieved? One can easily imagine benefits to self-regulation, especially because multiple elements attempting to transpose simultaneously may be particularly harmful to the host organism (recall the association in dysgenic crosses between high transposition rates and sterility). Even *in vitro*, the transposition rate of *Himar1* is reported to increase initially with the concentration of transposase, and then to decrease (Lampe et al. 1998), though the transposition rate for the related *Mos1* element from *D. mauritiana* simply plateaus at a high concentration (Tosi and Beverley 2000). The "overexpression inhibition" of *Himar1* could result from a concentration-dependent tendency of the transposase to form inactive aggregates (see also Heinlein et al. 1994). It would be interesting to see whether mutant transposons can be generated that do not show this self-inhibition, and to test whether they are responsive to their own copy number *in vivo*. Other possible mechanisms by which transposable elements could actively control their copy number are described in Box 7.3.

On the other hand, any reduction in the per capita transposition rate at high copy number may have nothing to do with the element itself, and everything to do with the host. Some host factor essential for transposition may be limiting. Or, hosts may be better able to suppress high copy number elements. Eukaryotes appear to have a number of mechanisms that suppress genes according to the number of copies in the genome (e.g., Mipping, Ripping, RNAi, and cosuppression; Galagan and Selker 2004, Schramke and Allshire 2004). In a particularly instructive study, a mutational screen was performed for increased transposition of *Tc1* elements in *C. elegans*. A mutation of large effect was discovered that also elevated the transposition rate of 3 other DNA transposons, and the mutant turned out to be a knockout of a gene involved in RNAi (Ketting et al. 1999; see also Tabara et al. 1999, Vastenhouw and Plasterk 2004). Similar results have been reported for RNAi and transposable elements in *Chlamydomonas* algae (Wu-Scharf et al. 2000). Transposable elements may become easier to detect and suppress

BOX 7.3

Candidate Mechanisms of Copy Number Control: Bacterial Plasmids

Bacterial plasmids actively control their copy number, and mechanistic understanding of how they accomplish this is quite advanced. Because similar mechanisms may be found in transposable elements, we review the 2 main classes of control mechanism (from Summers 1996).

Inhibitor dilution. Plasmids using this mechanism produce a *trans*-acting replication inhibitor whose concentration is proportional to copy number. Such proportionality can be achieved by constitutive synthesis of an unstable inhibitor; often this is an antisense RNA. As copy number increases, the concentration of the inhibitor increases, and replication rates fall, thus stabilizing copy number.

Autorepressor. Plasmids using this mechanism produce a protein that is rate-limiting for replication and that inhibits its own transcription (autorepression). This negative feedback loop means that the concentration of the protein (and therefore total replication rate) is constant, and independent of copy number. A constant total replication rate per cell means that the rate per element is inversely proportional to copy number.

as they become more frequent in the genome and, in this way, equilibrate in number.

One way that some host taxa suppress their transposable elements is by methylating them (i.e., by attaching a methyl group to cytosine residues; Bestor 2003). Methylation usually occurs to Cs followed by Gs (i.e., to CG dinucleotides or, in plants, to CNG trinucleotides, where N is any nucleotide). In principle, then, transposable elements could avoid methylation by losing their CG dinucleotides. Interestingly, *Ac* elements of maize have a deficit of CG dinucleotides in their coding region, but not in the inverted repeats, nor in the first 400bp untranslated region of the transcript (Kunze et al. 1988). Why have these been maintained? Have these been maintained in order to facilitate methylation-induced silencing, thereby allowing self-regulation? An alternative explanation is that elements that are normally methylated will be transiently hemimethylated after DNA replication (before the newly synthesized strand is methylated) and this may be a cue for excision, because *Ac* elements are selected to transpose immediately after DNA replication (Wang and Kunze 1998). It would be interesting to mutate these CG dinucleotides, to see how the behavior of the element is affected. Curiously, while methylation of an *Ac* element is associated with a lack of transcription, it does not interfere with the ability of the element to be transposed if other elements in the genome are transcribed (Kunze et al. 1988). This is surprising–because the transposase is likely to have come from an unmethylated element, we might have expected it to have evolved to recognize only unmethylated elements and thereby increase the probability of transposing the element from which it was derived.

Finally, copy numbers can be stabilized not only by the transposition rate decreasing with copy number, but also by the selective disadvantage of each insert increasing. This is possible because of the increasing chances of ectopic exchange or because of an overload of the cell's DNA repair system (Langley et al. 1988, Nuzhdin 1999). And all of these factors may be operating simultaneously.

Selection for Self-Recognition

When a particular insert is transcribed and translated, and the resulting protein is then involved in copying a different (unrelated) insert, the original insert gains nothing in terms of its own reproduction, but suffers whatever harm the transcription, translation, and transposition cause the host individual. It is worse than neutral. Hence elements will be selected to avoid this effect, and one way to do so is to produce proteins that preferentially copy the insert from which they are derived. LINEs have partly solved this problem by having their proteins bind to the mRNA from which they are derived, but even here the mechanism is not perfect. In humans, new *Alu* element insertions are about as frequent as new *L1* insertions, suggesting that the mechanism is only about 50% effective. In general, selection on LINEs should be to increase this value.

For DNA transposons and LTR retrotransposons, in which this form of *cis* preference is not possible, the same selection pressure could produce a different outcome. In particular, there may be selection for a mutant element that produced a transposase that recognizes only the mutant DNA sequence. This would require 2 genetic changes, in the transposase gene and in the DNA sequences recognized by the transposase. These changes could occur simultaneously but need not. For example, one can imagine that a mutant transposase could be selected because it was less active on the wild-type DNA sequences, and then mutant recognition sequences would be selected that are suitable substrates for the new transposase. Or, recognition sequences with greater affinity for the existing transposase might evolve, followed by a mutant transposase that only recognizes the new sequences. Note, though, that in terms of specificity, the *cis*-acting sequences and the transposases will be selected in opposite directions, the former to be recognized by all extant transposases and the latter selected to recognize only its own sequence.

This issue of self-recognition has implications for defining transposable element families (akin to defining species of more familiar organisms). Due to their *cis*-activity, different LINE inserts do not directly interact during transposition, and so the only way to group inserts into families based on function is to ask to what extent their future proliferation is limited by the same factors—in other words, to what extent they share the same genomic niche. The problem is analogous to defining species in wholly asexual taxa (Barraclough et al. 2003). For DNA transposons and LTR retrotransposons, there is the additional possibility of grouping inserts according to whether they are recognized by the same transposases.

Defective and Repressor Elements

If a particular transposable element with certain fixed properties is able to invade a gene pool when rare, as defined in the simple case by $R > 1$, it is *ecologically* stable. This does not ensure the long-term persistence of the transposable element in the face of mutations that may arise and be selected for—that is, the element may not be *evolutionarily* stable. Recall, for example, the DNA transposon's modus operandi: it contains a gene that is transcribed into RNA and then translated into protein in the cell's cytoplasm. The transposase returns to the nucleus, recognizes a particular DNA sequence, and is somehow involved in excising that sequence and inserting it else-where in the genome in a way that leads to a net increase in copy number. As we have seen, there is no way for the transposase to recognize which DNA sequence coded for its production. This means the system is open to parasites—DNA sequences that do not code for a transposase, but that still have the *cis*-acting sequences necessary to be recognized and acted on by transposase encoded by other elements. That is, defective or nonautono-mous elements can be complemented in *trans* and can accumulate in the ge-nome.

Such nonautonomous parasitic elements appear to be very common. For example, in a typical *D. melanogaster* strain with *P* elements, only one-third of the copies are full length and functional, whereas two-thirds are nonau-tonomous elements that have suffered 1 or more deletions but still have the *cis*-acting sequences necessary to be transposed (Engels 1989). In maize, *Ac* elements are functional transposons, and *Ds* elements are nonautonomous but mobilized by the transposase of *Ac* elements. Interestingly, *Ds* elements have multiple origins: some are simple deletion derivatives of *Ac*'s; *Ds2* ele-ments have internal regions that are derived from *Ac* but have sustained multiple deletions and substitutions; and *Ds1* elements have only 13bp at one end and 26bp at the other in common with *Ac*, along with about 400bp of internal sequence that has no apparent homology to *Ac* but still has transposase-binding sites (Kunze and Weil 2002). *Ds1* elements are also mo-bilized by an independent class of DNA transposons called *Uq* (*Ubiquitous*) elements. *Mu* elements of maize have many such classes of nonautono-mous elements that are homologous neither to it nor to each other (Walbot and Rudenko 2002). Genome-sequencing studies have uncovered a pleth-ora of these nonautonomous transposable elements, sometimes called MITEs (miniature inverted-repeat transposable elements), or at least their remnants, typically 100–1000bp long (Feschotte et al. 2002). For example, even in the relatively small *C. elegans* genome, there are fossils of some 27 families of MITEs, with a total copy number of about 7200, comprising sev-eral percent of the genome.

LTR retroelements can also mobilize in *trans* sequences that have the nec-essary *cis*-acting recognition sequences but do not encode the necessary pro-teins. For example, the *Bs1* LTR retroelement of maize does not have a *pol* gene, but instead carries a 654bp fragment of a host gene (a plasma mem-brane proton-translocating ATPase, *Zmpma1;* Bureau et al. 1994). The structure of *Bs1* is thus similar to that of vertebrate oncoviruses, retroviruses containing altered versions of host genes. *Bs1* is found in 1–5 copies in all maize lines analyzed and in its closest relative, teosinte. *TRIM* and *LARD* elements are more widespread nonautonomous LTR elements of plants (Witte et al. 2001, Kalendar et al. 2004). Such elements are also found in animals: in our own genome there are remnants of some 300,000 *MaLR* elements, mostly solo LTRs, constituting about 4% of the genome (Lander et al. 2001). Full-length elements have LTRs but no internal homology to retroelements, and they are thought to have been mobilized in *trans* by ERV-L endogenous retroviruses (Smit 1999).

Another variant class of transposon that can easily arise includes repressors of transposition. A mutant transposase that simply binds to a DNA transposon but is defective for excision will likely act as a repressor because it will interfere with the functioning of normal transposase. And if transposases function as dimers or multimers, a mutant transposase that binds to normal transposase but is otherwise defective will also act as a repressor ("multimer poisoning"). Because the net effect of transposable elements is harmful to the host, natural selection at the level of whole organisms can increase the frequency of repressor elements. Again, the best data are for *P* elements, for which at least some of the deletion derivatives act as repressors (Corish et al. 1996). One such element, called *KP*, is 1.2kb, less than half the length of intact elements, and it codes for a repressor that contains the DNA-binding domains but not others necessary for causing transposition (Lee et al. 1996, 1998). This element is very common in Eurasian populations but less so in American populations, and it has been observed to spread through experimental lab populations (Black et al. 1987, Gloor et al. 1993, Andrews and Gloor 1995). These repressor elements may be playing both ends of the stick: they reduce the overall number of transpositions, but if any is occurring, they are equally involved in the action. (Intact elements may also act as repressors, depending on their insertion site in the genome [Ronsseray et al. 1991, Gloor et al. 1993, Stuart et al. 2002], presumably because they stimulate some mechanism of host defense.) The *Spm* element of maize has also given rise to a deletion derivative that acts as a repressor (Cuypers et al. 1988). Somewhat analogously, there are endogenous retrovirus insertions in mice that appear to have gone to fixation because they are partially defective and protect the mouse from an exogenous retrovirus (Gardner et al. 1991).

Wildtype (+/+)   SD/+

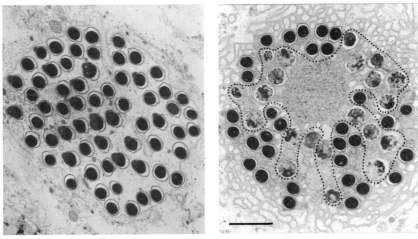

Plate 1. The effect of *SD* in *Drosophila*. Electron micrographs of developing spermatids. Note the wildtype (*left*) shows 64 well-condensed nuclei, while the *SD* male (*right*) shows 31 normal looking nuclei bearing *SD* and 10 poorly condensed nuclei carrying the wildtype chromosome (enclosed in dotted line; the missing ones have failed at an earlier stage of spermatogenesis). Bar, 1μm. From Tokuyasu et al. (1977) and Crow (1979).

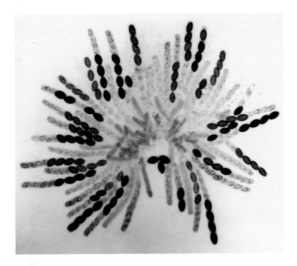

Plate 2. The effect of a spore killer in *Neurospora*. The rosette is derived from a cross of spore killer and wildtype strains. Mature asci would normally contain 8 viable black haploid spores, products of a single meiosis (followed by mitosis); here they have only 4 such spores, along with 4 tiny clear aborted spores. The viable spores all carry the *Spore killer-1* allele. The asci not showing the pattern are still immature. From Raju (1980).

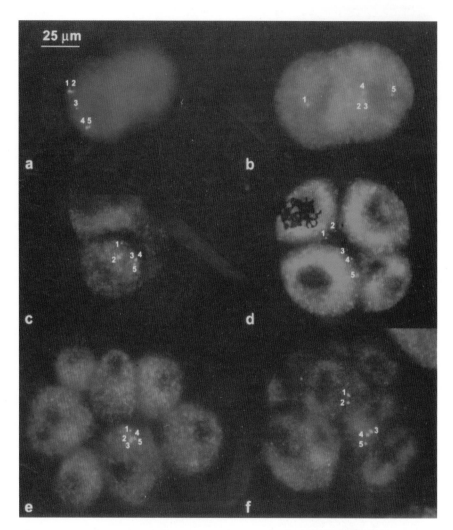

Plate 3. Paternal mitochondrial clumping in male mussels *Mytilus edulis*. Two (a, b), 4 (c, d), and 8 (e, f) cell embryos of 6 different mussels. Males are on the left and females on the right. The 5 paternal mitochondria are numbered. Note that in males they are always clumped into a single cell, while in females they disperse. In male two-celled embryos, they are known to be within the germinal cell. Courtesy of Bedford Institute of Oceanography, Dartmouth, Nova Scotia, Canada.

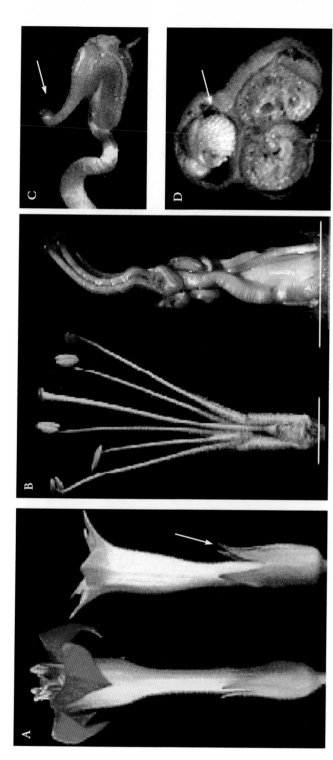

Plate 4. The effect of a CMS gene on flower morphology in tobacco. A. External morphology. Note the male-sterile flower (*right*) has a larger calyx (*arrow*) and a smaller and greenish corolla. B. Reproductive organs (sepals and petals removed). Note the male-sterile flower (*right*) shows fusion of some stamens with the carpel. C. Magnification of an anther from the male-sterile flower capped with a feminized structure (*arrow*). D. Transverse section of an ovary from the male-sterile flower showing an additional carpel chamber filled with ovules (*arrow*). Bars, 1 cm. From Bereterbide et al. (2002).

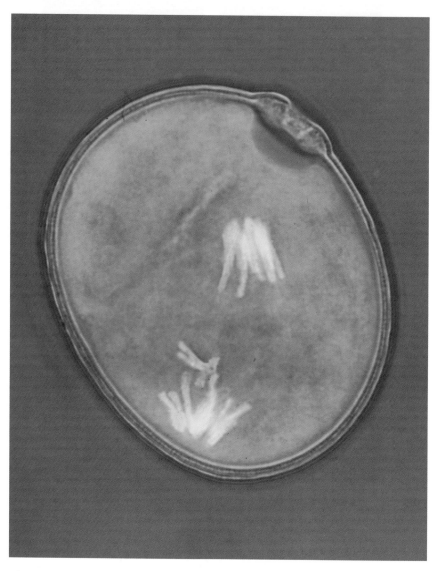

Plate 8. B chromosome nondisjunction in rye. Classic photo of first pollen grain mitosis in rye showing nondisjunction of two chromatids of a single B chromosome. The two are destined for the generative daughter nucleus, in other words, the germline. Courtesy of N. Jones.

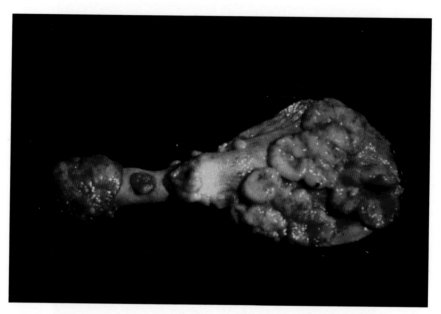

Plate 9. A pathogenic cell line. Canine transmissible venereal sarcoma (CTVS) is a cancerous cell line that escaped its original host and is now circulating among feral dog populations. Photo shows tumors on the penis and surrounding tissue. Courtesy of W. T. Weber.

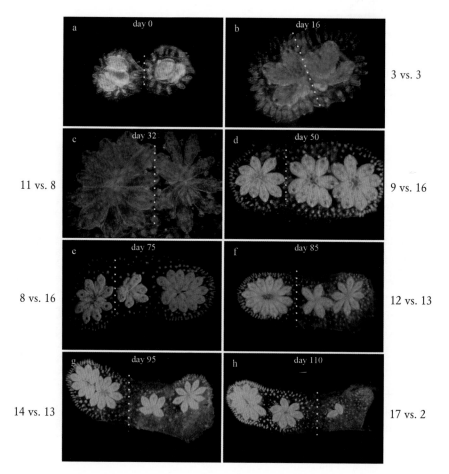

Plate 10. Chimeric resorption in the colonial ascidian *Botryllus schlosseri*. Photos show time series from a single interaction. Two young colonies that fused (a) developed within 16 days a 3- versus 3-zooids chimera (b). At first the left partner grew faster (c), and then the right partner began to develop (d, e). From day 85 the left partner began to resorb the other (f, g, h), until the right partner had disappeared at day 117. Dotted lines show the border between partners and numbers show the numbers of zooids in the two partners. Genotypic analysis of many such cases has shown that the surviving colony can be derived from either the apparent winning or apparent losing colony, and that its somatic cells may be derived from one while its germ cells are from the other. From Rinkevich (2002).

The evolution of such repressors may select for functional elements that avoid being repressed. If the repressor acts by binding to but not excising the DNA, there may be selection for elements that have different *cis*-acting regions that would not be bound by the repressor but would still be bound by their own transposase. If the repressor works by multimer poisoning, there may be selection for changes in the protein-protein binding domains, such that the functional transposase binds to other copies of itself but not to the repressor.

Extinction of Active Elements in Host Species

Despite their proliferative capabilities, there is abundant evidence from genome-sequencing studies that transposable elements often go extinct within a host species—that is, active copies of the element disappear from the species gene pool. In the human genome there are hundreds of thousands of inactive fossil DNA transposons, grouped into 63 families, all of which proliferated to varying extents, for varying lengths of time, in our ancestral gene pools and all of which are now completely inactive (Lander et al. 2001). Phylogenetic analysis for at least some of the families is consistent with a relatively short period of activity, followed by a long period of fossilization (Fig. 7.11). Similarly, there are remnants of at least 20 families of endogenous retroviruses in our genomes, each of which is thought to have invaded the germline over the last 100my and to have subsequently proliferated (copy numbers of full-length elements are estimated to range from 15–660, with many more solo LTRs), but most or all the families are now moribund (Tristem 2000).

The leading explanation for such extinctions is that DNA transposons and LTR retrotransposons often complement defective elements in *trans*, allowing the defective ones to transpose and accumulate in numbers, driving the functional ones extinct (Lohe et al. 1995, Robertson 2002). Consistent with this hypothesis, the *L1* LINEs, which show *cis*-preference in transposition and therefore purifying selection against defective elements, have been continuously active in our genomes for at least 150my (Smit et al. 1995, Lander et al. 2001). Repressor *L1* elements may also arise less readily, because the protein does not act as a multimer nor does it bind in *trans* to DNA or RNA. And the rDNA-specific *R1* and *R2* LINEs of arthropods appear to have been continuously active for some 500my (Eickbush 2002).

A   B

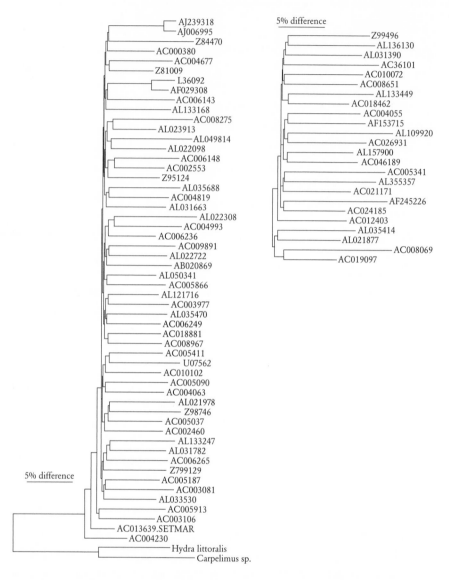

Figure 7.11 Phylogenetic history of 2 families of DNA transposons in the human genome. Each sequence is a different insert in the published genome sequence, identified by the accession number of the clone sequenced. For each family there is a burst of activity near the base of the tree and then long terminal branches to the tips. This is the pattern expected if the sequences are fossil remnants of a family that is no longer active. A. Relationships for 54 *Himar* inserts. B. Relationships for 24 *Charlie3* inserts. Adapted from Robertson (2002).

But this is not to say that LINEs never go extinct. In our own genome, there are ancient remnants of *L2* elements and 2 relatively low-copy-number elements *L3* and *CR1_Hs,* all of which are now moribund. Also, *L1* elements appear to be inactive in some South American rice rats and marsh rats (*Oryzomys* spp. and *Holochilus* spp.; Casavant et al. 2000). Curiously, these species can also have extremely variable karyotypes: Koop et al. (1983) found that, of 10 rice rats examined from a single isolated population, each one had a different karyotype, and Nachman and Myers (1989) found 26 different karyotypes among 42 marsh rats examined. Casavant et al. (2000) speculate that this may indicate a role for *L1* elements in chromosomal repair, but there is little independent data to confirm this. In insects too, LINE extinction does seem occasionally to happen. *I* elements, for example, appear to have been present in the *D. melanogaster* gene pool, to have gone extinct in at least some populations, and then to have reinvaded, either from other populations or perhaps from *D. simulans,* by introgression (Bucheton et al. 2002). And, the rDNA-specific LINE *R2* appears to have been lost from the common ancestor of *D. erecta* and *D. orena* (Lathe and Eickbush 1997).

Mathematical modeling of the spread of defective and repressor elements has confirmed that, under a wide range of conditions, nonautonomous and repressor elements can increase in frequency and drive intact elements extinct (Kaplan et al. 1985, Charlesworth and Langley 1986, Brookfield 1991, 1996). This process is facilitated under the following conditions:

- If there is a large cost of transposition for the host. Considering for simplicity just the DNA transposons, the cost to the host of a particular insertion can be divided into 3 components: (1) The effect of the DNA insert itself, on neighboring genes, on ectopic recombination, and so on. All elements will pay this cost. (2) The effect of the element being a substrate for transposase and thus introducing a double-strand gap and transposon insertions elsewhere in the genome. This cost depends on the affinity of the insertion for transposase. (3) The effect of any protein produced by the insertion on making elements (including itself) transpose. Nonautonomous elements avoid this cost, and repressor elements turn it into an advantage. The evolutionary fate of a transposable element family in a gene pool depends on the relative magnitude of these 3 costs. If the latter 2 costs are high, then, as elements accumulate in the genome, the production of transposase has more and more deleterious effects, thus selecting for nonautonomous and repressor elements (Charlesworth and Langley 1986, Brookfield 1991, 1996). As indicated earlier, there is no clear consensus about which of the many costs of an insertion are the most important. To help resolve the issue, it will be interesting to see whether new insertions that are unable to transpose, or unable to make a transposase (dead-on-arrival elements), typically drift to higher frequencies than functional inserts—if so, it would be evidence for the importance of costs associated with transposition itself (Yang and Nuzhdin 2003).

- If the variant element, either nonautonomous or repressor, has a transposition advantage compared to intact ones within the same host. For example, among artificially engineered *Tc1/mariner* elements, the frequency of transposition increases with smaller insert size (Lampe et al. 1998, Plasterk et al. 1999). Thus, deletion derivatives may have a transpositional advantage. More generally, elements that specialize at being a substrate for transposition—without the constraint of having to code for a transposase—may achieve higher rates of transposition.

- If mutation is biased such that intact elements give rise to nonautonomous (or repressor) elements more often than vice versa (Kaplan et al. 1985). This seems inevitable for all transposable elements. Among DNA transposons, those like *P* that induce recombinational repair of excision sites produce more deletion derivatives than those like *Tam3* that rely on end-joining. They may therefore be more susceptible to the accumulation of nonautonomous copies (Yamashita et al. 1999). As we have seen before, what is selected for in the short term may be harmful in the long term.

Note that if LTR retroelements evolve high ratios of gRNA:mRNA, as predicted by the model of Box 7.2, this will retard the spread of nonautonomous elements that effectively produce 100% gRNA—all else being equal, such elements are away from the optimum and will be selected against (Nee and Maynard Smith 1990). This may make LTR retroelements less prone to extinction than DNA transposons. Endogenous retroviruses in the human genome typically show a higher rate of synonymous than nonsynonymous nucleotide changes, consistent with some purifying selection for function and transposition rates being limited by the abundance of element-encoded proteins (Belshaw et al. 2004, 2005). This will most obviously be the case if there is often only a single active element in a cell, but could hold more generally.

Other possible causes of extinction (particularly for LINEs) are competition from other transposable elements (e.g., *L1* elements appear to have replaced *L2* elements in our genomes, although cause and effect is unclear) or the evolution of host repressors.

Horizontal Transmission and Long-Term Persistence

For DNA transposons, the general picture to emerge is as follows. When an intact element first invades a gene pool, it is rare and increases in frequency rapidly. Only functional elements are able to invade a new species, and elements producing a relatively large amount of transposase invade the fastest. Once there are several or many copies per genome, each producing a lot of transposase, transposase production becomes more and more costly, by causing more and more chromosome breaks. Nonautonomous and repressor elements then begin to accumulate. These may eventually drive the intact elements extinct. How then can intact elements persist over long evolutionary timespans? The answer seems to be horizontal transmission. The element can occasionally jump to a new gene pool, leaving the nonautonomous and repressor elements behind. As long as this happens (on average) at least once before the element goes extinct in the original species, it can persist over evolutionary time. Note that this condition for evolutionary persistence is analogous to the condition for ecological persistence, but at a new level. Just as a new insertion has a limited lifespan and the element must jump to a new site on average at least once before the insertion goes extinct, so a new gene pool colonization has a limited lifespan and the element must jump to a new species at least once before extinction.

This general picture presupposes a great deal of horizontal transmission of transposable elements among species—what is the evidence for this? We have already noted that *P* elements have recently invaded the *D. melanogaster* gene pool, an inference based on 3 observations (Engels 1992): (1) old strains tend not to have *P* elements, whereas all recently collected strains do (see Table 1.1); (2) *P* elements appear to be absent from the closest-known relatives, *D. simulans, D. sechellia, D. mauritiana,* and *D. yakuba* (Daniels et al. 1990, Clark et al. 1998); and (3) they are present in more distantly related species, including *D. willistoni*, from which a *P* element has been sequenced that is only 1bp different from the canonical *D. melanogaster* sequence (Daniels et al. 1990). These species are believed to have diverged about 50mya. Evidence similar to (2) and (3) also suggests a recent transfer of *P* elements from *Scaptomyza pallida* to the sibling species *D. bifasciata* and *D. imaii* (Hagemann et al. 1996). *P* elements in the 2 species are only 0.4% different, despite the species having diverged more than 35mya.

hobo is another DNA transposable element that appears to have invaded *D. melanogaster* sometime this century. The oldest strains show weak hybridization to a *hobo* probe, and no obvious activity, while all recently collected strains have the complete element (Pascual and Periquet 1991, Boussy and Daniels 1991). Moreover, elements in 3 different species (*melanogaster, simulans*, and *mauritiana*) differ by only 0.08% to 0.2% at all sites, compared to about 5% divergence at silent sites for ordinary loci (Simmons 1992). The direction of transfer is not yet clear (note these species can hybridize), nor is it clear whether the element was introduced to the species group from another species, perhaps outside the genus.

Apart from *D. melanogaster*, the best evidence of horizontal transmission of DNA transposons is for *mariner* elements:

- *mariner* elements from *D. mauritiana* and *Zaprionus tuberculatus* differ at only 3% of nucleotides, despite the hosts being thought to have diverged at least 50mya (Maruyama and Hartl 1991). A normal host gene (*Adh*) differs in the 2 species at 18% of sites. Also, some species that are more closely related to *D. mauritiana* than is *Z. tuberculatus* do not have such closely related *mariner* elements. Finally, a phylogeny of *mariner* elements from *Z. tuberculatus* and 5 *Drosophila* species is significantly different from that of the host species. The authors conclude that there has been a horizontal transfer event into an ancestor of *Z. tuberculatus*.
- *mariner* elements in *D. erecta* are only 3% different from those in the cat flea, *Ctenocephalides felis*, despite the hosts being in different orders (Lohe et al. 1995).
- In a PCR survey of 400 species of insects, a cluster of closely related *mariner* elements was found in the distantly related *Drosophila ananassae* (fruitfly), *Anopheles gambiae* (mosquito), *Haemotobia irritans* (hornfly), and *Chrysoperla plorabunda* (green lacewing) (Robertson and Lampe 1995). The consensus sequences of *mariners* from the green lacewing and horn-

fly differ by just 2bp out of 1044 (and 1 amino acid). The mosquito consensus differs from these two by 4 and 5 amino acids, respectively. The hornfly and fruitfly are in different families, estimated to have separated 100mya; both are in a different suborder from the mosquito, estimated to have separated at least 200mya; and all 3 are in a different order from the lacewing, estimated to have diverged over 265mya.

Other examples include horizontal transmission from a moth to a parasitoid wasp (Yoshiyama et al. 2001) and between fish and frogs (Leaver 2001).

Besides insects and vertebrates, *Tc1/mariner* elements are also found in other invertebrates, fungi, and plants. This wide taxonomic distribution, plus the relatively distant horizontal transfers, presumably reflects a low reliance on taxon-specific host factors for transposition. Indeed, transposition has been observed for elements artificially introduced into *Leishmania* protozoans, and into bacteria, as well as *in vitro* (Gueiros-Filho and Beverley 1997, Rubin et al. 1999). By contrast, *P* elements appear not to be active outside a relatively small set of dipterans (O'Brochta and Handler 1988, Perkins and Howells 1992). *mariner* elements are generalists, while *P* elements are specialists.

LTR retroelements can also move between species. The *copia* elements of *D. melanogaster* and *D. willistoni* are identical—a clear sign of recent transfer—despite the species having separated some 50mya (Jordan et al. 1999). These are the same 2 species involved in the transfer of *P* elements, but this time it was in the opposite direction: *copia* is abundant and widespread in *D. melanogaster* and relatives, but it is rare or patchily distributed in *D. willistoni*. The species' ranges are thought to have overlapped for only the last 200 years, implying that horizontal transmission need not be a rare event, given the opportunity, at least in some taxa. Phylogenetic analysis of *gypsy* elements in *Drosophila* species suggests a complex history of vertical descent and horizontal transfers (Terzian et al. 2000, Vázquez-Manrique et al. 2000), and similarly for LTR retroelements in some echinoderms (Gonzalez and Lessios 1999).

Some LINEs have also undergone at least occasional horizontal transmission (Župunski et al. 2001; Fig. 7.12), though this is not universal: *R1* and *R2* show no evidence of horizontal transmission among arthropods (Eickbush 2002). A reduced frequency of horizontal transmission compared to DNA transposons and LTR retroelements may be expected because the

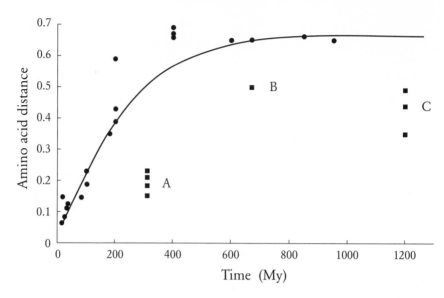

Figure 7.12 Sequence divergence of LINEs as a function of the estimated divergence time of their hosts. Data are for the RTE clade of LINEs. Note that 3 clusters of points fall below the curve passing through most other points, suggesting horizontal transmission: A (elements in cows and snakes), B (snakes and silk moths), and C (medaka fish and plants). Adapted from Župunski et al. (2001).

extrachromosomal phase of the replication cycle is entirely RNA, which is less stable than DNA. LINEs may also be less likely to insert into viral genomes (plausible vectors of horizontal transfer) because host DNA repair mechanisms are required for proper insertion, and these may not be associated with viral genomes (Malik et al. 1999). Nevertheless, we will be surprised if further instances of LINE horizontal transfer are not soon found, particularly among the protists and fungi, for horizontal transmission of HEGs has been documented in these taxa and they have no extrachromosomal phase at all in their replication cycle (see Chap. 6). Even infrequent cytoplasmic mixing between filamentous fungi of different species should allow LINE transfer.

The means by which all these horizontal transmission events occur is unknown. One plausible class of vectors are viruses, and an LTR retroelement from the host genome has been recovered from nuclear polyhedrosis viruses (NPVs) grown in insect cell cultures (Miller and Miller 1982, Fraser

1986). To the extent that NPVs are important vectors, we would expect horizontal transmission among Lepidopteran species (the most common hosts) to be more frequent than among other hosts. Other possible vectors are predatory mites (Houck et al. 1991) and parasitoid wasps, which may act as natural syringes for transferring DNA from one insect species to another, or aphids, which may transfer DNA from one plant species to another.

The importance of horizontal transmission in the distribution of transposable elements suggests that taxa with more easily accessible germlines are likely to have more families of transposable elements. Inaccessibility of the germline may explain, for example, why DNA transposons and LTR elements (other than endogenous retroviruses) appear to be rare in birds and mammals.

Transposable Elements in Inbred and Outcrossed Populations

As for other selfish genetic elements, a key factor likely to be important in the distribution and abundance of transposable elements is the host breeding system (Hickey 1982, Wright and Schoen 1999, Morgan 2001). In an inbreeding population, inserts will usually be homozygous, increasing their deleterious effects on the host (s), and therefore reducing their net reproductive rate (R). More generally, inbreeding increases the variance in copy number among individuals, and therefore the efficacy of natural selection in reducing their numbers. This may explain why DNA transposons and LINEs are absent from *S. cerevisiae*. This yeast is highly inbred and, as already noted, also has a dense genome (and so randomly inserting elements are likely to be particularly harmful). These 2 factors combined may prevent these elements from invading the gene pool and restrict the transposable element community to LTR retroelements that target safe havens. Other fungi have both DNA transposons and LINEs (Kempken and Kück 1998), and DNA transposons have been artificially introduced into yeast (Weil and Kunze 2000), and so their absence seems more likely to be due to high s than low t (transmission rate). Information on the transposable element content of outcrossing yeasts would be particularly interesting.

Host breeding system is also likely to affect transposable element evolution. In particular, in inbred hosts a daughter element (and any associated harmful effects) will remain associated with the parental element for more generations, and so selection for "prudent" transposition will be stronger, and evolutionarily stable transposition rates lower (Charlesworth and Langley 1986). The transposable elements of *C. elegans* and *A. thaliana* (both selfers) are apparently less active than those of *D. melanogaster* and *Z. mays*, consistent with this hypothesis, though closer comparisons would clearly be of interest.

Beneficial Inserts

So far we have assumed that almost all transposable element inserts are harmful to the host organism, and so each insert will tend to be driven extinct by natural selection on the host. And, at least for *Drosophila*, this seems to be a reasonable assumption. The best data come from insertion site frequencies, which show much the same pattern as they do for *P* elements, with insertions of any 1 transposable element at any 1 site being rare (Leigh Brown and Moss 1987, Charlesworth and Lapid 1989, Charlesworth et al. 1992, Biémont et al. 1994, Aulard et al. 1995; Fig. 7.13). These distributions are consistent with the estimates from laboratory studies of t and s being about 10^{-4}. Most of the transposons of *D. melanogaster* are also present in the sibling species *D. simulans,* and these also show low site occupancies (Nuzhdin 1995; for a rare exception for *copia*, see Vieira and Biémont 1996). These observations suggest the following conservative calculation: suppose the 2 species have a combined population of a million individuals, that they have been separated for a million generations, that each individual has a thousand active transposable elements, and that their rate of transposition is 10^{-4}. Then, there have been at least $10^6 \times 10^6 \times 10^3 \times 10^{-4} = 10^{11}$ insertion events since their separation. The genome itself is only 1.2×10^8bp long. Given this, it is striking (and even surprising) that so few beneficial mutations have occurred and fixed. That said, some fixed insertions are not visible by *in situ* hybridization, and it will be interesting to estimate the true rate of fixation from genome sequence data (Petrov et al. 2003, Bartolomé and Maside 2004).

Laboratory selection experiments appear, at face value, to give a different picture. In experiments with *D. melanogaster,* Torkamanzehi et al. (1992) found a significantly greater response to selection on bristle number in inbred lines with active *P* elements compared to lines without. Similar results were obtained by Frankham et al. (1991) in selection on inebriation times. These results suggest that particular insertions may be beneficial, at least in

272

some selective environments. The discrepancy presumably arises because the laboratory selection experiments are on much smaller populations, in which the supply of mutations will be more limiting than in natural populations.

The low site occupancies observed in *Drosophila* are by no means universal, and in other species many transposable element insertions have gone to fixation. In our own genome there are remnants of some 4 million transposable element insertions, the vast majority of which are fixed. One of the key unanswered questions in transposable element biology is what fraction of these (and others like them in other species) went to fixation because they were beneficial to the host and what fraction were effectively neutral, fixed by drift and selection at a linked locus (hitchhiking). In some cases, fragments of host genes (particularly the control regions) appear to be derived from transposable element insertions, suggesting the inserts may have been beneficial from the beginning. This pattern seems particularly common for MITEs. *Tourist-Zm11* provides the core promoter for the maize auxin-binding gene; the location of one rice MITE, *Ditto-Os2*, appears to correspond to the TATA box of the rice homeobox gene *Knotted-1;* and *Gaijin-Sol* probably supplies the polyadenylation signal and site for a sugarcane transporter gene (Bureau et al. 1996). Various contributions of transposable element sequences to the regulatory regions of human genes are reviewed by Jordan et al. (2003) and van de Lagemaat et al. (2003).

Whatever the reason for an insert going to fixation—whether it was beneficial to the host or hitchhiked along with a beneficial gene—this will (almost inevitably) have a dramatic effect on the population biology of the element. In the model presented earlier, each insert was found in an average of $1/s$ individuals before going extinct. But inserts with a beneficial effect s have a probability $2s$ of going to fixation, in which case it will go extinct as a transposable element only when a mutant arises and fixes that keeps the beneficial effect but is no longer able to transpose. An enormous number of individuals will then be carriers. Even a relatively low probability of this occurring (10^{-3}? 10^{-6}?) could have a dramatic effect on an element's net rate of increase (R). The only way this effect would not be important for R is if the only inserts that are beneficial are nonfunctional from the start (dead-on-arrival)—in other words, if the benefit caused was typically smaller than the cost to the organism of an active insert.

There is a link here between this disproportionate effect of a small pro-

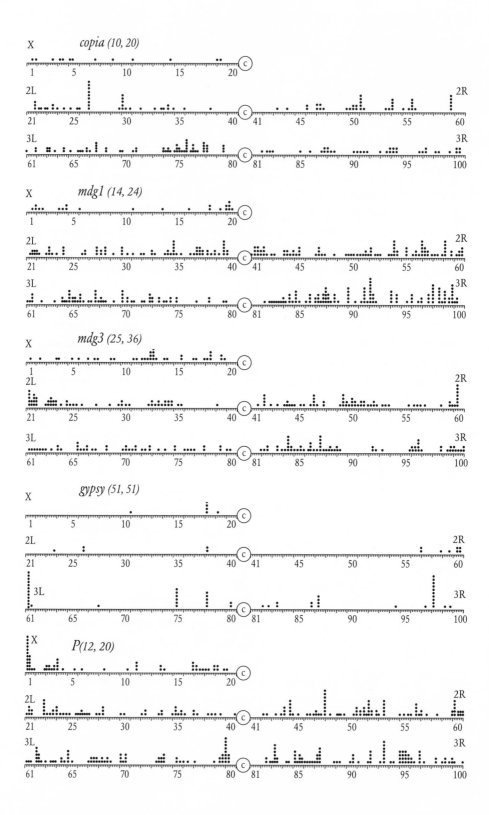

portion of beneficial mutations and Leigh's (1970, 1973) classic ideas on the evolution of mutator genes–genes that affect the rate at which other genes in the genome mutate. The key insight is that the evolution of mutators depends on whether the population is sexual or asexual. If the population is largely sexual, all that matters is the average effect of new mutations. Because most new mutations are deleterious, selection will be for ever-lower mutation rates. More precisely, an allele decreasing the mutation rate by an amount $\Delta\mu$ will have a selective advantage of $\Delta\mu s$, where s is the average selection coefficient against new mutations. Mutation rates will evolve ever lower, until this benefit no longer compensates for any physiological disadvantage (e.g., slower DNA replication). Alleles causing increased mutation rates cannot benefit from the small minority of beneficial mutations they cause because they recombine away from each other in an average of 2 generations (assuming free recombination between the mutation rate locus and the beneficial mutation), too short a time for selection of the beneficial mutation to increase the frequency of the mutator allele.

In asexual populations, on the other hand, an allele causing increased mutation rates remains linked to any beneficial mutations it causes, and goes to fixation along with them. If the environment changes sufficiently often that a mutator can produce a steady stream of beneficial mutations, mutation rates may evolve to be higher than in sexual populations. This is so even if the average effect of mutations is still negative. (Alternatively, if the environment is so constant that beneficial mutations never arise, mutation rates may evolve to be lower in asexual taxa, because of the permanent linkage between the mutator alleles and the deleterious mutations they cause.)

Transposable elements in sexual populations should behave like conventional mutators in asexual populations, because of the inherent linkage between the element and the (insertional) mutations it causes. Indeed, a purely cut-and-paste transposable element that has no mechanism for increasing in copy number when it transposes, and whose average effect is

---

Figure 7.13 The frequency of inserts is shown for *copia, mdg1, mdg3, gypsy,* and *P* elements along the 5 main chromosomal arms (X, 2L, 2R, 3L, 3R). Each dot represents a chromosome with an insert at that site; numbers in parentheses are the number of chromosomes analyzed for X chromosomes and autosomes, respectively. In general, inserts are well scattered but not completely random. Adapted from Biémont et al. (1994).

harmful to the host, could still spread through a population by hitchhiking with a minority of beneficial mutations. We are not aware of any transposable element that fits this description. All the ones we know of have an inherent tendency to increase in copy number when they transpose. But this may just reflect the fact that, if such an element were to exist, there would still be strong selection for a mutant element that over-replicated. If a transposable element in a sexual population did persist purely by the occasional creation of a beneficial mutation—in other words, they functioned as mutators, to increase the variation on which natural selection acts—their average effect would likely still be deleterious, and host genes would be selected to suppress their activity. There would be a conflict of interest over mutation rates, and it would still be appropriate to call them selfish genetic elements.

In wholly asexual taxa there can be no genetic conflict of interest over mutation rates. If transposable elements produce a small fraction of beneficial mutations, there may be no selection in asexual taxa to suppress their activity. Also, selection on transposable elements to be replicative should be weaker, and perhaps truly conservative transposition will be found in asexual eukaryotes. On the other hand, there would be no need for the transposase to be encoded by the transposing elements: it might just as well be encoded by a stable host gene that moves around unrelated segments of DNA that may have beneficial effects (e.g., promoters?). Bdelloid rotifers, long thought to be wholly asexual, have both DNA transposons and LTR retrotransposons (Arkhipova and Meselson 2000, Arkhipova et al. 2003), and these will be fascinating to study further.

Because the rare ability to cause beneficial insertions can have a dramatic effect on R, there may be selection for variants that are able to cause an increased frequency of beneficial mutations. Do transposable elements have adaptations to increase the probability of being useful to the host, even if of just one in a thousand insertions? Such adaptations are perhaps most likely to be found in nonautonomous elements like MITEs and SINEs, both because they are less constrained in terms of having to produce functional proteins and because they are less costly to the organism, and so more likely to go to fixation. Intriguingly, both the mouse *B2* SINE and the human *L1* LINE have antisense RNA polymerase II transcription promoters that do not serve any obvious purpose in increasing their own transposition but can drive the transcription of upstream host genes (Ferrigno et al. 2001,

Nigumann et al. 2002). Perhaps they have evolved because occasionally the novel transcript they generate is positively beneficial to the organism.

In principle, much the same considerations apply to the possible increase in frequency of a transposable element insertion by biased gene conversion (BGC; see Chap. 6). That is, in species in which insertions have a conversion bias over wildtype, transposable elements will have a much easier time spreading. If some insertions have a positive bias and others negative—for example, because of differences in genomic location—the former will have a disproportionate effect on R. And if there is any way for transposable elements to evolve a positive conversion bias, there will be strong selection for them to do so. But the interaction of transposable elements and BGC has barely been studied. Vincent and Petes (1989) showed that heterozygous *Ty* insertions in yeast have an overall gene conversion bias in favor of duplicating the *Ty* element to the homologous chromosome at both mitosis and meiosis, though the extent of bias appeared to vary for insertions at different locations. And recently, Ben-Aroya et al. (2004) have shown that *Ty* elements have a compact chromatin structure that extends some way into the flanking DNA. This compact structure may have evolved to reduce the frequency of ectopic exchange between *Ty* elements—if so, it may be an adaptation of the host, the element, or both. This compact structure may also protect the DNA from double-strand breaks, and so give the *Ty* element a gene conversion advantage, which would be of benefit to itself but not the host. We speculate that variation in the ability of transposable elements to spread by BGC may account in part for the otherwise-mysterious differences among species in their propensity to accumulate and fix transposable element insertions.

Rates of Fixation

Information on the rate of fixation of transposable element insertions in host genomes is beginning to accrue from genome-sequencing projects. Here we review recent studies on mammals and on maize.

Mammals. As we have already noted, there are some 4 million transposable element inserts detectable in our genome (Table 7.1). Together they make up about half of the genome. By comparison, there are about 30,000 protein-coding genes, and only about 1.5% of the genome is translated.

Table 7.1 Approximate transposable element composition of the human genome

	No. of Inserts	% of Genome Total	% of Genome Last 75my*
DNA transposons	400,000	3	1
LINEs	1,000,000	21	8
SINEs	2,000,000	14	11
LTR retroelements	600,000	9	4
Total	4,000,000	46	24

* Transposable element inserts acquired in the last 75my (i.e., not shared with mice).
From Lander et al. (2001) and Waterston et al. (2002).

Most of the inserts are inactive fossils and are fixed in the species—indeed, they range in age up to 150–200my. Presumably there are more that are older still but are undetectable because they have mutated beyond recognition. About half of these inserts (or a quarter of the whole genome) appeared since our last common ancestor with mice, about 75mya (Waterston et al. 2002). As a long-term average, this is equivalent to the new addition of 10kb every 1000 years. Mammalian genome sizes are relatively homogeneous and are not thought to have changed much over this time, and so an equivalent amount of DNA has probably been lost.

In the lineage leading to modern-day mice, transposable elements have been slightly more active on average, and the turnover slightly greater (about one-third of the genome new since the split with humans). At least some of this difference is presumably due to a shorter average generation time in the mouse lineage, and therefore greater number of generations. Currently, mice also have a greater diversity of active elements, including LTR retroelements. Over more recent times, *L1*s have shown much lower rates of accumulation in humans than in mice and rats per unit time, but approximately equal rates per generation (Table 7.2). The latter is somewhat surprising, because mice have about 30-fold more inserts that are active than humans do ($\approx$3000 vs. 100; Sassaman et al. 1997, DeBerardinis et al. 1998, Goodier et al. 2001), and the probability of a new mutation being due to an *L1* insertion is about 30-fold higher in mice (see earlier). Chimpanzees and bonobos apparently show a 2- to 3-fold higher accumulation rate of *L1* elements (per unit time) than do humans (Mathews et al. 2003).

A highly indirect estimate for the frequency of humans with a novel in-

Table 7.2 Accumulation rates for *L1* LINEs

Species	Element	Time Span Studied (my)	Accumulation Rate per 1000 Years	Accumulation Rate per Generation
Humans	*L1Ta*	3.5	0.2	0.004
Rats	*L1Rn*$_{mlvi2\text{-}rn}$	0.45	12	0.006
Mice	*L1Tf*	0.32	5	0.0025

From Boissinot et al. (2000).

sertion is about 1 in 16, or 0.06 (Kazazian 1999), about 10-fold higher than the rate of accumulation in the genome. Taken at face value, this would indicate that the rate of accumulation is only one-tenth of what it would be if inserts were selectively neutral—that is, the net effect of selection is to remove elements. Just as with conventional mutations, a minority of inserts may be beneficial, but the majority will not be. Selection in mice would appear to reduce the rate of accumulation even more, to 1/300 of the neutral rate. Perhaps this is due to a larger effective population size. Why mice have a greater number of active elements than humans is not clear.

A striking example of local variation in fixation rates occurs in *Microtus agrestis,* in which a nonautonomous LINE-like element has reached 200,000 copies per diploid genome, compared to 1000 copies in related species thought to have diverged 0.7mya (~3.5 million generations; Neitzel et al. 2002). For unknown reasons, many of these copies (40%) are in the sex chromosome heterochromatin.

Maize. Maize and sorghum are 2 species that diverged some 15–20mya and currently differ in genome size by a factor of about 3.5 (2400Mb vs. 750Mb, respectively; SanMiguel et al. 1998, Tikhonov et al. 1999). In a 240kb region of the maize genome near the *Adh-1* gene, there are 21 LTR retrotransposons and 2 solo LTRs, belonging to 11 different families; together they make up about two-thirds of the region. Many of the elements are nested, one inside the other: 14 of the 23 inserts are found in preexisting inserts, including 5 of the 6 most recent (Fig. 7.14). Preexisting inserts are likely to be "safe havens" for new inserts, because inserting there is unlikely to be harmful to the host. Not one LTR retrotransposable element was found in the homologous region of the sorghum genome. By comparing

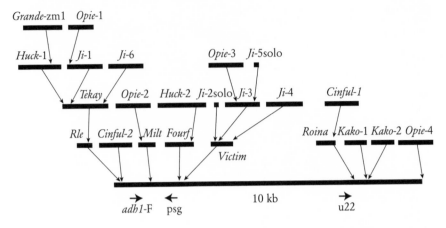

Figure 7.14 Nested arrangement of LTR retrotransposons in maize. More than half the inserts are into other inserts. Names indicate the family to which the inserts belong. The positions of 3 host genes are also indicated, at the bottom. Adapted from SanMiguel et al. (1998).

the LTRs of a single insert (the divergence of which gives an estimate of the time since insertion), it is estimated that at least 15 of them were inserted in the last 3my, as well as the 2 solo LTRs, totaling about 120kb (50% of the region analyzed). If one extrapolates to the whole genome, the results suggest that there have been some 170,000 inserts in the last 3my, or about 1 every 20 years, leading to a doubling of genome size. It would be interesting to know the genome-wide transposition rate, as a basis of comparison. And the relative contributions of drift, biased gene conversion, and natural selection to these fixation events remains unknown.

Transposable Elements and Host Evolution

We saw in the last two sections how transposable element insertions can provide new beneficial mutations for their hosts and can also be major players in the evolutionary dynamics of host genomes. We now review other ways they can be important for the host in providing mutational variation, both of chromosomal structure and molecular mechanisms. The importance of transposable elements as generators of mutational variability comes from the fact that they produce mutations that are unlikely to arise by any other means (e.g., during DNA replication).

Transposable Elements and Chromosomal Rearrangements

Transposable elements can cause duplications, deletions, and rearrangements of host genetic material in a bewildering number of ways. As with new inserts, most of these are expected to be harmful to the host, but some fraction will inevitably be beneficial. The position of genes along chromosomes often differs between closely related species (e.g., Carson 1983 for *Drosophila* species), indicating that some fraction of rearrangements do spread to fixation. And, at least some naturally occurring chromosomal rearrangements are associated with transposable elements. For example, in *D. melanogaster,* 3 low-frequency chromosomal inversions found in the Hawaiian islands were found to have *hobo* elements at each of their breakpoints, and a fourth had it at one end (at least to the level of resolution possible by *in situ* hybridization; Lyttle and Haymer 1992). More recent sequencing studies have associated transposable elements with rearrangements in *D. buzzatii,* mosquitoes, yeast, and humans (Kim et al. 1998, Mathiopoulos et al. 1998, Schwartz et al. 1998, Cáceres et al. 1999, Hughes and Coffin 2001).

The simplest mechanism by which transposable elements can cause a chromosomal rearrangement is by participating in an ectopic recombination event. That is, there can be homologous recombination between more-or-less identical sequences that are inserted at different locations in the genome. Depending on their relative positions and orientations, this can lead to a duplication, deletion, inversion, or translocation (Fig. 7.15).

As well as passively participating in ectopic recombination, DNA transposons can more actively cause chromosomal rearrangements in the following ways:

- *Alternative transposition.* Transposase normally cleaves at the 2 ends of a single insert, but occasionally it can "accidentally" cleave ends belonging to 2 different inserts. If the inserts are on the same chromatid, the entire intervening sequence will be transposed; if the inserts are on different chromatids, more complex rearrangements will result (Dowe et

DELETIONS AND DUPLICATIONS

1) Looping out

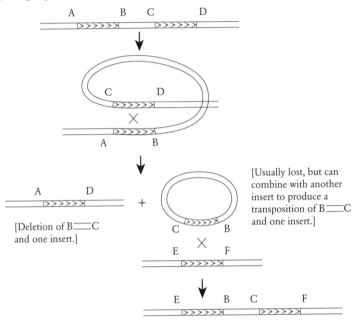

[Deletion of B⊐C and one insert.]

+

[Usually lost, but can combine with another insert to produce a transposition of B⊐C and one insert.]

2) Gene conversion

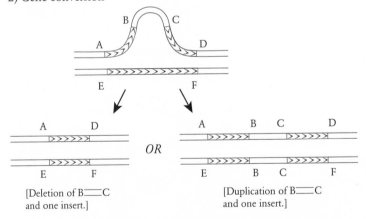

OR

[Deletion of B⊐C and one insert.]

[Duplication of B⊐C and one insert.]

3) Unequal crossing-over

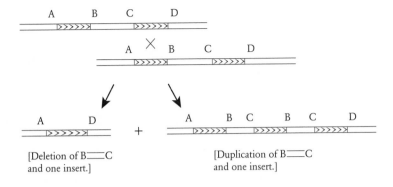

[Deletion of B⎓C
and one insert.]

[Duplication of B⎓C
and one insert.]

INVERSIONS

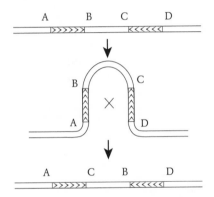

TRANSLOCATIONS

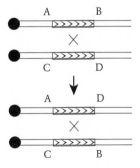

Figure 7.15 Chromosomal rearrangements caused by recombination between transposable element inserts at different locations in the genome. Each of these possibilities has been observed to occur in yeast. Adapted from Roeder and Fink (1983).

al. 1990, Harden and Ashburner 1990, Lovering et al. 1991, Zhang and Peterson 1999, Gray 2000).

• *Alternative end-joining.* Gaps left after excision of a DNA transposon are sometimes repaired simply by ligating the 2 ends back together, and if more than 1 element is excised simultaneously, the broken ends may then get rejoined in the "wrong" order, leading to inversions and translocations (Engels and Preston 1984).

• *Misrepair.* Repair of excision events is sometimes associated with complex DNA rearrangements. For example, Kloeckener-Gruissem and Freeling (1995) report on a *Mu* DNA transposon in the promoter region of the maize *Adh* gene, and an excision event that resulted in "promoter scrambling": the duplication, deletion, and inversion of flanking host sequences. This in turn resulted in novel tissue-specific patterns of expression. Because the transposon has excised, the role of a transposable element in generating this diversity would be difficult to detect just by examining the DNA sequence after the event.

In addition, the reverse transcriptase activity encoded by retroelements can produce the following large-effect mutations:

• *Gene duplication.* LINE proteins can "accidentally" reverse transcribe the wrong mRNA into the genome, leading to duplication of a host gene. These inserts will differ from the original host gene in having the structure of mRNA: untranscribed promoters and introns will be missing, and there will be an A-rich tail; it will also usually be bounded by direct repeats of the target site, created at the time of insertion. Usually, the new insert will not be functional, in which case it is called a processed pseudogene (Box 7.4). Very occasionally, the new gene will be functional, code for a protein, be selectively advantageous, and go to fixation. The testes-specific phosphoglycerate kinase (PGK) gene of mammals is probably one such example (McCarrey and Thomas 1987). PGK is an enzyme involved in glycolysis, the extraction of energy from simple sugars. Most mammals have 2 functional loci encoding the enzyme, *PGK-1* on the X chromosome and *PGK-2* on an autosome. *PGK-1* is active in somatic cells and premeiotic germ cells, but along with other X-linked genes is inactive during male meiosis (see Chap. 3). Fructose is

BOX 7.4

Processed Pseudogenes

Processed pseudogenes are nonfunctional copies of normal host genes that have been made by reverse transcription of mRNAs (Vanin 1985, Weiner et al. 1986, Wilde 1986). The reverse transcriptase used in pseudogene formation is thought to come from LINEs (Dornburg and Temin 1990, Tchénio et al. 1993, Esnault et al. 2000), and the predominance of LINEs over LTR retroelements in mammalian genomes may explain why processed pseudogenes are more common in mammals than in other taxa (Drouin and Dover 1987, Jeffs and Ashburner 1991). Estimates of the number of processed pseudogenes in the human genome range from 3600 to 13,000 (Zhang and Gerstein 2004). Genes that are highly transcribed in the germline should be more likely to form pseudogenes than those that are not. In our genomes there are many processed pseudogenes for β- and γ-actin, which are expressed in all cells, but none for α-cardiac or α-skeletal muscle actin, which are somatic tissue specific (Ponte et al. 1983).

Processed pseudogenes differ from SINEs (and LINEs) in typically not having internal promoters (most host genes do not have them), and so most will not be transcriptionally active—hence, they are dead ends in terms of retrotransposition. They are only one, or perhaps a few, retrotranspositional generations removed from the original host gene. In contrast to LINEs and SINEs, they have not had the opportunity to evolve adaptations for retrotransposition.

As with transposable elements, most processed pseudogenes are thought to be fixed in the human genome, though broad population surveys are rarely performed and examples of polymorphisms do exist (e.g., a dihydrofolate reductase pseudogene; Anagnou et al. 1984). As with the fixed transposable elements, the relative importance of drift, selection, and biased gene conversion is unclear. Once formed, processed pseudogenes may continue to interact with the original gene by nonallelic gene conversion—this inter-action could, for example, lead to the gain of intron sequences by the pseudogene or their loss from the parental gene (Wilde 1986). These interactions could also be a source of mutations in the func-tional genes, including, for rapidly evolving genes, atavistic back mutations.

present in semen and secreted in both the male and female reproductive tracts; to make use of this energy source, an autosomal copy of the gene has been selected that is expressed only in late spermatogenesis. This autosomal gene has all the hallmarks of having been reverse transcribed from mRNA: it has a poly(A) tail, bounding direct repeats, and no introns (compared to 10 in human *PGK-1*). Brosius (1999) reviews other genes likely to have been derived in the same way.

- *3′ transduction*. LINE mRNA can be processed "incorrectly" and there-fore contain a host sequence at the 3′ end, which is then reverse tran-scribed back into the genome, leading to duplication of the host se-quence. This is called 3′ transduction (Moran et al. 1999, Goodier et al. 2000). For example, *L1* elements of humans appear to have a relatively weak polyadenylation signal (poly(A)+ signal). This is the signal used by the cell to define where mRNAs are supposed to end: transcripts are cleaved at this site, and a poly(A) tail added. The consequence of a weak polyadenylation signal is that L1 mRNAs will often extend into the 3′-flanking DNA and this region will be reverse transcribed back into the genome along with the *L1* element. In this way, regulatory regions may get copied around the genome. A weak signal might also lead to exon shuffling: if an *L1* element is in an intron of a host gene, the host gene's polyadenylation signal might take precedence and all exons downstream of the *L1* insert would get copied (without the introns, which would be spliced out). If the element inserted into the intron of another gene, a chimeric gene could result. Boeke and Pickeral (1999) speculate that this process may explain why the last exon of host genes is often so much larger than internal exons. They also speculate that such processes might be the first step in the evolution of alternately spliced genes. Approximately one-quarter of *L1* retrotranspositions in humans involve copying 3′ sequences, typically several hundred base pairs (Goodier et al. 2000, Pickeral et al. 2000).

Why would *L1* have evolved such a weak polyadenylation signal? Conceivably, this process of 3′ transduction causes beneficial mutations linked to new inserts sufficiently often as to be selected for. An alternative explanation, suggested by Moran et al. (1999), is that it prevents *L1* elements in introns from terminating transcription of the host gene. This predicts that polyadenylation signals will be weaker (and 3′ transduction more common) in genomes with a higher percentage of intron sequences (Eickbush 1999). The euchromatic fraction of the human genome consists of 25–30% introns (Lander et al. 2001).

- *Intron removal.* The proteins encoded by LTR elements can "accidentally" encapsidate and make a cDNA copy of a host mRNA. This is unlikely to be integrated back into the genome (as this would require a second mistake, by the integrase protein) but could recombine with the host gene from which it is derived (Derr and Strathern 1993). This would result in the precise removal of any introns from the gene. Fink (1987) speculates that this process occurred thousands of times in the ancestors of baker's yeast, which now has very few introns.

It is possible, then, that transposable element activity increases the adaptation of their hosts over the long term, by inducing mutations that otherwise would not occur. But the size of the effect—if it exists—is unknown. And there may also be long-term costs. *Neurospora* fungi protect themselves against transposable elements by mutating any duplicated stretch of DNA longer than about 400bp, which has effectively halted adaptive evolution by gene duplication (Galagan and Selker 2004).

Transposable Elements and Genome Size

As well as producing variation in the arrangement of genetic material and in the functions encoded, transposable elements can also produce variation in the amount of DNA. Most obviously, they produce insertions that increase the size of the genome. As we have seen, the proliferation of transposable elements can also lead to deletions, by ectopic recombination between elements at different sites on the same chromosome (Fig. 7.15). On a smaller scale, the structure of LTR elements, with direct repeats at either end, can lead to looping out of 1 LTR and the intervening element. These transposable element-associated insertions and deletions play a key role in the evolution of genome size. We saw earlier that in the recent expansion of genome size in maize, it was predominantly the transposable element fraction that increased.

Another example is provided by barley. The *BARE-1* LTR retroelement in barley constitutes about 2.9% of total genome size and variation in genome size across *Hordeum* correlates positively with variation in the fraction made up of *BARE-1* (Vicient et al. 1999). More striking still, the excess of solo *BARE-1* LTRs over complete elements is negatively associated with both genome size and (more strongly) the fraction of genome made up of *BARE-1* elements. In other words, the apparent rate of removal of the LTR element through nonhomologous recombination is a better predictor of variation in genome size—and the fraction occupied by the element—than is copy number of the element itself. This result has been confirmed within a species of *Hordeum* (Kalendar et al. 2000). In addition, rates of solo LTRs to intact ones also varies along a cline.

At a broader scale, genome size is positively correlated with the fraction of the genome derived from transposable elements (Lynch and Conery 2003). That is, as genomes expand and contract over evolutionary time—and they appear to do both (Wendel et al. 2002, Jakob et al. 2004)—the fraction of the genome that is derived from transposable elements is disproportionately involved in the gains and losses. Again, the relative contributions of drift, biased gene conversion, and natural selection to the fixation of these insertions and deletions are unknown.

Genome size itself is known to be associated with a series of basic physiological parameters, such as rate of photosynthesis of plants, metabolic rates in birds and mammals and rate of development in amphibians (Gregory 2005). As yet, it has been difficult to disentangle cause and effect in these relationships. Particularly striking to us are instances in which large genomes appear to be imposing a cost. Vinogradov (2003) has shown that plants known to be of conservation concern have larger genomes than those of no concern. For globally threatened species, genome size is about twice that of unthreatened species, while it is 20% larger for those only locally threatened. The association also holds controlling for family and generic affiliation. Ploidy itself has no effect. Moreover, there is a negative correlation between species per genus and mean genome size, suggesting the association may be older than just the current wave of extinction. The effect runs counter to a long-term taxonomic bias toward larger genome size: the more recent the family of flowering plant in the paleontological record, the larger its genome size now. It will be interesting to test whether all of these associations are found in more formal phylogenetically controlled analyses. And the cause of the correlation remains completely unknown.

Vinogradov (2004) has repeated the analysis for vertebrates and found a more variable effect. For the groups with the smallest mean genome sizes (reptiles and birds), greater genome size predicts higher risk of extinction, while for fish, amphibians, and mammals there are no significant trends. On the other hand, large genome size may be inhibiting mental development in many species (Box 7.5).

Co-Option of Transposable Element Functions and Host Defenses

Transposable elements have evolved some sophisticated enzymatic capabilities (reverse transcription, cleaving and rearranging DNA, and so on) which, in some circumstances, may be useful to the rest of the genome (analogous to homing endonucleases and mating type switching in yeast; see Chap. 6). Here we review what appears to be a clear case of co-option of a transposable element in the evolution of the vertebrate immune system, the somewhat puzzling case of *Drosophila* telomeres, and telomeres more broadly. We also review suggestions regarding the co-option of host defenses.

The combinatorial immune system of vertebrates. Most vertebrates have immunoglobulin (*Ig*) and T-cell receptor (*TCR*) genes that are "split" and must be assembled by somatic recombination before they can be expressed. This assembly, called V(D)J recombination, occurs only in lymphocyte cells, and in some vertebrates is responsible for generating much of the diversity of antigen receptors within an individual organism, the assembly process resulting in slightly different genes in different cells (for a very brief outline of the immune system, see Chap. 11). The assembly process is initiated by proteins

BOX 7.5

Did Selfish Genes Drive Salamanders Stupid?

It has been argued that large genome size in salamanders has set sharp constraints on their mental evolution (Roth et al. 1997). The reason for this is a strong positive correlation between cell size and genome size across a great range of species and cell types (Roth et al. 1994). Even cells lacking genes, such as red blood cells in mammals, find their size strongly correlated with the genome size of the cells that gave rise to them (Gregory 2000). This means that a large genome will reduce the number of cells per unit brain size and the number of interconnections between them. Lower complexity is expected, along with lower mental acuity. Because the replication of selfish genes—in this case, transposable elements—is imagined to be a primary factor in genome expansion, the question can be put more sharply: did selfish genes drive salamanders stupid?

Salamanders and lungfishes have the largest genomes of any animals. Although polyploidy is frequent in salamanders, it does not make a strong contribution to the high genome sizes typical of the group. The parallel facts concerning mental development are equally striking. Salamanders (again, along with hagfish and some frogs) have the simplest nervous system of any vertebrate (Roth et al. 1993, Roth et al. 1997). Were it not taxonomically highly unlikely, the salamander system could easily be regarded as ancestral to all other vertebrate forms. Instead, comparative evidence shows that the brain and sense organs of salamanders have been simplified from a more complex ancestral state, and this is especially true of the more derived forms, the Plethodontidae.

These associations also work on a finer scale (Roth et al. 1997). The lepidosirenid lungfishes show greater reduction in neural complexity than do the lungfish *Neoceratodus* and also greater genome size. Desmognathine salamanders with relatively small genomes show greater morphological complexity of the brain than do the larger-genomed bolitoglossines, even though the latter are acrobatic salamanders operating in a 3-dimensional environment and using their tongues to capture prey. *Arenophryne rotunda* shows both the simplest brain and sense organs of any frog and also the largest genome. And so on.

And what happens when body size (and, therefore, brain size) is sharply reduced, as in miniaturized plethodontids (which are often small enough to seek refuge in earthworm holes)? Relative head size may be increased, as well as relative proportion of cells devoted to vision, but cell size is *not* reduced (Roth et al. 1995). There is no correlation between cell size and either head size or brain size. Instead, cell size seems largely resistant to adaptive variation, being strictly dependent on genome size. In short, salamander mental evolution, along with that of hagfish, caecilians, and some frogs, appears to be more a story about genome size expanding (for unknown reasons) and entraining nervous system changes than about the mental apparatus evolving in response to needs of the external environment. Certainly, when simple vertebrate nervous systems re-evolve, they appear almost always to be associated with unusually large genome sizes.

Comparative work makes it clear that genome size exerts a strong negative effect on rate of development, for example limb regeneration and embryonic time (but not egg size) in plethodontids (Sessions and Larson 1987, Jockusch 1997). In salamanders, genome size predicts rate of development, even when effects of nuclear and cell volume have been removed (Pagel and Johnstone 1992). This suggests that an important cost of extra (selfish) DNA is the time involved in replicating it every cell generation.

The matter can also be put the other way around. Lack of strong selection for mental acuity means weak selection opposing increases in genome size. But selection pressures are apt to be quite asymmetrical, with tiny increments in genome size due to transposable element increase having very small negative effects on intellectual ability.

encoded by the *RAG1* and *RAG2* genes cleaving the *Ig* and *TCR* genes at inverted repeats, and the mechanism of cleavage is in many ways similar to that which initiates DNA-based transposition. Moreover, the *RAG1* and *RAG2* genes, while not homologous, are themselves immediately adjacent to each other in the genome. These observations have led to the suggestion that RAG1, RAG2, and the inverted repeats they recognize in the *Ig* and *TCR* genes are descendents of an ancient transposon that has since been domesticated for host benefit (Agrawal et al. 1998, Hiom et al. 1998, Zhou et al. 2004). Indeed, the RAG1 and RAG2 proteins together can catalyze the transposition *in vitro* of a DNA fragment with the appropriate inverted repeats.

Note that in this example the linkage between the protein-coding genes and the inverted repeats they recognize is broken, so the *RAG* genes are not directly affected by their activity. Moreover, the genes are active only in somatic cells. These are features expected of genes acting for host benefit, not of selfish genes acting only for themselves.

The telomeres of *Drosophila*. Because of the way DNA is replicated (synthesis always in the 5' to 3' direction and initiated by an RNA primer), chromosomes cannot be replicated out to the last base pair, but instead are reduced by a couple base pairs at either end every cell cycle (Watson 1972). For example, a *Drosophila* chromosome loses 70–80bp from each end every generation (Levis et al. 1993). What prevents this process from gradually eating up the chromosome? It turns out that in *Drosophila* this continual shortening is counterbalanced by 2 LINEs, *Het-A* and *TART*, that specifically transpose to chromosome ends (Levis et al. 1993, Mason and Beissmann 1993, Pardue and DeBaryshe 2002). *Het-A* is the more abundant and active of the 2 transposons, and it has a number of unusual features. First, it encodes an ORF similar to the ORF1 of other LINEs but does not encode an ORF with any homology to reverse transcriptase. Thus, it is a nonautonomous element, and the RT must be supplied somehow in *trans*. Second, it does not have an internal promoter at its 5' end like other LINEs, but rather has a downstream-facing promoter at its 3' end. This means that an element that transposes to a chromosomal end cannot be transcribed, but instead must wait until another *Het-A* element transposes to its end, and then the first can be transcribed from the second element's promoter. Third, it has an unusually large 3' untranslated region downstream of the

ORF with no known function. The other element, *TART*, has its own surprises. It encodes 2 ORFs, and ORF2 has homology to RT, but also (like other LINEs) to an endonuclease domain, which should not be necessary for transpositions to chromosomal ends. Moreover, it has perfect direct repeats close to but not exactly at its ends, and it has an upstream-facing 3' promoter making antisense RNA transcripts that are 10 times more abundant than the sense transcripts. Like *Het-A*, it also has an unusually large 3' UTR.

These fascinating details await explanation. Are new telomeric inserts derived from other telomeric copies, or from one or a few master genes internal on some chromosome? No such sequences are apparent in the euchromatin, but they may exist in heterochromatin (the telomeres themselves are heterochromatic). What is supplying the RT for *Het-A* elements—is it a stable host gene, or is it another LINE element? Are they parasitizing *TART* elements? Why have *Het-A* elements been selected to promote the transcription of downstream sequences, and why have *TART* elements been selected to produce antisense transcripts?

The *Drosophila* mode of telomere reconstitution is unusual. In most other species, telomeres consist of hundreds or thousands of base pairs of a simple sequence 2–8bp long repeated many times (Alberts et al. 2002). These repeats are also synthesized by reverse transcription from an RNA template. The enzyme responsible, telomerase, is homologous to the reverse transcriptase of LINEs and LTR retroelements, and it may represent a much more ancient domestication of a selfish gene (Eickbush 1997).

Co-option of host defenses. We saw earlier that host organisms have mechanisms for suppressing their transposable elements, methylation being a well-studied example. Methylation is also used to regulate normal host genes (including imprinted genes; see Chap. 4). It is not known which came first, but it is possible that methylation evolved first as a means of controlling transposable elements and later was used to regulate host genes. Another possible example is Lyon's (1998) suggestion that X chromosome inactivation in female mammals may be an evolutionary elaboration of a system for controlling LINEs. According to this hypothesis, LINE insertion on the X chromosome could have been positively selected on the X at the time when the inactivation system was evolving. The age and distribution of LINEs on the X are consistent with this idea (Bailey et al. 2000).

Transposable Elements As Parasites, Not Host Adaptations or Mutualists

When Barbara McClintock (1952, 1956) discovered that certain maize loci could occasionally change their map position, it was only natural for her to wonder what function this served. The particular elements she observed (*Ac* and *Ds*) are active in somatic cells, and this fact–combined with the observation that inserts can affect the expression of neighboring genes–led McClintock to speculate that transposable elements normally function to turn genes on and off in somatic cells during development. Hence, she called them "controlling elements." This idea appears to be among the first recognitions that some mechanism for turning genes on and off during development must exist. The role of transposable elements in producing the variegated color patterns selected by horticulturalists seems to be an example of this (Plate 7), but decades of work on developmental genetics have shown that transposable elements do not play a significant role in development. Many years later, McClintock (1984) noted that "stress" appears to increase the activity of at least some transposable elements and suggested that they may function in "remodeling" the genome at times of difficulty (Box 7.6). In both of these suggestions, transposable elements were viewed as adaptations of the host (to control development or to respond to stress).

But the very structure of transposable elements suggests otherwise–in particular, the fact that the proteins carrying out the transposition reactions are always encoded by the transposable elements themselves. If, for example, transposable elements were the host's way of creating genetic variation, and so served to increase the efficacy of natural selection, there would be no need for these elements to contain a transposase gene: it could be stably integrated somewhere else in the genome, moving unrelated noncoding pieces of DNA (e.g., promoters, coding regions) around the genome. The contrast with the immune system is particularly instructive, for that is the arrangement there. Any finding of a MITE or SINE or other nonautonomous element that is mobilized by a stable host gene with no other function would be good evidence of transposition having evolved for host benefit. The fact that transposase proteins are encoded by transposable elements–and not by stable host genes–is inexplicable under the host adaptation hypothesis, and it only makes sense if they are selfish genetic elements. Other aspects of transposable element design also fail to support the hypothesis of host adaptation. For example, they do not carry host genes

BOX 7.6

Transposable Elements and Stress

McClintock (1984) noted that "stress" appears to increase the activity of at least some transposable elements and suggested that they may function in "remodeling" the genome at times of difficulty. But remodeling seems an overly optimistic description of what appear to be accidental chromosomal rearrangements. Why, then, are transposable elements activated by stress?

One possibility is that stress disrupts host mechanisms that suppress transposition. This cannot be the whole story, however, because transposable elements have been found with stress response promoters—they have evolved to be more active at times of stress (Wessler 1996, Grandbastien 1998, Takeda et al. 1999). Perhaps transposable elements have evolved to be active when host defenses are low. Or, perhaps at times of stress a new insert is more likely to be beneficial. Our own speculation is that times of stress are the safest time for a transposable element to be active because all the repair systems are active and damage to the host is least likely to occur. By contrast, when times are good and the host is making hay, transposition is particularly damaging to the host and, thus, damaging to the element.

whose optimal copy number varies. Nor are there mechanisms known for excision without integration (if the host sometimes wants more transposable elements, it should also sometimes want fewer).

If not host adaptations, perhaps transposable elements should be thought of as mutualists (e.g., Kidwell and Lisch 2001). After all, it is possible that the 4 million inserts in our own genome are beneficial, and transposable elements may even have evolved adaptations to increase the likelihood of an insert being beneficial. Almost certainly they are important sources of mutational variation that, combined with natural selection, can significantly increase the adaptedness of their hosts. However, we prefer to think of them as parasites because the average immediate effect on the host is negative. All the most obvious phenotypic effects of transposable elements are deleterious (insertion mutations, chromosomal rearrangements, and physiological costs). Moreover, transposition rates are orders of magnitude greater than excision rates, but transposable elements do not accumulate indefinitely, indicating that natural selection must be reducing their frequency, which can happen only if they are harmful. Like other classes of selfish genetic element discussed in this book, transposable elements are harmful to the host organism and persist in populations because of their non-Mendelian accumulation mechanisms (Doolittle and Sapienza 1980, Orgel and Crick 1980, Brookfield 1995, Zeyl and Bell 1996).

This does not mean that all insertions have been and will be harmful. It would be surprising if this were the case, and indeed there is evidence that some insertions are beneficial. There is no contradiction between calling transposable elements "selfish" or "parasitic" and entertaining the possibility that, in some species, the great majority of extant insertions may be beneficial, because natural selection ensures that the extant insertions are a highly biased sample of all those that have occurred. We define them as selfish based on selection pressures: the difference between transposition rates and excision rates means that the average effect of transposable element activity must be negative. Host genes will therefore be selected to suppress them, and transposons selected to circumvent host defenses. This is so even if a minority of insertions is beneficial and the host lineage is much better off with the elements. A similar distinction holds for more familiar organisms: under some conditions of nutrient cycling, herbivory may actually make a plant population more productive (de Mazancourt et al. 1998). But this does not mean that plants will be selected for palatability—rather, they will be selected to prevent herbivory, and only by considering the selection pressures will one be able to make sense of plant and herbivore design. Or, consider *Plasmodium*, the organism responsible for malaria. Normally, malaria is bad for an individual, but occasionally it can be beneficial: in the early decades of the 20th century, Julius Wagner-Jauregg intentionally infected thousands of patients suffering from syphilis with malaria, thereby inducing a fever and increasing syphilis remission rates from less than 1% to 30% (and in so doing earned a Nobel Prize; Nesse and Williams 1994). This does not, however, change our minds about whether *Plasmodium* is best considered a parasite. TEs may be important for the host in creating variation and thereby facilitate evolutionary change, but this is not their *raison d'être*, and the host is selected to suppress them.

Origins

Ancient, Chimeric, and Polyphyletic Origins

The proteins produced by transposable elements are complex, with multiple functions and multiple independent domains. Moreover, their domain structure varies widely, even within the 3 main classes (DNA, LINE, and LTR elements). Phylogenetic analysis indicates that, throughout transposable element evolution, there has been a strong theme of domains being gained and lost, with new acquisitions coming both from other transposable elements and from host genes. Among the best analyzed in this regard are the LINEs, all of which share a reverse transcription (RT) domain (Malik et al. 1999, Yang et al. 1999). If this domain is used to construct phylogenies, the ancestral LINE appears to have encoded a single ORF and have been site-specific, with a restriction-enzyme-like endonuclease (REL-endo) domain downstream of the RT domain (Fig. 7.16). In some lineages, this REL-endo domain was replaced by an AP endonuclease (APE) domain acquired from the DNA repair machinery of the host cell. Among those elements with an APE domain, 1 lineage (including *L1* from humans) appears to have gained a second ORF with a leucine zipper motif, while another lineage has gained an ORF with cysteine-histidine zinc finger motifs (similar to the gag protein of LTR retroelements). These are both nucleic acid–binding motifs. Cysteine-histidine motifs are also found sporadically in the original ORF both upstream and downstream of RT. Finally, among those with a *gag*-like ORF, RNase H domains appear sporadically on the RT phylogeny and were apparently gained at least once from the host cell and then lost several times. Recombination continues to be important in generating diversity of mammalian *L1* LINEs, particularly at the 5′ (front) end (Furano 2000, Moran and Gilbert 2002).

LTR retroelements also vary somewhat in domain structure (Eickbush and Malik 2002). Most have 2 ORFs, but vertebrate retroviruses have acquired a third ORF, encoding an envelope for intercell transfer. At least 6 other lineages of LTR retroelements have also acquired *env*-like genes, and in the 3 cases for which a likely source has been identified, it is in each case a

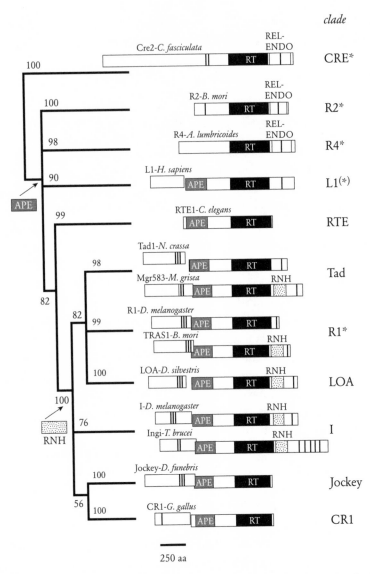

Figure 7.16 Gain and loss of enzymatic domains by LINEs over evolutionary time. For each of 11 clades of elements, representative molecular structures are shown as well as their phylogenetic relationships. RT = reverse transcriptase; APE = apurine/apyrimidinic endonuclease; REL-endo = restriction-enzyme-like endonuclease; RNH = RNase H domain. Vertical bars within an ORF represent the location and number of nucleic acid–binding domains. When 2 proteins are produced, they are slightly offset from each other. Numbers on branches are bootstrap support levels. * indicates that all known members of the clade are site-specific (i.e., target very specific sites in the genome), whereas (*) indicates that only some members are site-specific. Adapted from Malik et al. (1999).

virus (a baculovirus for *gypsy*-like elements, a phlebovirus for *CER7/13*-like elements, and a herpesvirus for *TAS*-like elements). LTR retroelements are also thought to be the source of the reverse transcriptase gene of plant caulimoviruses and vertebrate hepadnaviruses.

Most DNA transposable elements belong to one of a handful of families (e.g., the *hAT, En/Spm, piggyBac,* and *Mutator* families; Robertson 2002). The most widespread of these groupings is the *Tc1/mariner/pogo* superfamily, which includes transposable elements in animals, plants, fungi, ciliates, and bacteria. Within this superfamily, domain swapping also appears to have occurred (Plasterk et al. 1999). For example, *Impala,* an element of the fungus *Fusarium oxysporum,* is thought to be a chimeric element, one part derived from a *mariner*-like element and the other from a *Tc1*-like element (Langin et al. 1995). Some *Tc1*-like elements of fish are also thought to be chimeric (Ivics et al. 1996).

What about the more distant origins of the ancestral LINE, LTR retroelement, and DNA transposons? This is more speculative. Malik et al. (1999) suggest that the ancestral LINE evolved from a group II intron (see Chap. 6). These have an H-N-H type endonuclease domain downstream of RT. The ancestral LTR element with its 5 genes must itself be a chimeric element; the RT is obviously related to that of LINEs and most have an integrase that shares its catalytic core with that of *Tc1/mariner/pogo* transposase and a number of bacterial insertion sequences (Doak et al. 1994, Capy et al. 1996) as well as some bacterial host genes, including ribonuclease H and RuvC (Craig 1995). Thus LTR elements may be chimeras of a LINE and a DNA transposable element. For DNA transposons, as we have noted, the *Tc1/mariner/pogo* superfamily includes bacterial insertion sequences, as does the *Mutator* family (Robertson 2002), and some similarities have recently been identified between the active sites of these 2 families and the *hAT* and *piggyBac* families (Zhou et al. 2004).

It is striking that homology has yet to be detected between the *P* element and any other transposable element. *P* elements have thus far been found only in "muscamorphan and higher flies" (Robertson 2002). This tightly circumscribed distribution suggests the possibility that *P* elements have arisen relatively recently, perhaps somehow from 1 or more dipteran host genes. Further investigation is warranted, as this is the most promising opportunity for characterizing the birth of a transposable element. Interestingly, a stable host gene homologous to the first 3 exons of *P* elements has been discovered in some species of the *Drosophila montium* subgroup (reviewed in Miller et al. 1999). This gene has an additional untranslated exon and intron upstream of the *P*-homologous sequence and is missing the last exon, the presence of which distinguishes *P* transposase from *P* repressor. Its function is currently unknown. The authors interpret this gene as an example of transposable element domestication, and this does seem the most likely explanation. Nevertheless, it is only distantly related to *P* elements, and the possibility remains that it is the outgroup, and the source, of *P* elements.

Finally, we have already noted that SINEs are often derived in part from small untranslated RNAs, such as tRNAs or, in the case of human *Alu*s, 7SL RNA. Why tRNAs? Smit (1996) suggests that one predisposing factor is that tRNA genes have internal promoters, and so transcription is largely independent of genomic location. Consistent with this, *Alu* elements have also acquired an internal promoter, by a 2bp mutation of the original 7SL RNA gene. Why 7SL RNA? Boeke (1997) notes that this is the RNA scaffold of the signal recognition particle, which is involved in protein translation and thus is found near ribosomes. *Alu* RNA binds to some of the same proteins as the 7SL RNA, and Boeke (1997) speculates that it also accumulates near ribosomes and often is able to latch on to an *L1* protein as it is being synthesized, before the *L1* transcript does. tRNAs are also expected to congregate around ribosomes.

EIGHT

Female Drive

IN MALE MEIOSIS a diploid nucleus first replicates its DNA and then divides twice to create 4 haploid meiotic products, all of which are normally viable and develop into sperm or pollen. In female meiosis only 1 haploid cell normally survives to develop into an egg or ovule, with the others degenerating into "polar bodies." In the latter case, there is an obvious opportunity for drive: if a gene or chromosome can avoid the polar bodies and get into the egg or ovule more than 50% of the time, it will tend to increase in frequency. This is meiotic drive *sensu stricto,* as originally defined by Sandler and Novitski (1957). The movement of chromosomes at meiosis is mediated through the centromere, the region of the chromosome that attaches to the spindle, and centromere evolution figures prominently in this chapter. We begin by describing a well-studied supernumerary segment on chromosome 10 of maize (corn) that forms an extra "neocentromere" during meiosis, by which it attaches to the spindle and ratchets itself along—faster than its homolog—toward the spindle pole. Due to the geometry of female meiosis, this leads to it being transmitted to eggs at a greater than Mendelian rate—it drives. This behavior has allowed the segment to spread through maize populations, but it is a strategy that itself is open to parasitism; segments have evolved on other chromosomes that exploit the original segment and show drive in its presence.

No case of female drive has been studied as intensively as knobs in maize.

We review cases that may be similar but for which we lack key information on transmission through female meiosis. Female drive may also cause rapid evolutionary changes in the form, sequence, and number of normal centromeres—though again, direct observations of transmission rates are rare. And it may underlie one of the most striking aspects of female meiosis, namely, the high frequency with which eggs end up with the wrong number of chromosomes. More generally, differential transmission through female meiosis can drive the evolution of chromosome number, size, and shape—in other words, the karyotype. Finally, we review cases in which polar bodies have evolved to rejoin the germline and speculate whether this too may be the result of selfish genes.

Selfish Centromeres and Female Meiosis

Abnormal Chromosome 10 of Maize

In maize (as in many other species) the geometry of female meiosis is such that the 4 haploid meiotic products are produced in a row. The meiotic product at 1 end, nearest the base of the ovary, develops into the female gametophyte, while the other 3 degenerate. Any gene or chromosome region that can preferentially get itself into this basal meiotic product will thereby have a transmission advantage. There is a variant form of chromosome 10 that manages to do just that.

Chromosome 10 is the smallest of the maize chromosomes, and there are 2 structurally different forms: a common or normal type (*N10*) and an abnormal knobbed type (*Ab10*). These knobs function as neocentromeres at meiosis, meaning that the spindle attaches to them as well as to the normal centromere (Fig. 8.1). The knobs are at one end of the chromosome, and in plants that are heterozygous for the normal and knobbed chromosomes, there typically is a crossover between the normal centromere and the knob. This means that at the first meiotic division, chromatids with and without knobs go to each pole. Because the knob acts as a centromere, the chromatids carrying the knobs get to the poles first. This orientation is maintained for the second division, resulting in the knobbed chromosome being in the outside meiotic products, and the knobless chromosomes being on the inside. Because the basal-most meiotic product is 1 of the 2 outside meiotic products, the knobbed chromosome has been preferentially transmitted.

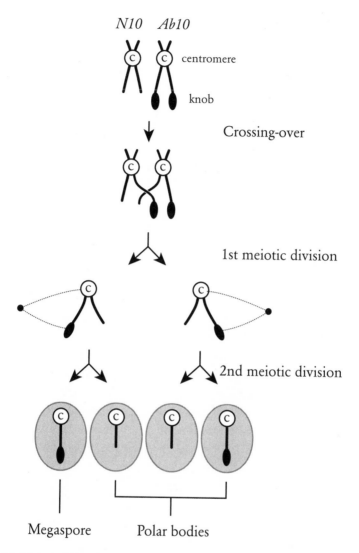

Figure 8.1 Meiotic drive of a chromosomal knob in maize. Normal and knob-bearing chromosomes pair up at meiosis, and typically 1 crossover occurs between the centromere and the knob. Consequently, at the first meiotic division, chromatids with and without a knob go to each pole. Because the knobs attach to the spindle and move along it, the knob-bearing chromatids lead the way. As a result, they tend to be found in the 2 outside meiotic products, one of which forms the megaspore. In this way the knob moves preferentially to the germline and away from the polar bodies.

Transmission rates average about 70%, varying in response to growth condi-tions and genetic background (Rhoades 1942, Kikudome 1959). The knobs are also active at male meiosis, but as all 4 meiotic products are viable there is no accumulation in the male line. They are not active at mitosis.

Details of how this mechanism works at the molecular level are beginning to emerge. Structurally, *Ab10* differs from *N10* in having extra DNA in-serted in at least 2 locations near the end of the long arm. It also has an in-version, which means that recombination rarely (if ever) occurs between *N10* and *Ab10* in this region (Fig. 8.2). The extra DNA is approximately equal to the length of the short arm of chromosome 10, or about 3% of the genome (Hiatt and Dawe 2003b). The proximal insertion has 3 prominent chromomeres, which consist mostly of tandem repeats of a 350bp motif called TR-1. The other, more distal insertion contains the large knob, which consists of about a million copies of a different, 180bp repeat (Peacock et al. 1981, Ananiev et al. 1998, Hiatt et al. 2002). Interestingly, the TR-1 and 180bp repeats have some homology, both to each other and to the centro-mere of chromosome 4, as well as to repeats on the B chromosome of maize (Alfenito and Birchler 1993, Hsu et al. 2003).

The TR-1 and 180bp repeats both show neocentromeric activity, and in both cases this activity depends on the *cis*-acting repeat sequences and on *trans*-acting genes in (or closely linked to) the unique regions of *Ab10*. Dif-ferent genes interact with the 2 types of repeat, and the latest studies suggest that at least 4 loci in the novel portion of *Ab10* contribute to drive (Hiatt et al. 2002, Hiatt and Dawe 2003a). Detailed cytogenetic studies have shown some differences between these neocentromeres and normal centromeres (Yu et al. 1997). For example, knobs interact with the sides of spindle fibers, in contrast to normal centromeres, which interact end-on (i.e., spindle fibers terminate at normal centromeres). Also, unlike normal centromeres, knobs are not microtubule-organizing centers. What they appear to have in com-mon is movement along the spindle by molecular motor protein(s). Appar-ently molecular motors (perhaps related to kinesins) interact with the TR-1 and 180bp repeats and with the spindle fibers, pulling the chromosome along the spindle.

Is there any significance to the fact that chromosome 10 is the shortest chromosome in maize? We speculate it is because short chromosomes tend to have relatively high rates of recombination per unit length, due to the re-quirement for at least 1 crossover event per bivalent to ensure proper segre-

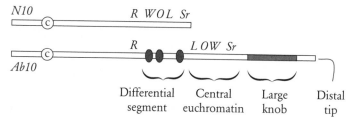

Figure 8.2 Normal *(N10)* and abnormal *(Ab10)* chromosome 10 of maize. The extra segment on *Ab10* has 4 distinct regions; in order from the centromere they are: (1) a differential segment that contains 3 prominent chromomeres; (2) central euchromatin, which contains a transposed and inverted portion of *N10* that spans at least 14 map units on *N10* and contains 3 known host genes (*W2, O7,* and *L13*); (3) a knob, which consists of deeply staining heterochromatin; and (4) the euchromatic distal tip. At least 5 independent breaks would be required to convert a normal chromosome 10 into *Ab10*. Adapted from Hiatt and Dawe (2003a).

gation. A high rate of recombination facilitates drive because at least 1 crossover is necessary between the normal centromere and the knob in order that chromatids with and without knobs go to each pole at the first division (i.e., the division is "equational" for the knobs). The ideal here would be exactly 1 crossover, which would give 100% second division segregation (SDS); for 2 crossovers there would be 50% SDS, assuming no chromatid interference (and for an infinite number, it would be 67%). It is interesting in this context that *Ab10* appears to encode factor(s) that strongly increase recombination (up to 5-fold) in some regions between knobs and centromeres (Kikudome 1959, Rhoades and Dempsey 1966, Hiatt and Dawe 2003a, 2003b). On the other hand, close physical proximity between centromere and neocentromere probably helps ensure that they are coordinated enough to be attached to the same spindle (Yu et al. 1997). In the absence of coordination, the 2 centromeres on the same chromosome could attach to different spindle poles, resulting in chromosome bridging, breakage, and loss. Thus knobs should do best where there is a high rate of recombination per unit length.

As well as being found in some domesticated maize races, *Ab10* is also found in the undomesticated sister taxa *Z. m. parviglumis* and *Z. m. mexicana* (teosinte). These latter also have a second cytologically distinguishable abnormal knobbed type, *Ab10-II,* which may be ancestral as it does not appear to contain the TR-1 repeat (Hiatt et al. 2002). In a survey of 51 populations of *Z. m. parviglumis* and *Z. m. mexicana, Ab10* (both types combined) was found in 37% of populations; in these populations, it had an average frequency of 14%, with a maximum frequency of 50% (data from Kato 1976, analyzed by Buckler et al. 1999).

Other Knobs in Maize

The emerging picture, then, is that there are protein-coding genes on *Ab10* that cause the TR-1 and 180bp repeats to act as neocentromeres and show drive. Because the protein-coding genes are linked to the repeats, they too show drive. But this system is open to exploitation: because the proteins are freely diffusible, they can act on repeats located anywhere in the genome. Thus, if repeats arise on a different chromosome, they too can drive in the presence of *Ab10*, even though the protein-coding genes do not gain anything by this, and probably even lose. The maize genome does, indeed, have many knobs (smaller than those on *Ab10*) that show neocentromeric behavior and drive in the presence of *Ab10*. Knobs have been found at 22 different places on all 10 chromosomes in the maize genome, and at a total of 34 different locations if one includes the wild relatives *Z. m. parviglumis* and *Z. m. mexicana* (Dawe and Cande 1996; Fig. 8.3). In a survey of 28 knobs, 14 consisted mostly of the 180bp repeat, 4 consisted mostly of the TR-1 repeat, and 10 had both types of repeat (Hiatt et al. 2002). All are thought to show meiotic drive in the presence of *Ab10*, with transmission rates from 59% to 82% for different knob sizes and positions (Longley 1945).

There appears to be a striking, positive association between knob size and drive. For example, transmission rates for knobs on the small arm of chromosome 9 are 69%, 65%, and 59%, respectively, for large, medium, and small knobs (data from Kikudome 1959, analyzed by Buckler et al. 1999). And, when a locus is heterozygous for knobs of different sizes, the larger knob appears to "win." For example, if an individual is heterozygous for a small and medium knob on chromosome 9, the latter segregates to 65–70% of the ovules. Recent microscopical studies show that larger knobs move faster on the spindle than smaller ones, perhaps because more molecular motors are attached to them (Yu et al. 1997). These results may have implications also for the growth of centromeric repeats, associated with drive.

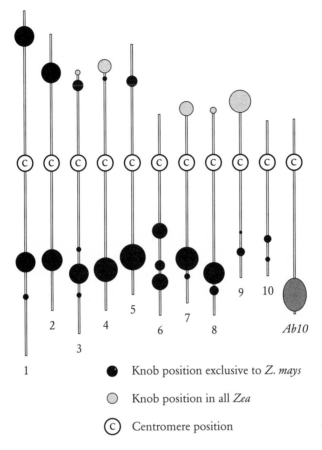

Figure 8.3 Location of knobs on maize chromosomes. The area of the circle is proportional to the size and frequency of the knob. The black internal knobs are unique to maize, while the gray terminal knobs are found in all *Zea* species. *Ab10*'s knob is larger than representation in this figure would allow. Adapted from Buckler et al. (1999).

Not surprisingly, the incidence of these other knobs is associated with the presence of *Ab10*. In a survey of 51 populations of *Z. m. parviglumis* and *Z. m. mexicana*, a composite "knob index," based on the size and frequency of knobs, was positively correlated with the frequency of *Ab10* across populations ($r_s = 0.37$, $p<0.01$; data from Kato 1976, analyzed by Buckler et al. 1999). The correlation is also significant in an analysis of phylogenetically independent contrasts.

For each chromosome arm there is likely to be an optimal position for a knob. As we have seen for *Ab10,* knobs work best if they are far enough away from the normal centromere that they are separated by at least 1 crossover event; on the other hand, they have to be close enough to coordinate attachment to the same spindle pole. Experimental data give some support to the idea of an optimum: a knob on the long arm of chromosome 3, found 25μm from the centromere at the pachytene stage of meiosis, is inherited in 71% of ovules. An inversion bringing it closer, 19μm from the centromere, resulted in 67% inheritance, and another inversion taking it more distal, 31μm from the centromere, resulted in 62% inheritance; both reductions were very highly significant (Buckler et al. 1999). Further support for the idea of an optimum position comes from a frequency distribution of knob positions, showing they tend to be clustered around 25μm from the centromere (Fig. 8.4). Only 28% of the genome is between 19μm and 31μm from the centromere, but 71% of knob sites and 82% of the "knob index" (which incorporates knob size) fall in this region. It is also worth noting that knobs are significantly overdispersed among chromosome arms, with generally 1 frequent position per arm (Fig. 8.3). This may reflect competition between knobs on the same arm.

There is good evidence, then, that meiotic drive has substantially reshaped the maize genome and has done so relatively rapidly. Most knobs are thought to have evolved within the last 100,000 years (Buckler et al. 1999).

Deleterious Effects of Knobs in Maize

If the only effect of *Ab10* or any other knob was to distort segregation ratios in its own favor, one would expect it to go to fixation. The fact that no knob has done so suggests they have deleterious side effects. There is suggestive evidence of this on several counts:

- In experimental crosses *Ab10* is transmitted to only about 45% of offspring through pollen, apparently because *Ab10*-bearing pollen is less effective than pollen with *N10* in fertilizing ovules (Rhoades 1942), indicating a cost of knobs on pollen fitness. This is an example of a sex-antagonistic effect in the predicted direction, the cost being borne by

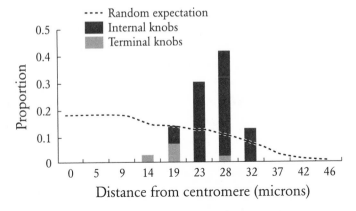

Figure 8.4 Position of maize knobs relative to the centromere. The Y axis shows the proportion of the total "knob index" at a given distance from the centromere; the knob index is a composite measure incorporating both the size and frequency of the knob. Distances were measured during pachytene. The random expectation is the expected distribution of knobs if they were randomly distributed along chromosome arms. Adapted from Buckler et al. (1999).

the nondriving sex (see Chap. 2). No such cost was observed for *Ab10-II* in the one maize background in which it has been tested (Rhoades and Dempsey 1988).

• Multiple B chromosomes can cause the loss of knobbed chromosomes (Rhoades and Dempsey 1972, 1973), indicating a cost of knobs in the presence of Bs.

• Knob DNA is the last to be replicated, and its presence probably lengthens the synthesis phase of the cell cycle (Pryor et al. 1980). This late replication can apparently also lead to mitotic abnormalities (Fluminhan and Kameya 1997).

In addition to these observations, there is also presumably the risk that the centromere and neocentromere on the same chromatid occasionally get attached to opposite spindles, leading to chromosome breakage and reduced fertility. Any of these costs could in principle slow the spread of a knob through a population and prevent it from going to fixation; whether any of them is quantitatively sufficient is not yet clear.

Knobs, Supernumerary Segments, and Neocentromeres in Other Species

Do similar elements exist in other species? Certainly there is none that is as well studied as *Ab10*. The closest parallel we are aware of is in the dicot *Rumex acetosa* (Polygonaceae). There is a heterochromatic knob near the end of the short arm of chromosome 1 that shows a transmission advantage through females of 63% (p<0.01), but not through males (43%, n.s.; Wilby and Parker 1988). The mechanism is not known, but it must be at least slightly different from that of *Ab10*, because the knob does not show neocentromeric activity in male meiosis (female meiosis was not examined). In a survey of British populations, the frequency of the knob was low (ca. 5%), with no obvious geographical variation.

More generally, knobs are found in many plant species (Hiatt et al. 2002, Dawe and Hiatt 2004). Usually they are polymorphic (e.g., *Vicia*, *Scilla*, *Anemone/Hepatica*, *Secale*, *Trillium*, and *Cetacea*) and seem to consist of long tandem arrays of DNA sequences. Neocentromeres have been observed in at least 14 species of plants, and at least in some the neocentromeres form at knobs. Interestingly, in half the cases the neocentromeres form only in species hybrids. This is a common feature of selfish genetic elements—that they become especially active in species hybrids—and suggests that perhaps suppressors have evolved that prevent neocentromeres from being active within species, and it is only by being dissociated from the suppressors in hybrids that they can become active.

Among animals, polymorphic supernumerary segments are common in some taxa such as the Orthoptera and may be analogous to knobs, though, again, transmission has rarely been studied. In one exception, in the grasshopper *Chorthippus jacobsi*, there are polymorphic supernumerary segments on 7 out of 8 autosomes; and at least 2 of them (both distal on the smallest chromosome) show significant drive, of about 80%, in both females *and* males (López-León et al. 1992a). The authors note that these supernumerary segments repel chiasmata in heterozygotes, so coadapted gene complexes can be created in regions close to a segment (if it remains polymorphic for a significant time).

Interestingly, supernumerary segments in orthopterans also show interactions with B chromosomes. We have already seen that the knob repeats of maize have some homology to repeats on the maize B, and that there is a competitive or antagonistic interaction between knobs and Bs in the sense that the presence of Bs is likely to lead to the loss of knob-bearing chromosomes. The same is true in the grasshopper *Eyprepocnemis plorans:* a supernumerary segment proximally located on the smallest autosome is composed of the same DNA repeats that make up a major proportion of the Bs in this species (Cabrero et al. 2003) and usually is transmitted normally, except that it is significantly eliminated by heterozygous females carrying Bs (López-León et al. 1991). Despite this apparent competitive exclusion, supernumerary chromosome segments in the acridid grasshoppers are significantly more likely in species that have Bs than those without, though there are many species that have only one or the other (data of Hewitt 1979, analyzed by López-León et al. 1991). It is possible that a B could translocate to an A and become a segment, and vice versa, but it is also possible that centromeric repeats colonize either location from each other and from A centromeres.

Meiosis-Specific Centromeres and Holocentric Chromosomes

As we have already indicated, there is presumably a risk when a chromatid has 2 centromeres that they attach to opposite spindles, leading to chromosome breakage and reduced fertility. As long as the effect is smaller than the advantage due to drive, a selfish neocentromere will still be able to spread. But as it does so, it generates selection pressures on unlinked genes to respond in some way. One response would be to suppress the neocentromeres and, as we have seen, there is suggestive evidence of this in the appearance of neocentromeric activity in species hybrids. Another possible response would be to suppress the original centromere and establish the neocentromere as the normal one. If, as in maize, the neocentromere is active only in meiosis, and the original centromere is suppressed only in meiosis, the species would have functionally different centromeres at mitosis and meiosis. Something like this is seen in species with "holocentric" chromosomes, in which the centromere is diffuse and the spindle attaches along the whole length of the chromosome. These have been best studied in nematodes (Goday and Pimpinelli 1989). Interestingly, they are only holocentric at mitosis; at meiosis, they show a standard localized centromere.

Localized centromeres may be necessary at meiosis so chromosomes can recombine without ending up attached to 2 spindles and being torn apart. Centromeres that are purely mitotic have no such constraint, and they are free to expand along the whole chromosome; perhaps this expansion even increases mitotic fidelity. In some parasitic nematodes, the meiosis-specific centromeric sequences are actually lost from the somatic cells, as part of their "chromatin diminution" (described further in Chap. 11). Some moths and bugs also have chromosomes that are holocentric in mitosis but have a localized centromere at meiosis (Holm and Rasmussen 1980, John 1990). We speculate that it was the evolution of meiosis-specific selfish centromeres that led to the evolution of distinct mitosis- and meiosis-specific centromeres, which in turn allowed the evolution of holocentric chromosomes.

More generally, meiosis and mitosis may differ in how actively the chromosomes participate in the 2 processes (Dawe and Hiatt 2004). Meiotic chromosomes in maize, *Drosophila*, and mice have all been shown to actively initiate spindle formation, whereas mitotic chromosomes are typically inert.

Selfish Centromeres and Meiosis I

The neocentromeres of maize are active at the first meiotic division in a way that they show drive at the second meiotic division. Normal centromeres segregate at the first meiotic division, and so if they are to show drive, they must recognize and exploit some asymmetry at this division. Many B chromosomes do just this (see Chap. 9). We are not aware of any direct evidence of driving centromeres on normal A chromosomes, but the possibility has been raised many times, and it could help explain some otherwise-puzzling observations.

The first puzzling observation is that centromeric sequences change rapidly over evolutionary time, even though the basic structure remains the same. The core of a centromere, where the kinetochore forms and the spindle microtubules attach, typically consists of at least 500kb of tandem repeats, in which the repeat unit size is remarkably constant across a wide range of species: 171bp in primates, 168 in rice, 175 in maize, 155 in a fungus gnat, 186 in a fish, and so on (reviewed in Henikoff et al. 2001). This is the same range of sizes as found for nucleosomes. These sequences show concerted evolution, with the same sequence evolving on different chromosomes (analogous to the same repeat being found in multiple knobs on different maize chromosomes), and the same sequences are also often found on B chromosomes (see Chap. 9). Within the centromere cores there may also be islands of complex DNA, and attached to the repeats is a centromere-specific histone protein, CenH3. The surprising observation is that, although the unit size is conserved, the centromeric repeats are among the fastest-evolving sequences yet found in eukaryotic genomes, differing even between closely related species (Henikoff et al. 2001). Moreover, its associated histone, CenH3, is also relatively fast evolving, even though the noncentromeric histone from which it is derived, H3, is among the most conservative proteins known. In *Drosophila* and *Arabidopsis* the centromeric histone has recently been under positive selection (Malik and Henikoff 2001, Talbert et al. 2002). Another centromeric protein (CENP-C) has been under positive selection in plants more generally and in mammals (Talbert et al. 2004). Why should centromeric sequences and associated proteins continually evolve while apparently not changing in function and while the spindle apparatus to which they attach hardly evolves at all (Henikoff and Malik 2002)?

On the assumption that such rapid coevolution indicates an evolutionary struggle–analogous to the rapid evolution associated with parasite-host interactions or sperm-egg protein interactions–Henikoff et al. (2001) suggest that selection has continually favored centromeres that bind more strongly or quickly to the part of the spindle apparatus directed toward the egg rather than the polar body in female meiosis. Moreover, they suggest that the centromeric proteins CenH3 and CENP-C counterevolve so as to make such biases less pronounced. Drive is imagined to be relatively weak in each generation, and it is weakly opposed by negative side effects, including a possible reduction in male fertility (Daniel 2002).

There is as yet no direct evidence for this idea–for example, we lack a demonstration that chromosomal transmission in females depends on the sequence of the centromeric repeats and how they interact with CenH3. The large number of repeats found in centromeres is circumstantial evidence that strength of binding to the spindle is selectively important, and deletion studies point in the same direction. In a *Drosophila melanogaster* minichromosome, the size of the centromere affects its chance of transmission, especially, as expected, in females (Murphy and Karpen 1995, Sun et al. 1997). *Dp1187* is a fully functional minichromosome (1.3Mb) that is not a normal part of the genome and has no known phenotypic effects. Only about one-fourth of the chromosome is necessary for normal transmission;

and in all such deletions, there is a small but significant drive of univalents in females (53%) but not males (49%). But once portions of the key core are deleted, transmission is progressively reduced, especially in females. Similarly in maize, misdivisions of a B chromosome centromere generate a range of centromeric sizes from 200kb to 4Mb and, across this range, decreasing size is associated with lower transmission (Kaszas and Birchler 1998). As with the minichromosome, excision of particular regions has major effects. The situation is made even more intriguing by observations that centromeres may be defined epigenetically, rather than by primary sequence (Amor et al. 2004). In addition, recent evidence from maize shows that mRNA produced by the centromeric repeats forms an important part of the protein matrix that surrounds the centromere (Topp et al. 2004).

Note that while these processes of centromeric competition may be strong in outbreeding, sexual species, they are expected to be rare or absent in asexual ones. In this regard, it is noteworthy that a centromeric repeat makes up 15–20% of the total genome of a sexual species of stick insect *Bacillus grandii*, but only 2–5% in a closely related asexual form *B. atticus* (Mantovani et al. 1997). Also, there is no diversification across a wide geographical range (Italy to Greece) in the repeat found in the asexual species, but clear diversity within Sicily of the sexual one. It would be most interesting if this relationship between breeding system and centromere size was generally true. One might also expect centromeres from outcrossed species to be "dominant" over those from inbred species and show preferential segregation in species hybrids (Fishman and Willis 2005).

The second puzzling observation is the high frequency with which female meiosis goes wrong. In humans, an estimated 10–25% of pregnancies have too many or too few chromosomes because of errors during female meiosis (Hassold et al. 1996). This frequency of nondisjunction is much higher than in males, and most errors are at the first meiotic division. How can a character so closely related to fitness be so error-prone?

One possible explanation is that the errors are due to selfish centromeres trying to get into the egg and avoid the polar body and that sometimes both homologs end up there (Axelrod and Hamilton 1981; see also Day and Taylor 1998, Zwick et al. 1999). It is an intriguing idea, well worth testing. Ideally, one would want to find some *cis*-acting sequence that, when mutated or deleted, caused drag when heterozygous with the wildtype but that, when homozygous, resulted in normal segregation with reduced levels of nondisjunction.

The Importance of Centromere Number:
Robertsonian Translocations in Mammals

Female drive may also affect the number of centromeres, as indicated by the transmission of Robertsonian translocations. This is a relatively common class of chromosomal rearrangement in which 2 chromosomes with terminal centromeres (acrocentrics) combine to give a large metacentric chromosome and a small centric fragment without any genes, which is then lost, without harmful consequences. In humans (who normally have 5 acrocentric chromosomes) Robertsonian translocations arise at a frequency of about 4×10^{-4} per gamete per generation, the majority of which are of maternal origin (Pardo-Manuel de Villena and Sapienza 2001c). In some lines of mice (which normally have 20 acrocentric chromosomes), the frequency can be as high as 10^{-3} per gamete per generation (Nachman and Searle 1995).

In females that are heterozygous for a Robertsonian translocation, the 2 shorter chromosomes pair up with the single larger chromosome at meiosis, and there is often a bias in which ones are transmitted to the egg. The bias seems common to the various translocations found within a species but the direction of the bias differs between species. In mice, metacentrics arising from Robertsonian translocations of various chromosomes tend to show drag, on average being inherited by about 40% of the progeny of heterozygotes, though there can be differences according to the chromosomes involved and the genetic background (Gropp and Winking 1981, Nachman and Searle 1995). This only occurs in females; inheritance is Mendelian through males, and direct observations show that the drive is due to preferential segregation at the first meiotic division (Pardo-Manuel de Villena and Sapienza 2001c). By contrast, Robertsonian translocations in humans show drive, typically being inherited by about 59% of progeny. Again, this effect is found only in females; inheritance through males is Mendelian. All else being equal, this drive would cause the metacentric chromosomes to spread through the human population. But heterozygotes are at a high risk of producing aneuploid gametes and so the metacentric chromosomes (which will always be in the heterozygous state when rare) are selected out, and the meiotic drive merely increases the average number of people who suffer from each translocation event.

As expected from these opposing transmission biases of Robertsonian translocations, the majority of mice chromosomes are acrocentric (in fact, all are) while the majority of human chromosomes are metacentric. A larger

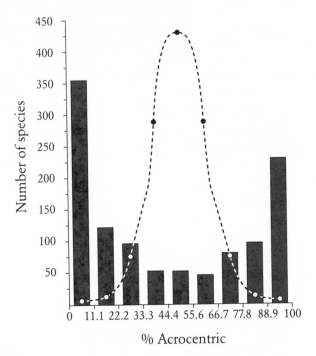

Figure 8.5A Distribution of acrocentric chromosomes in mammals. The number of species with various percentages of acrocentric chromosomes (divided into 9 equal-sized categories) is plotted for all mammals (black bars). Also plotted is the expected frequency if chromosomes were randomly distributed around the overall mean frequency of acrocentrics of 50.6% (dotted lines). Note the extraordinary contrast between the random and the actual. Adapted from Pardo-Manuel de Villena and Sapienza (2001a).

review of mammalian karyotypes reveals that, in the great majority of species, one type of chromosome greatly predominates, even though overall metacentric chromosomes are almost exactly as frequent as acrocentrics (Fig. 8.5A). About 50% of species have less than 10% or more than 90% acrocentric chromosomes, and only 14% of species have between 33% and 66% acrocentrics. Indeed, there is no order of mammals in which this central interval contains the largest number of species, and many of them individually have a bimodal distribution of chromosome shapes (Fig. 8.5B). This holds for some families and genera, as well. Evidence from *Mus musculus* reveals some recently derived populations that show a reversal in

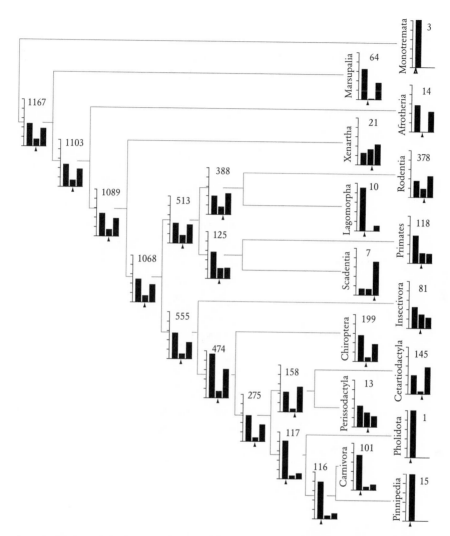

Figure 8.5B Distribution of acrocentric chromosomes in mammalian orders. Same as 8.5A except that 3 equal-sized categories of percent acrocentrics are used for orders of mammals, phylogenetically organized. Bars represent the proportion of species in each category, arrowheads indicate the mean, and numbers are the total number of species analyzed. Note that most orders lack the middle category almost entirely. Adapted from Pardo-Manuel de Villena and Sapienza (2001a).

the predominant kind of chromosome (i.e., mostly metacentric; references in Pardo-Manuel de Villena and Sapienza 2001a). (There is an interesting, if unexplained, negative correlation between metacentric length, as a fraction of total length, and transmission ratio against the 2 acrocentrics in a hybrid zone between races that differ strongly in number of metacentrics; Castiglia and Capanna 2000.) These observations indicate that the polarity of drive—or whatever else determines chromosome shape—easily and often reverses in mammals. This could itself be a genome maneuver to counter maladaptive effects of centromere competition.

Pardo-Manuel de Villena and Sapienza (2001a, 2001b) suggest that the relevant parameter determining the transmission of Robertsonian transloca-tions is the number of centromeres (2 vs. 1), with the "stronger" spindle tending to attach to the 2 acrocentrics and the "weaker" spindle to the metacentric. By hypothesis, the stronger spindle in some species (e.g., mice) leads to the egg, whereas in others (e.g., humans) it leads to the polar body. Consistent with the idea that centromere number is important, in XO fe-male mice (which are fertile) the X (1 centromere) drives against its absence (no centromere; LeMaire-Adkins and Hunt 2000). Data on B chromosome distributions is also supportive. Because most of these drive as a univalent, a B chromosome typically brings an extra centromere to 1 side of the meta-phase plate, just as do 2 acrocentrics compared to 1 metacentric. As ex-pected, B chromosomes are significantly more frequent in species of mam-mals with a higher proportion of acrocentric chromosomes (see Chap. 9). This is a robust finding: it holds separately for rodents, all other mammals, and in a taxonomic contrasts test for all mammals. It also holds within a species of grasshopper (Bidau and Marti 2004).

Sperm-Dependent Female Drive?

Quite a different form of female-specific meiotic drive has been suggested for 2 loci in mice, a homogeneously staining region (HSR) called *In* on chromosome 1 and the *Om* locus on chromosome 11 (Agulnik et al. 1993, Pardo-Manuel de Villena et al. 2000). Both reports come with the extraordi-nary claim that meiotic drive only occurs when the fertilizing sperm is wildtype; if the sperm carries a copy of the driving element, segregation at female meiosis is Mendelian. This is not impossible, as the second meiotic division in mice occurs after fertilization. But the results do need confirmation. For *In*, the original observations were for *Mus musculus musculus,* and subsequent research on a similar element in *Mus musculus domesticus* again showed non-Mendelian inheritance, but due to zygotic lethality rather than meiotic drive (Weichenhan et al. 1996; see Chap. 2). For *Om*, the data are indirect. Compelling evidence of sperm-dependent segregation would be very exciting and would raise a slew of evolutionary questions about the interaction.

Female Drive and Karyotype Evolution

As we have seen, the geometry of female meiosis can select for selfish genetic elements that actively exploit the asymmetry to enhance their transmission to the next generation. And this can have profound effects on the species karyotype—at least in maize. We have also seen that the mechanics of female meiosis may select for particular chromosomal morphologies—a more passive, less obviously selfish kind of adaptation. We now review further examples along this latter theme, in *Drosophila* and relatives. In these species, female meiosis is much like that of maize, in that the 4 meiotic products end up in a row and the egg nucleus always arises from the meiotic product at 1 end of the row (Rhoades 1952).

The best example of the mechanics of female meiosis affecting karyotype evolution comes from paracentric inversions (i.e., those that do not include the centromere), though drive itself is not involved. In *Drosophila* females heterozygous for such inversions, crossing-over within the inversion leads to 1 recombinant chromatid having 2 centromeres attached to different spindle poles, and 1 having no centromeres. Both of these tend to end up in the middle meiotic products, with the nonrecombinant chromosomes in the outside meiotic products, and therefore in the egg. This selection for nonrecombinant chromosomes underlies the use of inversions as balancer chromosomes (i.e., chromosomes that will not recombine with wildtype chromosomes) in *Drosophila* genetics. The fact that heterozygosity for a paracentric inversion reduces neither female fertility nor male fertility (because there is no crossing-over in male meiosis) means that rare inversions are not automatically selected against, as they would be in other taxa. As a consequence *Drosophila* populations are often segregating for paracentric inversions, apparently maintained by selection for reduced rates of recombination.

Other chromosomal differences can show drive at female meiosis. If 2 homologous chromosomes differ substantially in length (e.g., through a deletion), the shorter one has the transmission advantage (Novitski 1951, 1967). This arises for the following reason: if there is a crossover between the centromere and the deletion, the first division is "equational" for the deletion, with short and long chromatids going to each pole. Just by being bigger, the longer chromatid tends to drag more at the first division, and so the shorter chromatid leads the way. This is exactly the same as for *Ab10* of maize (Fig. 8.1), except in this case the longer chromatid does not have a neocentromere and moves slower rather than faster. This orientation is maintained for the second divisions, with the result that the shorter chromosomes are more likely to end up in the 2 meiotic products at the ends of the array and the longer chromosomes in the middle. Because the egg derives from 1 of the 2 outer meiotic products, the shorter chromosome has the transmission advantage. There is thus a bias in favor of deletions, and against insertions, particularly as one moves away from the centromere (or, more precisely, as second division segregation increases).

A similar mechanism also gives an inherent bias to chromosomal inversions that move the centromere closer to the middle of the chromosome (Heemert 1977, Foster and Whitten 1991). If females are heterozygous for an inversion across the centromere (a "pericentric" inversion), crossing-over within the inversion will produce reciprocal duplication-deficiency (*Dp-Df*) chromosomes that will be lethal if included in the egg. If the chromosomes differ substantially in the position of their centromeres, there will be a tendency at the first division for the more metacentric chromosome to segregate with a long *Dp-Df* chromatid (which will drag), and the more acrocentric chromosome to segregate with a short *Dp-Df* chromatid (which will precede it to the pole). Among functional eggs, the more metacentric chromosome will therefore have a transmission advantage. In a study of 3 blowfly *(Lucilia cuprina)* inversions with large differences in relative centromere positions, the more metacentric chromosome was inherited by about 60% of the progeny when the female was heterozygous (Foster and Whitten 1991). Transmission through males did not differ from the Mendelian 50%. Less extreme pericentric inversions showed no detectable drive, nor did paracentric inversions (i.e., those not including the centromere).

It is an open question to what extent such biases have been important in molding dipteran karyotypes—whether, for example, the advantage of shorter chromosomes has constrained the accumulation of "junk" DNA and kept the genome of *Drosophila* relatively small compared to some other insect groups (e.g., grasshoppers). If this effect has been important, we would expect the density of junk to decrease with distance from the centromere. It does seem that large supernumerary segments are rarer in dipterans than in some other taxa such as grasshoppers. As for the bias in favor of more metacentric chromosomes, the reduced fertility associated with pericentric inversion heterozygosity will be a selective barrier to karyotypic change. We would expect the barrier to be lower (and therefore drive more important) in species with small population sizes and with strong competition among siblings, so zygotes that die because they inherit a *Dp* or *Df* chromosome can be replaced by a close relative.

Karyotypes have been described in species after species for more than a hundred years. But until recently there has been little theoretical conception about how variation in simple parameters like the number, size, and shape of chromosomes could possibly be accounted for. Perhaps there is something idiosyncratic about every chromosomal rearrangement that determines its fate (e.g., changes in the sequence or expression of genes near the junctions), and one should therefore not expect to find general trends. But such trends are found, as suggested by White's (1973) survey of a large number of cases in which the same type of rearrangement has gone to fixation many times in the same lineage—what he calls "karyotypic ortho-selection" (Foster and Whitten 1991). The bimodal distribution of chromosome morphologies in mammals is another example of such concerted evolution. Perhaps now we can (dimly) see that part of the answer is likely to lie in the mechanics of female meiosis.

Polar Bodies Rejoining the Germline

We have emphasized in this chapter that the asymmetry of female meiosis presents an opportunity for a selfish genetic element to drive by arranging to be transmitted to the "correct" meiotic product. In principle, another way to exploit this asymmetry is to contrive, somehow, to convert the meiotic product the element ends up in into an egg or ovule—not to avoid polar bodies, but to transform them, and shift them back into the germline. In some species meiotic products that would normally be polar bodies do rejoin the germline. This can occur in two ways: the polar body can take the place of the sperm and fuse with the egg nucleus to restore diploidy, a form of automictic reproduction; or it can develop in parallel with the normal egg, and like it be fertilized by a sperm cell and then develop into a normal embryo. What is not yet clear in either case is whether the change of fate is due to genes active in the polar bodies (in which case they are selfish), or whether it is due to genes in the diploid tissue (in which case they are not).

Automixis. In some animal species, females go through 2 more-or-less normal meiotic divisions, producing 4 haploid meiotic products. Then diploidy is restored not by fertilization but by the second polar body fusing with the egg nucleus ("terminal fusion"; Fig 8.6A). In other species, the second polar body fuses instead with a derivative of the first polar body, leaving the egg nucleus to degenerate ("centric fusion"; Fig. 8.6B); and in yet other species the 2 products of the first meiotic division fuse, followed by a single second division, producing a diploid egg and a single polar body. Such automictic reproduction is the norm in some species, and reproduction is essentially clonal; in other species it seems to occur sporadically, as if by accident. In modeling the evolution of these unusual modes of reproduction, the convention is to consider the fusions as properties of the mother, and so genes for them are expected to increase if the fitness of the resulting offspring is more than half that of an outcrossed offspring (the "twofold cost of sex"). But perhaps these fusions are controlled instead by genes expressed in the polar body itself. The conditions for such a gene to spread would be considerably relaxed, and such a gene could legitimately be called "selfish," as it would be acting against the interests of genes in the mother, the egg nucleus, and other genes in the polar body that are linked to different degrees with the centromere. Such activity would appear to require either transcription in the polar bodies or, conceivably, the sequestration of active molecules by attachment to the chromosomes (analogous to *SD* in *Drosophila;* see Chap. 2).

Sexual progeny. In some plants and animals, 2 or even all 4 of the products of female meiosis can be fully functional, able to be fertilized and to develop into a normal organism. For example, in the plant *Sedum chrysanthum* (Crassulaceae), all 4 meiotic products appear to be functional, forming lateral tubes that grow through the maternal nucellus in the direction from which the pollen tube is going to arrive (Subramanyam 1967). Because 2 or 3 maternal cells may undergo meiosis in the same nucellus and all the products are active, there can be quite a tangle of haploid tubes (Fig. 8.7). Eventually, 1 of them penetrates through the nucellar epidermis, and this is the 1 that develops into the female gametophyte and is fertilized. The impression

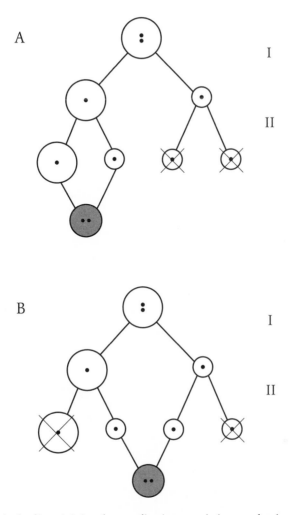

Figure 8.6 Polar bodies rejoining the germline in automictic reproduction. A. Terminal fusion. The crossed polar bodies degenerate and the shaded cell gives rise to the embryo. B. Central fusion. The crossed egg cell and polar body degenerate, and the shaded cell gives rise to the embryo. Adapted from Suomalainen et al. (1987).

one gets is of a race. That sister meiotic products might compete is not surprising; what is not clear is why in this and some other species (Maheshwari 1950, Johri 1963), multiple meiotic products are active while in others they are not. Is this due to gene expression in maternal tissues or in the polar bodies? It would also be interesting to know the evolutionary lifespan of

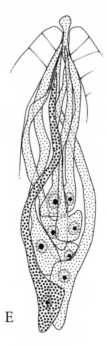

E

Figure 8.7 Competition among polar bodies in *Sedum chrysanthum (= Rosularia pallida)*. Two or 3 cells in a single nucellus can go through meiosis, each one producing 4 viable meiotic products that appear to race up to meet the incoming pollen. Only 1 of them will be fertilized and develop into a seed. Adapted from Maheshwari (1950).

such an arrangement—we might predict that competition will evolve to be sufficiently destructive that maternal modifiers reimposing a single winner might eventually be selected.

Development of multiple female meiotic products has occasionally been reported in animals as well. For example, in the polyclad flatworms *Prostheceraeus vittatus* and *Leptoplana*, the first meiotic division can produce 2 cells of equal size; in the second division, each of these produces a small polar body and then is fertilized to give rise to 2 independent embryos (Wilson 1928). The production of 2 embryos from a single meiotic precursor is reported as a "spontaneous abnormality"; we wonder whether there is genetic variation for this behavior and at which stage (pre- or postmeiotic) it is expressed.

NINE

B Chromosomes

B CHROMOSOMES ARE CHROMOSOMES that are additional to the normal set (called the As) and that are regularly found in some—but not all—individuals within a population. By definition then, they are not a *necessary* part of the genotype. Indeed, as we shall see, they are often harmful in their phenotypic effects (especially as they increase in number) and are maintained because they possess a "self-accumulation mechanism," that is, they give themselves an advantage in propagation not normally shared by any of the As. They drive. A minority of Bs, however, are either neutral or positive in their effects, the latter showing drag and the former near-Mendelian transmission.

At meiosis—when there is more than 1 B present—B chromosomes sometimes pair with each other and form chiasmata but not with the As. In other words, they are not trisomics, additional copies of 1 or more of the regular chromosomes that pair with their homologous chromosomes (e.g., all three 21st chromosomes causing Down's syndrome pair as a trivalent). They have escaped the discipline of the diploid set (for a typical A, 2 copies and 2 copies only) and can vary in number between (and within) individuals: 0, 1, 2, 3, 4, and so on. Although presumably all B chromosomes ultimately come from A material, they may be viewed now as independently evolving genetic parasites that are selected to increase their own rates of transmission while minimizing associated negative phenotypic effects (Östergren 1945, based in part on the work of Müntzing 1945). Because Bs are found in only some individuals, selection on a B to drive is usually stronger than it is on As to stop it (Östergren 1945). Regarding phenotypic traits, it is notable that in roughly half of all species with B chromosomes, Bs are highly heterochromatic (Jones and Rees 1982), a staining characteristic associated in other contexts with lack of genetic expression (e.g., Y chromosomes, inactivated Xs in female mammals, inactivated paternal chromosomes in male scale insects). There is independent evidence that Bs are usually mostly genetically inactive, perhaps having their main effects on transmission ratio distortion and (inadvertently) genome size and, therefore, rates of cell division (major gene effects being, with a few exceptions, conspicuously absent). As a result of the absence of coding genes and high frequency of repetitive elements, the study of the genetics and molecular biology of Bs is still in its infancy.

Although less widespread than transposable elements, Bs are the easiest kind of selfish element to detect, showing up usually as extra chromosomes visible under the microscope, often staining heterochromatic or with unique heterochromatic bands. Because they are so easy to detect, Bs were discovered early in the history of cytogenetics, about 100 years ago (Wilson 1906, 1907). They are now known from almost 2000 species (Jones and Rees 1982, Camacho et al. 2000, Jones pers. comm.) and are discovered regularly whenever cytogenetic work is done; a recent group shown to have Bs are the thorny-headed worms (Spakulova et al. 2002). Although there is large variation in the number of species with Bs between families, some of this variation is easily explained by differences in intensity of study. For example, only 3 plant families have more than 100 species with Bs and each has been studied intensively cytogenetically: the Gramineae (economic importance) and the Compositae and Liliaceae (amenable to study). Bs are present in a significant minority of species (2–15%) in many animal and plant groups. In a particularly well studied flora, British angiosperms, fully 15% of species have B chromosomes (Jones 1995, Burt and Trivers 1998b).

The chief advantage of this high frequency of known cases of selfish elements is the possibility of making comparative statements with some empirical foundation. Is inbreeding associated with a decrease in the frequency of the t haplotype on islands and SR in laboratory populations? Evidence may be partial or equivocal. Regarding the parallel assertion concerning B chromosomes—namely, that they are restricted to outbred populations (Jones and Rees 1982)—we shall see that there is clear and convincing positive evidence from a detailed study of the British flora. Comparative work also shows that the presence of Bs is associated with a variety of A characters, including genome size, chromosome number, ploidy level, and typical A chromosome shape. Unfortunately, because of the absence of genetic markers on B chromosomes, the phylogeny of the Bs themselves is nearly completely unknown.

The literature on B chromosomes is very large and has benefited from excellent review work: on all aspects of Bs (Jones and Rees 1982), the first international conference on Bs (Beukeboom 1994), Bs in plants (Jones 1995), and recent work on Bs (Camacho et al. 2000, Puertas 2002, Jones and Houben 2003, Camacho 2004, 2005).

We begin our review by describing the various modes of non-Mendelian inheritance. We describe the transmission parameters of several well-studied species, the phenotypic effects of B chromosomes, and the environmental correlates of their frequency in nature. In most species, Bs apparently exist in spite of negative phenotypic effects because their net effects on transmission (measured across the 2 sexes) is positive; but in a small minority of species, their net effect on transmission is negative (or neutral), while their net phenotypic effects are positive (or neutral). We show that B chromosomes are rare in inbred systems and that they appear to cause the A chromosomes to show heightened recombination, perhaps as an adaptive response by the As. B chromosomes are also more frequent in species with larger genomes, fewer chromosomes, and lower ploidy. Especially interesting is the fact that, in mammals, Bs are more frequent when a higher frequency of As are acrocentric.

A bizarre variety of B chromosomes have evolved: from Bs that are so small they consist only of centromeric material (Wolf et al. 1991, 1996) to Bs that gain their advantage by causing the destruction of all other paternal chromosomes (Werren and Stouthamer 2003). B chromosomes are a set of renegade As that have set themselves adrift from the regular set and the discipline of the diploid state and now evolve according to relatively simple principles of self-accumulation, as impeded by counteradaptations on the As and the growing phenotypic costs as they increase in number. They have evolved an impressive array of means to achieve drive—both mitotic and meiotic—and to these we turn next.

Drive

Drive is the key to understanding Bs and it occurs in a great variety of ways, but it is noteworthy that for none of these are the molecular mechanisms known. In what follows, we rely heavily on the excellent reviews found in Jones and Rees (1982), Jones (1991), and Jones (1995). We concentrate first on the various kinds of drive and then review what little is known about the genetics of drive. We review parameters of drive in well-studied species and emphasize that in a significant minority of species Bs fail to show drive and may even drag. We conclude by describing the way in which a greater degree of outbreeding is expected to increase the success of Bs, and we review supporting evidence from plants.

Types of Drive

Bs can drive during female meiosis by moving toward the egg pole, and in some very unusual ways, such as destroying paternal A chromosomes, but in all other cases, B chromosomes use nondisjunction as part of their mechanism of self-accumulation. When a B chromosome in a cell replicates itself, these two may fail to detach and this nondisjunction causes one daughter cell to inherit 2 B chromosomes and the other 0–rather than the faithful 1 copy each that regular disjunction gives (Plate 8). There has been no increase in *average* copy number, of course, only an increase in the within-individual variation; but if nondisjunction is either directional–with a bias toward the germline–or is followed by more rapid reproduction of the B-containing germinal cells, the B will increase in number in the relevant lineage (Fig. 9.1). Imagine, for example, the cell prior to the first cell of a gonad (or germline-soma split). When it divides, one cell initiates the gonad and the other daughter cell becomes a helper, in other words, a somatic cell. Directed nondisjunction at this crucial cell division will give the B an advantage in propagation. There are a series of such nodal points in development, and each gives an opportunity for directed nondisjunction.

In addition to directed nondisjunction, at any time in the development of the germline, the nondisjoined B may cause its cell to outreproduce sister cells lacking the B. This is not expected to be an intrinsic tendency of B chromosomes; quite the contrary, by adding DNA, Bs tend to slow cell development, increase cell size, and decrease rates of mitosis (see later). These effects are stronger (in rye) per unit B chromosome added than per unit A, perhaps because Bs are more compacted (e.g., Evans et al. 1972). Thus, over-reproduction of cells with Bs must result from some directed activity of the B, including possibly hostile actions toward neighboring cells lacking Bs. In

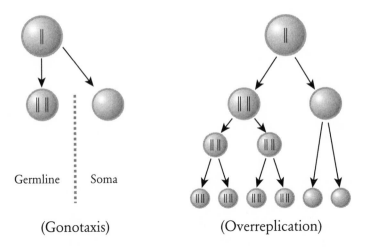

Germline ┊ Soma

(Gonotaxis)   (Overreplication)

Figure 9.1 Two primary ways by which B nondisjunction gives an advantage. Left: Gonotaxis, or preferential movement to the germline. Right: Overreplication (in germinal cells) compared to cells not containing the Bs.

any case, nondisjunction creates the variance in B number per cell that can then select for drive.

Because it is sometimes unclear whether nondisjunction is directional or is associated with overreproduction of B cells, we organize the following examples according to the time in development when the Bs gain their reproductive advantage. The gain can be:

1. Mitotic (before meiosis)
2. Meiotic
3. Gametophytic (after meiosis)
4. At fertilization

Like A drive, the selfish advantage usually occurs in one sex only.

Mitotic (before meiosis). In the plant *Crepis capillaris*, a single B replicates itself faithfully in vegetative tissue (meristem, root, caudicle, leaf); but 1 to 4 days prior to the transition to reproductive shoots, nondisjunction of Bs apparently results in stem cells with no Bs and 2 Bs and cells with 2 Bs soon predominate (Fig. 9.2). Although self-accumulation of B

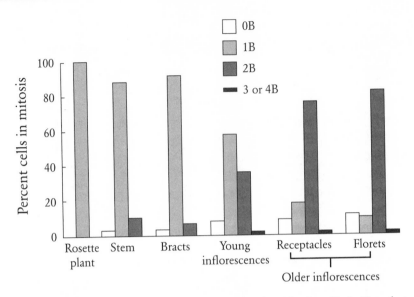

Figure 9.2 Distribution of Bs in different parts of the plant *Crepis capillaris*. Note that Bs increase in frequency as they approach the flowers. Adapted from Jones and Rees (1982), based on Rutishauser and Röthlisberger (1966).

chromosomes prior to meiosis is relatively rare in plants, it appears to be common in some animals. Nondisjunction always appears to be involved but to what degree this is directional is usually not known. In grasshoppers there are several examples of B chromosome increase during mitoses in testicular tissue (Nur 1969a). In *Calliptamus palaestinensis* this increase is achieved in early embryos during the differentiation of germ cells from somatic tissues, apparently by a form of directed nondisjunction (in 1B individuals) such that 2B germ cells outnumber 0B cells by 15:1 while somatic tissue remains 1B (Nur 1963). In *Camnula pellucida* Bs are 37% more frequent in spermatocytes than in the gastric caeca (Nur 1969b). The same kind of premeiotic drive also occurs in the locust *Locusta migratoria* (Kayano 1971, Viseras et al. 1990).

Meiotic. Meiotic drive for a B chromosome was first shown in the lily *Lilium callosum* (Kayano 1957). 1B females generate 80% 1B progeny (when crossed with 0B males) instead of the Mendelian 50%, and this is achieved by directional movement toward the micropylar pole in 80% of meioses. The developing embryo sac is broadest at the micropylar end, where the egg cell lies. Univalent Bs tend to lie outside the metaphase plate, usually on the micropylar side of the spindle. This drive appears to maintain the B in *Lilium* in spite of its considerable negative effects on survival and fertility (Kimura and Kayano 1961). Such drive is analogous to the drive of heterochromatic knobs on the A chromosomes of maize, in which knobs end up preferentially in the cells that become eggs (see Chap. 8).

A closely parallel animal case occurs in the grasshopper *Myrmeleotettix maculatus* (Hewitt 1973, 1976). The spindle is asymmetrical and, in both 1B and 2B oocytes, Bs tend to be found off the metaphase plate and on the side of the egg rather than the polar body (Fig. 9.3). The degree of asymmetrical distribution on each side of the metaphase plate correlates closely with the degree of B drive in females (Hewitt 1976). Bs show drag in males but the net effect is positive. There are many additional examples from grasshoppers (reviewed in Camacho 2005).

In males of the scale insect *Pseudococcus affinis* paternal chromosomes are heterochromatized early in development and kept that way until spermiogenesis, when they degenerate (see Chap. 10). A B is heterochromatic throughout life until late prophase I, when it decondenses, becomes euchromatic, and, whatever its origin, joins the maternal haploid set, which is passed on to offspring in unreduced form (Nur 1962). Transmission rates exceed 90% in some crosses. How B chromosomes revert to euchromatic at the key moment is unknown. Note that nondisjunction is not involved.

Gametophytic. By far the most common type of B drive in flowering plants takes place almost immediately after meiosis during the development of the male gametophyte (Jones and Rees 1982). The usual mechanism is directed nondisjunction of the Bs into the generative nucleus as opposed to the pollen tube nucleus (Fig. 9.4A). This is a clever trick because the (doubled) B not only passes into the next generation but it also avoids exerting any debilitating effects on itself during the key pollen tube competition preceding fertilization! Strangely enough, rye pollen from plants with Bs outcompetes pollen from plants without Bs in pollen tube competition (Puertas and Carmona 1976).

This mechanism is common in grasses but is also found more widely. In grasses the spindle is asymmetrical at first pollen mitosis, being blunt at the

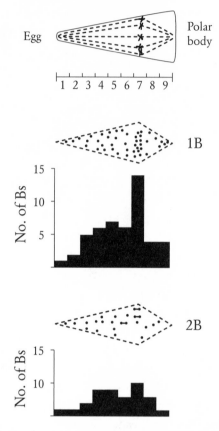

Egg — Polar body

1 2 3 4 5 6 7 8 9

No. of Bs

15

10

5

1B

No. of Bs

15

10

2B

Figure 9.3 B drive during female meiosis in a grasshopper. Distribution of B chromosomes on the spindle in oocytes of the grasshopper *Myrmeleotettix maculatus*. Top: division of the spindle into nine equal segments for scoring. Middle: B distribution in 1B individuals. Bottom: same for 2B. The asymmetry of the spindle means that more Bs end up on the egg side of the metaphase plate than on the polar body side. Adapted from Jones (1991), based on Hewitt (1976).

generative pole and longer and pointed at the vegetative; but whether this facilitates a move to the generative pole is unclear. Rye is unique among known grasses in adding drive through female reproduction to the usual pattern in males (Fig. 9.4B). The B chromosome system in rye is unusually well studied (reviewed in Jones and Puertas 1993).

A   B

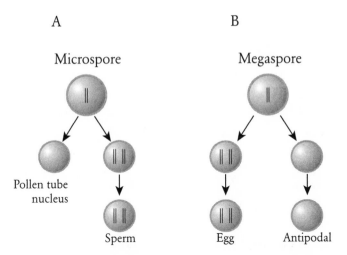

Figure 9.4 B drive in rye. B chromosome reproduction is shown for both male and female sporogenesis in rye. A. Microsporogenesis (male). The B shows nondisjunction, avoids the pollen tube nucleus during pollen grain mitosis, and lands instead in the generative nucleus that rises to the sperm cells. B. Megasporogenesis. The B shows nondisjunction and moves toward the egg cell, avoiding the antipodal. Adapted from Jones and Rees (1982).

At fertilization. The only known case of nondisjunction at the second pollen mitosis (in maize) appears to be nondirectional, but sperm with Bs are more likely to fertilize the ovule than those without, which are more likely to become part of the endosperm (Roman 1948), in which they are often lost during subsequent development (Alfenito and Birchler 1990).

Note that there are no cases of B drive in the male sex accomplished by disabling non-B-containing sperm, as there are for many cases of classic autosomal and X-Y drive (see Chaps. 2 and 3). At first glance, this seems puzzling. Meiosis itself will produce variation in B number (e.g., 1 and 0), and this can easily be augmented by nondisjunction (2 and 0). The B-containing cells could produce some developmental poison to which they possess the antidote. We believe that their escape from the diploid set may have limited this kind of spiteful action. An A chromosome (or a nonrecombining section of one) can attack its homolog and "know" that, because of the diploid system, it will not be found with its disabled homolog in the same cell—in other words, it will not suffer this kind of return cost. By contrast, a B attacking an A risks, absent any biased association at meiosis, being found with the injured A half the time. Put another way, a poison-antidote analogy is itself misleading. In the only case of A drive in males in which the mechanism is known (*SD* in *Drosophila*), the driver is inert to its poison not because it has a detoxifying agent but because it lacks the genomic material under attack in the homolog (see Chap. 2). Bs have plenty of targets to attack; but having set themselves free of diploidy, they must find it difficult to move opposite particular A chromosomes they have just disabled. Absent this ability, a B suffers the cost of the damaged A half the time, just as do the 0B sperm cells, for a net loss in transmission compared to a nonspiteful B.

Unusual modes of drive. PSR is a B chromosome that in a haplodiploid wasp enters the egg via a sperm and promptly destroys the A chromosomes with which it arrived, thereby converting what would normally have developed as a female into a male. The B gains a benefit if the sex ratio is female-biased (reviewed in Werren and Stouthamer 2003). In 2 other species, Bs also arrive alone from the male parent but in a different manner than does PSR; they arrive with sperm that otherwise serves only to initiate development in the egg (pseudogamy). The flatworm *Polycelis nigra* is a hermaphrodite that can reproduce either sexually or asexually (reviewed in Beukeboom et al. 1998). In the latter case, sperm are necessary to initiate development but their chromosomes are eliminated from the egg. Not so, apparently, for a B chromosome, which is transmitted paternally in pseudogamous couplings as well as maternally (Beukeboom et al. 1998). Rates of infestation (with at least 1 B) can reach 90%. This high frequency may, in part, be explained by the fact that each worm has a functional male and female reproductive system, so Bs can presumably be transmitted from each individual to the progeny of any other; they can, thus, jump between asexual lineages. Phenotypic costs of Bs are modest and found only in some populations—a decreasing rate of cocoon production with increasing B numbers and a decrease in the growth rates of offspring of B-carrying mothers. All other variables are unaffected, including adult body size, cocoon fertility and size, and number of young per cocoon (Beukeboom et al. 1998).

A somewhat similar system, involving microchromosomes, is found in an ameiotic fish, the Amazon molly *Poecilia formosa* (Schartl et al. 1995,

Table 9.1 Major nodal points for drive by Bs

Stage	2B	0B	Groups
Early in development	Germline	Soma	Plants/animals
Meiosis	Egg cell	Polar body	Plants/animals
1st pollen grain mitosis	Generative nucleus	Pollen grain nucleus	Plants
2nd pollen grain mitosis	Zygote	Endosperm	Plants

Schlupp et al. 1998). The species requires sperm from a closely related species to initiate development, but the sperm contribute no genes except those found on microchromosomes, which are occasionally passed by the male. These can never return to the species of origin; hence drive is not selected, which may be why such Bs are found in only about 5% of sexual individuals in the wild.

Nodal points in development. There are surprisingly few nodal points in development during which directed movement can give a B chromosome an advantage (Table 9.1). In animals there are usually only 2, one a consequence of multicellularity and the other of anisogamy. It could easily have been otherwise. For example our gonads could differentiate in place, successive waves of gonadal somatic tissue breaking off from the germinal core (e.g., in the testes, Sertoli cells, or endocrine cells). Each successive somatic differentiation would give another nodal point at which directed nondisjunction could give a B chromosome an advantage.

This is exactly the opposite of what actually occurs. To take only tetrapods: the germinal tissue invariably migrates some distance—by its own actions and with the help of intervening cells—into a particular region of somatic cells ("genital ridge") that it induces to differentiate into the various categories of helper cells. At this point the single nodal point connecting these tissues is many cell generations in the past, so that even nondirectional disjunction in somatic cells, followed by reactivation of 2B cells into germinal tissue, may be effectively foreclosed by too many generations of somatic differentiation since the nodal split. In short, the actual plan of development in complex, multicellular creatures such as humans requires only 1 nodal point between somatic and germinal tissue, itself distant in space and time from the differentiating gonadal tissue.

Genetics of A and B Factors Affecting B Drive

The genetic factors affecting B drive (located both on Bs and on A chromosomes) are poorly known. Except for rye and maize, nothing is known in detail about the genetics of B effects on B drive, and work on A genetic effects on B drive is confined to a few species. Like other selfish elements, the phenotype of an individual (high or low B drive) is known only from a study of the individual's progeny, making genetic studies more difficult. More importantly, Bs are not found in model organisms such as mice, *Drosophila*, yeast, and so on. Regarding evidence, it has long been known that rates of accumulation vary among rye cultivars (Müntzing 1949, Matthews and Jones 1983), and Carlson (1969) described an inbred maize line in which Bs did not accumulate. These results suggest, but do not prove, A variation in genes affecting B drive. Likewise, accumulation rates vary among individuals of *Hypochoeris maculata* (Compositae), but no genetic analysis has been performed (Parker et al. 1982). Genetic studies are reviewed here.

The scale insect *Pseudococcus obscurus*. The genetics of the A-B interaction has been studied with unusual care in the mealybug *Pseudococcus obscurus* (Nur and Brett 1985, reviewed in Bell and Burt 1990). The species shows paternal genome loss, which greatly facilitates genetic analysis. Males are functionally haploid because the paternal set of A chromosomes is made heterochromatic and inactive, and only the maternal set is transmitted in sperm (see Chap. 10). B chromosomes are transmitted regardless of their immediate ancestry and they thus drive. Paternal genome loss makes it possible to transfer a given B from one A background onto another. When outcrossing a male with Bs into a new population, with a female lacking Bs, he will inevitably produce some sons and these will transmit only their maternal A genes. If we were to back-cross these sons to their maternal lineage, we would in 2 generations introduce a B chromosome into a completely novel A background. Lines developed from singly inseminated females captured from nature vary widely in their frequency of Bs, some having as many as 4, on average, and some having none. In some lines, the number of Bs increased or decreased sharply in a few generations of lab culture, as if B success was being molded by A background. Similar variation has been found in the progeny of an uninfected individual mating with an infected individual within the same population: some sublines have eliminated Bs entirely;

and in some, they have proliferated. When a standard B found in 1 line was transferred into 3 other lines lacking Bs, it accumulated rapidly in 1 line but not in the other 2. A more elaborate series of crosses confirmed these results and also demonstrated variation within lines, so natural populations seem to be segregating for alleles at not less than 2 A loci that affect resistance to B drive (Nur and Brett 1987). There is no evidence in this species that Bs of different origin vary in their degree of drive against a constant A background (Nur and Brett 1985, 1987). Drive suppressor genes apparently work by interrupting the decondensation of the Bs so they no longer segregate preferentially with the euchromatic set (Nur and Brett 1988).

The grasshopper *Myrmeleotettix maculatus*. Crosses of individuals from 2 source populations of *M. maculatus*, one with Bs and one without, created families with which to measure variance in drive due to A and B factors (Shaw and Hewitt 1985, Shaw et al. 1985). These data suggested additive genetic variance for accumulation in the female line, perhaps due to a single A locus.

Rye. Extensive crosses have been performed in rye lines selected for high and low rates of B transmission. Genes affecting B transmission on the male side in controlled crosses of rye are found on the Bs themselves (Puertas et al. 1998; see also Jiménez et al. 1997). These "genes" are hypothesized to be sites for chiasmata formation, because in lines selected for high transmission, pairing of 2 Bs occurs 88% of the time (and the B is inherited 83%) while in low lines, Bs form bivalents 19% of the time (and are inherited 44%; Jiménez et al. 1997). It is interesting that, while rye Bs introduced into wheat undergo normal nondisjunction in both the male and female gametophyte (Müntzing 1970), they are less efficient at pairing together at meiosis and so are prone to elimination (Müntzing et al. 1969). Lower transmission through females in lines selected for low transmission could be the result of the same trait or it could be an independent one, located either on the As or on the B (Romera et al. 1991). Early observations suggested that the maternal parent affected B drive: pollen grains formed in 2B parents whose maternal parental was also 2B showed high B transmission, while 0B plants whose maternal parent was 2B were highly accepting of B pollen (Puertas et al. 1990). These intriguing observations have never been repeated.

Figure 9.5 B regions controlling B nondisjunction in maize. The B chromosome in maize, with 4 regions numbered that control nondisjunction of the B. The fourth contains the centromere (open circle). Adapted from Carlson and Chou (1981).

Maize. The genetics of the various B factors that determine B chromosome accumulation are known only in some detail from maize, and even here, no protein-coding genes have been isolated, only regions of the B. Recall that in maize the Bs show nondisjunction at the second pollen grain mitosis, and then the sperm with the 2 Bs is disproportionately likely to fertilize the egg rather than contribute to the polar body. Three regions of the B are crucial for the occurrence of nondisjunction, and a fourth enhances the frequency of nondisjunction (Lin 1978, 1979, Carlson and Chou 1981, reviewed in Carlson and Roseman 1992; Fig. 9.5). In addition, there is at least 1 gene on the A chromosomes that is active in the haploid egg, determining whether it is preferentially fertilized by B-containing sperm, and another that controls whether the B shows drag at female meiosis (Chiavarino et al. 1998, Chiavarino et al. 2001, González-Sánchez et al. 2003).

Transmission Rates in Well-Studied Species

Measures of B drive as a function of B number and sex of individual are known in some detail for about 15 species (Table 9.2). Several features of these data are worth mentioning. First, drive usually occurs in one sex and not the other; indeed, the B may suffer drag in the nondriving sex. Drive is always strongest for the univalent and typical values range from 0.7 to 0.9. Drive usually drops steadily with increasing B number. Because individuals with 1 B are also usually the most common B-bearing individuals, drive as a univalent is usually the critical drive parameter affecting frequency. If the B shows drag, drag is strongest in the univalent and transmission increases with B number so as to become nearly Mendelian at the highest numbers.

In other words, whether Bs drive or drag, they always show nearly 50% transmission at high B numbers.

Absence of Drive

Transmission data are available for Bs in about 70 species of plants and, according to Jones (1995), of these only about 60% show detectable drive. Half of these are in the Gramineae and typically show drive at second pollen mitosis via nondisjunction, with or without accumulation in the female germline. The remaining 40% either show no detectable drive or show a net loss of Bs. It is not clear how many of these are characterized by modest drive, rather than none at all. Bs showing drag must be beneficial to be maintained in a stable polymorphism. Jones (1995) suggests that it would be "grossly misleading" to regard all plant Bs as being selfish elements, maintained by drive. And, as we will see, in the grasshopper *Eyprepocnemis* there is clear evidence that successive B chromosomes have been neutralized, losing their drive and being partly replaced by new variants that reestablish drive, that are, in turn, neutralized.

Degree of Outcrossing and Drive

It has long been noted that Bs are preferentially found in outcrossing species (e.g., Jones and Rees 1982). Moss (1969) reports doing a comparative study of plant families (but presents no details) and concludes that Bs are largely excluded from inbred and asexual species. This, in turn, is expected on theoretical grounds (Puertas et al. 1987, Bell and Burt 1990, Shaw and Hewitt 1990). In inbred or asexual species, all natural selection is between competing lines of descent and lines without Bs are expected to outcompete those with Bs (which suffer the phenotypic costs). By contrast, in outcrossed species, uninfected lines of descent can be continually reinfected, and Bs can persist because they drive. Inbreeding increases the variance of B number among individuals and therefore the efficacy of natural selection in reducing their number. It also increases the power of selection because high B number is disproportionately associated with phenotypic cost and also with reduced drive.

This logic was addressed by Burt and Trivers (1998b) in a model in which

Table 9.2 Transmission rates of B chromosomes

Number of Bs:	Male 1	Male 2	Male 3	Male 4	Female 1	Female 2	Female 3	Female 4	Female 5	Reference
Meiotic										
Lilium callosum	0.54	0.44			0.83	0.54				Kimura and Kayano 1961
Chortoicetes terminifera	0.45	0.45/0.47*			0.77	0.61/0.80*				Gregg et al. 1984
Heteracris littoralis	0.52				0.61					Cano and Santos 1989
Melanoplus femur-rubrum	0.53				0.82					Lucov and Nur 1973
	0.50				0.75					Nur 1977
Myrmeleotettix maculatus										
Talybont	0.50	0.44			0.90					Nur 1977
Foxhole	0.37	0.30			0.57	0.62				Nur 1977
Other	0.44				0.80					Shaw 1984
Omocestus burri	0.50				0.76					Santos et al. 1993
Pseudococcus obscurus	0.84	0.71	0.63	0.51	0.34	0.48	0.46	0.50	0.47	Nur 1962, 1969b, 1977
Rattus fuscipes	0.65	0.50			0.87	0.63				Thomson 1984
Rattus rattus						0.74	0.61			Yosida 1978
Rattus rattus	0.43				0.67	0.55				Stitou et al. 2004
Premeiotic										
Crepis capillaris	0.90				0.83					Parker et al. 1989
Calliptamus palaestinensis					"Follicles with 2 Bs outnumber those with 0 Bs by 15:1"					Jones 1985
Postmeiotic										
Briza media	0.75									Bosemark 1957
Secale cereale–Östgöta		0.47				0.46				Jones 1991
–Vasa II		0.85				0.94				Jones 1991
Zea mays	0.56	0.67	0.51	0.46	0.29	0.45	0.45	0.47		Jones 1991
Unknown										
Hypochoeris maculata	0.46	0.4	0.4		0.62	0.43	0.53			Parker et al. 1982
Locusta migratoria	0.37†				0.82†					Pardo et al. 1994
Drag										
Allium schoenoprasum	0.43	0.40	0.50	0.49	0.34	0.29	0.35	0.46	0.50	Bougourd and Plowman 1996
	5: 0.48, 6: 0.41, 7: 0.30				6: 0.30, 7: 0.41					

*Transmission if Bs are the same/transmission if they are different.
†Average value not separated for different B chromosome classes.

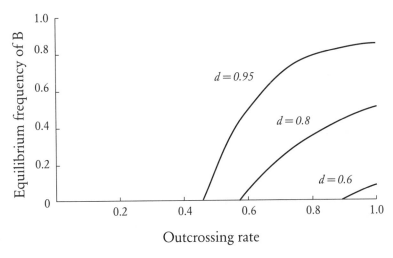

Figure 9.6 B frequency as a function of outcrossing. The expected frequency of B-containing individuals is plotted as a function of outcrossing rate. Calculations assume fitness values appropriate to *Lilium callosum* (0.95 for 1B individuals and 0.4 for 2B individuals) and that drive occurs only in 1B females ($d = 0.5$ for Mendelian transmission). Note that as a population becomes more inbred, the frequency of Bs declines until eventually they are lost from the population. Adapted from Burt and Trivers (1998b).

B number was permitted to vary from 0 to 2 and drive occurred only as a univalent in females (Fig. 9.6). The model confirms that higher degrees of outbreeding are associated with higher equilibrium frequencies of Bs, with exact values depending on the degree of B drive and the harm done to the host. Across 353 species of British flowering plants, outbreeding is, indeed, strongly associated with the presence of Bs. This is true for all plants (Table 9.3), separately for Gramineae, Compositae, and all others, and in a taxonomic contrasts test ($p \approx 0.002$). Experimental studies also support the relationship between Bs and outcrossing. Müntzing (1954) inbred the normally outbreeding rye (*Secale cereale*), and B frequency declined. Similarly, Bs experimentally introduced into *S. vavilovii*, which inbreeds, rapidly declined in frequency (Puertas et al. 1987).

Burt and Trivers (1998b) also noted that the connection between breeding system and B presence is inverted for beneficial Bs. That is, inbreeding promotes the spread of beneficial Bs. Naturally the relationship with breed-

Table 9.3 B chromosomes and breeding system in British plants

Breeding System	No. Species	% with Bs
Selfed	55	5.5
Mixed	205	6.8
Outcrossed	93	27.9
Total	353	12.5

From Burt and Trivers (1998b).

ing system may be more complex when Bs are beneficial at low number but harmful at higher.

Effects on the Phenotype

As in most cases of selfish genetic elements, phenotypic effects of Bs are often very modest. Effects on growth and development are usually slight. A striking exception, at least in plants, is the strong and widespread effect on fertility (Jones and Rees 1982). One universal effect of B chromosomes is that they increase genome size, which, in turn, has correlated effects on cell size and cycle. Increasing numbers of Bs have increasing effects and these obey an odd-even rule, whereby odd numbers of Bs (1 and 3) have larger effects on the phenotype than even numbers (e.g., 2 and 4). Here we review these and other phenotypic correlates of Bs, following closely Jones and Rees (1982) and Jones (1995) as updated wherever possible.

Effects on Genome Size, Cell Size, and Cell Cycle

Perhaps the most direct effect of Bs on the phenotype is their effect on genome size and correlated variables, such as cell size and duration of cell cycle. That is, Bs add DNA. They also appear to do so more than their relative lengths suggest (that is, they are more compacted). We might easily expect these effects on cell size and slowed-down cell cycle to be reflected in retarded growth and development; yet for plants, Jones (1995) notes that such effects are "slight and difficult to detect." It is not clear whether these may reflect an adaptation of the A genotype to the presence of Bs—in other words, whether the As are somehow compensating systematically for increases in cell size and length of cell cycle.

One effect of Bs lengthening cell cycles is foreclosure of a major avenue of drive. In *Crepis capillaris* 2B cells (generated by nondisjunction) may outreplicate 0B cells in above-ground plant tissues (Rutishauser and Röthlisberger 1966). Occasional examples may also occur in grasshopper testes but these have not been demonstrated. The tendency of 2Bs to lengthen cell cycles automatically makes their cells less competitive than those without Bs; but note that in fully 30% of plant species, the B is a tiny microchromosome and in another 25% the B is not tiny but is smaller than the smallest A. In *Crepis capillaris,* the B is one-third the size of the smallest A (Abraham et al. 1968). Across all organisms, Bs make up between 10% of the total genome and as little as 0.1% (with very occasional examples beyond these extremes).

Effects on the External Phenotype

This subject was last reviewed by Jones and Rees (1982), from which the following can be concluded. Most conspicuous are effects on flowering and fertility in plants, which are invariably negative. In contrast, effects on growth and vigor in plants are as often positive (for low B numbers) as they are negative. Two cases of positive effects on germination are known, one of which appears to be caused by a B whose overall phenotypic effect is positive. The negative effects on fertility in plants are also suggested by the fact that, for example, Bs are found in almost all wild populations of rye but in no domesticated cultivars, apparently because selection for high fertility removed them (Jones and Puertas 1993). Similar absences are true of other domesticated grasses, such as maize. Animal Bs show a more uniformly negative phenotypic profile. Effects on hatching, vigor, parasite resistance, fertility, and embryo and later development are all negative. Occasionally a B is known to increase the size of a phenotypic trait (where relation of size to fitness is unknown).

Disappearance from Somatic Tissue

In *Poa alpina* Bs are found in primary but not adventitious roots (Müntzing and Nygren 1955), while they are progressively lost from all roots in *Xanthisma texanum* (Berger and Witkus 1954, Semple 1976). More recently, Bs have been shown to be missing in adventitious roots of *Agropyrum mongolium* (Chen et al. 1993) and totally eliminated from stems and leaves in *Sor-*

ghum stipoideum (Wu 1992). In the 8 species summarized in Jones and Rees (1982) in which Bs disappear from main roots or adventitious ones, 4 species also show variation in B number in other somatic or reproductive tissue. Bs are also absent from the roots of 2 species of *Erianthus* (Sreenivaasan 1981). Why are there not many more species in which Bs disappear from roots or other tissue not destined to become germinal? One possibility is that any mutation on a B that caused it to disappear from somatic tissue would be vulnerable to manipulation by the A chromosomes so it malfunctions in the germinal tissue, thereby leading to B elimination. Jones and Rees list at least 12 species in which Bs vary in number in germinal tissue, as well as usually in some somatic tissue. Of course, plants are constrained by the fact that their dispersed gonads reduce the amount of strictly somatic tissue; but the infrequency of B elimination in roots and its complete absence in animal somatic cells invites the question, Why not more? It is noteworthy that Jones and Rees (1982) cite several cases in which Bs disappear from pollen mother cells but none in which they do so in egg mother cells. Why should this be true?

B Number and the Odd-Even Effect

The phenotypic effect of Bs on the host organism usually depends upon the number of Bs it carries. In a few cases, Bs have beneficial effects at low number but always negative effects at high numbers. For example, vegetative growth in the grass *Festuca pratensis* is greater with 1 and 2 Bs than with none, but lower at higher numbers (Bosemark 1957). In 2 species of the composite *Achillea*, growth and vigor is positive with low numbers of Bs (Stebbins 1971). And examples of beneficial Bs involve benefit at low numbers only. This, of course, is an expected feature of any system with drive: higher numbers than are useful to the organism will inevitably be generated by drive, even if some lower number is neutral or even useful. In most species, it is uncommon for individuals to have more than 3 Bs; but in a few, very large numbers are regularly found, up to 24 in the yellow-necked mouse *Apodemus peninsulae* (for other examples, see Camacho 2005).

Phenotypic effects also obey a very interesting odd-even rule: negative effects are almost always more severe in odd numbers (see Fig. 9.7). Indeed, almost all B effects obey an odd-even rule so graphs, whether negative or positive, nuclear or exophenotypic, have a saw-toothed appearance. In some cases, such as vegetative traits in rye, the general trajectory (with increasing

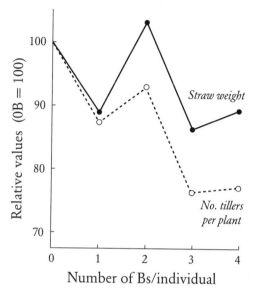

Figure 9.7 The odd-even effect on morphology of rye. Straw weight and number of tillers per plant (measures of phenotypic quality) are plotted as a function of number of B chromosomes per individual. Note that effects in plants with an odd number of Bs are always worse than in adjacent categories of even-numbered individuals. Adapted from Jones and Rees (1982), data from Müntzing (1963).

Bs) is downward but the 2B category rebounds above the 0 category (see Fig. 9.7). (Given drive through both pollen and ovules in rye, the 2B category is more nearly modal than in species with sex-limited drive.) In the harvestman spider *Metagagrella tenuipes*, there is even an odd-even effect on the chances of the spider being parasitized by hemogregarine protozoa: B-even individuals are much more likely to be parasitized in the summer than B-odds and they have lower body weight than those without Bs, while the body weight of B-odds is unchanged (Gorlov and Tsurusaki 2000a). Note this runs opposite to the usual pattern because here B-even individuals are harmed.

It is noteworthy that the odd-even rule applies even at the nuclear level, for example, in the association with reduced nuclear protein and RNA per nucleus in rye (Kirk and Jones 1970; Fig. 9.8). Why should there be this effect? One possibility is that, in a species long selected for a diploid genotype (mitochondrial and chloroplast genes excepted), a single unpaired chromo-

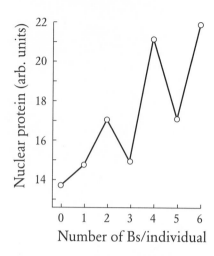

Figure 9.8 Nuclear protein amounts in root meristem cells of maize with different numbers of Bs. Note the striking odd-even effect. Adapted from Jones and Rees (1982), based on data in Ayonoadu and Rees (1971).

some is inherently more disruptive to normal mitotic and meiotic events than is a paired set. As we shall see, the odd-even rule applies also to effects of B number on chiasmata (for a recent example, see Ghaffari and Bidmeshkipoor 2002). If a single B is inherently more disruptive to normal diploid cell and genetic operations, this might especially be expected for a cell-wide genetic parameter such as number of chiasmata. Odd numbers of Bs typically have larger effects so our initial presumption is that these effects on the A chromosomes are negative to the As. The odd-even rule by which Bs affect the frequency of aberrant meiotic products has been explained by the negative effects of unpaired Bs lagging during meiosis, a defect corrected when Bs pair, as they often do when in even numbers (Camacho et al. 2004).

Negative Effects of Bs More Pronounced under Harsher Conditions

There is indirect evidence that the phenotypic effects of Bs are usually more severe when external conditions are relatively more negative for individuals of the species. Two lines of evidence are suggestive. Bs are often found at the center of a species' range or under conditions regarded as near-optimal, but are missing from peripheral areas or environments regarded as more challenging, and in experimental work, Bs are usually more harmful at higher densities (reviewed in Jones and Rees 1982). Regarding the former assertion, *Myrmeleotettix maculatus* is the classic case. Bs are more abundant in British populations in warmer, drier habitats, which are considered ideal for the species (Barker 1966, Hewitt and John 1967, 1970). Marginal populations have few or no Bs. A steep cline of B frequency from the coast to the inland appears to correlate best with average summer temperature (Hewitt and Brown 1970). Bs also show a steep decreasing cline with altitude in Wales (Hewitt and Ruscoe 1971). Together, the evidence suggests that Bs flourish where the species flourishes and are absent where it struggles, presumably reflecting the increasing phenotypic cost of Bs when external conditions are harsher. Note, however, that we are implicitly assuming that drive is not affected by ecological variables—which may, at least sometimes, be a bad assumption. Temperature alone can presumably affect cytogenetic processes directly, and, indeed, drive in this species appears to be reduced at low temperatures, so some of the population variation can be explained by variation in drive (Shaw and Hewitt 1984). With this caution in mind, we review here recent examples of similar associations on the assumption that they usually reflect varying phenotypic costs.

In the fish *Astyanax scabripinnis* a macro-B chromosome is found in the headwaters of numerous Brazilian rivers but not at lower elevations (Néo et al. 2000). Because migration is from higher to lower elevations, this distribution suggests strong selection against Bs at lower altitudes. The species is defined as a headwater species whose abundance declines downstream. Predators are less common upstream and fish species diversity is lower, but the precise ecological variable associated with B presence has not been identified.

In *Crepis capillaris* Bs are found at mid-altitudes in large populations, declining both toward the summit and toward the small, isolated lowland populations (Parker et al. 1991; Fig. 9.9). B chromosome numbers appear to be relatively constant over time. A similar pattern of B chromosome distribution is found in the plant *Ranunculus ficaria* (Gill et al. 1972). And in the Pacific giant salamander *Dicamptodon tenebrosus*, Bs range in number from 0 to 10 per individual and reach their highest average values in the middle of the species' range (Brinkman et al. 2000).

The density with which individuals are planted in experimental plots affects B survival in a way that reflects the phenotypic effects of the Bs (Jones

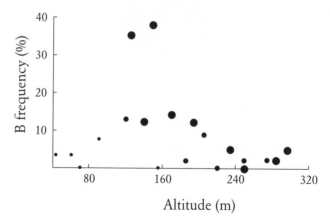

Figure 9.9 Proportion of *Crepis capillaris* in nature with Bs as a function of altitude and size of plant population. The percentage of individuals with at least 1 B is plotted as a function of altitude. Size of black dots is proportional to size of the local population. Note that Bs are found at intermediate altitudes and in larger populations. Adapted from Parker et al. (1991).

and Rees 1982). When these are positive, as in *Lolium*, increasing densities in experimental plantings result in greater survival of B-carrying individuals; or they result in no effects, as in *Allium schoenoprasum* (Holmes and Bougourd 1991), while the reverse is true in rye, a species with negative phenotypic effects of Bs. In experimental plantings of rye in nature, individuals with Bs survive better in less dense plantings and, by the same amount, in the preferable of 2 environments (Rees and Ayonoadu 1973). More detailed experiments confirmed the effect of density over a wide range, producing a strong, positive correlation between mean B frequency and percentage survival (Rees and Hutchinson 1973; Fig. 9.10). Parallel data from maple trees in nature show that the frequency of plants with Bs is inversely related to the severity of the maple stand die-back (Gervais et al. 1989). In pearl millet Bs are advantageous at low experimental densities and disadvantageous at high (Jayalakshmi and Pantulu 1982).

Is the Sex of Drive Associated with the Sex of Phenotypic Effect?

It is interesting to ask whether the sex in which drive occurs is associated with the sex in which the phenotypic effect occurs, assuming there are such

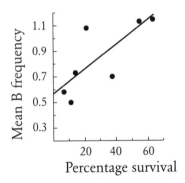

Figure 9.10 The higher the mortality of rye plants, the lower the frequency of Bs among the survivors. Mean frequency of Bs among survivors is plotted against percentage that survived in experimental patches. Adapted from Jones and Rees (1982) based on Rees and Hutchinson (1973).

sex differences. Two views can be argued. First, the sex in which B drive occurs is the sex whose phenotype selection on Bs will tend to improve (because beneficial effects are augmented by drive). This view leads to the expectation of sex-antagonistic effects, with negative effects being more often registered in the sex not showing drive (see Chap. 2). Second, drive itself is likely to disrupt meiosis (or mitosis), especially at odd and higher numbers, so the sex showing B drive is the one more likely to show a B-associated cost. Most plant species are hermaphrodites, so there is little scope for sex-specific selection during development but ample opportunity for sex-antagonistic selection on fertility. When the size or height of a plant affects male and female function differently, selection on Bs could favor the plant form most conducive to B spread through either pollen or ovules.

The evidence on phenotypic effects and drive, analyzed by sex, has been reviewed (Palestis et al. 2004b) and, so far, shows no clear trends (Table 9.4). Overall, drive occurs predominantly in males instead of females (34 vs. 9) but so do phenotypic effects (13 vs. 5). There is also significant heterogeneity by group. In the composites, drive occurs overwhelmingly in males and negative phenotypic effects only in them, while in grasses drive also occurs overwhelmingly in males but negative phenotypic effects slightly more often in females. In grasshoppers, drive occurs slightly more often in females but negative effects are registered in males only. The striking effect of Bs in grasshoppers is a near-universal negative effect on male sperm production

Table 9.4 Number of species in each group in which drive is through males or females or in which a negative phenotypic effect has been registered in either males or females

Group	Male	Female
Composites		
drive	6	1
phenotype (-)	5	0
Grasses		
drive	24	2
phenotype (-)	4	5
Grasshoppers		
drive	4	6
phenotype (-)	18	0

From Palestis et al. (2004b).

and, therefore, fertility. The problem is that Bs often interfere with normal cell division. For example, they lag, impeding normal cell separation, leading to the formation of restitution nuclei and a displaced spindle apparatus, so large diploid sperm are produced, along with a small anucleated bud (Bidau 1986). Also, sperm cells, being very small, may more easily suffer negative effects from the increase in nuclear DNA. In this case, the strong negative effect in males seems unrelated to the sex in which the B drives.

It would also be interesting to know whether the mode of drive was associated with some phenotypic costs and not others. For example, it might be that drive during meiosis is more likely to produce negative side-effects on fertility than is drive during first pollen mitosis (or, the other way around). Data on animals are currently too scarce, but it might be possible to combine data on mode of drive in plants with form of phenotypic cost.

B Effects on Recombination Among the As

One of the most striking effects of B chromosomes is their tendency, in about a half of all cases, to exert influence on recombination among the As, usually increasing chiasmata number (or variance in their number) but sometimes decreasing chiasmata number (or variance in their number; see Table 9.5 for studies of flowering plants and grasshoppers). There are 3 obvious possibilities for explaining the data: the increase in A recombination

Table 9.5 Effects of Bs on chiasmata in As

Effect	No. Plant Species	No. Grasshopper Species
Mean Xta frequency		
Increases	11	9
Decreases	6	1
No effect	21	10
Mixed	4	1
Between-cell variance		
Increases	8	5
Decreases	1	0
No effect	2	8
Between-bivalent variance		
Increases	5	
Decreases	0	
No effect	2	
Mixed	1	

From B. Palestis (pers. comm.).

benefits the B genes, it benefits the A genes, or it benefits neither set of genes but is an incidental by-product of another character (such as hetero-chromatin) that selection for some reason fails to eliminate. Assuming the recombination rate of As without a B is their natural optimum, an increase associated with Bs suggests either that the Bs benefit from breaking up coadapted gene complexes arrayed against them (with the cost of breaking up some beneficial complexes serving other phenotypic needs) or that the As benefit from increasing their own recombination when Bs are present, the better to rid their progeny of the Bs by recombining new coadapted complexes arrayed against them (but with the cost of breaking up some ben-eficial complexes; Bell and Burt 1990). Bs may also have been selected to in-crease their own rates of chiasmata—in order, in 2B individuals, to avoid uni-valent loss by pairing up—and this trait increases chiasmata in As (Carlson 1994). Having Bs in only 1 copy results in especially large A effects, because there is no paired B with which to form B chiasmata.

This last argument does not apply when Bs drive in 2B individuals as univalents, because failure to pair will conserve drive. A possible case occurs in *Myrmeleotettix maculatus,* in which Bs pair less often during female meio-sis, in which they drive, than during male meiosis, in which they do not

(Hewitt 1976). Cases are known in which Bs have a strong effect on their own chiasmata rates, as in pearl millet, but have little effect on the As (Pantulu and Manga 1975, Sabba Rao and Pantulu 1978). In *Crepis capillaris*, Bs increase chiasmata on both As and Bs (Brown and Jones 1976), while in *Eyprepocnemis plorans*, greater B chiasmata are associated with fewer A, and vice versa (Henriques-Gil et al. 1982).

The simplest incidental by-product theory fails to explain why, if so many relationships are possible between Bs and A recombination, they have not been eliminated in favor of no effect at all. Indeed, there is substantial variation within species in B effects on A chiasmata. In rye, Bs are associated with higher chiasmata frequency in some cultivars (Zecevic and Paunovic 1969) and not others (Jones and Rees 1967). Similar within-species variation in B effects on A chiasmata also occurs in several grasshoppers (John and Hewitt 1965, John and Freeman 1975, Westerman and Dempsey 1977, Camacho et al. 1980). If no genetic entity benefits from these effects and all are harmed (both Bs and As), why are they not quickly eliminated?

In Carlson's theory, A effects are a consequence of B attempts to affect their own tendency to pair (and form chiasma). When selected to pair, Bs encourage recombination, and this effect spills over on to the As. Bs and As are in conflict over B recombination, leading to a variety of possible outcomes. For example, As could be selected to induce Bs to recombine (to prevent drive), with their own increase being an unselected side effect.

The two adaptive explanations also easily accommodate the available facts. The tendency for A recombination to increase with B number is caused, in one theory, by their increasing power to disrupt the As and, in the other, from an increasingly desperate defensive response by the As as B infestation mounts! In one view, the Bs win most of the time (when recombination increases); in the other, the Bs win only a minority of the time (when recombination decreases). In either view, smaller effects with even numbers of Bs are consistent with a general tendency for negative phenotypic consequences to be less severe with even number of Bs. Because drive is also weaker at even numbers, relatively more offspring of odd-numbered B individuals will contain Bs, requiring a relatively larger recombinational response by the As. It is noteworthy that the odd-even rule may even work within an individual. In 3 individuals of the grasshopper *Locusta migratoria*, which had 2Bs in some testicular cells and 1B in others, the mean chiasma frequencies were higher in the 1B cells (Cabrero et al. 1984).

There is one fact that is not congenial to the theory that Bs induce an

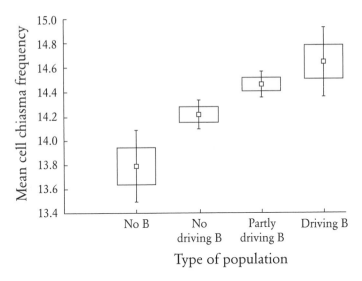

Figure 9.11 Mean chiasmata frequency as a function of different B categories in a grasshopper *Eyprepocnemis plorans*. The stronger the drive of the B, the greater are the chiasmata among the A chromosomes. Adapted from Camacho (2005).

adaptive increase by the A chromosomes, the better to combat the Bs. Why are positive effects on variance in A chiasmata at least as strong as the effects on frequency of chiasmata (see Table 9.5)? If the changes in variance have no effect on average rates of recombination (by no means certain), where is the benefit in this response? On the other hand, this fact is consistent with the incidental by-product view, including Carlson's theory.

Recently, Camacho et al. (2002) have shown an association within a species between degree of B drive, among 3 types, and degree of effect on A recombination (Fig. 9.11). Although cases are few, the pattern is certainly striking: the more drive a B shows in *Eyprepocnemis plorans*, the more it is associated with high rates of recombination in the As. And the lowest rates of A recombination are associated with no Bs at all. If this trend indicates that A recombination is adaptively adjusted to the presence of harmful Bs, it raises the question of how those Bs are recognized.

Pairing of A Chromosomes in Hybrids

Finally, B chromosomes have a very curious effect on the pairing behavior of A chromosomes in interspecific hybrids. In various interspecific hybrids–for example, between wheat and rye or species of *Lolium*–B chromosomes cause homologous (within species) chromosomes to form bivalents at metaphase I, instead of homologous and homeologous (interspecific) ones forming quadrivalents. In this sense, Bs are said to diploidize the genome. The benefit to the plant is that offspring are guaranteed to inherit similar numbers of genes from each species. In 2 species, the B seems to confer effects that are also produced by genes on the As, for example, the *ph* locus on wheat and the pairing genes located on As of *Lolium* (Jones 1995).

Ultrastructural details are available for *Lolium* (Jenkins 1986). In any case, the effect of Bs on the pairing behavior of the As challenges our understanding, because it is unclear "why the Bs of *Lolium* should have evolved properties which they are unlikely to express in nature" (Jones 1995).

Neutral and Beneficial Bs

Beneficial B Chromosomes

In a minority of species with Bs, the Bs are not harmful to the creatures they inhabit, nor are they maintained by drive. They either show no transmission ratio distortion or they show modest drag. They often have the usual negative effects on phenotypic characters (especially with a high B number) but typically show strong positive effects on early survival, often under extreme conditions (e.g., of competition or drought). It is not clear whether beneficial Bs evolved from Bs that once showed drive or, indeed, whether they have now given up entirely on drive. As always, evidence is less detailed than we would wish. Likely examples of beneficial Bs can be found in plants *(Allium, Lolium)* while an apparently neutral B (both in transmission and net phenotypic effects) is found in most populations of the grasshopper *Eyprepocnemis*.

Allium schoenoprasum. Perhaps the best evidence of a beneficial B comes from the chive (reviewed in Bougourd and Jones 1997). A small section of its habitat in the United Kingdom shows B chromosomes at surprisingly high frequencies: 55% of seeds collected in July had Bs while 64% of seedlings in September had B chromosomes. This increase occurs during a time of about 97% mortality as seedlings first become established (Holmes and Bougourd 1989). Careful experiments show that the first 2 weeks are critical and, while initial density increases mortality, this is not selective regarding Bs (Holmes and Bougourd 1991). Watering conditions are important to the survival of Bs. Under "drought," Bs enjoy the same survival benefit as in nature (roughly 12%), while under normal and excessive watering Bs are neutral, or nearly so, in their effects. Experiments also show that Bs are associated with rapid germination under all conditions but that this characteristic confers a large advantage only under "drought" conditions (Plowman and Bougourd 1994). Seeds were characterized for a series of traits that affected germination rate (such as linear dimensions and shape), but none of these was associated with B number. *Allium porrum* plants with Bs germinate on average a day earlier than those without (Gray and Thomas 1985). It is important to emphasize that, in other phenotypic characters, Bs have negative effects, especially with high B number, for example, on plant height, number of seeds/flower, and seed weight (Bougourd and Parker 1979b).

Controlled crosses show that the B drags through both pollen and ovules, with a mean transmission of about 40%, and this transmission rate is unaffected by B number (including odd-even; Bougourd and Plowman 1996). But studies of transmission rates in nature have revealed a remarkable fact: there are significant differences in transmission between nearby riverine populations, with one showing the usual drag at all B numbers and the other showing drive at low numbers and neutrality or drag at high numbers (Fig. 9.12). The B is also highly polymorphic in this species, with at least 12 types. *A. schoenoprasum* is self-compatible and reproduces both sexually via seeds and asexually via bulbs from a short rhizome (Stevens and Bougourd 1988). Inbreeding probably increases B frequency, as does asexual reproduction (by deleting meiosis, with its drag). Selfing is associated with large inbreeding depression, and degree of outcrossing has been tentatively estimated at about 80%.

Lolium perenne. It is not known whether Bs in *Lolium perenne* drive, but several studies have shown that the more densely plants with B chromosomes are planted, the better they survive. Because overall survival is reduced, the B is giving a benefit under adverse conditions (Hutchinson 1975, Teoh et al. 1976). This is the reverse of the pattern found in rye (Hutchinson 1975; see Fig. 9.10), a species well known for its drive in both sexes (balanced by negative phenotypic effects). *Lolium* is also unusual in showing re-

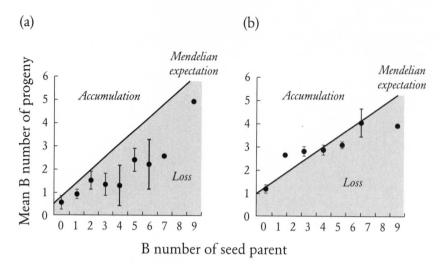

Figure 9.12 Transmission of B chromosomes in 2 natural populations of chives as a function of B number. The Mendelian or 50:50 line is given. Note that in population (a) from River Wye, the Bs always show drag (points below the line) while at (b) River Infon, Bs appear to drive at low numbers and then drag. Adapted from Bougourd and Plowman (1996).

duced chiasmata frequency with increasing number of Bs (Cameron and Rees 1967), which is opposite of the usual pattern.

Nectria haematococca. Mating population Vl of the fungus *N. haematococca* is capable of attacking at least 9 species of plants and 1 animal, in addition to being able to live as a saprophyte (VanEtten et al. 1994). Key to its attack on plant species is the ability to detoxify (e.g., by demethylating) host compounds specifically designed to deter it (e.g., phytoalexin pisatin in the pea). Pisatin demethylating ability (Pda) genes are located on chromosomes that are sometimes lost during meiosis, with the resulting isolates being viable so long as they do not need to detoxify host compounds. In that sense, these chromosomes are dispensable and hence analogous to B chromosomes (Miao et al. 1991a, 1991b, VanEtten et al. 1994, Kim et al. 1995, Han et al. 2001, Funnell et al. 2002). Although *Nectria* is haploid, individuals may get 2 Bs via the usual route of nondisjunction but no drive is known (Miao et al. 1991a). *Nectria* is briefly diploid after sexual union and all products of meiosis are functional. A dispensable 1Mb chromosome containing at least 3 genes also controls host-specific pathogenicity in the fungal plant pathogen *Alternaria alternata* (Hatta et al. 2002).

Evolutionary logic. Beneficial genes located on B chromosomes have special properties compared to beneficial genes on the regular set and to negative genes on Bs. The most important of these are the greater variation in copy number of Bs, their linkage relationships, and interaction with degree of inbreeding.

- Variation in copy number. An important difference between A and B chromosomes is that the latter vary in copy number, giving 1, 2, 3, and so on doses of any active gene, while diploidy in the As gives only 1 or 2. This difference could cause problems in dosage compensation. It is well known that negative phenotypic traits are more pronounced as the number of Bs increases, but this could easily be due to increases in genome size or in disruption of normal mitosis and meiosis. In actual fact, the extra copies of the beneficial genes are likely to cause negative net effects at high numbers, if we can judge by the general effect of B numbers in all other species. It almost seems perverse that when Bs are beneficial they drag, but this, of course, is implied if B numbers stay relatively constant over time. We could easily imagine that, for everyone concerned, it would be better to combine benefit with drive. But Bs that are beneficial at low numbers and harmful at high may have a better net effect on the phenotype if they drag. The effect depends on asymmetries in B phenotypic effect by B number: if increases above the mean are more costly than the benefits of decreases below the mean, drag will lower the Bs' chances of reaching the high numbers at which their phenotypic effects are disproportionately negative. This theory predicts that the degree of drag increases with the B number, but the opposite appears to be true (see Table 9.2).
- Interaction with inbreeding. While inbreeding selects against selfish B propagation, inbreeding has nearly the opposite effect on beneficial Bs. Inbreeding increases benefits in progeny while decreasing drag; so when these effects are large, inbreeding increases B frequency. Under more nearly neutral effects, the relationship is more complex but still shows positive effects over some of the range. Understanding how changes in

breeding system affect whether Bs are beneficial or negative would be very valuable. It would also be useful to know what such selection implies regarding their structure.

• Linkage. B genes give up linkage with A genes for linkage with other B genes. But there is much less recombination on Bs, because in most species the majority of individuals have only 1 B (and hence no recombination), while those with 2 may show chiasmata, but less so than on the As. Bs evolve, therefore, more like asexual elements in a coevolving world of sexual elements (the A chromosomes). This is almost certainly a disadvantage. In any case, there are very few genes on the Bs.

B Chromosomes in *Eyprepocnemis plorans:* A Case of Continuous Neutralization?

Probably the best studied B chromosome system in any animal is the widespread B system of the grasshopper *E. plorans*. It is suitably complex and, in some ways, unusual. First, it has a very large number of B chromosomes, more than 50 at last count (Camacho et al. 2003), some of which arise in the lab at appreciable frequencies (e.g., 0.05% and 0.21%), often from each other. Novel B forms also occur regularly in natural populations (López-León et al. 1993). These Bs are clearly derived from each other and sometimes compete with each other for transmission. Usually 1 B predominates and several additional Bs are found in a population. Rate of infection with at least 1 B is often higher than 30% and sometimes reaches 70% (Camacho et al. 1980). Second, early analyses uncovered no evidence of drive. In particular, the 3 most frequent Bs—B_1, B_2, and B_3 (presumed also to be the oldest)—showed no drive (López-León et al. 1992b), while infrequent variants sometimes showed drag (López-León et al. 1993). Associated with this was an absence of any phenotypic effects, at least at low B number, on such traits as body size and size-based somatic condition (Martin-Alganza et al. 1997) or on several fitness components, such as mating frequency (López-León et al. 1992c) and clutch size, egg fertility, and embryo-to-adult viability (Camacho et al. 1997).

A growing body of more recent evidence suggests a cyclical view of B evolution, in which Bs first appear with drive (or soon develop it); are neutralized, along with any phenotypic effects; are reinvaded by new driving Bs, with negative phenotypic effects, which are once again neutralized; and so

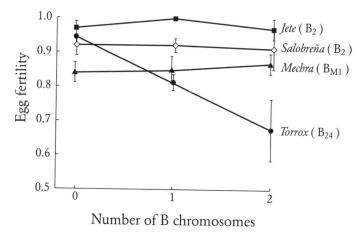

Figure 9.13 Egg fertility as a function of B chromosome number for 4 different Bs in *E. plorans*. Egg fertility is the proportion of eggs containing an embryo. Note that B number affects egg fertility only for the one B still showing drive, B_{24} Locality of B is also given. Adapted from Camacho et al (2003).

on. B_2 shows drive when outbred to populations lacking it, and B_{24} regularly drives in its own population (Zurita et al. 1998). In turn, only B_{24} shows any phenotypic cost (Camacho et al. 2003; Fig. 9.13), while increasing effects on A recombination are associated with an increasing degree of drive (see Fig. 9.11). In one Spanish population, the mean number of B_{24}s per individual increased from 0.34 in 1984 to 0.98 in 1992 to 1.53 in 1994, completely re-placing B_2 (Zurita et al. 1998). In turn, drive of B_{24} has decreased from 0.7 in 1992 to 0.52 in 1998 (Perfectti et al. 2004). B_2 drives in females, but this drive is prevented when females mate with males from the same population (Herrera et al. 1996). This in itself is a most unusual effect because it is not the sperm of the male that alters drive in the egg (which is still undergoing meiosis at the time of fertilization) but his accessory chemicals.

Copulations last about 24 hours, permitting large samples of those copu-lating in nature to be compared with noncopulators (López-León et al. 1992c). Such comparisons reveal no effect of B constitution (0, 1, 2+) on mating frequencies of either males or females, nor is there any evidence of assortative mating (and, hence, none of inbreeding). This still permits some sexual selection in double matings, through sperm competition, with Bs expected, if anything, to increase numbers of aberrant sperm and hence decrease competitive ability; but *E. plorans*, in fact, shows strong second male sperm precedence (≈90%), apparently without regard to B constitution (López-León et al. 1993). In the lab, with cages of 10 males and 15 females, Bs lower male mating success and postpone the age of first mating in both sexes (Martin et al. 1996).

As in plants, there are probably many species of animals with Bs that do not drive or are nearly neutral. In the arachnid harvestman *Metagagrella tenuipes*, there is no evidence of a B accumulation mechanism, nor sex differences in presence of Bs, nor effect on survival from juvenile to adult (Gorlov and Tsurusaki 2000a, 2000b). At present, we have no idea how many such species in which Bs do not appear to drive have, in fact, traveled the route that the Bs in *E. plorans* have taken. Of course, most species show Bs with little or no variation, in contrast to the striking polymorphism in *E. plorans*.

Structure and Content

Size

Jones (1995) has recently reviewed evidence on the size distribution of Bs in plants. In no plant species is the B chromosome larger than the largest A and in only a few species is the B roughly the size of the largest As. In fully 40% of the species, the B is one-fourth to three-fourths the size of the average A. In 26% of cases, the B is smaller than the smallest A (but not tiny) and in 30% it is a tiny microchromosome (Fig. 9.14). In short, Bs are generally small and in more than one-half of all cases are smaller than the smallest A. In more than half the latter cases, they are tiny. The smallest in any plant species is found in the fern *Ophioglossum* (Goswami and Khandelwal 1980); and in animals, the smallest B is in 2 species of the fly *Megaselia*, in which the B is little more than a centromere (Wolf et al. 1991, 1996). Unfortunately, there is no phylogenetic evidence that would allow us to order the size of Bs. Are Bs initially larger and selected to become smaller as genes are shut down that are redundant on the As (Green 1990)? Or do Bs often start very tiny and sometimes grow in size, as when they add repeated sequences useful in drive? In animals, the B is sometimes as large as the largest A and, in one case, larger: the cyprinid fish *Alburnus alburnus* has a B that is infested

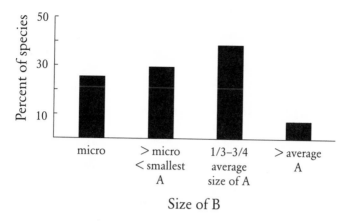

Figure 9.14 Size of B chromosomes in plant species. Percentage of plant species that have Bs for 4 categories of B size: tiny, dot-chromosomes (micro); larger than micro but smaller than smallest A; somewhat smaller than the average size of As (one-third to three-fourths of A size); larger than the average size of the As. Data from Jones (1995).

with retrotransposon-derived repetitive DNA not found on the As (Ziegler et al. 2003). As pointed out by Hewitt (1979) and reaffirmed since then (Camacho 2005), large Bs are more likely to be mitotically stable, while small ones may be unstable and, thus, vary in number within an individual. This variation, when it occurs in germinal tissue, may make drive more likely, while in somatic tissue it could make B elimination from selected tissues, for example, root cells, more likely.

Polymorphism

Jones's database reveals at least 65 species with 2 or more types of B chromosome. Many show only 2 forms, such as the large and the microsized one in *Brachycome dichromosomatica* (Smith-White and Carter 1970). Sometimes there are 3 and occasionally there are many more. The origin of many of these polymorphisms can be explained by reference to centromeric misdivision, in which a single B at meiosis may give rise to 2 unequally sized acrocentrics, and then their isochromosomes, and then further derivates by deletions. Jones cites rye as an example; it has a single widely distributed B chromosome type that sometimes generates these derivative forms but none of them seems selected over the dominant form (Jones and Puertas 1993). Chives *(Allium schoenoprasum)* provides an example of the opposite, namely high structural polymorphism of Bs. In Wales 7 different forms combine to give as many as 29 different B karyotypes (Bougourd and Parker 1979a, 1979b). It may be noteworthy that the B appears to be beneficial in *A. schoenoprasum*, and in the equally polymorphic grasshopper *Eyprepocnemis plorans* the B is often neutral. Polymorphism may also result when drive at female meiosis is strong for univalents, but weak or absent in 2B individuals due to meiotic pairing. Then as 1 B form increases in frequency, thereby increasing the number of 2B individuals, new mutants are favored that do not pair with the old B but instead continue to act as univalents, with drive. This is a form of frequency-dependent selection, with rare variants favored because they will rarely pair with themselves, and common ones disfavored for the opposite reason.

Heterochromatin

Bs are often assumed to be mostly heterochromatic, but this is truer of animals than plants. Perhaps only one-half of all plant Bs are heterochromatic (Jones 1975, 1995). In plants, Bs may often be euchromatic or contain major euchromatic segments, apparently about as often as A chromosomes. There is even one species, *Scilla vedenskyi*, in which all 5 autosomes show heterochromatin but the B does not (Greihuber and Speta 1976). Heterochromatin is presumably associated with repetitive DNA. C-banding is supposed to reveal constitutive heterochromatin but "in terms of structure and organization of Bs in plants there seems to be no specific useful information to be had so far from C-banding studies" (Jones 1995).

Genes

The absence of genes on B chromosomes is very striking. In such an intensively studied species as maize, the complete absence of B genes with a growing list of A genes was clear more than 20 years ago (Jones and Rees 1982). In Jones' (1995) review, the only important phenotypic genes listed are the virulence factors of fungal B chromosomes. The other host characters are leaf-striping (in maize), male fertility, achene color, crown rust resistance, and a regulatory effect on an A esterase gene, but for some of these the evidence is weak. It is impressive how short the list is when compared with lists of A genes.

One possible exception is rRNA genes, though the data are not yet conclusive. On the one hand, Jones (1995) lists 4 plant Bs with rRNA genes and 20 with a nucleolus-organizing region (NOR), which typically contain rRNA sequences. Similar findings come from animals. Green (1990) lists 9 species of animals and Camacho (2005) provides additional references on NORs located on Bs. For example, Bs in 2 species of *Cnephia* blackflies have NORs (Procunier 1975). And inactive ribosomal cistrons are spread throughout the B of *Rattus rattus* (Stitou et al. 2000). The B originated from one of the smaller NOR-carrying A chromosomes, after which there were expansions of repetitive and ribosomal DNA. These are methylated, heterochromatized, and silenced. On the other hand, the cytologically detectable chromosomal constriction that is usually associated with active NORs is underrepresented on Bs compared to random expectations (Jones and Rees 1982).

Why might rDNA be so common on Bs? Camacho et al. (2000) suggest that the extra rDNA may increase transcription rates and, thereby, growth; but if this were beneficial, why do they so quickly become inactive? It is perhaps noteworthy that copy number of rDNA on the As increases strongly with increasing genome size, even though the latter is not associated with an increase in protein-coding genes (Prokopowich et al. 2003). Because B chromosomes are more frequent in plant species with larger genomes, this effect may contribute to rDNA copy number on Bs. Alternatively, the recent discovery in yeast that rDNA repeats are among the last regions of the genome to segregate at mitosis, and that special enzymatic functions are needed for their segregation (D'Amours et al. 2004, Sullivan et al. 2004), suggests that rDNA repeats on Bs may sometimes be involved in nondisjunction and drive.

Jones (1995) has called attention to a key fact. If Bs really evolved from As by degenerating according to Muller's ratchet (Green 1990), we should see many "young" Bs in the process of degeneracy, but this does not seem to be the case. Almost all Bs seem to be stripped of genes. We believe this may be because Bs often evolve from the centromeres outward, beginning with very few genes and having such genes rapidly invaded by centromeric and pericentromeric repeats, which give an advantage in drive. Thus, even B drive itself may not be controlled by protein-coding genes as much as by repetitive DNA.

Tandem Repeats

Modern sequencing techniques applied to B chromosomes have uncovered a rich array of tandem repeats while identifying almost no functional genes. The characterization of these repeats has, in turn, given us our first unambiguous evidence regarding genetic homology to sequences elsewhere in the genome. B chromosomes are usually composed of repeated DNA sequences that vary in copy number and repeat type in very interesting patterns (Camacho et al. 2000). In some species (e.g., *Drosophila subsilvestris*) repeat elements appear to characterize the entire B (Gutknecht et al. 1995). As already indicated, half of all Bs in plants are heterochromatic (Jones 1975, 1995), a trait associated with blocks of repetitive DNA (Bigot et al. 1990, Charlesworth et al. 1994). At the same time, Bs that are euchromatic may harbor large blocks of tandem repeats, as in the Australian daisy *Brachycome dichromosomatica* (Leach et al. 1995). These are located near the centromere and consist either entirely of blocks of repeated DNA (175bp, as is typical of centromeric repeats) or interspersed between retrotransposons and fragments of chloroplast DNA (Franks et al. 1996). (In addition, there is no evidence of other B-specific repeats.) Because tandem arrays of repeated sequences may be involved in chromosome separation, it is tempting to suggest that the centromeric B repeat has evolved to increase nondisjunction and, thereby, subsequent drive during pollen meiosis (Leach et al. 1995).

Detailed study of the genetic structure of a micro B chromosome in *B. dichromosomatica* reveals a complex structure that immediately suggests a hybrid origin for the B (Houben et al. 2001). Because the genomic organization of the B is nothing like that of any of the As, it could not have originated by simple excision, nor did it contribute to the polymorphic heterochromatic segments also found in the species (Houben et al. 2001). Instead, the B appears to be a heteromorphic assemblage of 2 different repeat elements, along with additional DNA. The chief repeat is not organized as an extended and uninterrupted tandem array but rather as clusters of repeats, interspersed with other repeats. These are atypical of most repeats in animals and plants, in which long, uninterrupted arrays, sometimes many kilobases in length, are commonly observed (references in Zinic et al. 2000). Parallel evidence from rye again suggests that the B chromosome is a complex assemblage of 2 repeat families that, in turn, have been generated from fragments of a variety of sequence elements (Langdon et al. 2000).

Evidence that B-specific repetitive DNA is often centromeric or pericentromeric comes from various sources. In *Crepis capillaris* a repeat family found throughout As and Bs is most concentrated in B pericentromeric heterochromatin (Jamilena et al. 1994). In the maize B chromosome, many sequences are highly repetitive (and found on A chromosomes at lower copy number but not near their centromeres; Alfenito and Birchler 1993). A B chromosome–specific sequence contains a repeat unit conserved in plant A chromosome telomeres but is localized to both telomeric and centromeric regions of the B (Qi et al. 2002). One B-specific repetitive element bears genetic similarity to a retrotransposon (Stark et al. 1996). In the fish *Prochilodus lineatus*, Bs are largely characterized by a pericentromeric repeat found in lower copy number on almost half of the A chromosomes (de Jesus et al. 2003). In the field mouse *Apodemus peninsulae*, all B chromosomes contain a large amount of repeated DNA sequences; one predominates on the Bs and is found in lower density on the As (Karamysheva et al. 2002). This is the general pattern: the apparent invasion of the Bs by some A repeats, particularly centromeric and pericentromeric, which then are greatly amplified on the Bs.

The importance of repeat elements to the constitution of B chromosomes suggests a novel perspective on B evolution. The classical view is that B chromosomes may begin large, the size of sex chromosomes, but they then diminish in size to limit phenotypic costs, especially those resulting from triploid expression. Thus, genes decay and are silenced, the noncoding DNA is invaded by repetitive elements, phenotypic effects are minimized, and B DNA is lost (Green 1990). An alternative view is that B chromosomes increase from an initial small size—sometimes not much larger than a centromere and 2 telomeres—through the successive addition of repetitive sequences donated from the A chromosomes or through unequal crossing-over among repeats already present. Especially likely to be useful to the B are centromeric repeats, which may give an advantage in meiotic drive, and pericentromeric repeats, which may affect rates of nondisjunction (Henikoff and Malik 2002). The latter is a key factor in many forms of B drive. Repetitive elements are a major determinant of variation in genome size across species, so species with large genomes have a disproportionate amount of repetitive DNA that can be contributed to aid B development. The smallest animal Bs are found in 2 species of the humpbacked fly *Megaselia,* and they appear to be little more than centromeres, completely lacking arms (Wolf et al. 1991, 1996). It is as easy to imagine that they began this way as that they decayed into their present state. Centromere-sized Bs could result from Robertsonian fusions that also generate small centromeric fragments (also containing telomeres).

Where do the B chromosome tandem repeats originate? There is now reliable evidence for several species. The evidence from *B. dichromosomatica* forms an interesting pattern. The sequence is found on at least 1 A chromosome of the species at low copy number (Leach et al. 1995). Multiple copies are found in 4 related species not known to harbor B chromosomes. Thus, it seems likely that these repeats originated on A chromosomes, from which they colonized the B and then greatly expanded in copy number. Whether the repeat came from within the species or outside is not known. Low copy number on an A could also be the result of a high copy number repeat on a B making frequent contributions to the A.

The Origin of Bs

We have not yet identified the source of Bs in any B-containing species. Nor do we understand in general how they originate. By logic they must come from the A chromosomes or from outside the species. There is evidence for both. The key prerequisite is a centromere because acentric fragments are rapidly lost in both meiosis and mitosis. Of course, telomeres are also necessary but these could, in principle, evolve after a centromeric fragment appeared, as long as the centromere provided stable inheritance or drive. Centromeres do arise *de novo* in humans (Amor and Choo 2002), but how important this may be in B formation is completely unknown. Centromeric fragments can arise by Robertsonian fusions of 2 acrocentric chromosomes, and Bs in *Paspalum stoloniferum* (Avdulov and Titova 1933) and *Haplopappus gracilis* (Jackson 1960) are possible cases. Although this mechanism has the twin feature of generating a centromere and very little else, we have no idea how important it may have been in generating B chromosomes. (As we will see, Bs in mammals are associated with fewer metacentric chromosomes, not more.)

Interspecific hybridization is also thought to generate B chromosomes easily because new chromosomes can be introduced that fail to pair with any of the other chromosomes yet regularly replicate themselves during mitosis and meiosis (Sapre and Deshpande 1987). Naturally occurring hybrids between *Coix gigantea* (N = 20) and *C. aquatica* (C = 10) and their backcrosses generate N = 11 individuals, some of whom have 10 *aquatica* chromosomes and 1 small *gigantea*, while others have 9 *aquatica* and 2 small *gigantea*. In either case, the *gigantea* chromosomes may take on an independent existence within the *aquatica* genome, pairing with none of the other chromosomes. If it is nonhomologous, it remains as a univalent, does not interfere with meiosis, and carries out its own division and distribution independently, but erratically, like a B chromosome. Note that failure to pair with the As is of critical importance to the neo-B because pairing interferes with meiosis, resulting in a variable number of laggards and the production of abnormal and probably nonfunctional gametes. Thus, in any scheme for the evolution of early B chromosomes, failure to pair with As should evolve at once (while pairing may be favored between Bs to avoid univalent meiotic loss).

There are several other ways in which interspecific movement may be congenial to Bs. They may enter a new genome already possessing drive and without any coadapted A genes opposing their spread. Thus, like other driving elements, they should easily pass species borders. In addition, in some cases they can even bypass the initial block of lowered hybrid viability. For example, the PSR chromosome in *Nasonia* enters via a male and converts a diploid female into a haploid male by destroying the paternal genome; it thus destroys the half of the genome that might cause hybrid inviability (Werren and Stouthamer 2003). A molecular phylogeny of the retrotransposon NATE shows that PSR is, indeed, a recent transfer into *Nasonia* from a species of *Trichomalopsis* (McAllister and Werren 1997). Finally, Bs that are advantageous, such as those in *Nectria* that confer resistance to a pea antifungal, may confer an immediate advantage when introduced interspecifically (Camacho 2005).

Dhar et al. (2002) have recently described in detail how a centric fragment originated in *Plantago lagopus* and within a few generations acquired many of the features of a B chromosome. It arose from a spontaneous trisomic for chromosome 2 and has lost almost all the coding region of that chromosome, but it is massively amplified for 5S rDNA. Indeed, it consists almost entirely of 5S rDNA repeats. 5S rDNA is also known for amplification on A chromosomes: in flax it accounts for 3–6% of the total genome (copy number varying by strain from 50,000 to 115,000 copies; Schneeberger and Cullis 1992, Goldsbrough et al. 1981). In the case of the neo-B in *P. lagopus*, the 5S rDNA sequences are not expressed, though there is a functioning NOR at each end of the chromosome. The neo-B is an isochromosome (i.e., has 2 identical arms), formed by misdivision of the centromere sometime after the rDNA amplification. Although the As are mostly euchromatic, the neo-B is completely heterochromatic. It also has telomeres. In fewer than 10 generations this neo-B acquired many typical B characters.

Another origin of a B chromosome in the lab occurred interspecifically. A centromeric fragment appeared after several generations of introgressing a *Nasonia giraulti* region into an *N. vitripennis* background (Perfectti and Werren 2001). The centromeric fragment steadily improved its inheritance through the female (and showed a decrease in mitotic instability) while still showing drag. Inheritance was nearly perfect through males.

The fact that different B variants within a species are closely related suggests that the rate of appearance of successful novel elements is small, compared to the rate of diversification of the existing B (Cabrero et al. 1999, Houben et al. 1999). Sometimes Bs appear to endure longer than speciation events. They also appear to cross species borders, as suggested by the similarity between Bs in closely related species—for example, in *Rhamdia* fish (Fenocchio et al. 2000)—or the fact that they pair in hybrids, as in *Secale cereale* and *S. segetale* (Niwa and Sakamoto 1995). Similar evidence of common origin in closely related species can be found in rodents (reviewed in Camacho 2005). Likewise, a reanalysis of grasshopper data supplied by Hewitt (1979) shows homogeneity of B characters (e.g. size, degree of polymorphism) in 17 genera and heterogeneity in only 4, a significant effect that suggests common origin but may also reflect common forces acting on related species (Camacho 2005).

The frequency with which B chromosomes originate and the length of time that they persist within host lineages together determine the overall frequency of Bs within a host clade. This frequency appears to vary taxonomically. In a dataset of 23,652 plant species that have had some sort of karyotypic study, of which 979 had Bs (4%), heterogeneity in the frequency of Bs was found at all ranks above the species level (Levin et al. 2005). About 8% of monocots have Bs versus 3% for eudicots; and they are mostly in nonmonocot basal angiosperms. Significant heterogeneity in B frequency occurs among related orders, families within orders, and major taxa within families. There are many B chromosomes hotspots, including Liliales and Commelinales at the order level.

A Factors Associated with B Presence

Genome Size

In a review of factors associated with variation in genome size, we happened to notice that, whenever genome size increased, Bs were more frequent (plants vs. animals, monocots vs. dicots, Orthoptera vs. Diptera). With no particular theoretical bias, we performed a detailed test of this observation across 226 species of British flowering plants for which we could also get data on possibly related variables, such as breeding system, chromosome number, and ploidy level (Palestis et al. 2004c, Trivers et al. 2004). Most critically, we made an effort to measure intensity of study and to correct our data with this measure. In the process, we discovered that B chromosomes are much more likely in species with large genomes and in species with relatively few chromosomes. Why this should be true is unclear.

Correcting for intensity of study. B chromosomes are typically small and absent from most individuals. They easily invite nondetection. To correct for possible biases thereby introduced, we need a measure of study effort. The most direct measure would be to tabulate number of karyotypic studies per species, because these alone will lead to B chromosomes being identified. For flowering plants this would be a herculean task. Darlington and Wylie (1956) list only "the most recent references" and these come to more than 2400. Recent search engines do not access the literature efficiently prior to 1980, leaving a long period of time that would need to be searched by hand. We have adopted an intermediate posture. We assume that species are relatively well studied genetically or poorly studied in such a way that the number of studies of genome size, karyotypes, isozymes, and so on are positively correlated per species. To test this assumption, we measured the number of karyotypic studies and studies of genome size for a subsample of 25 species and found they were strongly positively correlated (Trivers et al. 2004). The number of studies of genome size can be measured easily and precisely. Species with larger genomes turn out to be studied more intensively and study effort is positively correlated with the presence of B chromosomes, but this is not true when the data are tested by independent taxonomic contrasts and the genome size of species with Bs varies very little with intensity of study. When a logistic regression analysis is applied, study effort drops out as a significant variable affecting the interaction (Table 9.6). Indeed, major correlations between B chromosome frequency and genome size, as well as chromosome number, are strengthened when study effort is corrected for.

Genome size and presence of B chromosomes. B chromosomes are much more likely to be found in large-genomed plants, the average difference between genome size with and without Bs being 60%. This is true when the data are corrected for ploidy and is true in an independent taxonomic contrast. B chromosomes are completely absent from the species with the smallest genomes. But the correlation does not hold across all flowering plants. Although monocots as a group have larger genomes and more species with B chromosomes than do dicots, within monocots (and also grasses) there is no relationship between genome size and presence of Bs, while relationships are significant within dicots (and Compositae). (Both monocots and dicots have a class of species with small genomes but only monocots also have a class with large ones.)

The meaning of these results is obscure. On the one hand, they suggest a general, positive relationship between genome size and presence of Bs. Where selection is, in general, reduced against selfish elements, species may harbor both B chromosomes and more A chromosome DNA, due to the spread of such selfish elements as retrotransposons. Indeed, outbreeding is associated with both the presence of Bs and larger genomes (though, in the latter case, not by independent taxonomic contrasts); but there is a strong positive association ($p = 0.002$) after correcting for breeding system and we do not know whether other factors may explain this association.

An alternative effect considers A chromosomes as possible donors of genetic material to Bs. Most of the DNA in large-genomed species is repetitive. Bs may originate from centric A fragments consisting largely of repetitive DNA or Bs, after origination by whatever means, may disproportionately accumulate useful repetitive DNA from repetitive-rich As found in species with large genomes.

There may also be a methodological explanation for the results. Differen-

Table 9.6 Logistic regression analysis for factors associated with the presence of B chromosomes across 226 species of British plants

Variable	Coefficient	s.e.	Partial r	Logistic Likelihood Ratio Tests		
				Chi-Square	df	Probability
Intercept	−1.79	0.90	−0.10			
Genome size[1]	1.55	0.54	0.18	9.48	1	0.002
Breeding system[2]	1.67	0.45	0.24	14.83	1	0.0001
Chromosome no.[3]	−0.09	0.03	−0.19	13.11	1	0.0003
No. estimates	0.30	0.21	0.00	2.10	1	0.15
Variable ploidy?[4]	0.36	0.44	0.00	0.68	1	0.41

[1]Log (4C/ploidy)
[2]Selfing + mixed versus outcrossing.
[3]Minimum reported number of chromosomes.
[4]Constant versus variable ploidy.
From Trivers et al. (2004).

tial rates of DNA excision may explain much variation in genome size (Petrov et al. 2000, Petrov 2002). If rates of excision on Bs are similar to those on As, the size of the A genome and the size of any Bs will be correlated, so Bs may be especially difficult to detect in species with small A genomes. We know of no way of falsifying this assertion. Beyond these few suggestions, we can only invite others to try to explain these (and the following) results.

Chromosome Number

Because it was convenient, we also noted chromosome number and discovered a strong negative association between chromosome number and the presence of Bs (after logistic correction, $p < 0.0003$; Trivers et al. 2004). Why this should be true is also obscure to us. It may be that species with many chromosomes have a more strongly developed meiotic machinery for fair segregation, though they will also have more units to segregate. The effect, if any, is perhaps most likely in species with a few large A chromosomes, which generate small Bs. The correlation with chromosome number is also consistent with the following bias. Fewer numbers mean greater size of each A chromosome. If B size is a constant fraction of mean A size, in species with relatively few A chromosomes, Bs will be physically larger and, therefore, easier to detect.

Ploidy

One might think that variable ploidy—that is, more than 1 ploidy level within a species—would be positively associated with frequency of B chromosomes, because crossing between individuals differing in ploidy might easily generate B chromosomes. But in our logistic analysis, variable ploidy has no effect on the presence of B chromosomes, while ploidy itself shows a weak, negative correlation with the presence of Bs. This could result from regularization of single Bs into paired ones under increase of ploidy. An analogous phenomenon has been described for a haplodiploid species (Araujo et al. 2001, 2002).

Shape of A Chromosomes

On the theory of centromeric drive (see Chap. 8), there is a frequent switch in polarity regarding the relative success of different numbers of centromeres on the metaphase plate during meiosis. Thus, a Robertsonian fusion of 2 acrocentrics into 1 metacentric pits 2 centromeres against 1. If relatively more centromeres are being favored by drive (as in mice), the karyotype becomes mostly acrocentric, and vice versa (as in humans). The same kind of selection may apply to the Bs. In the typical B chromosome that drives during female meiosis, most of the drive is achieved when the B appears as a univalent and there may be no drive as a bivalent. In this case, the single extra B amounts to an extra centromere on one side of the metaphase plate. Where extra centromeres are favored for A chromosomes, they may likewise be favored as B chromosomes, leading to a correlation between the percentage of A chromosomes that are acrocentric and the presence of B chromosomes.

Data from mammals strongly support this supposition (Palestis et al. 2004a). For mammals as a whole (p <0.0001), rodents (p = 0.004) and all other mammals (p = 0.04), B chromosomes are more likely to be found in species with a larger fraction of acrocentric A chromosomes (Table 9.7). The effect is also significant in an independent taxonomic contrasts test for all mammals (p = 0.03). These data are the strongest evidence to date in support of the meiotic drive theory of chromosomal evolution in mammals (Pardo-Manuel de Villena and Sapienza 2001a; see Chap. 8).

Table 9.7 Average percentage of A chromosomes that are acrocentric for mammals with and without B chromosomes.

	No. Species	% Acrocentric Chromosomes	Probability
All mammals			
Bs present	57	67	
Bs absent	1116	43	< 0.0001
Rodents			
Bs present	45	68	
Bs absent	346	50	0.004
Non-rodents			
Bs present	12	64	
Bs absent	669	41	0.04

From Palestis et al. (2004a).

In addition, the shape of the B chromosome is positively correlated with the typical shape of the A chromosomes (p = 0.04). This is consistent with the view that most Bs within species are recently derived from conspecific As or, at least, that similar forces have acted on their shapes after originating. A similar association appears to be true of grasshoppers (Camacho 2005).

There is still a dearth of information on how drive actually occurs in mammals. In one well-studied case, *Rattus fuscipes,* Bs drive strongly during female meiosis and show weak or no drive in males (Thomson 1984), and in *R. rattus* too there is drive in female meiosis (Stitou et al. 2004). In the only other cases known, drive may occur by mitotic instability in males but nothing is known about drive in females, and transmission data are lacking for both sexes (Patton 1977, Volobujev 1980). There are, of course, alternative explanations for the pattern we have described. If acrocentric As are more likely to donate Bs than are metacentric As, per arm, more acrocentrics will be associated with higher donation of B material. In addition, any other selection pressure favoring more over fewer centromeres will generate the same correlation. Not all species obey the general rule, of course, and exceptions may occur for many reasons: no drive through female meiosis or the fusion of 2 chromosomes to form a metacentric, which generates a centric fragment that becomes a B (Patton 1977).

Bs and the Sex Ratio

Because B chromosomes almost always drive in one sex and not in the other, they are expected to bias the sex ratio toward the sex in which they drive, while the As would be selected to have the opposite effect. In addition, under certain conditions Bs may be selected to pair with and move opposite to the X chromosome, leading the B to become a Y chromosome. Sometimes by simply changing sex an individual (or a genetic element) can take advantage of locally unbalanced sex ratios. This is known to occur in the wasp *Nasonia vitripennis*, in which the PSR B chromosome destroys the paternal A chromosomes, thereby converting diploid females into haploid males. Such an element can flourish when the sex ratio is female-biased for other reasons. We begin our account of sex ratio effects with PSR, which has been studied in some depth.

Paternal Sex Ratio (PSR) in *Nasonia*

PSR is a B chromosome with an unusually destructive form of self-propagation (Nur et al. 1988, Werren and Stouthamer 2003). Found in a parasitoid wasp *Nasonia vitripennis* (males haploid, females diploid), it is inherited from the father and arrives with a haploid set of A chromosomes in a sperm cell. Upon fertilization of an egg, it immediately causes these same A chromosomes to condense into a chromatin mass that replicates once during the first cell division but not thereafter and is lost, usually by the fifth division. Thus, it converts diploid females into haploid males. At first glance this would appear to induce drive, because as adults these males will produce sperm cells by mitosis ($r = 1$) instead of egg cells by meiosis ($r = 1/2$); but in a haplodiploid species, only females produce sons. At equilibrium under outbreeding, the expected sex ratio is 1:1 (better put: the expected ratio of investment in offspring is 1:1; hereafter, we implicitly assume male cost equals female cost), so a female can expect to produce twice as many offspring as a male, exactly balancing the greater transmission rate in males. In short, PSR requires a female-biased sex ratio in order to gain an advantage. Such biased sex ratios are expected under at least two conditions: local mate competition and the spread of selfish female-biasing genetic elements.

As is well known, any tendency toward sib-mating favors a female-biased sex ratio (Hamilton 1967). Male *Nasonia* have only vestigial wings and usually mate with females emerging from the same parasitized host (fly pupa) or a nearby set of such hosts. Often these are sisters. This kind of breeding structure is both congenial and uncongenial to the spread of PSR (Werren and Beukeboom 1993). The female-biased sex ratio is congenial because the expected reproductive success of a male is then higher than that of his sister, but the very small size of the demes accentuates the negative effect of producing excess males. When the 2 opposing forces are compared, PSR is seen to invade the natural system to an equilibrial frequency of no more than 3% of males. Incidentally, about 10% of PSR males do produce some daughters that, in turn, appear to lack PSR (Beukeboom and Werren 1993a). This may result from the B being lost in some spermatogonia.

When the sex ratio is female-biased without sib-mating or highly limited demes, PSR is expected to invade to much higher frequencies (Werren and Beukeboom 1993). A factor that will produce a female-biased sex ratio in large demes is well known from *Nasonia:* MSR (maternal sex ratio) is a cytoplasmic factor that causes all eggs to be fertilized (all-female families). These expectations have been confirmed in lab experiments that varied factors such as deme size and MSR frequencies (Beukeboom and Werren 1993b). In Utah where 17% of females have MSR, 11% of females are mated to PSR males. (Presumably in matings between the two, PSR wins out and all-male broods are produced.)

The phenotypic effects of PSR form an interesting pattern. In general, PSR males are as fit, or fitter, than standard males (Beukeboom 1994). No significant differences in longevity were found (except in one sample in which PSR individuals survived better), nor in ability to compete for males and sperm depletion rates. PSR males produced larger family sizes (11–22%), but this at least partly reflects savings in producing all-male broods. PSR males have one major disadvantage: in multiple matings, they do poorly, especially if they are the last to mate. This disability may select for more rapid development, the one trait in which they show a consistent, if slight, advantage. Altogether, Beukeboom (1994) concludes that the phenotypic effects of PSR are unlikely to exert much effect on its numbers.

PSR is an example of an ultra-destructive B chromosome that, under most conditions in nature, cannot flourish but that is aided by the spread of another selfish element, biasing the sex ratio in the opposite direction. Unlike most other B chromosomes, it does not spread by causing directed nondisjunction or biased meiosis in its own favor. It is rare across the range of its host, being restricted to the Wyoming Basin within the United States, where usually 1–5% of females are mated to PSR males (Beukeboom and Werren 2000).

A chromosome with a similar effect appears to have evolved in the tiny parasitoid wasp *Trichogramma kaykai* (Stouthamer et al. 2001). Indeed, the mechanism of action appears to be the same as in *Nasonia:* all paternal chromosomes except the B condense into a chromatin mass and eventually fail to replicate (van Vugt et al. 2003). Like many other *Trichogramma* and other micro-Hymenoptera (Stouthamer 1997), many or all females are infected with the bacterium *Wolbachia,* which, by modifying the first mitotic division, converts haploid males into diploid females (Stouthamer and Kazmer 1994). The B chromosome exactly reverses this action and turns diploid females into haploid males. Because many of these species are inbred and have female-biased sex ratios anyway, the spread of feminizing *Wolbachia* is able to make many species completely parthenogenetic (Stouthamer 1997). Spread of this B keeps *Wolbachia* infection levels low at 6–25%.

PSR-like elements may be much more widespread in the haplodiploid Hymenoptera than is commonly supposed (Werren and Stouthamer 2003). Although PSR usually produces all-male broods, these are also expected of virgin females and are, thus, easily overlooked. Many species produce female-biased sex ratios, not only under inbreeding and local mate competition but also with frequent feminizing cytoplasmic elements. Finally, in both *Nasonia* and *Trichogramma*, PSR has easily been introduced into related species (reviewed in Werren and Stouthamer 2003). The Bs' ultra-destructive behavior should make interspecific movement easier because they destroy the A genes with which they arrived. Thus, there is no hybrid inviability to impede movement into the new species; the B arrives alone.

X–B Associations in Orthoptera

In most of the Orthoptera, the female is XX and the male XO and B drive occurs in one sex only. This presents a special opportunity for a B chromosome. When B drive occurs in the female, it can gain an additional advantage by associating (and segregating) with the X chromosome in male meiosis: it will pass preferentially into daughters, where it will later drive, instead of sons, where it will not. The potential net increase in drive is substantial. If

B transmission in males is Mendelian, a gene causing exclusive X–B association would result in all daughters with a B and all sons without, or a doubling of drive a generation later. With drag in males, the potential benefit can be even greater. Notice that persistent X–B associations might be expected to slightly depress the overall adult sex ratio because it would cause a higher percentage of females than males to carry B chromosomes, with their negative phenotypic effects (although these may be weaker than comparable effects in males). The argument goes the other way around if drive occurs in males. The B is then selected to go to the opposite pole of the X, in order to reappear in males. Whether this predisposes male XO systems to become XY is unclear.

The key piece of evidence–that daughters with Bs outnumber sons with Bs in species showing drive in females, and the reverse for male drive–is, in fact, missing, even in species in which X–B association continues throughout meiosis. Instead, there is evidence of very strong, nonrandom pairing of the X and B chromosomes, usually together, but sometimes in opposition, often changing during meiosis; but there is no evidence that the final outcome (progeny sex ratio of B-containing individuals) is other than 50/50 (reviewed in Camacho 2005). We would expect this result given conflict over B drive, but only if we assume the As have the final control. Under B drive in females, the B is selected to grab hold of the X during male meiosis and move across the spindle apparatus with it, while the As are selected to pry apart the B and the X and force them to assort at random or, better yet, make them move in opposition to each other so Bs are forced into males where they would later show drag. (The optimal strategy for the As may also depend on a sex difference in phenotypic costs: if the cost of a B is considerably greater in a male than in a female, As might be selected to keep the B in females, even though they drive there.)

Has the *Drosophila* Y Evolved from a B?

XO/XX systems may be vulnerable to the invasion of a B chromosome that comes to segregate against the X, thus becoming a Y chromosome (Hackstein et al. 1996). Any B with a sex-antagonistic effect favoring males (whether through survival, fertility, or drive) would naturally benefit from segregating against the lone X, because this would ensure passage to males, where the benefit is found. Likewise, male fertility factors benefit by being Y-linked. Reasons for supposing that in *D. melanogaster*, unlike in mammals, the Y is not a degenerate X chromosome have been reviewed by Hackstein et al. (1996) and Carvalho (2002). In particular, genetic studies have failed to identify any X paralogs for Y-linked genes, whether these are male fertility genes or others (Brosseau 1960, Lohe et al. 1993, Carvalho 2002). Almost all appear to be derived from autosomal genes. The few other Y open-reading frames appear to be remnants of pseudogenes and retrotransposons. It is noteworthy that there are a number of XO species of *Drosophila*. Unlike in mammals, the Y does not determine sex; rather, it is the relative number of Xs to autosomes. Neither the Y chromosome fertility genes nor the associated repetitive elements are conserved between species. At the molecular level, both Bs and *Drosophila* Y chromosomes are packed with repetitive elements; likewise, the Y is largely heterochromatic, as are often B chromosomes. In summary, it is as easy to imagine Y chromosomes growing in size from small B chromosomes as decreasing in size from large Xs. In the *repleta* group the primitive condition seems to have been dot-like Ys that increased in size, instead of nearly X-sized ones (Wasserman 1982). It has recently been shown that the Y in *D. pseudoobscura* originated 18mya not from a B but as a result of an autosomal translocation onto the X, whose paired segment became a neo-Y (Carvalho and Clark 2005).

In the psyllid bug *Psylla foersteri*, a B chromosome appears to have all but become a Y chromosome in some populations in which all males have 1 X and 1 B chromosome and the 2 assort oppositely of each other some 98% of the time (Nokkala et al. 2000, Nokkala and Nokkala 2004). In other populations, a few individuals lack Bs or have 3, so the incipient Y is clearly a B chromosome. In some populations X and Bs cosegregate in 25% of male meioses and a lower frequency of individuals have Bs. In a related species, Bs vary in number between cysts of a male's testes, so drive may occur in males, a factor that would predispose a B toward segregating opposite the X and thereby showing up in sons.

An additional possible example of a B rejoining the regular chromosomes occurs in the haplodiploid wasp *Trypoxylon albitarse* in which, in some populations, all males have 1 B and females, 2, while in other populations, there are a few females with 0 or 3 Bs, apparently after a rapid B invasion (Araujo et al. 2001, 2002). In principle, Bs may also join the A genome by translocating onto an A as an extranumerary chromosomal segment (Camacho 2005).

Other Effects of Bs on the Sex Ratio

There is tantalizing evidence of sex ratio effects that make sense on the assumption that a B chromosome wants to bias the sex ratio in the direction of the sex within which it drives. For example, in the characid fish *Astyanax scabripinnis*, there is a highly significant association between B chromosomes and sex, with a disproportionately high frequency of males without a B chromosome and a high frequency of females with 1 B (Vicente et al. 1996; see also Néo et al. 2000). Five intersexes were discovered, all of which had Bs, so Bs may have a direct effect on the sexual phenotype (Néo et al. 2000). But nothing is known about the sex in which drive occurs. (The B is, incidentally, very large.) In the fairy shrimp *Branchipus schaefferi*, Bs are only found in males. The more Bs are found in a male, the more male-biased is the population sex ratio (Beladjal et al. 2002). Presumably drive occurs in males, though it is unclear why Bs are not found in females. Significantly more Bs are found in female than in male marsupial frogs, but nothing is known about the causes of this difference (Schmid et al. 2002). In *Distonyx groenlandicus*, a lemming with X*Y females (see Chap. 3), a B chromosome appears to have arisen from the Y chromosome (Berend et al. 2001). This is very interesting because, if it could bias the sex ratio toward males (by reversing the feminizing effects of X*Y), it would immediately spread because X* creates a dearth of males. For another example of Y derivation, see the frog *Leiopelma hochstetteri* (Sharbel et al. 1998). In the zebra finch there appears to be a germline-limited accessory chromosome in all males and females examined (Pigozzi and Solari 1998). Because it appears to be eliminated in males, it presumably shows complete drive in females.

Jones and Rees (1982) pointed out that when Bs are not small chromosomes, they are often the size of sex chromosomes, with some evidence of derivation from them. This may be because aneuploidy is more easily tolerated for sex chromosomes, as it is for small autosomes, giving the B chromosome a protected status during which to perfect drive. An example of probable X derivation comes from *Eyprepocnemis plorans* in which one B, B_2, has the same centromeric location and the same order of its 2 major constituents (a 185bp repeat and rDNA) as does the X chromosome (and none of the autosomes). Note that Haig (1993d) has suggested a similar argument for the way in which sex chromosomal tolerance of aneuploidy led to the evolution of germline-limited chromosomes in the sciarids.

Male Sterility in *Plantago*

In *Plantago coronopus,* a single chromosome is associated with male sterility and is never found in male-fertile plants (Paliwal and Hyde 1959; but more recent work calls this into question, Raghuvanshi and Kumar 1983). This B chromosome is much smaller than any of the As, and it is heterochromatic and resembles the centromeric region of a typical A chromosome. In the related *P. serraria,* B drive occurs during meiotic segregation in egg mother cells (Fröst 1959). If the same is true in *P. coronopus* and there is sufficient reproductive compensation, selection may favor the B switching development of a hermaphrodite into pure-female, in order to benefit from drive there. In addition, there is evidence that the male-sterile plants often reproduce apomictically. Bs are known to increase A chromosome instability in tapetal cells of maize, so they could presumably easily evolve the ability to cause early destruction of tapetal tissue, as do CMS genes (Chiavarino et al. 2000).

TEN

Genomic Exclusion

IN A VARIETY OF UNUSUAL—and, sometimes, bizarre—genetic systems, individuals discard the chromosomes they inherited from one parent and only transmit those from the other parent. The chromosomes that are transmitted may thereby enjoy a 2-fold advantage; they are analogous to autosomal drivers, except that drive applies to an entire haploid set of chromosomes, and drive may be based on parent-of-origin, not on a specific DNA sequence. But the extent to which these chromosomes act together to produce drive is completely unknown, so it is entirely possible that 1 or a few genes direct the entire process.

We discuss here 3 systems of genomic exclusion that differ fundamentally in their mechanisms, phylogenetic distribution, and underlying logic (Table 10.1). The most important and widespread system is paternal genome loss (PGL) in males: males begin life as diploids but they only transmit their maternal chromosomes to offspring. Sometimes, the paternal chromosomes are lost early in development so males spend most of their lives as haploids. In others, paternal chromosomes are inactivated early in development—so males are functionally haploid—but the paternal genes are carried throughout life, only to be jettisoned during spermatogenesis. Whatever the mechanism and time of action during development, the key feature is that all maternal genes show complete drive in males and (with one exception) the male functions throughout life as a haploid. These genes are like autosomal driving elements such as *t* and *SD*, except that transmission rate differences are determined not by specific nucleotide sequences but by the parental derivation of the chromosomes (maternal vs. paternal). The resulting pattern of inheritance is the same as that seen in haplodiploid species (e.g., monogonont rotifers, some mites, thrips, whiteflies, ants, bees, and wasps) and for the X chromosome in male heterogametic species. Hence, the system is sometimes called parahaplodiploidy. Males are identically related to offspring, and full sisters are related by three-fourths (Hamilton 1972, Trivers and Hare 1976). PGL in males has been described in some mites, scale insects, a beetle, and probably a suborder of springtails (Collembola). It is also found as a component of the more complicated genetic systems of fungus gnats and gall midges. Some of these are very successful groups. More than half of the 6000 species of scale insects show paternal genome loss and these species are widespread pests of plants worldwide. There are thousands and perhaps tens of thousands of species of fungal gnats and gall midges, all or most of which show PGL in males. Phytoseiid mites, in turn, are an important family of predatory mites. Why PGL in these species and not in others? In this chapter we review ideas that inbreeding and selection for female-biased sex ratios are predisposing factors, and that PGL may have evolved as a consequence of bacterial endosymbionts manipulating their host's chromosomes.

The second system of genomic exclusion is very different: when individuals from 2 different species mate, the hybrid offspring may be fertile; but when it reproduces, it passes on only the haploid genome from one parental species, discarding the haploid genome from the other species during gametogenesis. That is, the genes from a species are used somatically for one generation and then replaced. This asexual form of reproduction may benefit from repeated hybridization with a sexual species whose genes are used but not passed on. Sometimes only female hybrids are produced and are fertile, and so the system is analogous to a driving Y chromosome in a female heterogametic species. This system is called hybridogenesis, or hemiclonal reproduction, and is much more occasional than PGL, with examples found so far in single species complexes of topminnows, water frogs, and stick insects.

Finally, we have maternal genome loss, or androgenesis, in which the maternal genome is ejected from its own egg after fertilization and the progeny contain only the paternal genome. It is the forced reproduction of paternal genes by maternal tissue. This system is found in some hermaphroditic clams and other species and is a particular puzzle to understand, because it

Table 10.1 Systems of genome exclusion

Paternal genome loss (PGL) in males or parahaplodiploidy
>*Distribution.* Very important system, widespread in some groups, such as scale insects, fungus gnats, and phytoseiid mites. Also found in a beetle and in springtails. May have frequently given rise to haplodiploid species.
>
>*Key features.* Associated with very small body size, inbreeding, and local mate competition, with strongly female-biased sex ratios. The latter favor a driving X, joined by maternal autosomes, with little or no resistance from the paternal genes: in short, PGL itself. Also associated with bacteria transmitted only through eggs that may have been selected to haploidize males.

Hybridogenesis or hemiclonal reproduction
>*Distribution.* Found only in interspecific hybrids. Known from topminnows (*Poeciliopsis*), stick insects (*Bacillus*) and water frogs (*Rana*).
>
>*Key features.* Asexual ("hemiclonal") females borrow paternal genes every generation from closely related sexual form, which may often give a heterotic advantage over pure parthenogenesis.

Androgenesis or maternal genome loss
>*Distribution.* Found only in a cypress tree (*Cupressus*), a clam (*Corbicula*), and a stick insect (*Bacillus*).
>
>*Key features.* Infrequent in the stick insect, which is dioecious and hybridogenetic. Common in the other two, which are hermaphroditic. This latter trait helps prevent extinction and, if combined with selfing, reduces (or eliminates) resistance from the maternal genes.

imposes such a severe cost on the investing sex (females). Indeed, one species of cypress tree is now near extinction. Although associated with hermaphroditism, androgenesis also occurs sporadically in the very species of stick insects that also shows hybridogenesis. (The opposite system of gynogenesis, or pseudogamy, in which the paternal genome is excluded after fertilization, is similar to conventional asexual reproduction, and we do not discuss it further; see Stenseth et al. 1985.)

Each system has a very different logic and, except for stick insects, there is no overlap in the kinds of species found with the different systems of genomic exclusion. We begin our treatment with PGL in males.

Paternal Genome Loss in Males, or Parahaplodiploidy

Relatively simple forms of PGL occur in some mites, scale insects, a beetle, and probably also in some springtails (Fig. 10.1). We review the biology of PGL in each of these taxa in turn. We also review those aspects of the spe-

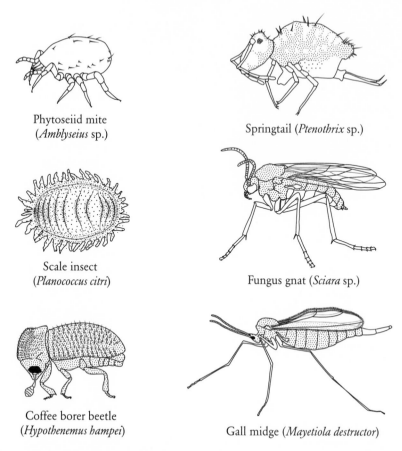

Phytoseiid mite
(*Amblyseius* sp.)

Springtail (*Ptenothrix* sp.)

Scale insect
(*Planococcus citri*)

Fungus gnat (*Sciara* sp.)

Coffee borer beetle
(*Hypothenemus hampei*)

Gall midge (*Mayetiola destructor*)

Figure 10.1 Representatives of 6 taxa with paternal genome loss. All of them are small arthropods, but otherwise unremarkable.

cies' natural histories that are most relevant to the various ideas that have been put forward about why PGL may have evolved—for example, facts about the mating systems, sex-determining systems, and the presence of maternally inherited endosymbiotic bacteria (Table 10.2). We then discuss in more detail these various ideas about the evolution of PGL, and about the links between PGL and haplodiploid systems. PGL also occurs as a component of more complicated chromosomal systems in some dipterans (fungus gnats and gall midges), but discussion of these cases is deferred until later in the chapter.

PGL in Mites

The Phytoseiidae are predatory mites that commonly attack herbivorous mites. There are about 1000 described species, and presumably many more undescribed (Oliver 1983). Several species are used in the biological control of spider mites and so are relatively well studied (Hoy 1985). Individuals are small, typically about 0.4mm long, with no eyes, antennae, or wings, and generations are short (about a week from egg to adult). In all sexual species that have been examined, males are haploid and females need to mate before laying eggs–presumably they all show PGL, though this has been confirmed in only a few species (Nelson-Rees et al. 1980). In addition, there are wholly asexual species. Chromosomes are few in number, generally 6 or 8 in females and half that in males (Hoy 1985). Intracellular symbionts–which are presumably vertically transmitted through females–have been found in at least 1 species (Hess and Hoy 1982).

Cytological observations on one species, *Typhlodromus* (= *Metaseiulus*) *occidentalis,* show that normal mitotic cell divisions occur in all eggs for the first 24 hours of development (Nelson-Rees et al. 1980). During the second day, half the chromosomes are ejected from most or all cells of about 40% of the eggs (presumably the males). First there is a "reductional" division in which homologous chromosomes pair and then separate; one set is slightly more condensed than the other and there is no indication of crossing-over. After separation the condensed set is then ejected out of the cell, where it stains very darkly. Evidence that the ejected chromosomes are paternally derived comes from crosses that show that pesticide resistance of males is inherited only from the mother, both in *T. occidentalis* and *Phytoseiulus persimilis* (Helle et al. 1978, Nelson-Rees et al. 1980). In addition, males that have been exposed to high levels of X-irradiation produce no daughters (presumably because of dominant lethal mutations) but do produce sons. Curiously, though, these sons are sterile, suggesting perhaps some residual function for the paternal genome–this is true of *T. occidentalis, P. persimilis,* and *Amblyseius bibens* (Helle et al. 1978, Hoy 1979).

PGL is probably also found in the Anoetidae and Dermanyssidae families of mites, though the evidence here is indirect: males are haploid and females must mate before laying any eggs (Nelson-Rees et al. 1980, Oliver 1983, Hoy 1985). In each case the ancestral state was male heterogamety (either XO or XY; Oliver 1983).

Table 10.2 Factors associated with PGL in males

Group	Breeding System	Sex Ratio	Control of Sex Ratio	Achiasmate Meiosis in Males	Ancestral Sex Chromosome System
Phytoseiid mites	Inbreeding, LMC	Female-biased	Maternal	???	XO or XY
Scale insects	LMC?	Female-biased	Maternal	Yes	XO
Scolytid beetle	Inbreeding, LMC	Strongly female-biased	Presumably maternal	No	XY
Collembola	Inbreeding, LMC	Female-biased (1:10)	Presumably maternal	No	???
Sciaridae and Cecidomyidae	???[a]	???[a]	Maternal	Yes	XY

For evidence, see text.
LMC=local mate competition
[a]Sib-mating not possible in monogenic species and sex ratio presumably equal.

Behavioral and population studies suggest high levels of inbreeding and a good fit to Hamilton's (1967) local mate competition model, at least in some species (Sabelis and Nagelkerke 1987). Progeny of a single female developing on a leaf are likely to mate with each other because males develop faster than females but do not disperse; rather, they make and maintain contact with juvenile females, waiting until they emerge. Mating occurs immediately afterwards (Hoy and Smilanick 1979, Hoy 1985). Dispersal is primarily by mated females, either walking or in the wind (Johnson and Croft 1976, Sabelis and van der Meer 1986). Furthermore, the sex ratio of phytoseiids is typically female-biased, as expected under local mate competition (Sabelis and Nagelkerke 1987).

The primary sex ratio is under maternal control, as indicated by many observations of adaptive sex ratio adjustments. For example, the proportion of males declines between the first day or two of egg laying and the remainder of the egg-laying period in both *P. persimilis* and *A. bibens* (Schulten et al. 1978). Increasing the amount of food available to a female increases the proportion of daughters in *A. californicus*, *T. occidentalis*, and *P. persimilis*, while crowding females increases the proportion of sons in *A. anonymomys*, *A. idaeus*, and *T. occidentalis* (reviewed in Sabelis and Nagelkerke 1987). In *P. persimilis*, female eggs are 15% larger than male eggs, and this requires maternal control of sex of the egg. Finally, females of *T. occidentalis* and *P. persimilis* adjust the sex ratio they produce to the presence of conspecifics or their cues (Nagelkerke and Sabelis 1998). Sex ratio adjustment is precise in that its variance is less than binomial and improves over time as successive eggs are laid. Indeed, sex ratio control appears to be as precise in parahaplodiploid mites as it is in haplodiploid species, even though control is more straightforward in the latter.

PGL in Scale Insects

PGL in males is widespread, though not universal, in scale insects (Homoptera: superfamily Coccoidea), which is a widespread and important group of insects. There are about 6000 species of scale insects, which live as external parasites of perennial plants (Miller and Kosztarab 1979). Their primary adaptation is for a female to settle at one spot on the stem of a plant and to insert her mouthparts down into the phloem, from which nutrients are derived. From this adaptation comes a strong trend toward sedentary habits, with a scale, or covering, evolving as protection for mother and offspring. Specialization in the Coccoidea usually involves reduction in size and in mobility, reduction or loss of structures (e.g., legs), and, most importantly, the development of various mechanisms to protect the eggs or the young larvae. These mechanisms include the retention of the developing eggs inside the female until sometime prior to hatching, the secretion of wax filaments to form an ovisac, the hardening of the body to form a scale, or the formation of a true scale as in the armored scale insects (Diaspididae; Nur 1980). Thus, larval development is commonly gregarious in scale insects. As with many other sap feeders, endosymbiotic bacteria that supply vitamins and amino acids are common in scale insects. These are found in an unusual set of host cells and passed on from mother to offspring (for more details, see Chap. 11).

As in all Homoptera, chromosomes are holocentric, lacking a localized centromere (Nur 1980). That is, the spindle apparatus attaches to each chromosome throughout its length. Each fragment is thus "centric" and much less likely to be lost in mitosis and meiosis than are acentric fragments in a conventional species. This behavior may help explain the enormous variety of karyotypes within the Coccoidea. To take the extreme, among 17 species in a single genus (*Apiomorpha*) of gall-inducing scale insects, the chromosome number varies by a factor of 48, from $N = 4$ to $N \sim 192$ (Cook 2000). Less extreme but still substantial variation in chromosome number characterizes many other coccoid taxa, with increases in number suggested to be more frequent than decreases (Nur 1980). By contrast, almost all species in the largest family, Diaspididae, show a uniform $N = 4$ (Brown 1965). Coccoids also show inverted meiosis, in which the first meiotic division is equational and the second reductional, instead of vice versa as in the usual system. Finally, the group is unusually rich in genetic systems, including conventional diploid outbred, diploid parthenogenetic, haplodiploid, and PGL species. The most likely ancestral state for the evolution of PGL in scale insects is XO male heterogamety, as seen in the outgroup *Puto* and in various coccoid genera. And meiosis is achiasmatic in males.

The 3 systems of PGL. In scale insects there are 3 systems of paternal genome loss: the lecanoid, the Comstockiella, and the diaspid (Kitchin 1975, Nur 1980, 1982). In the first 2 systems, the paternal genome is made heterochromatic early in development and kept inactive throughout development, until it is eliminated in meiosis. The 2 systems differ only in the details of the elimination. Heterochromatinization occurs after 6 or 7 cleavage divisions (i.e., in 64–128 cell blastulas), when zygotic gene transcription first occurs and when the paternal set would normally have first become active (Herrick and Seger 1999). In artificially induced haploid embryos or tissues (i.e., in the absence of the maternal genome), the paternal genome is heterochromatinized as usual (suggesting maternal transcripts or proteins are responsible for induction), but then the chromosomes revert to being euchromatic (suggesting the maternally derived homologs are responsible for maintenance; Brown and Nur 1964). In the lecanoid system the paternal chromosomes go through meiosis with the euchromatic set: there is no crossing-over and the spermatids receiving the heterochromatic set simply fail to develop into functional sperm. Recent cytogenetic studies on the mealybug *Planococcus citri* confirm that the first meiotic division is equational (i.e., separating the 2 chromatids of each chromosome), and then at the second meiotic division a monopolar spindle forms, to which the euchromatic chromosomes attach but not the heterochromatic ones, thus allowing the cosegregation of the maternally derived chromosomes (Bongiorni et al. 2004).

In the Comstockiella system, the heterochromatic paternal chromosomes are destroyed in whole or in part just prior to meiosis. Sometimes all of the paternal chromosomes are destroyed, and there is no second division; at other times, 1 chromosome survives and it is "ejected" from the telophase nuclei. In yet other cases, 2 or more chromosomes survive and there is a second meiotic division; and at the limit, all survive and one has a lecanoid system. Thus the number of paternal chromosomes eliminated before meiosis can vary from all to none, and this range of variation can be seen among species, among individuals of the same species, and among testicular cysts of the same individual.

Finally, in the third system, the diaspidid, the paternal chromosomes are eliminated at the same blastocyst stage at which they are inactivated in the other 2 systems. This is similar to what is seen in phytoseiid mites and can be seen as an evolutionary advance—saving the cost of replicating inactivated DNA (and also preventing reactivation of the paternal set). This system is seen only in some species of a single highly specialized family

(Diaspididae–armored scale insects). It has been suggested that this trend towards earlier ejection of the paternal chromosomes may reflect a history of resistance on the part of some or all of the paternal chromosomes (Herrick and Seger 1999).

At the molecular level, virtually nothing is known except that maternal and paternal genomes in both sexes may differ in methylation patterns (Nur 1990b, Bongiorni et al. 1999).

Inbreeding and local mate competition? Adult males usually develop wings, but they feed neither during the prepupal (third) or pupal (fourth) stages nor as adults (Nur 1980). They usually live only a few hours or days, are poor fliers, and often mate near where they hatch, presumably with sisters and other close relatives (Comstock 1940, Miller and Kosztarab 1979). Recent work on sex ratio of eggs of *P. citri* confirms that the primary sex ratio for that species is female-biased (Varndell and Godfray 1996). Whether these traits are ancestral to the group is not known, but the male traits cited characterize scale insects generally, not just those with PGL.

Maternal control of the sex ratio. As with the phytoseiid mites, there are a number of observations suggesting maternal control of the sex ratio. In *P. citri*, females that are aged before being allowed to mate produce a higher proportion of males (reviewed in Werren and Charnov 1978). Several other species show the same effect, and temperature and photoperiod effects have also been observed (Nur 1990a). The sex ratio of eggs in *P. citri* is sensitive to variables associated with degree of competition between females (adult crowding) and availability of males in the environment (time without contact with males before breeding), though in a pattern that is difficult to interpret (Varndell and Godfray 1996). In *Stictococcus* (a 2n-2n species), eggs that will turn into females contain endosymbionts and those that will become males do not (Brown and Chandra 1977). While the cause and effect are not certain, it is clear that sex is determined before fertilization. Finally, in the diaspid *Pseudaulacaspis pentagona*, eggs containing female embryos are coral colored and are laid first; eggs containing male embryos are whitish and laid later, again implying maternal control. The pattern of pigment formation is not known specifically for this species, but in 3 other families of scale insects, egg pigments are present before fertilization. (On the other hand, embryos are laid at a fairly advanced stage in *P. pentagona;* Nur 1990a.)

PGL in the Coffee Borer Beetle

The coffee borer beetle *Hypothenemus hampei* (Scolytidae) shows paternal genome loss in males, as inferred from absence of paternal inheritance by sons (as in haplodiploid species) and failure of unfertilized eggs to develop (unlike haplodiploid species). There is cytological evidence that males remain diploid throughout life but, from early on, one set of chromosomes appears to be inactivated and is later discarded during spermatogenesis (Brun et al. 1995a, 1995b, Borsa and Kjellberg 1996). The species has the classic marks of local mate competition. A single female enters a coffee berry, digs a gallery, and deposits her eggs. She produces a strongly female-biased sex ratio (~1:10), but this bias decreases (as in parasitic wasps) with increasing competition between mothers (Borsa and Kjellberg 1996). The dwarf and flightless males have one fewer molts than females and they usually mate with their sisters. The ancestral sex-determining system for beetles is XY/XX, the small Y not being involved in sex determination.

PGL in Springtails?

Recent studies have shown that in 3 species of springtails (Collembola, suborder Symphypleona), spermatogenesis is aberrant in a manner that strongly suggests paternal genome loss (Dallai et al. 1999, 2000). As usual, meiosis is normal in females, with pairing and crossing-over, while in males $2n = 10$ spermatogonia give way to daughter cells of $n = 6$ and $n = 4$. The $n = 4$ cells have very little cytoplasm and do not divide again. The nuclear material becomes heavily compacted and the cell degenerates. The $n = 6$ cell divides once to give 2 sperm cells. Females are $2n = 12$ and also produce sex cells with $n = 6$. Thus, offspring of both sexes begin with $n = 12$, but 2 of these chromosomes are then lost in the male very early in development—before soma and germ cells differentiate and possibly as early as the first mitotic division. In spermatogenesis paternal and maternal chromosomes do not pair and meiosis is thus achiasmatic, but whether this was original to the group is not clear. Preliminary evidence suggests that the aberrant system is found in 3 additional species and may well be general for the Symphypleona. Because both sexes are wingless with low mobility, inbreeding is assumed to be common and the production of nearly all-female broods is known to be common. Population sex ratios are often strongly female-biased in the Collembola (Hopkin 1997) and, in particular, in the species with the aberrant system. What is not yet known is whether the chromosomes eliminated are, as expected, the paternal ones.

Evolution of PGL

The question of how PGL systems may have evolved has long been a matter of theorizing, attracting some of the finest minds in evolutionary biology (Brown 1964, Bull 1979, Haig 1993c, Hamilton 1993, Normark 2004). There is as yet no consensus, but several factors are likely to have been important.

Drive. If a maternal effect mutation arose that caused paternally derived chromosomes in sons to be silenced and ejected, it would have a transmission advantage and could thereby spread through a population. The gene might, for example, produce RNAs or proteins that are introduced into the egg and attack any incoming paternal genome that contains a Y chromosome (or does not contain an X). Alternatively, a mutation expressed in the offspring that, when maternally derived, attacked paternal genomes with a Y (or no X) would also show a transmission advantage. Such a mutation could be imprinted or could simply be X-linked. Crucially for these mechanistic speculations, the ancestral state for PGL is typically male heterogametic. This kind of drive is selected in all male heterogametic species—why then does PGL only occur in some mites and small-bodied insects?

Inbreeding. Inbreeding affects the evolution of PGL in several ways. First, inbreeding leads to homozygosity and the expression of deleterious recessive mutations. A history of inbreeding will purge these mutations from the population, and so make the switch to the functional haploidy of PGL more likely. Such a switch would not easily be tolerated in outcrossed species. Second, inbreeding reduces the importance of drive (and drag)—in inbred species the maternal and paternal genomes are related and so the conflict between them is less. Finally, in a species with a 50:50 sex ratio, inbreeding would usually be associated with intense competition among brothers for mates (local mate competition), and so selects for a female-biased sex ratio. This is exactly what PGL produces in male heterogametic species: males that produce X-bearing sperm only. Thus an autosomal mutation that was expressed independently of parental origin and that attacked paternal genomes with a Y chromosome could still spread because of the sex ratio bias. Indeed, an autosomal "suicide" mutation that was only active when paternally derived and caused the elimination of Y-bearing paternal genomes could also spread, if inbreeding was sufficiently intense—not so much a selfish gene as a self-sacrificial gene. The only chromosome unlikely to evolve PGL—and indeed to resist it—is a Y, as it will not be found in sisters. Y chromosomes are likely to have been present in the ancestors of some PGL taxa but not others (Table 10.2).

Whether or not PGL evolves in order to bias the sex ratio, the fact that it does so initially means that selection on the PGL mutation should be frequency-dependent. That is, the selective advantage of the mutation will decline as it becomes more common and the population becomes more female biased. The mutation may therefore come to some intermediate equilibrium frequency, depending on such factors as where it is located (X vs. autosome), its mode of action (maternal vs. paternal vs. zygotic), any pleiotropic effects it may have, and the degree of inbreeding in the population. If it does stabilize at an intermediate frequency (rather than going to fixation and driving the population extinct), there will then be 2 types of males in the population: wildtype males producing a 50:50 sex ratio and PGL males producing all daughters. Females will then be selected to control the sex of their progeny so their broods may have the optimal sex ratio and so they can make facultative adjustments in the sex ratio. As we have seen, there is considerable evidence for such maternal control in both phytoseiid mites and scale insects.

The evidence on inbreeding in taxa with PGL is suggestive. Many species of phytoseiid mites fit the pattern of local mate competition, though PGL is so ancient that it is difficult to infer the ancestral mating system with confidence. Scale insects are mixed: adult males are winged unlike the usual case for sib-mating (Hamilton 1967), yet they have very short lives, are poor fliers, and appear often to mate near their place of birth. Again, it is difficult to be confident about what mating systems were like when PGL first arose. The bark beetle genus *Hypothenemus* is particularly interesting in this respect, as it is probably the most recently evolved case of PGL, and it fits the classic local mate competition mold very well. The study of other, even more recently evolved cases will be especially valuable for inferring the general principles of PGL evolution. There may be tens or even hundreds of small twigs on the tree of life with PGL awaiting discovery.

Egg-transmitted bacteria. Perhaps the original PGL mutation did not occur in the nucleus but instead in a maternally transmitted bacterial endosymbiont (Hamilton 1993, Normark 2004). This is not impossible: maternally transmitted *Wolbachia* endosymbionts are well known to affect the chromosome dynamics of their hosts (see Box 5.3). Such a mutation may be positively selected for two reasons. First, in an inbreeding population, the bacteria in a male will benefit from the female-biased sex ratio produced by PGL because the bacteria in him will be related to those in his daughters, who will, in turn, pass on the bacteria. Alternatively, in an outcrossed population, bacteria that caused males to be functionally haploid would, at least initially, often kill the males. If progeny were gregarious, resources would be freed for the males' sisters, which would again contain closely related endosymbionts, so the mutant bacteria would increase in frequency. Bacterial endosymbionts have been reported from phytoseiid mites and are common in scale insects. Larvae from both taxa are often gregarious, as are those of the coffee borer beetle.

Achiasmatic meiosis in males. After crossing-over and forming chiasmata, each daughter chromosome is a mixture of paternal and maternal parts. These occur more or less at random and after this time there is no possibility of whole-genome drive based on parental origin, unless we imagine "reverse recombination," a highly unlikely achievement. Thus, systems that for other reasons have evolved achiasmatic meiosis (i.e., without crossing-over) should be more likely to see PGL evolve. Achiasmatic meiosis is almost always found in one sex only and usually in the heterogametic sex. Thus, many flies (Diptera) are male heterogametic and male achiasmatic, while butterflies and moths (Lepidoptera) are female heterogametic and female achiasmatic.

Time of action. In taxa with PGL the paternal genome is not silenced or eliminated immediately after fertilization, but instead after several mitotic divisions. This delayed action contrasts with several other PGL-like phenomena. For example, the PSR chromosome of *Nasonia* wasps causes paternally inherited autosomes to be lost and does so from the first mitotic division (Reed and Werren 1995; see Chap. 9). Similarly, the *Wolbachia* bacteria that cause cytoplasmic incompatibility in *Drosophila* do so by interfering with the separation of paternal chromosomes and act from the first mitotic division onward (Tram and Sullivan 2002). In *Drosophila* both maternally active and paternally active nuclear mutations are known that interfere with the paternally derived chromosomes in the fertilized egg (Loppin et al. 2000, 2001, Ohsako et al. 2003). And in gynogenetic or pseudogamous asexuality, the paternal genome is excluded from the beginning. The silencing (or ejection) of chromosomes in PGL taxa is, by comparison, delayed. The timing may be linked to when the progeny begins to express its own genes (rather than relying on transcripts and proteins inherited from the mother) and may be linked to the fact that silencing occurs only in males rather than in all progeny (so that sex identification may be necessary), but otherwise the significance of the delay is obscure.

PGL and Haplodiploidy

PGL systems are sometimes described as parahaplodiploid in order to emphasize that inheritance is the same as in haplodiploid species. Under PGL the male begins life as a diploid–and may remain diploid–but he only transmits his mother's genes. It is natural to wonder whether the two systems tend to evolve under the same circumstances. Certainly, they are often found together in the same groups: thus, both kinds of inheritance are frequently found in mites and coccoids. There are several reasons to expect this coincidence, including the possibility that parahaplodiploidy may provide one route to the evolution of haplodiploidy (Schrader and Hughes-Schrader 1931). Once PGL is achieved, one final step is all that is needed for haplodiploidy to emerge. Evidence from mites suggests that this may have happened. In dermanyssine mites (Acari: Mesostigmata), molecular taxonomy reveals that the haplodiploid clade evolved from a PGL–rather than a conventional diploid–system (Cruickshank and Thomas 1999). (By contrast, in scolytid beetles, a haplodiploid clade has apparently evolved directly from a diploid species; Normark et al. 1999.) Other reasons for PGL and haplodiploidy to co-occur include common predisposing factors, such as inbreeding and selection for a female-biased sex ratio.

Sciarid Chromosome System

Fungus gnats of the genus *Sciara* have a chromosomal system among the most complicated of any animal, one so complex as easily to deter further study (but reviewed in Gerbi 1986). The system shows evidence of several major kinds of selfish genes, including those involved in paternal genome loss, germline-limited chromosomes, and selfish X interactions. Haig (1993d) has provided a hypothetical account of how the complex sciarid system we see today may have resulted from a long history of intragenomic conflict centering on sex determination and the sex ratio. Of particular interest is Haig's account of how sex ratio conflict may have spun off a set of germline-linked chromosomes (Ls), that is, chromosomes absent from somatic cells. We begin our account with as simple an overview of the sciarid chromosomal system as we can muster, based primarily on *Sciara coprophila*. We then present Haig's ideas about how the system may have evolved. Finally, we review the cytogenetic basis for the various aberrant chromosomal behaviors.

Notable Features of the Sciarid System

The sciarid chromosome system has 3 separate elements: the autosomes; the X chromosome, which has an unusually complicated lifecycle in the 2 sexes; and the germline-limited chromosomes, which may vary in number within species. Germline transmission of these elements is entirely normal in adult females, with each autosome, X, and L having a one-half chance of transmission. Spermatogenesis is highly aberrant in males. Autosomes show paternal loss, as does the X chromosome, while all L chromosomes are transmitted (i.e., they drive). An added complexity is that the maternal X shows nondisjunction and each sperm cell has 2 identical maternal Xs. This, as we shall see, is a very temporary gain.

Thus, the zygote has 3 X chromosomes, it has 2n autosomes, and it has a varying number of L chromosomes, typically 3. But the numbers quickly change in the embryo. By the seventh round of cell divisions, 1 of the 2 paternal X chromosomes is eliminated from the germ and soma of both sexes, as is 1 of the 2 Ls (parental origin uncertain). The remaining L chromosomes are retained only in the germline. The second paternal X is eliminated in the soma of individuals destined to become males. That is, males are somatically XO, with the lone X being maternal! Females retain both Xs in the soma.

Several features of this genetic system are worth noting. In spermatogenesis, the maternal X fails to disjoin and 2 copies are retained in each sperm, yet this temporary gain is almost immediately reversed in the embryo when 1 of the 2 paternal Xs is eliminated in both sexes. This temporary gain, followed by reversal, with no net gain, suggests a possible history of intra-genomic conflict. Note that the second paternal X is eliminated from the soma of individuals who become males. This suggests conflict over the sex ratio and leaves the maternal X with a possibly controlling influence over later loss in transmission of the paternal X. Note that (unlike other taxa with PGL) paternal genes in males are not inactivated early in life but remain active throughout life, and so males are functionally diploid.

Note also that, in spite of paternal genome loss among the regular chromosomes of males, the L chromosomes are all transmitted, although 1 paternal L is promptly eliminated from the embryo. This is reminiscent of the B chromosome in *Pseudococcus,* which escapes paternal genome loss and instead shows drive (Nur 1962, Nur and Brett 1988) or PSR in *Nasonia,* a B that drives by causing paternal genome loss (Nur et al. 1988; see Chap. 9).

An unusual feature of some sciarids, including *S. coprophila,* is that females produce only one sex, sons or daughters, and are said to be monogenic. This trait is controlled by a large paracentric inversion on the X chromosome such that those females who are heterozygous for the inversion produce only daughters while homozygous wildtype (no inversion) produce only sons.

An Evolutionary Hypothesis

Here we summarize Haig's (1993d) theory, describe some of its predictions, and briefly outline some difficulties and alternatives.

A driving X selects for driving autosomes. The evolutionary trajectory is imagined to begin with a driving X chromosome, like those found in other dipterans (see Chap. 3). But instead of the autosomes evolving to suppress the driving X, they evolve to drive *with* the X. The natural set of chromosomes to drive is *maternal.* They are the first to dwell in the new cytoplasm, itself maternally supplied. The arriving paternal chromosomes can be marked or "imprinted" for later preferential treatment (e.g., ejection from gametes). The only surprise perhaps is that the imprint should be retained throughout life, despite robust paternal genome expression. But, as we shall see, there is also a genetic bias that could bridge this gap: in males the soma contains only a maternal X. Thus, from early on there is a cytoplasmic bias in favor of maternal genes and a genetic bias predisposed in the same direction. It would be interesting if X chromosomes were especially likely to harbor genes biasing the cytoplasm toward maternal interests.

The female-biased sex ratio selects for maternal control. With X drive now accompanied by autosomal drive and no Y standing in the way, the sex ratio is expected to become female biased. Mothers are then selected to control the sex ratio, or at least to increase the number of males they produce. A plausible means is suggested by the current X chromosome constitution of somatic cells: females have both Xs and males only the maternal. It is easy to imagine a gene that converts females into males by engineering the elimination of the paternal X in somatic cells.

The doubling of the maternal X by nondisjunction in spermatogenesis, followed by its early elimination in the embryo, is puzzling even if we imagine their system to be a vestige of an earlier sex ratio conflict. Haig notes that PGL selects for males to produce daughters (because sons are an evolutionary dead end for their genes), and that inclusion of an extra X chromosome in each sperm might help accomplish that. In eggs that would normally develop into males, by eliminating 1 X from the soma, the trick might just work, if the rule was "eliminate 1 paternal X" (rather than "eliminate all paternal Xs"). But in order for the strategy to be selected, the sex ratio must be *male* biased, so the advantage of converting what would normally be sons into daughters outweighs the cost of converting what would normally be fertile daughters into XXX aneuploids that would probably be lethal or sterile.

Xs evolve into Ls. Haig argues that L chromosomes evolved from X chromosomes. Once paternal genome loss is in place, a paternal chromosome that resisted exclusion from the germline would normally generate an individual with 3 homologous chromosomes instead of 2. This kind of aneuploidy is usually fatal. X chromosomes, however, will tolerate aneuploidy whenever there is a process in place in which all but a set number of Xs (1 or 2) is eliminated from the soma—for example, as in the rule "delete all paternal Xs" (assuming meiosis is normal in females).

Such a process permits the following kind of story to be told. Imagine a mutant X* that in males resisted elimination when paternal, so sperm were XX* instead of X. (This is before the evolution of maternal X nondisjunction in spermatogenesis and subsequent ejection in the offspring.) Daughters are inviable because they are aneuploid (XXX*), but in males both paternal Xs are eliminated from soma and they are viable, with XXX* germlines. Because *Sciara* tolerates germline aneuploidy, offspring inherit a paternal X* and maternal X, with the paternal X marked for elimination.

X* now acts as a sperm-borne pathogen that kills daughters and is transmitted from father to son. If it stays in the male line, it is destined to disappear. But if it manages to lose whatever functions cause female aneuploid inviability before its disappearance, it could persist. Initial success in females will select for increased rates of loss of function. And it will be selected for exclusion from the female soma, with all elements agreeing on the benefit of such action.

In short, an X has now evolved into an L, showing drive in males, normal segregation in females, and elimination from the soma in both sexes. But the evolution of L chromosomes brings new selection pressures on the sex ratio. Because they drive more in males than in females, they are expected to bias the sex ratio toward sons. This is the first time in the evolution of the system in which an element favors a male-biased sex ratio and, unimpeded, the sex ratio is expected to reach equilibrium at 2:1, assuming twice the propagation of Ls through males as through females. But any male bias selects for countermeasures by the autosomes and the X. One such countermeasure is monogenic egg production.

The existence of Ls selects for monogenic (single sex) reproduction.
Imagine a mutant X chromosome, X′, which resists the action of the Ls and causes itself to produce only female-determining eggs. Such an X will spread because females enjoy higher reproductive success when the sex ratio is male biased. The spread of this gene, in turn, favors regular Xs that produce more sons when XX. The system should rapidly stabilize at 50% XX′ females that produce only daughters and 50% XX females that produce only sons. Note that X′ never appears in males and, thus, is expected to lose male functions. When X′O males are of low viability, there is no longer any selection for autosomes to reverse the action of X′ because this would result in the production of low-viability males, an effect that should help stabilize the monogenic system further.

The key assumption is that the L chromosome evolves to control the sex ratio, thus favoring the monogenic counterstrategy. L chromosomes may more easily evolve control of the sex ratio because, being germline-limited, there is no negative pleiotropic covariation with other phenotypic traits. A mutant gene on a regular sex chromosome causing a sex ratio effect may very well sacrifice the gene's original phenotypic benefit or have some other negative phenotypic effect, while no such effect burdens the spread of novel sex ratio mutants on the L. An entire chromosome is free to evolve narrow genetic functions with little or no constraint from pleiotropic phenotypic effects.

Predictions. Haig's theory makes a number of mechanistic predictions, including: there is a distant homology between X and L chromosomes; Ls compete with each other for retention in the germline; multiple independent elements (and not a single locus) control the primary sex ratio; elimination of paternal chromosomes in spermatogenesis is caused by expression of maternal genes; the imprint of paternal chromosomes is established after–rather than before–fertilization; the elimination of paternal X chromosomes is controlled by the expression of L chromosomes; and this effect is suppressed in the oocytes or nurse cells of XX′ females.

There is very little evidence on these predictions. That Ls came from Xs is consistent with the fact that in the first spermatocyte the L remains condensed and heteropycnotic (Metz 1938), just as do the sex chromosomes in many species. Likewise, the mechanism of L elimination–failure of sister chromatids to separate–appears to be the same as for somatic X elimination. *Sciara* also has 2 species that lack Ls entirely (*ocellaris* and *reynoldsi*), and these show no evidence of the XX, XX′ system of monogenic reproduction assumed to evolve in response to Ls (Metz and Lawrence 1938). Sex ratios in *ocellaris* are digenic, while those of *reynoldsi* appear to be a mixture of di- and monogenic.

Difficulties and alternatives. As Haig (1993d) is the first to admit, his story is a post-hoc attempt to fit theory to the facts that goes well beyond the available facts and permits many alternative scenarios not yet explored. Yet it is the first argument of any sort to explain the evolution of an extremely aberrant genetic system, itself apparently made up of a variety of selfish genetic elements. Especially striking are arguments that link the evolution of one subsystem to another, as in: X drive → L chromosomes → monogenic eggs. But there are many steps left unexplained. For example, it is not obvious how, mechanistically, a driving X chromosome can be "joined" by the maternal autosomes. Paternal genome loss may instead have evolved directly, as we hypothesized earlier for the simpler PGL systems. Also, Ls could easily have evolved from standard B chromosomes rather than from Xs, a possibility that removes the difficulties of aneuploidy. It is also easy to imagine that bacterial endosymbionts played a leading role in the evolution of the system (Normark 2004). Bacterial endosymbionts are associated with monogeny in wood lice (*Armadillidium* spp.; Rigaud et al. 1997).

Further progress will clearly require much more empirical work, both evolutionary and mechanistic, on the species concerned. It is also worth emphasizing that, regardless of how exactly the system evolved, there are abundant opportunities for conflict in the extant species. Autosomes, Xs, and Ls all have different patterns of inheritance, and so different optima regarding sex ratios and behavior toward kin. Investigations into these current conflicts have barely begun.

Mechanisms

Little is known about the mechanisms underlying the sciarid chromosomal system. In particular, almost nothing is known at the molecular level. We provide a review of the major findings for each of the genetic elements in this section.

Elimination of 1 of the 2 paternal Xs in embryonic germline tissue. In most cases of chromosome elimination (certainly paternal genome loss), elimination is achieved by a failure to pair at the metaphase plate or the pulling into a dead-end corner by filaments attached to a nongenerative pole. By contrast, the disappearance of 1 of the 2 paternal Xs from the primordial germ cells of both sexes occurs in a most unusual way, with 1 X moving quickly toward the nuclear membrane; becoming attached, engulfed, and evaginated; and reappearing as a vesicle in the cytoplasm, a lone chromosome surrounded by a double membrane (Perondini and Ribeiro 1997 and references cited therein). Noteworthy is the complete absence of spindles. Instead, there is clear evidence that the membrane is primed for regenerating itself: ribosomes on the membrane become numerous near the point of attachment, and endoplasmic reticulum and mitochondria are frequent in the nearby cytoplasm. All 3 are used to regenerate membranes. The transformation from independent nuclear chromosome to cytoplasmic vesicle takes about 3 hours, and higher temperature leads to quicker transformation.

Paternal X elimination in somatic tissue. Elimination of paternal Xs from somatic tissue (1 lost in females, 2 in males) proceeds by a different mechanism than elimination in the germline (de Saint Phalle and Sullivan 1996). The first abnormality shown by a soon-to-be-eliminated X is a failure of sister chromatids to separate. The chromatids separate at the centromeres and the telomeres but remain attached along a region of the long arm of the X. The centromeres remain attached to the spindle apparatus and under tension at the metaphase plate, stretched in opposite directions so, if one centromere should become detached, the entire complex would migrate to the opposite pole. This work also confirmed classic findings that the process requires a *cis*-acting region of the X chromosome, near the centromere, known as the "controlling element" (Crouse 1960). If this region is translocated to an autosome, the autosome is eliminated in embryos when it is of paternal origin. As in regular X-elimination, this elimination is due to the failure of the long arms of the autosome to separate. The controlling element is in the region of the proximal X heterochromatin that contains most of *S. coprophila*'s rDNA repeats (Crouse 1979). Perhaps this location is important: it has recently been shown in yeast that rDNA separates later in mitosis than other genomic regions and that its separation depends on a different mechanism than is used by much of the rest of the genome (D'Amours et al. 2004, Sullivan et al. 2004). For unknown reasons, Xs are eliminated earlier in males than in females. The elimination occurs in cycles 7, 8, and 9 in males and almost exclusively in cycle 9 in females.

Sanchez and Perondini (1999) have developed a model for sex determination in sciarids based on control of X chromosome elimination. They argue that the evidence is best accommodated by assuming a 2-factor system in which a chromosomal factor interacts with the X chromosomes causing their elimination. A maternal factor, in turn, regulates the chromosomal factor and maternal imprint of chromosomes makes them impervious to the chromosomal factor.

Elimination of the Ls. L chromosomes are eliminated from somatic cells during the fifth cell cycle. Virtually nothing is known about the process except that, as in X elimination, L sister chromatids fail to separate on the metaphase plate (de Saint Phalle and Sullivan 1996). It would not be surprising if this failure were controlled by an element on the L, because the L would benefit from its own somatic exclusion.

Paternal genome loss in spermatogenesis. Paternal chromosome elimination in *S. coprophila* has been described in a series of papers by Metz (summarized in Metz 1938). The process appears to be the same in related species (Esteban et al. 1997). A monopolar spindle is formed in meiosis I (that is, spindles radiate out from a single centrosome), the chromosomes remain scattered and fail to pair, and they are not oriented in a metaphase-like array. Maternal chromosomes move toward the single pole while paternal chromosomes attach to the spindle and move in the opposite direction into a cytoplasmic bud to be eliminated (Kubai 1982). Because the spindles from the pole extend all the way to the nongenerative side and because they attach to the paternal chromatin cluster and their microtubules interact with kinetochores of paternal chromosomes, it has been speculated that they are involved as much in taking the paternal chromosomes away as they are in drawing the maternal chromosomes to the pole (Kubai 1982, Esteban et al. 1997, Fuge 1997). While maternal chromosomes are drawn to the pole, with the centromeres leading, the reverse is the case with paternal chromosomes, which often move as if the ends are being pulled toward the nongenerative side—in other words, their centromeres lag (Abbott et al. 1981). Filaments are found where the paternal chromosomes end up, but they are not oriented in an astral shape, and they contain both actin and myosin (Esteban et al. 1997).

How these traits may interact with any imprints applied to the paternal chromosomes is obscure. Some evidence suggests that relative histone acetylation may play a major role in chromosomal imprinting (Goday and Ruiz 2002). In early germ cells, the paternal chromosomes are highly acetylated on histones H3 and H4. An exception is the paternal X that is eliminated from germ nuclei. At later stages, before the beginning of gonial mitoses, both sets of chromosomes show high levels of H3/H4 acetylation; but in male meiosis, only the maternal chromosomes are highly acetylated. The paternal set that is eliminated appears to be underacetylated.

Maternal X nondisjunction in male meiosis II. In the second meiotic division in males, the spindle is bipolar, but while the maternal autosomes and L chromosomes align on the metaphase plate, the maternal X is already near one pole. For autosomes and Ls, the sister chromatids segregate to opposite poles, but both X chromatids go to the same pole. The chromosomes at this pole are incorporated into the functional sperm, and the chromosomes at the other pole are discarded in a secondary bud. This nondisjunction and accumulation of the maternal X at male meiosis appears to involve the same *cis*-acting controlling element as nondisjunction and loss of paternal Xs in somatic cells (Crouse 1960).

PGL in Gall Midges

Like the Sciaridae, the Cecidomyidae (gall midges) are a family of dipteran insects with highly unusual chromosome systems. It is not yet clear whether the 2 families are closely related. Perhaps the best-studied case is the Hessian fly *Mayetiola destructor*. It shares with the sciarids a suite of traits. It has germline-limited chromosomes (Es for "eliminated"), paternal genome loss, and monogenic broods in which the sex of offspring is determined by maternal genotype (Hatchett and Gallun 1970, Stuart and Hatchett 1988). There is somatic elimination of X_1X_2 in males (presumably the paternal pair). The E chromosomes are necessary only for fertility (Bantock 1961). That they should acquire fertility genes—or have such effects—is not surprising: they are only found in germinal cells and they are so much more numerous than the somatic chromosomes (28 vs. 8 in *M. destructor*) that their absence could easily impede normal meiosis. Unlike L chromosomes, E chromosomes are eliminated in spermatogenesis and have predominantly maternal inheritance (White 1973).

Hybridogenesis, or Hemiclonal Reproduction

In hybridogenesis, individuals from 2 species mate to form a hybrid that is fertile but only transmits the haploid genome of one parent—the other haploid genome is discarded during gametogenesis. Thus haploid genomes from the donor species can invade the recipient host species. Often the hybrids are of one sex only (female), and so the system amounts to mother-daughter transmission of an entire haploid genome with the other half borrowed anew each generation and then discarded. Thus the system is also called hemiclonal reproduction: offspring are produced that combine a usually invariant maternal genome (hemiclone) with a recombinant, sexual paternal genome.

Hybridogenesis is best known from the fish *Poeciliopsis*, the frog *Rana esculenta*, and the stick insect *Bacillus* (Table 10.3). It is also known in the Ibe-

Table 10.3 Key features of hybridogenetic species

Feature	Poeciliopsis	Rana	Bacillus
Within-species mating	Impossible (no males)	Possible between hemiclones; mostly inviable within hemiclones	
Males	Absent	Regular but low sperm quality and unattractive	Rare and sterile
F₁ hybrids of parental species	Mostly inviable	Often better than either parental species (and the existing hybridogens)	
Cross to one's own sexual species	Inviable, or severely disabled and sterile	Usually viable	Completely viable and fertile
Age of hemiclone	120,000 generations	Recent (?)	Recent

rian fish *Rutilus* (= *Tropidophoxinellus*) *alburnoides,* but it has barely been studied there (Carmona et al. 1997, Alves et al. 1998). A possible case of incipient hybridogenesis is found in a midge (Polukonova and Belianina 2002). We begin with *Poeciliopsis,* which was the first case to be studied in some detail.

The Topminnow *Poeciliopsis*

Poeciliopsis is a genus of small freshwater fish (Box 10.1). Crosses of *P. monacha* × *P. lucida* occasionally produce offspring that survive to maturity and are fertile (Schultz 1973). Due to some unknown peculiarity of sex determination, these are all females, and so F2 progeny cannot be produced. But if these hybrid females are backcrossed to *lucida* males, a striking result is observed: the offspring are no more *lucida*-like than the F₁ hybrids (reviewed in Schultz 1977). One can repeat this backcross for as many generations as one wishes, and still the *monacha* traits are not diluted out. The implication is that the hybrid females only pass on the *monacha* genome in their eggs and exclude the *lucida* genome, and this conclusion has been confirmed with allozyme polymorphisms—in each case, hybrids only transmit the *monacha* allele, not the *lucida* allele. Very much the same observations also apply to crosses of a *P. monacha* female × *P. occidentalis* male—all the progeny are female and they only pass on the *monacha* genome (Wetherington et al. 1987). Or, crossing a *P. monacha-lucida* hybrid female ×

P. occidentalis male produces a *monacha-occidentalis* hybrid, with complete re-placement of the *lucida* genome. (The reciprocal crosses, of *P. lucida* or *P. occidentalis* female × *P. monacha* male, have never been successful; Schultz 1977.)

Cytological studies of oogenesis in these hybrid females indicate that at some point there is a mitosis with a monopolar spindle, to which only the *monacha* chromosomes attach; the others are lost in the cytoplasm (Cimino 1972). There is no synapsis of homologs and no crossing-over. This aberrant mitosis is followed by an abbreviated 1-step meiosis, in which the *monacha* chromatids separate into chromosomes. The *monacha* genome is thus a driv-ing haplotype. If introduced into a *P. lucida* or *P. occidentalis* (or *P. latidens;* Schultz 1977) population, then it will be transmitted intact from mother to daughter, analogous to a driving Y chromosome in a female-heterogametic species.

The ranges of the 2 host species, *P. occidentalis* and *lucida,* do not overlap, and each one has a very narrow zone of overlap with *P. monacha,* which is currently restricted to the headwaters of just 3 river systems. The frequency and diversity of *monacha-occidentalis* hemiclonal females is much higher where *occidentalis* and *monacha* overlap and can hybridize (Vrijenhoek 1979, Angus 1980, Quattro et al. 1992). By contrast, streams with low clonal diversity are thought to be derived from a single hybridization event. The frequency of *monacha-lucida* hemiclones is less variable, with hybrid females found al-most everywhere that *lucida* is found.

In principle, the driving *monacha* hemiclone has a 2-fold fitness advan-tage over *occidentalis* or *lucida* genotypes, and so it might be expected to in-crease in frequency and go to fixation, at which point it would drive both the host species and itself extinct. As with driving sex chromosomes, there must be some countervailing frequency-dependent selection pressure pre-venting it from going to fixation. In *Poeciliopsis,* the dominant force ap-pears to be the strong avoidance of hybrid females by *lucida* or *occidentalis* males. Laboratory mate choice experiments show that male *lucida* have a great reluctance to mate with *monacha-lucida* females (which should not be surprising, because the hybrid is an evolutionary dead end for their genes; McKay 1971). Furthermore, this mate choice seems to produce the sort of frequency-dependent effects that would keep hemiclones at an intermediate frequency. Field collections show that the probability of a hybrid female be-ing pregnant (i.e., having successfully mated) decreases as the relative fre-

BOX 10.1

Poeciliopsis

This is a genus of about 20 species of fish found in Pacific drainages from the southern United States to Colombia and in Atlantic drainages from southern Mexico to Honduras. Adults are typically 2–5 cm long and are omnivores. As with other members of the Poeciliidae (including guppies, killifish, platyfish, swordtails, topminnows, and mollies), there is internal fertilization and live birth (males have an intromittent organ called a gonopodium). *Poeciliopsis* females usually carry 2–3 separate broods at different stages of development at any one time ("superfetation"), possibly fathered by different males. In *P. monacha* there is no obvious postfertilization transfer of resources from mother to zygote, which instead lives off of yolk supplies ("lecithotrophy"); in all other species there is such resource transfer across a placental barrier ("matrotrophy"). The genetics of sex determination are not well understood: there are no obvious sex chromosomes and within the family Poeciliidae there are species with male heterogamety, female heterogamety, and a mixture of the two. As well as diploid sexual and hemiclonal taxa, there are also triploid asexual, all-female taxa that must mate with a sexual male in order for development to begin ("gynogenesis" or "pseudogamy"). From Meffe and Snelson (1989).

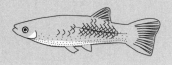

quency of hemiclonal females increases (Fig. 10.2A). In addition, there appears to be a "rare hemiclone" advantage in attracting males, which would contribute to the maintenance of hemiclonal diversity within a single site (Keegan-Rogers 1984).

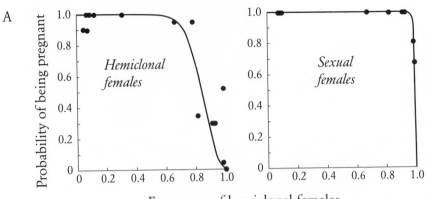

A

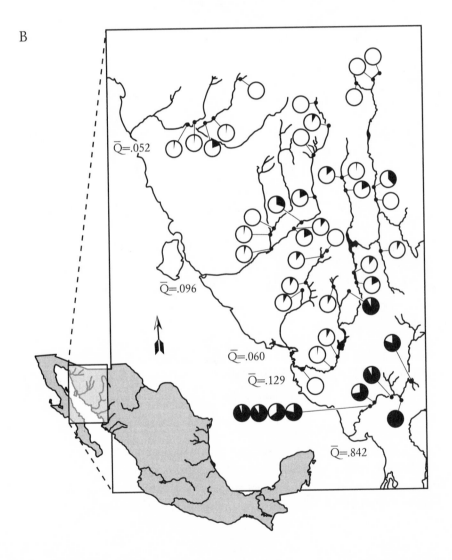

B

If the only fitness differences between sexual and hemiclonal females are the 2-fold drive and the probability of being pregnant, simple calculations show that the equilibrium frequency of hemiclones should be about 85% (Moore 1976). In fact, the frequency of hemiclones is usually less than this (Fig. 10.2B), indicating some other, additional cost of hemiclonal reproduction (e.g., more intense competition or mutational degeneration). And male mate choice is probably not the only factor preserving stability (and as we shall see, it apparently does not act at all in the parallel case of *Rana esculenta*). Ecological differences also exist between sexuals and hemiclones that could allow "niche partitioning" (Weeks et al. 1992). This means that competition among sexuals, or among hemiclones, is more intense than that between sexuals and hemiclones, and this competition also gives the required frequency-dependent selection. Something like this may also contribute to the coexistence of multiple clones in a single locale.

Hemiclones and *P. monacha*. What happens in the zone of overlap with *P. monacha*? One possibility is that, in this zone, hemiclonal hybrid females are constantly being produced and these females often mate with *monacha* males to produce normal *monacha* progeny—that is, that a substantial fraction of the *monacha* gene pool is transmitted through hemiclonal hybrid females. If this were the case, one would expect selection in the *monacha* gene pool for genes that increase hybrid fitness, and *monacha* genotypes from the hybrid zone should produce fitter hybrids than those from outside the hybrid zone. Furthermore, if the hybrids were sufficiently fit, one might expect both hybrid females and *monacha* females to prefer to mate with *lucida*

Figure 10.2 Frequency-dependent selection and distribution of *Poeciliopsis monacha-occidentalis*. A. The probability of sexual and hemiclonal females having mated and being pregnant is plotted as a function of the frequency of hemiclonal females. Each point is a different natural population. Note that the probability of a hemiclonal female being pregnant is more sensitive to an increase in hemiclonal frequency than is the probability of a sexual female being pregnant; this difference tends to stabilize the frequency of hemiclonal females at an intermediate equilibrium. B. Frequency of hemiclonal females in different populations, and the average for each river system ($\overline{Q}$). Note that in most populations the frequency is less than the 85% expected from the difference in probability of being fertilized. Adapted from Moore (1976).

or *occidentalis* males rather than *monacha* males because their genes would get a transmission advantage in the progeny. At the limit, we might expect all hybrid and *monacha* females to mate with *lucida* or *occidentalis* males, and *monacha* itself might go extinct, leaving only the hemiclonal hybrids and the host species.

It is not clear how much opportunity there has been for such selection. The zone of overlap between *monacha* and either *lucida* or *occidentalis* is very narrow and has been little studied. What we do know can be summarized as follows. Allozyme and mitochondrial RFLP studies show that hybridogenetic females arose more than once. This is particularly well supported for *monacha-lucida* hemiclones (Quattro et al. 1991). However, the success rate of hybridization is low, even in the lab. Only 30% of *monacha* × *occidentalis* crosses give at least 1 surviving progeny, and only 7% of *monacha* × *lucida* crosses do (Schultz 1973, Wetherington et al. 1987). Thus hemiclones may not arise frequently, even in areas of overlap. At least in *monacha* × *lucida* crosses, mortality is mainly due to the fact that *monacha* eggs and embryos are much larger than those of *lucida* (5.6 vs. 1.4mm^3) and hybrid embryos, which are intermediate in size, have not used up all the yolk provided by the *monacha* mother by the time they are born, and so the heart ends up outside the body. This is not a problem for hybrid progeny of hybrid females, because they provide an intermediate amount of yolk. Why the reciprocal cross (*lucida* female × *monacha* male) is never successful is not known.

Hemiclonal decay? Though it is clear that hemiclonal reproduction can be ecologically successful over the short term, it is not clear that individual clones are very successful on an evolutionary timescale. Because hemiclones are transmitted intact, with no recombination, natural selection does not work as efficiently on them as on normal, sexual gene pools. Hemiclones are not expected to adapt as rapidly as sexual species to changes in the environment, nor to purge themselves of deleterious mutations as efficiently, and for both reasons might be expected to degenerate over evolutionary time and go extinct (in the same way that obligately asexual taxa go extinct; Burt 2000). Several observations are consistent with mutational degeneration (Vrijenhoek 1984). Many naturally occurring hemiclones cannot mate successfully with *monacha* males: of 12 *monacha-lucida* and 2 *monacha-occidentalis* hemiclones tested, only 7 gave viable progeny (Leslie and Vrijenhoek 1980). By contrast, hemiclones generated in the laboratory do give viable progeny on crossing with *monacha* males, though unfortunately only 2 have been studied (Vrijenhoek 1984). Similarly, *monacha-lucida* × *monacha* hybrid males, when backcrossed to *monacha-lucida* females, produce very few offspring (Leslie and Vrijenhoek 1978). In addition, some hemiclones have silenced enzyme alleles (Spinella and Vrijenhoek 1982).

All these observations are consistent with mutational decay, though it is difficult to exclude the possibility that these features have evolved as a by-product of adaptation to hemiclonal reproduction. For example, selection on hybrid females to mimic *lucida* or *occidentalis* females, and so increase mating success, might conceivably select for the silencing of *monacha* genes (Vrijenhoek 1984). And hemiclonal decay, if it does happen, can be slow: the *monacha-occidentalis* clones that are thought to have colonized the 4 most northerly river systems are estimated, from mitochondrial divergence within the clones, to be about 60,000 years old (about 120,000 generations; Quattro et al. 1992).

Table 10.3 summarizes key reproductive parameters for the 3 systems of hybridogenesis. What is noteworthy for *Poeciliopsis* is the degree to which the *monacha* hemiclone becomes isolated after it becomes part of the *P. monacha-lucida* hybridogen. No males are produced, matings with itself are fatal or soon become so, and new clones probably enter relatively infrequently. These factors should increase the importance over time of processes of hemiclonal decay.

The Water Frog *Rana esculenta*

The European water frog *Rana esculenta* is a hybrid of *R. ridibunda* and *R. lessonae*. It can breed with either parental species, but over a large section of Europe it coexists with only one parental species, *R. lessonae* (reviewed in Vorburger 2001a). Unlike *Poeciliopsis*, both sexes occur in *R. esculenta* and in both sexes only the *ridibunda* genome is transmitted to subsequent generations. The *lessonae* genome is eliminated during gametogenesis, before premeiotic DNA synthesis. The *R. ridibunda* genome then undergoes reduplication and after meiosis is passed unchanged into gametes (Vinogradov et al. 1990).

Although both sexes of *R. esculenta* are common, females appear to be more successful than males (Graf and Polls-Pelaz 1989). Low reproduction of *R. esculenta* males partly reflects female choice against them in favor of *R.*

lessonae males. Indeed, both species favor *R. lessonae* males—at long range (frog calls) and in response to amplexus (decreased clutch size; Reyer et al. 1999, Roesli and Reyer 2000). In addition the sperm of *R. esculenta* males are ejaculated in smaller numbers and are less competitive in mixed ejaculations (corrected for density) than are sperm of the other species (Reyer et al. 2003). Testes of *R. esculenta* males are smaller, often of abnormal structure and with fewer spermatogonia, which take longer to mature (Ogielska and Bartmanska 1999). Matings between *R. esculenta* typically produce only females (consistent with the Y chromosome of males usually being derived from *R. lessonae* and therefore lost). These females have *R. ridibunda* genotypes, so recombination between clones in such females may occur (but has not yet been demonstrated; Vorburger 2001c). And *R. esculenta* × *R. ridibunda* matings lead to balanced, viable young, again allowing recombination before returning to asexual, hemiclonal life in *R. esculenta* (Schmidt 1993).

Although females are choosy, males do not appear to be (Engeler and Reyer 2001). This is surprising, because mating with *R. esculenta* females is a genetic dead end for *R. lessonae* males, just as it is for *P. lucida* male fish, which are highly discriminating (McKay 1971). Engeler and Reyer argue that the anomaly can be explained by male preference for larger, more fecund females within his own species. Female *R. esculenta* are larger than *R. lessonae*, perhaps confusing the males with an apparent offer of higher fecundity.

There is clear evidence of hemiclonal decline, in the form of deleterious recessives that are more than offset by strong heterotic benefits from hybridization (Hotz et al. 1999, Vorburger 2001a). If individuals within a *R. esculenta* clone mate with each other, all progeny are usually inviable, perishing early in larval life. This mortality presumably reflects the accumulation of deleterious, recessive alleles in a local, asexually reproducing population. By contrast, matings between members of different clones, for example, from different geographic regions, often give viable young, suggesting that in isolation they have accrued a different set of lethal recessives (Vorburger 2001b). Alternatively, the initial instantaneous formation of separate strains may have isolated differing sets of recessive lethals. Different hemiclones also appear to evolve adaptations to the local, outgroup males. In experimental crosses between frogs from two different regions, within-population matings between *R. esculenta* females and *R. lessonae* males produce larger frogs at metamorphosis than do between-population matings (Semlitsch et

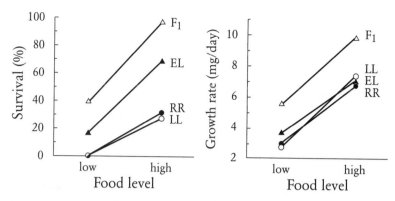

Figure 10.3 Fitness-related parameters as a function of food levels for 4 kinds of *Rana* crosses. Percentage survival and growth rate until metamorphosis (in mg/day) as a function of low or high food availability for the following crosses: F_1 = first-generation hybrid *R. ridibunda* × *R. lessonae;* EL = natural hybrid lineages, *R. esculenta;* LL = *R. lessonae;* and RR = *R. ridibunda.* Note that first-generation hybrids shows superior performance compared to either parental strain. Adapted from Hotz et al. (1999).

al. 1996). At the same time, clonal variation is locally selected because mixtures of clones outperform monoclones (Semlitsch et al. 1997).

The success of *R. esculenta* may have 3 causes: heterosis, a broader niche, and a specialized niche. Perhaps most importantly, *R. esculenta* shows strong heterotic effects. For example, individuals of this species are superior to either *R. ridibunda* or *R. lessonae* in a variety of larval life history traits, such as survival, growth rate, and early age of metamorphosis (Hotz et al. 1999). Especially striking is the fact that first-generation hybrids show greater heterosis than established ones (which presumably show some hemiclonal decay; Fig. 10.3). *Poeciliopsis* is the complete opposite. New hybrids between *P. lucida* and *P. monacha* are usually inviable.

The weaker, more general heterosis of later-generation *R. esculenta* appears largely to be behavioral. That is, growth efficiency is not heterotic in *R. esculenta* but is intermediate between the 2 parental species. The growth effect seems to be mediated through increased feeding time and consumption (Rist et al. 1997). In addition, several studies have shown that *R. esculenta* has a wider environmental tolerance than does *R. lessonae*. Especially in harsh environments, such as low food, high larval density, or the presence of agricultural chemicals, it does better than the parental species (reviewed in

Negovetic et al. 2001). This may reflect heterosis as well as selection to occupy a broader niche. Finally, there is evidence for strong niche segregation, at least by water temperature and oxygen levels. As measured by survival and body mass at metamorphosis, *R. esculenta* performs much better at lower temperatures, while *R. lessonae* does better at higher (Negovetic et al. 2001). For unknown reasons, *R. esculenta* females appear to be relatively more numerous where there is less variation in water temperature (Peter et al. 2002). *R. esculenta* is intermediate between the other 2 species in the breeding habitat they occupy (Pagano et al. 2001), itself characterized by intermediate levels of oxygen (Plenet et al. 2000). In dispersal behavior, *R. esculenta* falls exactly between the parental species (Peter 2001). The ability of the 2 genomes in *R. esculenta* to work together to produce such a superior product is especially striking because they differ so much in genome size: *R. ridibunda*'s genome is 16% larger than that of *R. lessonae* (Vinogradov and Chubinishvili 1999).

The hybridogenetic drive of *ridibunda* genotypes may be an important feature in the success of introduced *R. ridibunda*, which have replaced the native *R. esculenta* and *R. lessonae* in several areas of central Europe (Vorburger and Reyer 2003). The *R. ridibunda* frog can breed with both native species, producing *R. esculenta* offspring with *R. lessonae*, and *R. ridibunda* with *R. esculenta*. Matings between hybrids are, in turn, viable because they are unlikely to share the deleterious mutations that characterize the *ridibunda* portion of the *R. esculenta* genome.

It is noteworthy that 2 Balkan species of *Rana* (*shqiperica* and *epeirotica*) form natural hybrids with *R. ridibunda*, but these fail to show genome exclusion (Guerrini et al. 1997). This could reflect a weakness in the local *R. ridibunda* genome, but even when the males are imported from regions where they do induce genome exclusion in local hybrids, they fail to induce exclusion of the Balkan genomes. Whether this resistance is an evolved capacity or a preexisting condition is unknown.

In summary, *R. esculenta* provides a clear case of a species in which a genome gains an immediate benefit in transmission that is then augmented by the advantages of outbreeding with a closely related species. By parasitizing a bisexual, closely related species every generation, the *ridibunda* genome enjoys a largely asexual (or hemiclonal) life without suffering the usual costs.

The Stick Insect *Bacillus rossius-grandii*

Two strains of hybridogenetic stick insects are found in Sicily (reviewed in Tinti and Scali 1993). Males are rare and sterile (due to severely impaired gametogenesis). The hybridogenetic females transmit only their *rossius* genes, discarding the *grandii* genome. Contrary to earlier conclusions (Tinti and Scali 1992), recent work shows that genome loss occurs before meiosis in *Bacillus,* just as it does in *Poeciliopsis* and *Rana* (Marescalchi and Scali 2001). This loss requires an extra round of DNA replication during meiosis in order to produce hemiclonal eggs after 2 cytologically normal meiotic divisions. As we will see, the species also shows androgenesis: in as many as 20% of a female's progeny, both sons and daughters, all genes arise from the syngamy of 2 sperm nuclei (Mantovani and Scali 1992, Tinti and Scali 1992).

Bacillus is also unusual in that hybridogens coexist with parthenogens, *B. whitei,* formed from the same 2 parental species (in each case with *B. rossius* acting as the female parent; Mantovani et al. 2001). The evidence suggests that the hybridogenetic species may have given rise to the parthenogenetic one. On the one hand, the peculiar features of egg maturation are shared by the hemiclonal and clonal eggs (Scali et al. 1995). On the other hand, hybridogenetic female *B. rossius-grandii* occasionally produce parthenogens. The reason for the extraordinary genetic diversity within *Bacillus* is completely unknown. What is certain is that it provides unusual opportunities for genomes to cycle between the various systems.

Evolution of Hybridogenesis

The evolutionary origins of hybridogenetic drive are unknown. That is to say, while we know that it arises by hybridization, we do not know anything about the differences between species that give rise to genomic exclusion in F_1 hybrids, and how those differences evolved. Perhaps hybridogenesis reflects coevolution between the mitotic machinery as expressed in premeiotic cells and *cis*-acting sequences at centromeres. Also, many selfish genetic elements are manifest only in population or species hybrids (e.g., driving sex chromosomes and cytoplasmic male sterility; see Chaps. 3 and 5). We wonder, then, whether some driving selfish genetic element has swept through *P. monacha*, or *R. ridibunda*, or *B. rossius*, and has now gone to fixation in those species, and the only evidence for it is in species hybrids, when introduced into a population in which it has not spread. The alternative would be that some other sort of genetic change has swept through one species or another, due to conventional natural selection, and the behavior in hybrid genotypes is only an accident or coincidence. It will be interesting to find out.

Androgenesis, or Maternal Genome Loss

"Perhaps it is just a matter of time before someone discovers (or invents in the laboratory) an all-male species. It makes diploid sperm that inseminate eggs of a related species and give rise to diploid nuclei that exclude the egg pronuclei. The exclusively sperm derived genes then direct the development of a male animal with the father's genotype" (Williams 1988, 294).

Among the most bizarre genetic systems ever to have evolved are those in which only paternal genes are passed on even though the female or female tissue does all the parental investment. Androgenesis is unusual in displaying a form of selfish action that was predicted to exist before actually being reported. It is now known from 3 groups of organisms (McKone and Halpern 2003) and typically occurs within a species, but may also occur between species, as Williams imagined. Two of the cases, a conifer and several clam species, are hermaphroditic. This characteristic, as we shall see, may both facilitate the evolution of androgenesis and make subsequent extinction less likely. The third group, stick insects, is dioecious, and androgenesis occurs only in relatively infrequent hybrid, triploid forms. One of the 2 hermaphroditic cases, the conifer, is a very rare species and we begin with it.

The Conifer *Cupressus dupreziana*

C. dupreziana is a very rare, hermaphroditic conifer, confined to a section of a desert in Algeria (Pichot and El Maâtaoui 2000). Only 231 individuals were alive in 1970 (Pichot et al. 2001). It reproduces by androgenesis, that is, diploid pollen provides all of the nuclear genes for the next generation.

Androgenesis occurs in the following way. Megasporogenesis (leading to seeds) is normal, but microsporogenesis (leading to pollen) is not. Premeiotic development conforms to the usual coniferous patterns, in which microcytes separate and begin meiosis, but unexpected irregularities then appear (El Maâtaoui and Pichot 2001). These include abnormal chromosome separation and cytokinesis, followed by fusion of the meiotic products. It is not entirely clear whether these occur at random, in which case heterozygosity decreases, or whether heterozygosity is maintained by having unlike products reunite. What evidence is available suggests that the species lacks genetic diversity in nature (El Maâtaoui pers. comm.). It is thought that the diploid pollen travels to a female cone and develops into an embryo without first fusing with a female cell.

The conifer is part of a family, the Cupresseaceae, all members of which show paternal inheritance of both mitochondria and chloroplasts, a rare trait in plants, found only in some related conifer families (reviewed in Mogensen 1996). While not a prerequisite for androgenesis, this probably facilitates its evolution, because all the requisite DNA can arrive from paternal tissue alone.

It is hard to escape the impression that androgenesis has put this species on the fast road to extinction. Certainly, it is associated with a major extinction-biasing factor, namely, low fecundity. Only 10% of seeds are normal—in other words, an embryo surrounded by a reserve, endosperm-like tissue (Pichot et al. 1998). The remaining seeds lack an embryo or lack both embryo and endosperm. A reduction of female fecundity by a factor of 10 must place it at an enormous competitive disadvantage when overlapping with conventional species. It is not known whether selfing occurs in nature. If it did, the hermaphrodite's genes would be passed at the same rate whether through paternal tissue, maternal, or both.

It has been argued, instead, that androgenesis represents an evolutionary response to the threat of extinction, preventing inbreeding through uniparental inheritance. But it is not clear whether diploid pollen is the product of selfing and a 90% cost is a high price to pay for inbreeding avoidance. The only way we see an advantage for androgenesis is if it can be used to colonize other species. Controlled crosses between *C. sempervirens* and *C. dupreziana* produce only viable *dupreziana* offspring (Pichot et al. 2001). This means if it often overlaps another *Cupressus* species, it may be able to propagate itself by using heterospecific "surrogate mothers." Recent work does show colonization of a closely related species in nature (El Maâtaoui pers. comm.).

Why should female fecundity be so low? A drop in female fecundity in response to androgenesis makes sense if some paternity is not androgenetic, so in aborting the androgenetic, the maternal tissues save resources for nonandrogenetic. But no evidence has been produced to support this possibility. Another possibility is that androgenesis has selected for male-biased investment in reproductive function, because androgenetic genomes are passed preferentially by males (McKone and Halpern 2003). The less selfing there is, the stronger such selection, so androgenetic genomes are expected to be outbreeders, perpetually on the lookout for new maternal tissue to colonize.

The Clam *Corbicula*

Several species of the freshwater clam *Corbicula*–diploid, triploid, and tetraploid–appear to reproduce entirely by androgenesis (Komaru et al. 1998, Byrne et al. 2000, Qiu et al. 2001). The sperm cells have the same DNA content as somatic cells and presumably arise by some kind of ameiotic process whose details are unknown (Komaru and Konishi 1999). What is known is that the entire female nuclear genome is ejected in 2 polar bodies shortly after the sperm enters the egg (Komaru et al. 1998, 2000). Comparative data on *Corbicula* and related genera show that androgenesis occurs in species with biflagellate sperm, brooding of young, and hermaphroditism (reviewed in McKone and Halpern 2003). The significance of the first is unclear. Brooding of young may facilitate androgenesis because, at least initially, such a radical change in the genetic system may increase the frequency of aneuploid progeny. In brooding species there is usually some natural abortion of offspring, and so aneuploid progeny can be replaced by normal euploid progeny with little or no effect on a female's fecundity. Parthenogenesis is also associated with brooding, probably for the same reason (Lively and Johnson 1994). Hermaphroditism makes long-term survival more likely and, if associated with selfing, sharply reduces the cost to the maternal genome (because it will be closely related to the paternal).

The Stick Insect *Bacillus rossius-grandii*

As we have noted, in this hybrid species as many as 20% of a female's progeny, both sons and daughters, can arise by androgenesis. In this case, 2 haploid sperm of a male fuse inside the egg to generate a diploid genome, which then displaces the maternal set (Tinti and Scali 1992).

Androgenetics have also been observed in both intra- and interspecific crosses of pure species and in *Bacillus* androgenetics can produce fertile progeny of both sexes (Tinti et al. 1995). The ability of *Bacillus* stick insects to produce fertile androgenetics of both sexes permits hybridogenetic hemiclones to reenter the ancestral population. A *B. grandii* androgen is produced when the entire hybridogenetic egg nucleus degenerates and the offspring receive only paternal genes. These arise from fusion of 2 sperm nuclei of the many found in the polyspermic eggs. (The possible effects of multiple matings on this system are completely unexplored.) *B. rossius* androgens can be generated experimentally when *B. rossius* males fertilize hemiclonal females, producing fully fertile androgens of both sexes (Tinti and Scali 1993, 1995). Likewise, androgens can be imposed on the otherwise parthenogenetic *B. whitei.*

ELEVEN

Selfish Cell Lineages

YET ANOTHER KIND of within-individual conflict can result from the fact that cells within a multicellular individual are not quite genetically identical to each other. On the one hand, clonally generated cells from a single progenitor cell (e.g., a zygote) become less alike genetically with time, through spontaneous mutation and mitotic recombination during each cell cycle. This effect is inevitable for any multicellular species, but it is expected to vary in intensity, being strongest in large, long-lived ones. Certainly all humans are, in this sense, genetic *mosaics*. Within our bodies, variant cell lineages can expand or contract according to their tendency to proliferate. Both cancer and the adaptive immune system are fundamentally based on the principle of differential cell reproduction. We review the evidence of such differential reproduction and the resulting cell lineage selection in both somatic and germline tissues, as well as the idea that controlling cell lineage selection may have been an important selective agent in the evolution of development.

Genetic variation among cells and cell lineage selection can also arise because organisms are *chimeras*—that is, derived from the amalgamation of cells of different genotypes. In our own species, low-level chimerism between mother and offspring is not uncommon, and in marmosets and tamarins chimerism between twins is the norm. In colonial invertebrates, genetically distinct colonies often fuse, and competition for representation in the germline is expected to be particularly intense. In mushrooms, chimerism is an essential component of the sexual cycle, with genetically unrelated nuclei both cooperating and competing in a common cytoplasm. And in several taxa, while the germline is uniform, the soma is chimeric, and this can lead to differences among tissues and among genes in coefficients of relatedness to family members, and thus conflicts over optimal patterns of behavior. The disunity of the organism is manifest once more.

Mosaics

Spontaneous mutation is an inevitable source of genetic variation within any multicellular organism, particularly those that are large and long-lived. There are, for example, about 10^{13} cells in the human body, and about 10^{12} cell divisions per day. Because the mutation rate per nucleotide is about 10^{-9}, this means every possible single nucleotide mutation occurs in our genome hundreds of times every day, and within our lifetime the whole range of Mendelian genetic diseases probably arises at one time or another, in one cell or another. In addition, mitotic recombination (which arises as a consequence of DNA repair; see Chap. 6) can convert heterozygous cells into homozygotes. Rates of mitotic recombination are 10^{-7} to 10^{-5} per cell cycle in yeast, and, in the germline, 10^{-4} to 10^{-2} per individual generation in *Drosophila*, and 10^{-5} to 10^{-4} per individual generation in plants (reviewed in Otto and Hastings 1998). Loss of heterozygosity can also occur by duplication of an entire chromosome, followed by loss of the homolog (Cervantes et al. 2002).

Many times these genetic changes have no effect—for example, a brain-specific gene mutates in the kidney and has no effect. Or an essential gene is lost, the cell dies, and another cell divides to take its place. In terms of cell proliferation, such mutations are, respectively, neutral or deleterious. For the organism, most such mutations are probably harmless. Instead, the danger lies in the small minority of mutations that give the cell a proliferative advantage, for these can lead to cancer.

Somatic Cell Lineage Selection: Cancer and the Adaptive Immune System

Cancers are selfish cell lineages: clones of cells that, within an individual's lifetime, have evolved high rates of replication compared to other clones, at the expense of host fitness. They arise and spread by the conventional Darwinian process of repeated rounds of mutation and selection (Nowell 1976, Vogelstein and Kinzler 1993). This realization helps to make sense of much of the phenomenology of cancer. For example, environmental contributors to cancer can include mutagens or anything that increases the rate of cell division, and thus the opportunity for mutations to arise or selection to occur (Ames et al. 1995). Also, cancers are more of a problem in mitotically active tissues in which the need for regeneration is common (e.g., epithelial tissues), and less common in mitotically inactive tissues in which the need for regeneration is rare (e.g., the heart, where trauma is usually fatal; Williams and Nesse 1991). Somatic mutation and cell lineage selection are also important in the medical treatment of cancers, as tumors can evolve resistance to chemotherapy. Mutations that allow cells to produce their own mitosis-inducing signals, suppress contact inhibition, evade apoptosis, attract the vascular system, or disperse to other locations in the body (metastasize) can all have a selective advantage.

Because cancers are harmful to the organism, there will be selection for adaptations that prevent or delay their occurrence—anticancer adaptations. An indication of the importance of such adaptations in our own species is suggested by the fact that in wild mice raised under benign conditions in the laboratory, some 46% have gross tumors at death (Andervort and Dunn 1962). Humans are some 3000 times larger and live 20–30 times longer than mice, so if the probability of a cell becoming cancerous was the same per unit time in us as in mice, none of us would make it out of the womb alive, let alone reach puberty (Peto et al. 1975). Selection to prevent cancer must have been ever-present in the evolution of animal development (Graham 1992).

Anticancer adaptations must work by attacking the evolutionary process of somatic mutation and cell-lineage selection, and there are 3 ways to do this (Leroi et al. 2003). First, they can reduce the mutation rate. Some stem cells seem to arrange DNA replication and cell division to retain an "immortal strand" of DNA within the stem cell—that is, an unreplicated strand from which all copies are derived—and this should reduce the mutation rate (Cairns 1975, Potten et al. 2002). Some stem cells also do not seem to repair certain forms of DNA damage, preferring instead to die, which can also reduce the mutation rate (Cairns 2002). Second, anticancer adaptations can reduce the selective advantage (at the level of competing cell lineages) of each intermediary step toward tumor formation, so the mutations do not accumulate as much, or as fast. This can occur by modifiers evolving that make the cancer-contributing mutations more recessive. Stem cell dynamics and tissue architecture are also important. Cairns (1975) suggests that interposing a series of transiently amplifying cells between stem cells and terminally differentiated cells reduces the number of stem cells required and so also reduces the frequency of cancer (assuming that only stem cells can become cancerous; see also Frank et al. 2003). Similarly, having separate patches of stem cells, between which migration is difficult or impossible (as is likely for those in the colonic crypts), also reduces the selective advantage of an oncogenic mutation.

Finally, anticancer adaptations can work by adding another set of controls on cell proliferation, so more mutations are required to turn a cell cancerous (Nunney 1999). Cells in the retina (a relatively small and nonproliferative tissue) can become cancerous by the inactivation of only 1 tumor-suppressor locus, whereas cells in the lower gastrointestinal tract (a large, constantly proliferating tissue) require knockouts of 3 loci as well as the activation of an oncogene (Vogelstein and Kinzler 1993). It appears that we have evolved added redundancy in those tissues most requiring it.

Cell lineage selection is not always harmful to the organism. As we have indicated, it can mean that a defective mutant cell is replaced by a non-mutant neighbor. It even sometimes happens that a patient with an inherited disorder (e.g., of the immune system) has a back mutation in a cell that allows the cell to proliferate and cause a spontaneous cure (Hirschorn et al. 1996, Erickson 2003). Most impressive of all, vertebrates have evolved an adaptive immune system that is fundamentally based upon the principle of cell lineage selection (Alberts et al. 2002). Briefly, there are about 2×10^{12} lymphocytes (B cells and T cells) in the human body (together weighing about the same as the brain or liver). They mostly differ one from another due to rearrangements of the antibody- or T cell receptor-encoding genes. (The mechanism responsible for these genome rearrangements is thought to have evolved from a domesticated transposable element; see Chap. 7.) These rearrangements are more or less random with respect to function, and this variation is acted on by 2 types of selection. First, lymphocytes that recognize "self" are preferentially killed, inactivated, or induced to change. Second, lymphocytes that bind to pathogens are stimulated to replicate. Moreover, in B cells, this replication is associated with hypermutation of the antibody-encoding gene, creating more variation, and cells that bind tightest to the antigen are induced to replicate fastest. In this way our immune system can mount a response to almost any foreign antigen and produce antibodies with increasingly higher affinity as an infection persists, while not attacking our own bodies.

Cell Lineage Selection in the Germline

Though novel genotypes may evolve by cell lineage selection during the lifetime of an organism, those genotypes almost always disappear when the organism dies. There is one wonderfully bizarre exception of a cancer that does not die with its host and is now circulating among feral dog populations (Box 11.1; Plate 9). The other, more general, exception is for cell lineage selection that occurs within the germline—the precise genotype may be lost by recombination, but changes in allele frequencies can be passed on to subsequent generations.

Selection in the germline has been experimentally demonstrated in classic experiments on *Drosophila* (Abrahamson et al. 1966). These experiments involve irradiating males and then measuring the frequency of recessive lethal mutations transmitted by those males on their X chromosome and on one of their autosomes. If one looks at progeny from matings immediately after the irradiation, when there has been no opportunity for any selection, the ratio of lethals transmitted on the X and on the autosome is about 0.38:1. This is not too different from the relative size of the 2 chromosomes. But if one looks at progeny from matings 17–21 days after the irradiation, the relative frequency of X-linked lethals is only 0.18:1, approximately half of what it was previously. The sperm cells transferred in these matings would have been derived from cells that were primary spermatogonia at the time of irradiation. These form a small population of asynchronously dividing cells that function as a reservoir for the secondary cells, which undergo 4 synchronous mitotic divisions to produce a cyst of 16 cells, each of which then undergoes meiosis to form a total of 64 haploid sperm. In the development of primary spermatogonia into sperm, selection is expected to act more intensely against X-linked mutations because there is only a single copy of the X, whereas mutations on the autosomes are masked by the other copy. This selection is positive for the individual. Cell lineage selection in the germline has also been observed in mosaic flies generated by inducing mitotic recombination at particular mutant loci (Extavour and García-Bellido 2001).

424

BOX 11.1

Canine Transmissible Venereal Sarcoma: The Origin of a Highly Degenerate Mammal

Mammals are continually sloughing off cells (e.g., skin), but these cells usually die once they are separated from the organism. Even the most aggressive and invasive cancer cells usually die when their host dies. But there is a fascinating exception: canine transmissible venereal sarcoma (CTVS) is an infectious disease of dogs that is caused by a pathogenic lineage of cancerous cells (Das and Das 2000). This cell lineage is transmitted from one dog to another, usually during coitus. Once on a new host, the cells reproduce to form a tumor-like growth, usually around the genitals, and the cell lineage can then be transmitted to another host (Plate 9). It is this continuity of the cell lineage (as opposed to mere continuity of an infectious virus) that distinguishes CTVS from other transmissible cancers (e.g., human cervical cancer).

CTVS can be found in many parts of the world, and in some regions it is the most common dog tumor. It is thought to have originated only once and spread worldwide, and a LINE retrotransposable element insertion upstream of the c-*MYC* oncogene (Choi et al. 1999) was presumably important in its genesis. There are no obvious differences in susceptibility among breeds of dogs, and the cells can also be transmitted to foxes. Even without treatment, tumors usually regress after 1–3 months; if regression is complete, the host is immune to subsequent reinfection. This "naturally occurring allograft" has become a true pathogen. One could even think of it as a highly degenerate mammal.

The opportunity for selection in the germline increases with the number of germline cell divisions from zygote to zygote. Even in species with segregated germlines, this is often more than 20; in a 40-year-old man it is 600–700; and in long-lived plants and animals without a segregated germline, it

Table 11.1   Number of germline cell divisions from zygote to zygote in different species

Species	Number of Cell Divisions
Maize	50
Drosophila	35 (at age 18 days for males & 25 days for females)
Mice (female)	25
(male)	62 (at age 9 months)
Humans (female)	23
(male)	36 (at age 13; + 23 per year thereafter)

Age refers to the time at which the zygote is produced.
From Otto and Hastings (1998).

is presumably much more (Table 11.1). Only genes that are expressed in the germline are subject to germline selection, but this may be a substantial number of genes. Germline development is a complex process involving proliferation, active and passive migration, contacts and signaling with various cell types, and gamete differentiation as well as basic cellular metabolism (Saffman and Lasko 1999). One can reasonably suppose that it is open to influence by a substantial fraction of the genome. Consistent with this expectation, mosaic analysis in *Drosophila* has shown that most zygotic lethal mutations are homozygous lethal in the germline (67% for point mutations, 88% for chromosomal deletions; García-Bellido and Robbins 1983). These numbers are expected to vary among cell types and are known to be different in tergite cells (20% and 58%, respectively).

In humans, cell lineage selection in the germline has been implicated in the transmission of 2 genetic diseases. First, myotonic dystrophy type 1 (DM1) is an autosomal dominant disorder caused by the expansion of a CTG triplet repeat in the 3′ untranslated region of the *DMPK* gene. Normal alleles have 5–30 repeats, mildly affected individuals from 50–80, and severely affected individuals 2000 or more. The disease shows a phenomenon known as "anticipation," in which the symptoms become increasingly severe and appear at a progressively younger age in successive generations. This is because offspring tend to have more repeats than parents. The conventional explanation for this difference is mutation bias, but a positive association has been found in lymphoblastoid cell lines between the number of repeats and cell replication rate (Khajavi et al. 2001). The same association occurring in the germline could contribute to (or cause) the increase in number of repeats between parents and offspring.

Second, Apert syndrome is another autosomal dominant disorder; approximately two-thirds of the cases are due to a C→G mutation at position 755 in the *FGFR2* gene, which causes a Ser→Trp change in the protein (Goriely et al. 2003). This is a male-specific mutation hotspot: in a study of 57 cases, the mutation always occurred on the paternally derived allele (Moloney et al. 1996). On the basis of the observed birth prevalence of the disease (~1 in 70,000), the apparent rate of C→G mutations at this site is about 10^{-5}, which is 200- to 800-fold higher than the usual rate for C→G mutations at CG dinucleotides. Moreover, the incidence rises sharply with the age of the father (Fig. 11.1). Goriely et al. (2003) analyzed the allelic distribution of mutations in sperm samples from men of different ages and concluded that the simplest explanation for the data is that the C→G mutation gives the cell an advantage in the male germline. In particular, they suggest that in a testicular stem cell population of 10^7 to 10^8, the mutation may cause a clonal expansion to about 10^3 cells (essentially a small neoplastic lesion) and this could account for the elevated mutation rate.

In both these examples of human diseases the mutations are so deleterious at the individual level that they cannot possibly go to fixation. Rather, the population comes to have an intermediate equilibrium frequency of the mutation, and the effect of cell lineage selection is to increase the effective mutation rate, and thus the equilibrium frequency of the disease. In principle, mitotic recombination and cell lineage selection in the germline can also lead to "mitotic" drive and the spread and fixation of a gene if heterozygous individuals give rise to homozygous cells in their germlines at an appreciable rate, and if one of those homozygous types has a proliferative advantage (Otto and Hastings 1998). The opportunity for this effect is greatest in species without a segregated germline and in which there are many germline cell divisions per generation.

Evolution of the Germline

We saw in our discussion of cancer that the evolution of development has almost certainly been influenced by the need to prevent the proliferation of variant cell lineages. This need applies not only to the construction of somatic tissues like the colon but also to the way the organism makes its gametes. Indeed, the dangers of cell lineage selection will be particularly acute in germinal tissue because not only is there the risk of cancer but there is

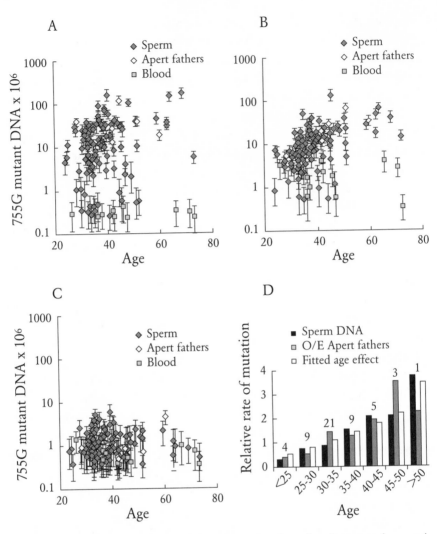

Figure 11.1 Frequency of mutant sperm cells as a function of age in men and comparison with the paternal age effect for Apert syndrome. At position 755 of the *FGFR2* locus, the wildtype is a C, but mutants can be found in sperm samples. A. Frequency of C→G mutants increases with male age in sperm but not blood cells. B. Frequency of C→T mutations increases in sperm but not blood cells. C. Frequency of C→A mutants increases in neither sperm nor blood cells. D. Relative levels of C→G in sperm (black); observed/expected rates of Apert syndrome births; and rates predicted from a best fit model of observed birth data (white) each in relation to the father's age. Adapted from Goriely et al. (2003).

also the danger, perhaps even more acute, of creating a genetically inferior gamete carrying a deleterious mutation. There are probably many mutations that give the cell a small proliferative advantage and do not typically harm the individual in which they occur but that would be harmful if passed on to the next generation. Just one such mutation could be enough to kill an offspring, and therefore the strategy of adding extra layers of redundancy does not work as well in the germline. Other strategies such as restricting the migration of cells also seem likely to be less effective in reducing the transmission of deleterious mutations than in reducing the incidence of cancer deaths.

So, how can an organism produce gametes in a way that reduces the probability of transmitting a new deleterious mutation? There are several possibilities. First, one might expect gametes to be derived from a small population of specialized cells that are distinct from the rest of the body. This pattern would reduce the number of cell divisions and therefore the mutation rate. Perhaps more importantly, it would also reduce the number of germline cells and therefore the probability that a mutation that increases the cell's rate of proliferation would get established and be transmitted. Second, one might expect cell division in the germline to be under the control of both the cell itself and neighboring somatic cells, such that only if both "agree" would division take place. The reliance on the cell itself would give some degree of purifying selection, which, as in somatic cells, would probably be a good thing; the simultaneous reliance on a somatic cell that is distantly related and unlikely to share the same somatic mutations might help prevent proliferative mutations from spreading. That is, one expects cell division to be under both cell-autonomous and nonautonomous control. Finally, one might expect cell divisions in the germline to be synchronous, but with the option of a cell not dividing and dying. Again, this possibility would allow purifying selection but reduce the opportunity for a selfish lineage to proliferate.

It is easy to point to features of the germline consistent with these expectations. Many animal phyla segregate their germline early in development. After segregation, germ cells often migrate considerable distances to reach the gonad, where they are surrounded by somatic cells (Saffman and Lasko 1999). Somatic cells often play a key role in the proliferation and differentition of the germ cells. In plants it is the position of cells within the meristem that determines whether they will be part of the germline (Pineda-

Krch and Lehtilä 2002). As plant cells are immobile, they too have limited opportunities for selfish proliferation. Finally, some degree of replication synchrony is common in gametogenesis. Unfortunately, while it is easy to point to features of germline development that are consistent with expectations, the extent to which these features evolved in order to modulate germline selection is unknown.

Even these features do not entirely isolate germ cells from selfish somatic elements. In *Drosophila* a retrotransposon (412) is very active in the somatic mesoderm that gives rise to the testes and such cells surround and attach to newly arriving primordial germ cells (Brookman et al. 1992). 412 activity is not induced by the arrival of the germ cells but by prior-acting homeotic genes and 412 remains active while the germ cells are dividing. The suspicion is inescapable that 412 is pumping either copies of itself into the germ cells or other chemicals that aid 412 transposition later in spermatogenesis, but this has not been shown. 412 is inactive in ovaries but active in spermatocytes (Borie et al. 2002).

Direct evidence of a reduced frequency of mutants in the germline comes from studies of mice (Walter et al. 1998). Male germline cells have a significantly lower frequency of mutants than somatic tissues (brain, liver, and Sertoli cells, which are somatic cells in the testes; Fig. 11.2A). It is not clear whether this difference is due to a lower mutation rate or fewer cell divisions. Comparisons of DNA repair efficiencies in different tissues have shown that 1 mechanism (base excision repair) is more active in germline cells than in somatic cells (Intano et al. 2001, 2002, Olsen et al. 2001), while another mechanism (nucleotide excision repair) is inactive in the germline but active in somatic cells (Jansen et al. 2001). The interpretation of these latter results is not yet clear: as noted above, a cell choosing to die rather than repair its DNA could reduce the overall mutation rate.

Mutant frequencies also differ between germ cells at different stages of development, and between young and old mice (Fig. 11.2B). Tests made on purified populations of 8 different spermatogenic cell types show a significantly lower frequency of mutants than do the somatic cells. Moreover, mutant frequencies decline through spermatogenic stages, particularly from type A to type B spermatogonia. This transition appears to be selective for cells with fewer mutations. Consistent with such selection, there is significant apoptosis at this stage, though the mechanistic basis of the selection is unknown.

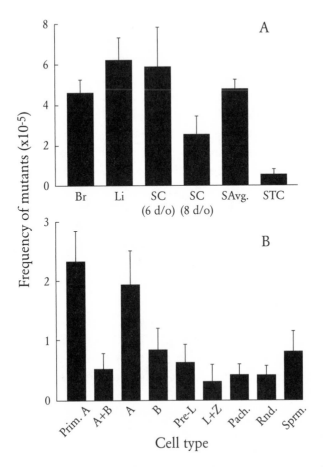

Figure 11.2 Mutant frequencies in different tissues of mice. Mice were genetically engineered to carry a silent *lacI* gene, and then the gene was recovered from various tissues and introduced into bacteria to measure the frequency of mutations. Bars show mean and standard errors. A. Comparison of somatic and germline tissues. Br, brain; Li, liver; SC, Sertoli cells; SAvg, somatic average; STC, seminiferous tubule cells. There was no significant difference among the somatic cell types, but significantly lower mutant frequencies in the seminiferous tubule cells. B. Comparison of different spermatogenic cell types. Prim. A, primitive type A spermatogonia; A, type A spermatogonia; B, type B spermatogonia; Pre-L, preleptotene spermatocytes; L + Z, leptotene plus zygotene spermatocytes; Pach., pachytene spermatocytes; Rnd, round spermatids; Sprm, spermatozoa. All cell types had a significantly lower mutant frequency than the average somatic frequency. Furthermore, there was a significant decline in mutant frequency between primitive type A spermatogonia and type B spermatogonia. Adapted from Walter et al. (1998).

Finally, we have assumed in our discussion that most mutations that increase the proliferation rate of a cell would be harmful if transmitted to the next generation. There are no hard data on the matter, and others appear to make the opposite assumption (e.g., Otto and Hastings 1998). Whitham and Slobodchikoff (1981) suggest that in trees and other long-lived plants, branches are genetically different, and branches that better fit the local environment (e.g., are more resistant to insect pests) grow and expand and make more pollen and ovules at the expense of other branches on the same plant that are less fit (see also Gill et al. 1995). In this way, a single individual may evolve over time and pass on the advantageous changes to its progeny. To some extent, such adaptive evolution must occur. Indeed, the prerequisites for such evolution have been shown in detail for 2 species of branching red algae, *Delisea pulchra* and *Asparagopsis armata* (Monro and Poore 2004). There is significant within-individual phenotypic variation in both species in such fitness-related traits as growth, secondary metabolic concentrations, and rates of tissue loss due to herbivory. Significant heritabilities were also shown for growth in *A. armata* and the other 2 traits in *D. pulchra*.

Though some aspects of cell lineage selection may be beneficial to the organism and its progeny, we are not aware of any evidence of plants, animals, or any other organism evolving mechanisms to increase the efficacy of cell lineage selection in the germline. No evidence of high mutation rates, either in general or at specific loci (e.g., those involved in pest resistance); no mechanism analogous to the vertebrate immune system that rearranges the genome at specific sites to create variation; and no high rates of gene conversion, which would be a relatively easy way to create variation. If such adaptations are indeed missing (as opposed to merely undiscovered), the implication is that positive selection in the germline is, on average, dangerous and to be avoided. Mutations that are beneficial at the cell lineage level may be more often deleterious than beneficial when passed on in a gamete, or the average effect of those that are deleterious may be larger than the average effect of those that are beneficial.

Selfish Genes and Germline-Limited DNA

Whatever the reasons for the evolution of a segregated germline, the fact that it has evolved is of considerable importance for genes that persist by drive and are harmful to the host organism. As we have seen, many B

Table 11.2 Occurrence of germline-limited DNA

Chromatin Loss	
Ciliophora (=Cilates)	Widespread but not universal; at least 1 species also shows loss of whole chromosomes
Nematoda	10 species in 2 orders in 1 subclass; not universal in those orders
Crustacea: Copepoda	6 species of *Cyclops* and 1 species in each of 3 related genera
Chromosome Loss	
Insects: Diptera	Widespread in 3 families: Cecidomyidae, Sciaridae, and Chironomidae: Orthocladiinae
Chordata: Myxinida	4 species of hagfish in 3 genera
Sex Chromosome Loss	
Insecta: Siphonaptera	*Ctenocephalides orientis*
Chordata: Mammalia	At least 10 species of marsupials

chromosomes accumulate by preferentially segregating away from somatic cells and toward germline cells (see Chap. 9). And many transposable elements are active only in the germline and are dormant in somatic cells (see Chap. 7). A more extreme manifestation of the same principle is for a selfish gene to excise itself and be lost from somatic tissues, only to persist in the germline. In fact, there are many examples of germline-limited DNA, some of which we have discussed in previous chapters (Table 11.2). In all cases our suspicion is that selfish genes must be involved, although in many cases the molecular details are poorly known and in no case is it clear why diminution has evolved. The evidence is most easily organized by taxonomic group.

Nematodes. Chromosome diminution was arguably the first selfish genetic phenomenon described, when Boveri observed in 1887 that chromosomes are sharply diminished in size in the somatic cells of *Parascaris* nematodes (for recent reviews, see Müller et al. 1996, Müller and Tobler 2000). Absent any other explanation, this suggests the differential accumulation of unnecessary DNA in the germline, followed by its excision from the soma. It is noteworthy that chromatin diminution is found in only a few species (7), all of which are parasitic (associated with very high egg number). A highly variable percentage of DNA is eliminated between closely related species, ranging from 0 to 95%, leaving a near-constant level of somatic

DNA. This contradicts any obvious germline function or function in the first few presomatic cell divisions. It also suggests that there is no net gain, that is, there appears to be no overall reduction in the (somatic) genome in species with genome diminution. Thus, germline limited DNA seems to invade and expand germline genome size without having any effect on amount of somatic DNA.

In *Parascaris univalens* the haploid genome consists of 1 compound chromosome with an internal euchromatic segment flanked by large, heterochromatic segments that are discarded from somatic cells. The single large euchromatic section fragments into as many as 50 separate chromosomes. The heterochromatic segments consist of blocks of two (5bp and 10bp) repeats and are highly variable in size. Why these short repeats should have become so common is unknown. It is also not clear why euchromatic fragmentation should be associated with heterochromatic segment loss.

It is also interesting that an rDNA gene seems to have invaded the germline-limited DNA in *Ascaris lumbricoides*. The variant ribosomal form predominates in oocytes but is eliminated from all presomatic cells and replaced by a somatic form (Etter et al. 1994).

Hagfish and copepods. In 8 species of hagfish chromosome elimination is known to occur during early embryogenesis in somatic cells (Kubota et al. 2001). As in nematodes, eliminated chromosomes are mostly heterochromatic and diminution eliminates most heterochromatic material. In *Eptatretus burgeri* 88% of the eliminated DNA consists of a 64bp repeat, tandemly and massively amplified.

Not much is known about chromosome diminution in copepods, except that it occurs in at least 9 species of *Cyclops* and its relatives, varies from 35% to more than 90% by amount, and occurs at about the time of germline-soma differentiation (Beerman 1977, Wyngaard and Gregory 2001). By one theory, the excess DNA plays a vital role in early development; but, as in nematodes, this hypothesis seems scarcely credible because the excess amount differs so sharply between closely related species. Somatic tissue in *Cyclops* is derived from only 4 or 5 segregational divisions. A similar observation has been made with *Parascaris*.

Ciliated protozoa. Because they are unicellular, the distinction between germ and soma in ciliated protozoa refers not to tissues or cells but to 2 different nuclei within a single cell, a small generative nucleus and a large somatic one derived from it. The micronucleus is small, diploid, divides by typical mitosis, and is mostly transcriptionally inactive. The macro is large, divides amitotically by budding, and is very active in transcription (reviewed in Prescott 1994, Coyne et al. 1996, Jahn and Klobutcher 2002). Only the micro is involved in the sexual cycle: after conjugation, the macro disappears and a new one is derived from the new micro. Thus, micros are germline and macros are somatic.

In all species studied, some micro DNA is eliminated from the somatic nucleus (10% to more than 95%, depending on the species). The remaining DNA in the macro is amplified many times. Micro chromosomes typically fragment into smaller macro chromosomes, and some DNA at the boundaries between fragments is usually lost. In addition, some sequences are spliced out of the micro DNA. These internal eliminated sequences (IESs) are divided into 2 classes: short (<100bp to several kb), noncoding sequences, which may be very numerous, and large (4–22kb) sequences with open reading frames and homology to known families of transposable elements (such as *Tc1/mariner*).

The details of these splicing reactions differ according to species. During the reorganization of a micro into a macro, the 5 pairs of *Tetrahymena thermophila* chromosomes are broken at some 200 positions to yield macro chromosomes with an average length of 600kb. During this process, 15% of the micro's DNA is eliminated from the macro and the remaining DNA is amplified about 45 times. About 6000 small sequences (600bp to a few kb), none of which reside within coding regions, are excised. Two transposable-like elements are also eliminated, one containing 15 open reading frames. Deletion boundaries for both groups are variable.

By contrast, in the spirotrichs, 200 or more chromosomes in the micro may fragment into 20,000 or more in the macro, almost all consisting of a single gene. Many of the small eliminated sequences are less than 100bp in length and they are very numerous (N~60,000). Their excision boundaries are precise, leaving only 1 copy of their short terminal repeat (2–6bp). They are often located within coding regions so their precise excision presumably permits the gene in which they were located to function properly. All the eliminated transposable element-like sequences are related to *Tc1/mariner* and as many as 5000 such sequences (some of which interrupt coding regions) are excised. The precise deletion of transposable elements from macro DNA is reminiscent of the silencing of transposable elements in the somatic cells of plants and animals, taken one step further (see Chap. 7). And the smaller, noncoding IESs may be analogous to the nonautonomous transposable elements that parasitize the functional elements. Indeed, some of the smaller IESs have similar inverted repeats at their ends as the transposable elements.

The ciliated protozoa diverged from other eukaryotes more than a billion years ago and the major subgroups diverged shortly thereafter (Wright and Lynn 1997). Only the most primitive group lacks a macronucleus (but has several small nuclei instead). Macros, along with chromosome diminution, may have evolved more than once within the ciliated protozoa, and it will be interesting to see the extent to which their extensive genome remodeling derives from the action and domestication of transposable element functions (Jahn and Klobutcher 2002).

Chimeras

We usually think of organisms as being made up of cells that are clonal descendants of a single progenitor cell, usually a zygote. As we have seen, cell lineage selection can work on genetic variants that arise by spontaneous mutation or mitotic recombination, but there is limited opportunity for such selection to produce complex adaptations. Mutation rates are too low, individual lifespans too short, and, most importantly, there is the bottleneck of passing through a single-cell stage every generation. But not all organisms are like this: in some taxa, individuals may be chimeric, their cells clonal descendents of 2 or more genetically distinct progenitors. If such chimerism is a regular occurrence, it greatly increases the opportunity for within-organism genetic conflicts. These will be of 2 types. First, cell lineages may compete for representation in the germline, to increase the chance of being transmitted to the next generation. Second, even if the chimerism is purely somatic, cells with different ancestries will have different coefficients of relatedness to family members, and the selective pressures they experience will be governed accordingly.

Taxonomic Survey of Chimerism

The incidence and extent of chimerism, and thus the opportunity for these conflicts, varies widely among taxa; probably in most species it never occurs. In our own species, chimeric individuals do occasionally arise *in utero*. The most common form of chimerism involves the blood: in some 8% of dizygotic twins, and 21% of triplets, sibling genotypes are detectable in blood samples (van Dijk et al. 1996). And fetal cells can persist in the mother and maternal cells in the offspring—sometimes for decades after pregnancy (Rinkevich 2001). This low-level "microchimerism" has yet to be associated with any biological effect, except sometimes with autoimmune disease; in principle, it could offer new routes to sibling rivalries and parent-offspring conflicts (Trivers 1974, 1985). Much more rarely, there can also be so-called whole-body chimeras—one such individual was discovered because she had a hazel and a brown eye. In another case, the apparent genetic profile of a woman did not match that of any of her 4 children, and the favored conclusion was that she was chimeric, with one cell population predominating in the soma and the other in the germline. In humans, whole-body chimerism appears to be so rare as to have a minimal evolutionary role.

Chimerism is more frequent and more extensive in marmosets and tamarins (callitrichid primates; reviewed in Haig 1999c). Females in these species usually give birth to 2 or more offspring per pregnancy. Twinning has secondarily reevolved from singleton births, and the fetuses develop with a connected circulatory system. Chimerism has been reported in blood, bone marrow, lymph nodes, and spleen, but it appears not to occur in lung or liver. Of particular interest is whether the germ cells are chimeric. This is known for opposite-sex twins but not for same-sex, in which it would be adaptive. If found in same-sex twins, there would be strong selection on genes for their ability to make germ cells that are good colonizers of their twin's gonads. It would also mean that a female from such twinships would naturally produce both offspring and nephews or nieces, so such an individual would be expected to be more likely to forego personal reproduction to help others than would an individual that produced only its own offspring.

Even if the germline is not chimeric, individuals are somatic chimeras of cells that are (ignoring mutation) genetically identical to those in its gonads and cells that are from the twin. If a chimeric marmoset reproduces, somatic cells that are genetically identical to the gonads will consider the offspring as related by one-half, whereas the other somatic cells will consider the offspring as related by only one-fourth to one-eighth (depending on whether the twins had the same father). If instead the marmoset helps its mother to reproduce, the coefficients of relatedness to the sibling will be one-half to one-fourth for both sets of cells (depending on whether the twins shared the same father with the new offspring). Thus, within every marmoset there are cells that, all else being equal, would prefer their mother to reproduce than the marmoset in which they reside. Haig (1999c) speculates that this may have been important in the social evolution of marmosets, which are unusual among primates in the degree to which reproduction within social groups is monopolized by a single female, with other group members (including progeny) helping to raise that female's offspring.

Competition among genetically distinct cell lineages for representation in the germline is apparently common in some colonial marine invertebrates—sponges, cnidarians, bryozoans, and ascidians—in which fusion of genetically distinct individuals is a normal part of the lifecycle. The best-studied species in this regard is the ascidian *Botryllus schlosseri* (e.g., Chadwick-Furman and Weissman 1995, Stoner and Weissman 1996, Stoner et al. 1999). Colonies that share at least 1 allele at a single multiallelic histocompatibility locus can fuse, but what happens next depends on the genotypes involved. At least in the laboratory, one colony usually seems to disappear over a period of about 2 months (Plate 10). Genetic analysis shows that sometimes the associated genotype has also disappeared; but in other pairings, the cells of the apparently disappeared colony actually replaced the cells of the colony that was thought to persist. Moreover, which genotype persists and which disappears can differ between somatic cells and germ cells, so the final colony can have somatic cells of one genotype and germline cells of the other. Thus the somatic cells of one genotype can be hijacked by the germline cells of another. Data from nature show 6–8% of individuals are chimeric (Karakashian and Milkman 1967, Ben-Shlomo et al. 2001). Much higher frequencies of chimerics, in some populations more than 60%, have been reported for the clonal ascidian *Diplosoma listerianum* (Sommerfeldt et al. 2003). In about 40% of these chimeras the 2 kinds of cells are thoroughly mixed and some chimerics contain as many as 6 different genotypes.

The fact that fusion in *B. schlosseri* is restricted to colonies sharing an allele at a histocompatibility locus (and that there can be more than 100 alleles segregating in a population; Rinkevich et al. 1995) indicates that fusion usually occurs with close kin. Presumably this requirement has evolved because fusion on average increases a colony's reproductive success but does not double it. (The only attempt to identify a benefit to chimerism was unsuccessful, though benefit has been noted in other colonial species; Rinkevich 1996, Sommerfeldt et al. 2003.) This limitation of fusion to close kin has the effect of attenuating, though not abolishing, the selection for germline colonization. It also provides the opportunity for genetic conflicts of interest over fusion, with loci tightly linked to the histocompatibility locus being selected to favor fusion more than those that are unlinked. Analogous conflicts of interest apply to loci linked and unlinked to self-incompatibility loci in plants and to the MHC locus in animals.

In mushrooms and their relatives, chimerism is an essential component of the sexual cycle. Briefly, haploid spores germinate and grow to make haploid "monokaryotic" hyphae. If 2 genetically distinct hyphae meet, they can fuse, and there is a reciprocal migration of nuclei from one hypha to the other. These nuclei replicate within the recipient hypha, which then comes to be populated by 2 genetically distinct nuclei. These "dikaryotic" hyphae continue to grow and are the main vegetative part of the lifecycle. Eventually a fruiting body is formed, within which 2 genetically distinct haploid nuclei fuse and then almost immediately go through meiosis to form haploid spores.

The dikaryon is therefore a chimeric individual, in which fusions between genetically identical hyphae are specifically prevented. This mating preference maximizes the opportunity for conflict and, as might be expected, close observations often indicate differential nuclear transmission—that is, drive (Ramsdale 1999). This can be manifest when the dikaryon makes asexual spores (conidia), if one nuclear genotype is transmitted more frequently than the other, or in matings between a dikaryon and a monokaryon, in which one nuclear genotype of the dikaryon is more likely to invade the monokaryon than the other. Occasionally it is even observed that both genotypes of the dikaryon invade and displace the resident monokaryon genotype.

Slime molds such as *Dictyostelium* provide another example of chimerism. In these species a multicellular slug-like pseudoplasmodium is formed by the aggregation of haploid amoebae. Amoebae of different genotypes can join together in the lab to form a chimeric slug, and the same probably also occurs at least sometimes in nature (Fortunato et al. 2003). Again, conflict is then expected over representation in the germline. Coalescence of genetically distinct spores, sporelings, and basal crusts has also been observed in red algae (Santelices et al. 2003).

Somatic Chimerism and Polar Bodies

A special kind of chimerism exists in taxa with polar bodies that persist and develop into specialized auxiliary tissues of the offspring. These are purely somatic tissues and do not form part of the germline, but they may nevertheless influence the organism's phenotype. For unknown reasons, polar body involvement in auxiliary tissues seems particularly likely to evolve at the interface between the organism proper and another organism. Conflicts between the auxiliary tissue and the organism proper can arise in interactions with kin, because the tissues will evolve according to different coefficients of relatedness. Conflicts between different loci within the auxiliary tissue are also possible, because relatedness to the rest of the organism can depend on the degree of linkage to the centromere. We briefly review the biology of these auxiliary tissues in 3 independent lineages.

Flowering plants and endosperm. In flowering plants a pollen grain produces 2 genetically identical haploid sperm, one that fertilizes the ovule to form the zygote and another that fertilizes a different maternally derived cell to form the endosperm (Maheshwari 1950, Johri 1963). The endosperm develops to varying degrees in different species and appears to act as a conduit or interface through which maternal nutrients are invested in the developing embryo and seed. In some species the endosperm includes the development of invasive haustoria (Fig. 11.3). In most plant species (~70%), the endosperm is triploid, containing 2 copies of the embryo's maternal genome and 1 copy of its paternal genome, but several other variants have evolved, apparently many times, usually involving the polar bodies (Table 11.3). As a consequence, there are maternally derived genes in the endosperm that are not found in the embryo proper, and the evolution of the endosperm is dictated by coefficients of relatedness different from those of the embryo. The greater the representation of the polar bodies, the less aggressive the endosperm is expected to be, both in extracting nutrients from the mother and in competing with neighboring embryos. There should therefore be a correlation across species between endosperm genetics and morphology. Note that if the endosperm becomes sufficiently passive in promoting its embryo, the latter may evolve to take matters into its own hands. In the Orchidaceae and Podostemaceae, for example, the endosperm is vestigial or absent, while the embryo itself makes extensive haustoria (Masand and Kapil 1966). We find it suggestive that, in both families, the endosperm (or female gametophyte) is often derived in part from the second polar body and so has lower relatedness to its own embryo (Table 11.3).

Also suggestive are reports from the mistletoe family (Loranthaceae). Female reproduction is unusual in this family in that several female gametophytes (from independent meioses) may develop—and get fertilized—within the same carpel. Only one of the resulting embryos will survive, but the endosperm is genetically chimeric, being derived from a fusion of the endosperms of the multiple embryos (Bhatnagar and Johri 1983). Again, this will lead to a comparatively low coefficient of relatedness with the embryo proper, which may help explain why the formation of endosperm haustoria is rare in mistletoes (unlike other families in the Santales). In one species *(Struthanthus vulgaris)* in which the endosperm does form a haustorium, this endosperm is not genetically chimeric but is derived from a single embryo sac (Venturelli 1981).

Parasitoid wasps and extraembryonic membranes. In many species of parasitoid wasp, specialized membranes form around the developing embryo that function to anchor the embryo to host tissue, protect it, translocate nutrients to it, and possibly also secrete chemicals into the host (Tremblay and Caltagirone 1973, Quicke 1997, Grbić et al. 1998). In addition, in some species cells or fragments from these membranes become separated from the embryo and float freely in the host hemolymph, swelling greatly as they absorb nutrients from the host, again possibly secreting chemicals, until finally they are consumed by the parasitoid larva.

The genetic constitution of these membranes is highly variable. In some species they are derived from cells of the embryo, and so are genetically identical to it; in other species they are derived from the polar bodies, and so are genetically distinct from the embryo. Among those that are derived from the polar bodies, the precise genetic constitution varies widely: in some species, all 3 polar bodies fuse together into one (triploid) nucleus, which then proliferates; in others the second polar body fuses with the innermost derivative of the first polar body to form a diploid nucleus, which then proliferates, while the outermost derivative degenerates; and in yet

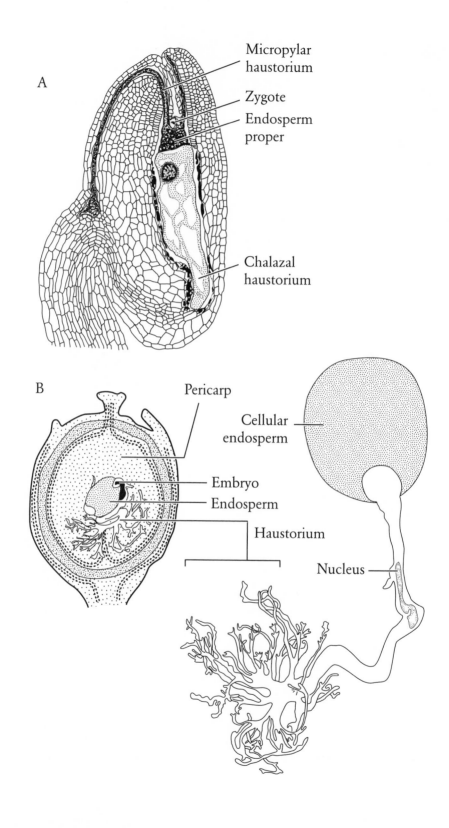

A

Micropylar haustorium

Zygote

Endosperm proper

Chalazal haustorium

B

Pericarp

Cellular endosperm

Embryo

Endosperm

Haustorium

Nucleus

Table 11.3 Comparative endosperm genetics

Type of Endosperm	Maternal Contribution*	Relative Relatedness to Own Embryo¶	Selected Families That Include Species with This Type of Endosperm
Monosporic			
Oenothera	a	4	Onagraceae (all?)
Polygonum	aa	3	~70% of flowering plants
Bisporic			
Allium, Endymion	ab	3−q	Balanophoraceae, Podostemaceae, Alismataceae, Butomaceae, Liliaceae, Orchidaceae
Tetrasporic			
Drusa	ab, ac, or ad†	2 ⅓	Compositae, Euphorbiaceae, Rubiaceae, Ulmaceae, Umbelliferae, Liliaceae
Adoxa	ac or ad†	2 + q/2	Adoxaceae, Caprifoliaceae, Ulmaceae, Liliaceae
Fritillaria, Penaea, Plumbago, Plumbagella	abcd	1 ½	Euphorbiaceae, Malpighiaceae, Penaeaceae, Plumbaginaceae, Liliaceae
Peperomia	8 cells, probably aabbccdd	1 ¼	Euphorbiaceae, Gunneraceae, Piperaceae

In each case the paternal contribution is a haploid genome identical to the paternal genome of the embryo.

* The 4 meiotic products and their descendants are labeled a to d, with a being the maternal contribution to the embryo. Note that a≡b and c≡d if segregation is at the first meiotic division and a≡c and b≡d if at the second division, where ≡ denotes identity by descent.

¶ The relatedness of the endosperm to its own embryo, divided by the relatedness to another embryo of the same mother. Calculations assume different fathers, and therefore are upper limits; to the extent that fathers are the same, the relative relatedness of endosperm to its own embryo will decrease. q is the frequency of second division segregation. q=0 if there is no recombination between the locus and the centromere; q=1 if there is always 1 crossover; q=0.5 if there are always 2; and q=0.667 if there are infinitely many (Fincham 1983). "Heterofertilization"–the derivation of the egg and endosperm from different pollen grains–will slightly lower the values; this is estimated to occur with a frequency of 1–2% in maize (Sarkar and Coe 1971).

†Probably with equal frequency.

From Maheshwari (1950), Johri (1963), and Bulmer (1986).

Figure 11.3 Examples of invasive haustoria formed by endosperm cells. A. *Melampyrum lineare* (Scrophulariaceae), showing the penetration of one branch of the micropylar haustorium as far as the funiculus; the endosperm proper; and the binucleate, 1-celled, chalazal haustorium. B. *Iodina rhombifolia* (Santalaceae), showing the cellular endosperm proper and the "aggressive" chalazal haustorium. Adapted from Vijayaraghavan and Prabhakar (1984).

other species, the first polar body does not divide, nor does it fuse with the second polar body, and instead the 2 nuclei proliferate separately in the same syncytium (Tremblay and Caltagirone 1973). In yet other species both polar bodies *and* embryonic nuclei contribute to the extraembryonic membranes: the polar bodies may fuse with an embryonic cell, and this fusion product then proliferates; or in one species the extraembryonic membrane consists of 8 large cells, each of which has 4 nuclei, 2 derived from a fusion of the 3 polar bodies and 2 derived from the embryo proper. After hatching, these cells transform into independent giant cells that seem to play an active role in the consumption of the host.

Individuals (embryo plus membranes) can thus be genetic chimeras. The key parameter determining the importance of this intra-individual genetic variance seems likely to be whether the mother typically lays more than 1 egg in the host. If she does, the evolution of the extraembryonic membranes is determined by different coefficients of relatedness than is that of the embryo itself—in particular, the extraembryonic membranes are expected to be more kindly and less aggressive toward maternal kin than the embryo proper. From this perspective, as well as protecting the embryo, the membranes may also be restraining it. The level of inbreeding is also important in determining just how different the embryo and polar bodies typically are.

It is also striking that, despite all the variability, in no case does only 1 polar body contribute to the extraembryonic membranes—there are always at least 2, and the second polar body is always involved. This guarantees that there will be no part of the genome in the membranes that is unrelated to the embryo proper. (For loci near the centromere, relatedness under outcrossing is $r = 0$ between the embryo and the first polar body.) Were such a genomic region to exist, it would be a prime location for a killer mutation that was active in extraembryonic membranes and attacked the embryo, assuming such attack benefited surviving siblings.

Scale insects and bacteriocytes. Polar bodies can also contribute to the specialized cells in scale insects that house endosymbiotic bacteria (Tremblay and Caltagirone 1973). These bacteria are vertically inherited through the egg from the mother. In mealy bugs (Pseudococcidae), these bacteria are contained in a large organ, the bacteriome (or mycetome), about one-third the length of the entire body. While in many species the cells of the bacteriome are derived from the embryo proper, in other species the first polar body does not divide, but instead fuses with the second polar body and then sometimes with 1 or 2 embryonic cells, to produce highly polyploid cells that engulf the bacteria and develop into the bacteriome. Something similar occurs in some armored scales (Diaspididae): the first polar body fuses with the second and with a cleavage nucleus to yield a pentaploid nucleus, which then proliferates. The cells are invaded by the endosymbiotic bacteria, giving rise to bacteriocytes (or mycetocytes) that do not conglomerate into a single organ but instead remain dispersed as single cells. In other species, 2 embryonic cells contribute to the fusion nucleus, which is therefore initially heptaploid. And, in yet other species, the bacteriocytes are derived purely from embryonic cells, with no polar bodies involved, though pentaploid cells without bacteria may still be present (Brown 1965).

One final diaspid phenomenon is worth mentioning. In *Gaga lutea* 3 of the 5 genomes of the bacteriome are heterochromatized. If, as Brown (1965) speculated, these are the polar body genomes, then individuality has been reestablished. In a parallel plant example, the 3 heterochromatized genomes of a pentaploid endosperm are known to be the polar bodies (Buzek et al. 1998). By contrast, in whiteflies (allied to scale insects), bacteriocytes invade ovarian cells and are transmitted thereby directly to the next generation, where they eventually break down and release their bacteria (Costa et al. 1996, Szklarzewicz and Moskal 2001). Their genetic constitution is unknown.

One notable feature of the scale insect bacteriomes and bacteriocytes is that they remain an important feature of the organism throughout its life, not just in early development, as in the endosperm and extraembryonic membranes. As before, conflict between the different tissues of a scale insect is expected to occur when there are interactions with kin, in which case these tissues evolve according to different coefficients of relatedness than the rest of the organism. Again, the bacteriocytes are more closely related to maternal kin than is the rest of the organism.

Bacteriocytes are also unique in that the paternal genome in males remains within the nucleus and is actively expressed—in other tissues, the paternal genome is silenced or ejected early in male development (see Chap. 10). Normark (2004) speculates that retention and activity of the paternal genome evolved to trick the bacteria into thinking they were inside females, and so prevent them from suicidally killing the male. Assuming the killing occurs early enough (e.g., while the scale insect is still under the brood chamber of his mother), bacteria may be selected to kill males (through which they will not be transmitted) in order to increase the space available for sisters containing clone-mates of the bacterial killer (see Box 5.3). The presence and activity of the male genome makes the bacteriocytes in males and females genetically equivalent. However, note that such activity is also a threat because the scale insect's paternal chromosomes are potential allies of the bacteria (they too are propagated only by females). Perhaps this is why they are outnumbered 4:1 by maternal haplotypes. Of course, how the system evolved and why there is so much variability remain unexplained.

TWELVE

Summary and Future Directions

IN THIS BOOK we have documented the diversity of selfish genetic elements in considerable detail, organizing the material by class of selfish genetic element. We now review the subject briefly by topic, to draw out some of the common themes that animate the subject and are likely to continue to do so in the future. For ease of reading, we have not repeated here the references for every fact and supposition cited, but we have made them available in the index, organized by topic.

Despite much recent progress, we are, in certain ways, just beginning to scratch the surface of the subject of selfish genetic elements. Our review reveals a whole series of general questions for which we typically have, at best, very partial answers. What is the prevalence of various kinds of selfish elements in different species and groups of species? Why are particular selfish elements found in some species instead of others? What is the evolutionary depth of particular selfish elements (e.g., B chromosomes)—shallow with repeated origins, or longer survival with fewer origins? What are the major factors that control the spread of selfish genetic elements and how do they act? What new forms of drive remain to be described? And what were the major selection pressures when the genome was first put together—a subject shrouded in mystery? Still, a number of conclusions are worth summarizing. We do so for selfish genetic elements generally, reviewing their:

- Logic and mode of action
- Molecular genetics
- Interaction with host breeding system
- Fate within species
- Movement between species
- Role in host evolution

Logic of Selfish Genetic Elements

Are selfish elements really selfish? This question has been answered definitively. Work on group after group has confirmed that selfish genetic elements really are (with some important exceptions) selfish. Despite many efforts to find benefits at the level of the individual—or, more commonly, at the level of the population or species—no general explanations have proved successful. For example, transposable elements have been described as an adaptive response to stress, a means of regulating gene expression, and a means of remodeling the genome; B chromosomes have been described as an organismic device to produce greater recombination in the As; and genomic imprinting has been described as a mechanism to protect against parthenogenesis or invasive trophoblast cancers or as a developmental rheostat (take your pick). Yet few of these explanations have survived even cursory inspection, and none a careful look at the evidence. Selfish genes appear to evolve because they benefit themselves directly, with all other, unlinked genes in the individual harmed or (at best) unaffected.

On the other hand, selfish elements have also sometimes "gone over to the other side," in other words, been co-opted to serve a useful function at the level of the individual, but the general situation conducive to such evolution has hardly been addressed.

Modes of selfish action. The vast majority of selfish genes spread by distorting their transmission from parent to offspring, contriving to be passed on to more than the Mendelian 50% of progeny (Table 12.1). This distortion can be a very powerful force when repeated generation after generation. Rates of increase vary in strength from almost 100% per generation (Bs in rye, spore killers in fungi) to about 50% per generation (nearly complete drive in one sex, as in *t*, *SD*, driving Xs and Ys) to only 1 in 1000 or 1 in 10,000 (common rates of transposition); but the latter are associated with colonizing new locations, so both copies can transpose and there is no obvious upper limit to copy number. As outlined in Chapter 1, these basic strategies have been identified for achieving drive: interfering with the competitor allele, replicating more than once per cell cycle, and preferential inclusion in the germline. Then there are the selfish genes that distort not their own transmission but how the host organism behaves toward relatives. The best examples are genes involved in genomic imprinting in mammals and (when interpreted this way) cytoplasmic male sterility (CMS) in plants.

Molecular Genetics

The full diversity of selfish genetic elements is only now emerging. For many selfish genes, we lack even rudimentary information about what they are and how they work. Nevertheless, some common features have become apparent.

Structure and size. Most selfish genetic elements encode 1 or more proteins. In some cases these proteins perform some novel function(s). HEGs and transposable elements encode proteins involved in DNA metabolism—nucleases, polymerases, ligases, and transposases. *Ab10* of maize somehow encodes the functions of neocentromeres. In other cases the selfish genetic element appears simply to interfere with normal host function: gametogenesis, sexual differentiation, fetal growth, and so on. The way they achieve this interference is sometimes reflected in the underlying molecular structure. *Sd* encodes a truncated host protein that is missing 40% of its length and consequently is mislocalized to the nucleus. There it causes drive of Rsp^i alleles that are themselves the absence of chromosomal repeats. At least 1 of the driving alleles in the *t* complex appears to be a deletion. Interestingly, the *t* haplotype *Responder* gene *(Tcr)* is a chimera of normal host genes, as are all CMS genes sequenced to date. Unfortunately we do not yet know how these proteins work.

Finally, some selected elements do not encode anything at all. Gs and Cs can spread because they are favored by the DNA mismatch repair system, and chromosomal rearrangements can spread because they reposition the centromere and so increase transmission at female meiosis. These noncoding elements obviously have more passive forms of drive than those that encode 1 or more proteins, without the same degree of adaptive complexity.

Table 12.1 The major classes of selfish genetic elements discovered thus far

Class (examples)	Logic / Mechanism	Distribution
Gamete killers (t haplotype in mice, SD in Drosophila, spore killers in fungi, SR in flies, D in mosquitoes)	Kills meiotic products (gametes, spores) to which it was not transmitted, thereby increasing the success of those to which it was. May be on autosomes or sex chromosomes; in latter case distorts sex ratio.	Animals, fungi, plants
Maternal effect killers (Medea in beetles, HSR and scat in mice)	Acts in mothers to kill offspring to which it was not transmitted, thereby increasing the success of those to which it was; can be rescued by paternal allele.	Animals; predicted in plants but not yet observed
Gametophyte factors (Ga1 in maize)	Acts in style to kill incoming pollen grains that do not carry a copy of the gene, thereby increasing the success of pollen grains that do.	Plants
Dominant feminizers (X* in lemmings)	Feminizes XY individuals and consequently is transmitted to 67% of the offspring assuming full compensation (because YY progeny die). In some species X* also shows germline nondisjunction and is transmitted to all progeny.	Rodents
Cytoplasmic male sterility (CMS) genes	Kills pollen (from which they are not transmitted) and thereby increases ovule success by avoiding inbreeding depression and/or reallocating resources.	Plants
PSR-like elements (PSR in Nasonia)	Excludes all other paternal chromosomes from the zygote, thereby converting a female into a male. Can only spread in female-biased populations of haplodiploid species.	Hymenoptera
Paternal genome loss (PGL)	Paternally derived chromosomes of males excluded from transmission, allowing maternally derived genes to drive. Now apparently under maternal nuclear control, but original genes of unknown location (nuclear or endosymbiotic) and unknown activity (maternal, paternal, or zygotic).	Insects, mites
Hybridogenetic hemiclones	In species hybrids, the haploid genome of one species excludes the haploid genome of the other from the gametes, thereby increasing its own transmission.	Animals

Element	Description	Distribution
Androgenetic elements	Genes that act in the zygote when paternally transmitted to exclude the maternal genome, thereby increasing their own transmission.	Animals, plants
Homing endonuclease genes (HEGs) (ω and VDE in yeast)	Cuts chromosomes not containing a copy of the gene and gets copied across as a by-product of the DNA repair system.	Fungi, plants, protists
Retrohoming group II introns	Makes a new copy of the gene via an RNA template.	Fungi, algae
Nuclear plasmids (2μm plasmid in yeast)	Circular plasmid that recombines while replicating, allowing many copies to be made despite replication originating only once.	Yeasts
DNA transposons (P elements in Drosophila, Ac elements in maize)	Makes a protein that cuts the gene out of the genome and inserts it elsewhere in such a way as to lead to an increase in copy number.	Animals, fungi, plants
LINE & SINE retrotransposons (L1 and Alu elements in humans)	Makes a new copy of the gene via an RNA template.	Animals, fungi, plants
LTR retrotransposons (Ty elements in yeast, copia elements in Drosophila)	Makes a new copy of the gene via an RNA template.	Animals, fungi, plants
Neocentromeres (Ab10 in maize)	Pulls the chromatid along the spindle at the 1st meiotic division, thereby increasing the probability of it being transmitted to the egg and avoiding the polar body.	Maize, others??
B chromosomes (2000 species)	Supernumerary chromosome that drives by gonotaxis and/or over-replication.	Animals, plants
Imprinted genes	Affects social behavior (especially toward mothers) in a way that is beneficial to themselves but harmful to most other genes in the same organism (unimprinted or oppositely imprinted). Conflicts arise both over gene expression and over the imposition or rejection of imprints.	Mammals, plants

In addition, there are sequences that are favored by biased gene conversion (e.g., Gs, Cs, minisatellite repeats), chromosomal rearrangements that have an increased chance of being transmitted to the egg at female meiosis, and mutations that increase proliferation of cell lineages or mitochondrial lineages at a cost to the organism (e.g., cancer).

Returning to the protein-coding selfish elements, some of these encode more than 1 protein, and most or all of them also have noncoding sequences that are essential for proper function. For example, DNA transposons consist, at a minimum, of a gene encoding a transposase and 2 ends that are recognized by the transposase. LTR retrotransposable elements encode up to 6 proteins and have *cis*-acting regions needed for packaging the RNA and other functions. *Ab10* of maize consists of both noncoding DNA repeats and protein-coding genes that make those repeats act as centromeres at meiosis, pulling them along the spindle apparatus. Autosomal killers like *t* and *SD* have multiple *trans*-acting driver elements that manipulate the transmission of a *cis*-acting responder allele, and sex chromosome drive also often involves multiple loci. In these latter examples, the different components are not located side by side on the host chromosome but instead are dispersed along it: distributed selfish genetic elements rather than unitary ones.

It is important for such distributed elements that the rate of recombination breaking up the various components not be too high: too much recombination and they cannot increase in frequency. Moreover, if there is any recombination, there will be selection to reduce it. The *t* complex spans a region that in wildtype chromosomes recombines 20% of the time. For *t* this rate is much reduced, to about 0.1%, by the evolution of 4 inversions. Inversions are also associated with many driving sex chromosomes, and some sort of block to recombination is found in the spore killer complexes of *Neurospora* fungi. The requirement for linkage between different components of a selfish element presumably explains why most autosomal killers are near centromeres, where recombination is often reduced. That is, centromeric regions are preadapted to evolve these complexes. Genes that are unlinked to the selfish element may be selected to impose recombination upon them, but this is a conflict that is often easily resolved in favor of the selfish elements by the simple evolution of inversions, making recombination costly for all parties.

Complex selfish genes with many components are expected to have evolved from simpler elements, and in some cases there has clearly been an evolutionary pathway of increasing size and complexity. The *t* haplotype has expanded by the sequential accretion of additional driving loci, tied together by inversions. *SD* too has acquired its enhancers and modifiers. LINEs with 2 genes are clearly derived from LINEs with only 1, and B chromosomes, we suggest, may often be expanded centromeres, with accretion of repetitive sequences. This complexity evolves because of selection for in-creased drive and decreased harm to the host organism.

The evolution of complexity also depends on the available mutations, which in turn should depend on the size of the selfish element and its link-age group (Fig. 12.1). The *t* haplotype, for example, currently encompasses some 30Mb and 300 genes. Mutations anywhere in this region that increase drive can be positively selected, as can those that accentuate the effects of male drive by giving males sex-antagonistic benefits in survival or reproduc-tion. The *t* can also increase in size over evolutionary time by acquiring ad-ditional inversions, with little immediate cost. Selfish sex chromosomes, particularly X chromosomes, also have a lot of genomic resources with which to work. And plant mitochondria, with up to 40 protein-coding genes, are apparently more likely to evolve CMS genes than animal mito-chondria, with their 12–13 genes, are to evolve the equivalent (CMS in her-maphrodites, male suicide in species with separate sexes). The upper limit to the size of a selfish "element" would appear to be one-half of the diploid ge-nome—the half that excludes the other half in systems such as paternal ge-nome loss and androgenesis.

At the other extreme are the HEGs, which have been under continuous selection for ever-smaller size to increase the probability of their getting copied successfully by the host DNA repair system. Transposable elements are also usually selected for small size, as judged by their generally compact organization. The considerable adaptive complexity of HEGs and trans-posable elements is instead due to the time available—perhaps a billion or more years, compared to 3my for the *t* haplotype and 60,000 years for a *Poeciliopsis monacha* hemiclone. In addition, because HEGs and transpos-able elements are small and effectively unlinked to the host genome, selec-tion can act more effectively on the mutations that do occur in them com-pared to those in the *t* haplotype, in which selection on the many host-benefiting genes contained in the region sometimes interferes with selection on drive. B chromosomes combine lack of linkage with weak constraints on size, and they might thereby have the greatest opportunity to increase in complexity; but it is unclear how old any B chromosome is and present evi-dence suggests that complexity largely depends on repeated, noncoding DNA.

Location. For some classes of selfish gene, their mode of action requires that they be at a certain location in the host genome. If we divide the ge-

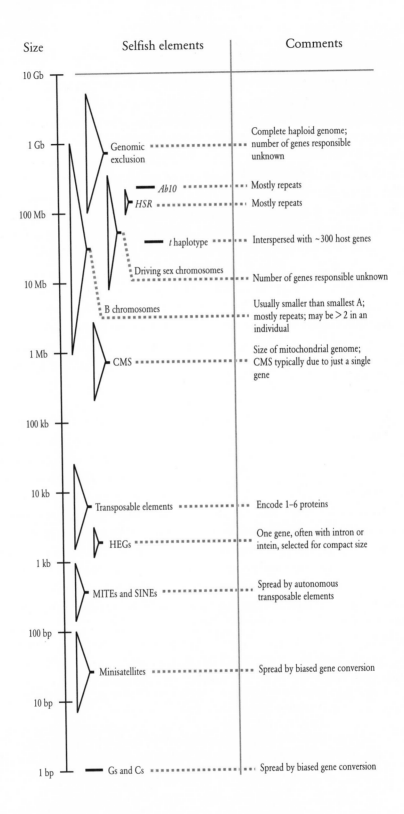

Size	Selfish elements	Comments

Size

10 Gb

1 Gb — Genomic exclusion — Complete haploid genome; number of genes responsible unknown

Ab10 — Mostly repeats

HSR — Mostly repeats

100 Mb

t haplotype — Interspersed with ~300 host genes

Driving sex chromosomes — Number of genes responsible unknown

10 Mb

B chromosomes — Usually smaller than smallest A; mostly repeats; may be > 2 in an individual

1 Mb — CMS — Size of mitochondrial genome; CMS typically due to just a single gene

100 kb

10 kb — Transposable elements — Encode 1–6 proteins

HEGs — One gene, often with intron or intein, selected for compact size

1 kb — MITEs and SINEs — Spread by autonomous transposable elements

100 bp

Minisatellites — Spread by biased gene conversion

10 bp

1 bp — Gs and Cs — Spread by biased gene conversion

nome into compartments (autosomes, X, Y, and cytoplasm), some genes spread only if they are in the correct compartment. In many plants a male sterility gene spreads if it occurs in the cytoplasmic compartment, but not if it occurs in the nucleus. A dominant feminizer like that found in lemmings spreads if it is on the X (or in the cytoplasm), but not if it is anywhere else. In other cases there is a bias, but it is not clear why. For example, why are HEGs and group II introns largely or completely restricted to organelles?

Even within a genomic compartment, where the selfish element is located can be important. HEGs must be in the middle of their own recognition sequence, or within a few base pairs of it. Move a HEG to another part of the genome without changing its recognition sequence, and it will not spread. *Ab10* and the other selfish knobs of maize have an increased chance of ending up in the egg only if they are not completely linked to the centromere. Indeed, they work best when there is exactly 1 crossover between themselves and the centromere, which gives them an optimal position on each chromosome. By contrast, autosomal killers typically form close to the centromere.

Mechanism. There has been tremendous progress over the past 20 years in our mechanistic understanding of how some selfish genes work. We now have atomic-level structural models of homing endonucleases bound to their recognition sites, and detailed molecular models of many steps in various transposition pathways. For imprinting, we now know that differential methylation of DNA plays a key role. But for many classes of selfish genes we have no idea at all how they work at the molecular level (e.g., *Medea* factors in beetles, gametophyte factors in plants, *SR*, X*, B chromosomes, and the various systems of genomic exclusion).

It is also true that the more we know about molecular mechanisms, the more we wish to know. Now that we know that genomic imprinting in-

---

Figure 12.1 The size of selfish genetic elements. The size (or range of sizes) shown indicate the linkage group that is transmitted selfishly as a unit, rather than the size of the causal genes (which is unknown for many cases). For example, in genomic exclusion the size of the haploid genome is given rather than the size of the genes within it that cause drive. Thus the size indicates the extent of DNA within which mutations increasing drive are positively selected.

volves the application, maintenance, and reading of epigenetic imprints, we want to know the interactions between the components of this system, and how they may conflict. Now that we know about the genes involved in *SD*, we want to know how they manage to affect only half the sperm, or for the genes involved in CMS, how they manage to affect only pollen production rather than any other aspect of development. Now that we know about HEGs binding to DNA, we want to know how to design novel HEGs, and perhaps use them for gene therapy or population genetic engineering. Understanding the mechanistic bases of selfish action is important both in its own right and for what it reveals about the evolutionary processes that created the element.

The last 20 years have also seen a great increase in the diversity of selfish genetic elements described. It is easy to predict that this trend too will continue. New subclasses of transposable elements continue to be defined, and determined searches will surely uncover new instances of autosomal killers, neocentromeres, genomic exclusion, and so on. Imprinting-based conflicts may be extended to new taxa–filamentous fungi, for example, or eusocial insects–and new phenotypes–dispersal or conscious mentation. And what about well-known phenomena with a selfish aspect that has not yet been appreciated? For example, we nominate rDNA as a prime candidate for functional DNA that may also have a selfish component. Multiple copies are spread around the genome and their number varies greatly between species (being positively associated with genome size). The effect on host function of a new mutation in one of those copies seems likely to be small, and biased gene conversion may therefore be particularly important in the fate of new mutations. And rDNA is also known to invade selfish elements such as B chromosomes and germline-limited DNA.

Are there any more completely new classes of selfish genes to be discovered? The answer to this question depends partly on what we mean by "completely new." The fact that some important new categories–such as maternal effect killers and androgenetic elements–were only discovered in the past 15 years suggests that many surprises await us in the future. We have suggested novel ways that sex chromosomes may drive by exploiting facultative sex ratio adjustments. To complement the gamete killers and maternal effect killers, perhaps someone will one day describe a sibling killer– a gene that kills diploid siblings that do not carry a copy of itself. And sex-linked maternal effect killers ought to be found in viviparous female-heterogametic species (e.g., snakes). It is also somewhat mysterious why no killer B chromosomes have yet been discovered with poison-antidote systems analogous to those found in bacterial plasmids, or why there are no transposable elements that self-splice at the RNA or protein levels. Perhaps one day these too will be described (or their absence explained). Likewise, we would be surprised if selfish X effects toward relatives were not discovered because the X chromosome is usually robust in size and gene content, unlike mtDNA and, often, Y chromosomes.

Selfish Genes and Sex

One feature that virtually all selfish genetic elements have in common is a reliance on the host being sexual—indeed, selfish genetic elements have been described as the original sexually transmitted diseases. The host mating system is also important, in particular whether inbreeding or outcrossing predominates, though the effect varies for different classes of selfish element: some types of selfish element spread faster and persist longer in outcrossed species, while others do better in partially inbred species. Selfish genes are also expected to affect the processes of sexual selection, though such effects have as yet been little studied.

Sex versus asex. For some types of selfish genetic element, their dependency on the host being sexual is obvious. One cannot exploit an asymmetry of meiosis if there is no meiosis. One cannot get ahead by killing gametes with the alternative allele if there are no gametes, and one cannot act on minority kinship relations if the genome is transmitted *en masse* and kinship relations are uniform across the genome.

But the importance of sex goes beyond these obvious effects. Even transposable elements and B chromosomes, which can accumulate in mitotic cells, will not be able to persist in a wholly asexual lineage if they are harmful to the host organism. In such a lineage the entire genome is transmitted intact from one generation to the next and acts as the unit of selection. If there are selfish genetic elements within a clonal species, different lineages will inevitably be differentially burdened and there will be selection among lineages for those that are less burdened. In lineages with particularly active elements, the frequency of the element may increase over generations, but those lineages will gradually disappear from the population, replaced by those with less active elements. It is possible for transposable elements to persist indefinitely in asexual species, but only if they are good for the lineage. Such elements would not attract the evolution of suppressors and it would be inappropriate to call them selfish. Sequences homologous to transposable elements have been found in the putatively ancient asexual bdelloid rotifers, but whether they are still mobile is not yet known.

The same logic also applies to the evolution of transposition mechanisms within sexual species. It is easy to imagine a gene encoding a protein that recognizes the gene and makes tandem copies of it, one beside the other. Such a gene might expand to tens, hundreds, or thousands of copies at a single site, but selection will drive the gene array extinct unless the copies are either useful to the organism or get separated from each other by meiotic recombination. Hence, the only transposable elements we see are those that make daughter copies that are dispersed away from the parent copy.

The inability of selfish elements to spread through an asexual population also probably explains why nuclear genes impose uniparental inheritance on their organelles, with one sex (usually the males) sometimes sabotaging their own mitochondria so they do not get transmitted to the offspring. Uniparental inheritance reduces the risk of transmitting a fast-replicating but otherwise defective organelle (though it generates conflicts over sex allocation as a side effect). A requirement for sex also explains why a selfish plasmid of slime mold mitochondria has evolved to reimpose biparental inheritance (Kawano et al. 1991).

Inbreeding versus outcrossing. In many respects complete selfing is like asexuality, and we expect it to be similarly inhospitable to selfish genetic elements. This is most easily seen for selfish elements like the *t* haplotype or HEGs that drive only in heterozygotes, and so anything that reduces the frequency of heterozygotes—like inbreeding—reduces the importance of drive. All else being equal, these types of selfish element will spread more rapidly through outcrossed populations than through inbred ones, and this effect has been experimentally demonstrated for a HEG and a plasmid in yeast. Reducing the risk of acquiring new selfish elements may even be an important reason for yeasts to prefer to inbreed. Ironically, the very efficient inbreeding system in yeast depends on a system of mating type switching that itself depends on a domesticated HEG. By reducing the importance of drive, inbreeding also selects for less active (i.e., more benign) elements, insofar as there is a trade-off between drive and the harm done to the host.

The distribution of selfish elements as a function of host breeding system has been best studied for B chromosomes in flowering plants. Outcrossed species are much more likely to carry a B than inbred species. Indeed, the theory and data are sufficiently compelling that any exception—any highly inbred species carrying a B—is a promising candidate for a beneficial B. More speculatively, we have suggested that DNA transposons and LINE elements are absent from baker's yeast because they are too harmful to persist in such an inbred species. The LTR retroelements that do exist in yeast are unusual in having complex adaptations to target safe havens in the genome where they are less likely to cause harm. For other classes of selfish genetic element, the association with breeding system is less clear, and there may even be suggestions of the opposite trend. *Podospora anserina* is a pseudo-homothallic fungus that readily selfs in the lab and presumably is highly inbred in nature, yet it has an abundance of HEGs and spore killers. Both these types of element usually have minimal effect on the host and so may be expected to go to fixation, even in highly (but not exclusively) inbred species. They may then degenerate for lack of anything to target. Perhaps by reducing the effectiveness of drive, inbreeding delays the degeneration of these selfish genes and increases their persistence times, making them more abundant.

The mating system also interacts in interesting ways with selfish elements that alter the sex ratio. Inbreeding typically selects for a female-biased sex ratio, and so in principle could select for autosomes that favor driving X chromosomes. The best-studied cases of driving sex chromosomes occur in outcrossed species (flies, mosquitoes, and lemmings), and so are counterselected by autosomes. But some inbred spiders have driving X chromosomes, though whether this drive is controlled by the X or the autosomes (or both) is unknown. Some taxa with paternal genome loss (PGL) are inbred, and the system may initially have evolved in male heterogametic ancestors precisely in order to produce more daughters. Indeed, it is theoretically possible in this case that the genes initially involved were on the paternal genome and excluded themselves—not so much selfish genes as self-sacrificial genes.

CMS genes are also expected to be more successful in (partially) inbred species, but not because they produce a female-biased sex ratio (though they do that as well). Rather, it is because they are selected to abort pollen production if there is any level of selfing and inbreeding depression, whereas nuclear genes only favor aborting pollen if the selfing rate and inbreeding depression are both substantial. Here we have a selfish gene manipulating the host breeding system, and this manipulation has knockon effects for the spread of virtually all other types of selfish genes. Consistent with expectations, CMS appears to be most common in partially inbred species and rare in self-incompatible species, though careful comparative work is needed to verify this conclusion. Gametophyte factors in plants also seem more likely to get established in partially inbred species than in obligately outcrossed ones, though there is no data yet to test this possibility.

Conflicts among the genes in an organism over inbreeding and outcrossing are expected to be common, though as yet we have no direct evidence. Thus, in animals, the costs and benefits of inbreeding are expected to differ between X, Y, and autosomes, and between maternally and paternally derived genes, with their relative proclivities reversing between males and females. Whether such conflicts have led in fact to interesting evolutionary dynamics is as yet unknown. The recent demonstration of imprinting being involved in MHC detection by women may give some clue about what is possible.

Prokaryotes versus eukaryotes. The importance of sex in the evolution of selfish genetic elements can also be seen by comparing prokaryotic and eukaryotic genomes. We have not reviewed in this book the bewildering array of mobile genetic elements found in bacteria, and evolutionary understanding of these elements lags behind that for eukaryotes. Nevertheless, one apparent difference is that the relatively clear distinction in eukaryotes between stable Mendelian genes that benefit the host and mobile driving genes that harm the host does not apply to prokaryotes. Bacterial plasmids not only carry genes needed for their own transmission but also contain genes that are clearly beneficial to the host organism. These tend not to be essential housekeeping genes, but instead those whose benefit is ecologically contingent. The same applies to bacterial transposable elements. It is possible that the relatively limited opportunities for these elements to transfer from one cell to another—the bacterial equivalent of drive—mean that purely selfish elements that never help the organism are often unable to spread and persist in populations. In eukaryotes, regular sex means that such elements are able to persist, and, we suggest, puts such a premium on drive that the increased bulk needed to encode a host-benefiting gene is usually selected against.

This distinction between prokaryotes and eukaryotes is not absolute. Some eukaryotic transposable elements may have evolved promoters that function to occasionally produce a beneficial mutation. Some pathogenic fungi have B chromosomes with host-specific virulence genes, and some yeasts have RNA replicons encoding toxins that kill other yeast cells. And bacteria have both retrohoming group II introns and HEGs, which may be truly selfish. Interestingly, the group II introns are typically associated with the more mobile fraction of the bacterial genome—the plasmids and transposons. And bacterial HEGs usually target stable housekeeping genes but are found in bacteria with unusual sexual systems—in archaea that have cell fusion, for example, and in mycobacteria with their unusual form of chromosomal conjugation. There is also some evidence that HEGs can be beneficial to bacterial cells in mixed cultures, perhaps by killing competing bacteria that do not contain the gene (rev. in Burt and Kaufopanou 2005).

The evolution of meiosis. Compared to the unidirectional modes of gene transfer typically found in bacteria—transformation, transduction, and conjugation—the preeminent feature of eukaryotic sex is the regimented symmetry and fairness (usually) of Mendelian segregation. It is easy to imagine that these features evolved to contain selfish genetic elements and promote genes that benefit the organism, but we know so little about the evolution of meiosis that it is impossible to be more definite until we learn more about the underlying molecular machinery. What we do know is that one fundamental component of meiosis, recombination, in fact produces its own weak form of drive: biased gene conversion. Sequences at which recombination is initiated are lost as a by-product of the process, giving recombinophobic sequences a transmission advantage. For this reason there can be conflicts between *cis*- and *trans*-acting components of the recombinational machinery, and recombination is likely to be something imposed on a DNA sequence rather than something it has evolved to attract. The resulting dynamics may explain the gain and loss of recombinational hotspots and coldspots over evolutionary time. Though the mechanism does lead to conflicts, one can imagine this may be a more stable situation and healthier for the host lineage than the opposite, in which recombinogenic sequences have a transmission advantage.

Sexual selection. With sex comes sexual selection, but its role in the evolution of selfish genetic elements remains mostly obscure. There are tantalizing hints of some biases in mate choice. Male and female *t* heterozygotes, who are most adversely affected by mating with another *t* mouse, are themselves most averse to doing so. In confined conditions of crowding and high male-male competition with no chance to emigrate, *t* males do very poorly. Female *D. pseudoobscura* do not discriminate in their first matings but in later ones appear to avoid *SR* males. In stalk-eyed flies, males with driving Xs do less well in competition for mates than do those without. On the other hand, it is surprising that we have no good examples of attractiveness or success in male-male competition being negatively biased by possession of B chromosomes. In plants, a major form of drive occurs at the first pollen tube mitosis in grasses: Bs avoid the nucleus that constructs the pollen tube and land instead in the generative nucleus, so the B contrives to spread during a time of sexual selection. Do individuals with relatively more transposable element insertions (especially males) do less well in competition for mates? No one knows.

Fate of a Selfish Gene within a Species

When a selfish genetic element first arises in a species—whether it originates *de novo* inside the species or is acquired from another one—it will increase in frequency if it drives more than it harms the host. Once established in the species, the selfish gene may continue to increase in frequency all the way to fixation, or it may stop at some intermediate equilibrium frequency. Either way, the long-term fate of most selfish genetic elements is to go extinct.

Fixation versus polymorphism. Elements go to fixation if, as they increase in frequency, the strength of drive continues to be greater than the harm they do to the host organism. Likely examples of such elements include fungal spore killers, maternal effect killers, gametophyte factors, and HEGs. Once an element is fixed in the population and there are no more susceptible alleles, the phenotype associated with drive (e.g., killing or cleavage) disappears. Only in crosses with populations or species in which it has not gone to fixation does drive reappear. Some gamete killers in plants are

Table 12.2   Sources of frequency-dependent selection that prevent selfish elements from going to fixation

Selfish Element	Source of Frequency-Dependent Selection
t haplotype of mice	Homozygous male sterile, and often homozygous lethal
X* of lemmings	Automatic negative frequency dependence because of sex ratio effect
SR of flies	The cost to the male of halving sperm number increases as males become less frequent; also negative homozygous fitness effects in females
CMS genes of plants	The cost to the plant of not being able to self-fertilize increases as pollen availability decreases
PSR in wasps	Automatic negative frequency dependence because of sex ratio effect
Hybridogenetic hemiclones of *Poeciliopsis*	Frequency dependent mate choice by the parasitized males

Note: The frequency-dependent mate choice found in *Poeciliopsis* may not be general—e.g., it is not found in hybridogenetic water frogs.

known only from interspecific hybridizations, and one possible explanation is that they have gone to fixation in one species but not the other.

By contrast, if the strength of drive decreases as an element increases in frequency or as the harm done to the organism increases, the element may not go to fixation and instead may stop at some intermediate frequency. The *t* haplotype, for example, is homozygous male sterile, and often homozygous lethal, and so cannot go to fixation. Most sex ratio distorters also fall into this category, though the precise details differ in each case (Table 12.2). Driving Y chromosomes in mosquitoes are an exception, for no such frequency-dependent effects have been found (in the absence of suppressors). And there is an ascertainment bias of unknown strength: were these elements to go to fixation, the population might go extinct, and so would not be available for study.

For B chromosomes and transposable elements, there is no such thing as fixation: in principle they could increase indefinitely in copy number, until eventually they impose such a burden on their hosts that they drive the species extinct. Whether this ever happens is unknown, but increased genome size in plants is associated with a high current risk of extinction. Certainly there are many instances in which Bs and transposable elements reach an intermediate abundance and do not increase in frequency any further. This failure to increase in numbers indefinitely must be due to the rate of drive decreasing with copy number, or the harm done to the host increasing. As it happens, there is evidence for both effects in both classes of element.

Selfish elements that do not go to fixation and instead come to an intermediate equilibrium often impose a significant genetic load on the population and attract the evolution of suppressors. Strikingly, the evolution of suppressors can, in some circumstances, lead to an *increase* in the frequency of the selfish element. In the case of CMS, the spread of a nuclear restorer gene increases pollen supplies, which could allow the CMS gene to increase further. Similarly, the spread of a suppressor of X chromosome drive would increase the number of males, which could allow the driving X to increase in frequency—even to fixation, with the suppressor then following behind. Alternatively, under certain conditions, both selfish genes and suppressors may remain polymorphic. For CMS, it is common to have several sterile cytotypes present in a single population, each with its different nuclear restorers, all maintained by complex frequency-dependent selection.

Extinction. Whether a selfish genetic element goes to fixation in a population or remains at some intermediate equilibrium frequency, over the long term it will almost certainly go extinct within that species. If such extinctions did not occur, all species would have many more selfish genetic elements than they do. The longest known cases of continuous activity within a host lineage are the *R1* and *R2* LINE elements of insect rDNA, which have been vertically transmitted with their hosts for hundreds of millions of years. But these are exceptional—most other selfish elements die out much more quickly.

The reasons these elements go extinct are varied and poorly understood, but may often be a by-product of how the element works. First, there is the problem of being too successful in the short term and going to fixation. An autosomal killer that goes to fixation has no susceptible alleles left to attack, and so there is no longer selection for killing. The killing function may then degenerate over evolutionary time, either because it imposes some cost on the organism or simply due to mutation pressure. Much the same considerations apply to HEGs, and defective elements are indeed commonly observed. In principle HEGs could persist by occasionally moving to a new location in the genome, but this would require a complex set of mutations and is apparently very rare. By contrast, the *t* haplotype cannot go to fixation, and so there are always susceptible alleles to attack, and the purifying selection maintaining the killing function continues. Had *t* been more "successful" in the short term and gone to fixation, it probably would not have lasted for 3my (and counting) as a functioning entity.

Another likely reason for selfish elements to go extinct is that they accumulate their own parasites–defective versions of the element–that swamp out the functional ones. This effect may be especially important for transposable elements. For example, in most populations of *D. melanogaster,* defective *P* elements outnumber intact ones. They may have a higher transposition rate than intact elements, and their accumulation raises the cost of the intact elements and could drive them extinct. LTR retrotransposable elements also spin off parasitic nonautonomous elements. LINE elements have partially solved the problem because the protein has evolved a tendency to copy the mRNA molecule from which it was synthesized, though this tendency is not perfect, and LINEs still have their parasitic SINEs. The accumulation of such parasites may explain why many genomes, including our own, contain fossil elements that once were active and no longer are. And the effect is not limited to transposable elements: *Ab10* of maize has presumably suffered from the proliferation of nonautonomous knobs throughout the genome.

Another possible cause of extinction is suggested by the *t* haplotype, which, as we have indicated, spans a region including some 300 host genes, bound together by 4 inversions. These inversions are beneficial in the short term because they tie together the various components of the element. In the longer term, because they restrict recombination with the wildtype chromosomes, selection on the 300 host genes is less effective, leading to the accumulation of deleterious mutations and the failure to incorporate new beneficial mutations. The sharp reduction of MHC variability on the *t* is one such possible cost. Over time, the reduced efficacy of selection could lead to the slow degeneration of the entire haplotype and its loss from the population.

Finally there are the host genes that are resistant to or suppress the activity of the selfish gene. As we have seen, in some circumstances the presence of one of these genes can lead to a short-term increase in the frequency of the selfish one. But if the suppressor completely abolishes drive and goes to fixation, regardless of the short-term effect, the long-term effect is for the selfish genetic element to disappear. Before then, there may be a phase in which it is only visible in hybridizations—as sometimes occurs with driving X chromosomes in flies and CMS genes in plants.

Movement between Species

Selfish genes can differ from their hosts not only in the direction of selection but also in their phylogenetic histories. This is because many selfish genetic elements are able to move between host species. They can do this by introgression from one species to another via species hybrids; in addition, some (small) elements are capable of much more distant horizontal transfers, though how they manage such a transfer is not usually known.

Hybridization. Because drive is such a powerful force, we imagine that a selfish gene that arises in one species will eventually colonize most or all other species that can hybridize with the original one. The *t* haplotype, for example, is found in 4 species of mice that diverged 0.5–1mya, but the *t*'s themselves diverged only about 0.1mya. Presumably a beneficial mutation occurred in the *t* of one species and then spread through the others; the mutation may have increased the rate of drive or it may have been in any of the 300 host-benefiting genes in the complex. In either case, movement would have occurred by hybridization between species, followed by backcrosses. For those selfish elements that attract the evolution of host suppressors, these suppressors will not initially be present in the recipient species but may be transferred in subsequent hybridization events.

For any gene, selfish or otherwise, the phylogenetic distance that it can move by hybridization is relatively restricted, as the hybrid must be viable and fertile in order for movement to be possible. One selfish element that is exceptional in this regard is PSR, the male-limited B chromosome of *Nasonia* and other haplodiploid wasps. This chromosome is active in the fertilized zygote and eliminates all paternally derived chromosomes except itself, thus causing the zygote to develop into a haploid PSR-bearing male. A PSR-bearing male that mates with a female of another species produces PSR-bearing males of the second species, not hybrids. The limits of its movement thus depend on the extent of (occasional) mating between species. One wonders whether PSR might even have evolved to make males less discriminating about the species affiliation of the females they approach.

In any case, tracing the evolutionary history of this chromosome (and others like it) should be very interesting.

Hybridization can be involved in the birth of a selfish element as well as its introduction into a new species. Some B chromosomes, for example, are thought to have originated from hybridization events. Hybridogenetic hemiclones currently rely on interspecific hybridization events to drive, but in this case it is unclear whether drive originated with the hybridization, or whether there was a previous phase of drive within one species that went to fixation.

Horizontal transfer. Transposable elements and HEGs are capable of moving between much more distantly related host taxa than would be possible by hybridization—between orders, phyla, even kingdoms. It is not known how they do this. In some cases, perhaps the element is occasionally released into the environment by a dead or dying cell and then taken up by another organism and incorporated into the genome; in others, some vector might be used such as a virus or a predatory mite. Whatever the mechanism, such transfers allow a selfish element to outlive the host species in which it first arose, and even to persist for hundreds of millions or billions of years. Within each host lineage there may be a perpetual evolutionary cycle of acquisition, fixation, degeneration, and loss, followed by reacquisition. As long as an element transfers to a new species at least once before degenerating in the old species, it will persist indefinitely.

The frequency of horizontal transmission varies widely among the different classes of element. DNA transposable elements might be expected to transfer readily because they have an extrachromosomal phase in the transposition reaction during which the element is associated with the transposase protein, at which time it can be inserted into any available chromosome. HEGs, by contrast, are never associated with their protein and are never separated from the host chromosome. Among retrotransposable elements, those with LTRs might transfer readily because they make a capsid (and some even an envelope), whereas LINE RNA is less well protected. Another important parameter is the specificity of requirements for host factors. *P* elements, for example, are active in only a restricted set of dipteran insects and so have more limited opportunities for horizontal transmission than *mariner* elements, which have been experimentally transferred across kingdoms. For HEGs, whether they work in a new species depends in part on whether they recognize the appropriate target sequence. This requirement explains why they preferentially target well-conserved parts of the genome and why some ignore synonymous sites in the target sequence.

One important factor affecting the likelihood of a potential host acquiring a selfish element by horizontal transmission is the accessibility of the germline. Organisms like yeast that take up DNA from the environment and incorporate it into their genome presumably have a high propensity for acquiring selfish genes. Species with distributed germlines (plants and some animals) are presumably more susceptible than those with segregated germlines. The combined difficulties of no extrachromosomal phase and segregated germlines may explain why HEGs have not been discovered in insects or vertebrates.

Wide horizontal transmission by some means other than hybridization may also occur in other selfish genetic elements, though there is as yet no direct evidence. The spore killers of ascomycete fungi are one plausible example, based on an easily accessible germline and an apparent need for horizontal transmission, given that they are expected to fix and then degenerate within species. CMS genes also might be expected to show horizontal transmission, given that normal host-benefiting mitochondrial genes do so. By contrast, it is difficult to see how a selfish element like the *t* haplotype, with its constituent genes scattered in among 300 host genes, could transfer to a fly, a fish, or even another *Mus* species if it was not able to hybridize. The same could be said for all but the smallest B chromosomes.

Besides allowing selfish elements to persist over long evolutionary timespans, horizontal transmission also allows them to diversify as they adapt to their separate gene pools. As a result, if these elements come back together again in the same species, they may coexist rather than exclude each other. As a consequence, there can be more than one active retrotransposable element, for example, in a single host species. Indeed, all selfish genes using reverse transcriptase are thought to be derived from a single element that diversified into the multiple types observed now (e.g., group II introns, LINEs, and LTR elements). Most DNA transposons fall into one of several ancient and widespread families, as do all HEGs.

Distribution among Species

Selfish elements are not uniformly distributed across species but typically occur in some groups rather than others. Only a small part of this variation is understood. As we noted earlier, obligately outcrossed plants are

468

more likely to suffer from B chromosomes but appear less likely to have CMS. Such breeding system effects can be expected for virtually all other types of selfish element. An easily accessible germline is probably a risk factor for selfish genes that move between species (HEGs and transposable elements). Somatic cells and tissues may afford some protection from these but allow gonotactic selfish genes (e.g., neocentromeres and B chromosomes) to spread. Yeasts are immune to such elements because they have no somatic cells to avoid—not even polar bodies at meiosis. By contrast, the first pollen grain mitosis, which produces a generative nucleus that is inactive during pollen tube competition and a somatic nucleus that is highly active, represents a perfect opportunity for B gonotaxis and it has been exploited numerous times, especially in the grasses. A deeper analysis of the opportunities for gonotaxis in different lineages would be most welcome.

For other classes of selfish gene, hosts with well-developed family lives are most susceptible. Maternal effect killers are expected whenever the death of a juvenile can help a parent or sibling. And if degrees of relatedness differ for the maternal and paternal genomes—as often they will—parent-specific gene expression and genomic imprinting is expected to evolve. Thus we expect to find imprinting in mammals and flowering plants (as we do) and also in ants, in species with diploid-on-haploid lifecycles, and in species with intimate male care of offspring (e.g., pipefishes and seahorses). For these latter groups we have no knowledge one way or the other. Imprinting can also evolve in other contexts, as in paternal genome loss in scale insects.

Much else remains unknown. For example, Bs are more common in plants with larger genomes and fewer chromosomes, and in mammals with a higher percentage of acrocentric chromosomes; only for the last do we have a theory (based on centromeric drive) that would predict this association.

Some types of selfish element are less widespread than might be expected. Why, for example, have nuclear plasmids only been reported in yeasts? And sex chromosome drive seems surprisingly uncommon, given the relative ease with which it is detected. Gamete killer sex chromosomes have thus far only been found in the Diptera. We have suggested that this may be because the sex chromosomes are often active in meiotic cells (unlike in mammals), but this is just a speculation. And we have no idea why some *Drosophila* species show X drive while others do not. The X* strategy in lemmings would seem to work in many other species with strong maternal compensation and it is perhaps surprising that we do not have more examples. On the other hand, the analogous Y* system in some other murids taxes our ability to explain its evolution anywhere.

For paternal genome loss in males, we have theories for how it may have been favored by inbreeding and local mate competition—or, conversely, outbreeding plus bacterial endosymbionts!—but evidence does not yet clearly favor either system of logic. For other systems of genomic exclusion, we have no clue as to what causes the known taxonomic distribution. Why do our examples of hybridogenesis come from particular groups of fish, frogs, and stick insects? Because they happened to be discovered first in these groups (but are, in fact, broadly distributed) or because these groups share something else in common?

Humans and mice make for an interesting comparison, as 2 well-studied mammals. For reasons that are not clear, mice have more selfish genetic elements than humans: 1 gamete killer (*t*) versus 0; 3 likely maternal effect killers (*HSR*, *scat*[+], and *Om*[DDK]) versus 0; at least 3 different families of LTR retrotransposons known to be active (*IAPs*, *ETns*, and *MaLRs*) versus 0; and about 30 times more *L1* LINE inserts that are active. One can only speculate at this point on an explanation for the difference—possibilities include the larger population size of mice; or the fact that they are a complex of 4 occasionally hybridizing sibling species compared to 1 relatively homogeneous species. Or, maybe selection at the host level is stronger against selfish genes in humans—perhaps because our complex nervous system is especially sensitive to the genetic defects, such as aneuploidy, that selfish elements may generate (Crow 1982), or perhaps because, being larger and longer-lived, we are more at risk from somatic mutations and cancer (Gifford and Tristem 2003, Leroi et al. 2003).

Some selfish genetic elements are relatively easy to detect, and therefore suitable for obtaining large datasets on their distribution. Traditionally, B chromosomes and CMS genes have been the easiest. More recently, PCR and complete genome sequencing have made transposable elements and HEGs relatively easy to find. Others are more difficult. It is not usually obvious whether half the sperm are defective in 5% of the males of a population, or whether there is a neocentromere active only at meiosis in 5% of the females. And we do not yet know what gamete killers, maternal effect killers, or gametophyte factors look like in a genome sequence—neither active ones nor fossil remnants.

Role in Host Evolution

The immediate, short-term effects of selfish genetic elements on their host organisms are negative, but the amount of harm varies widely. *P* elements of *D. melanogaster* are particularly active, and our best estimate is that, if all *P* elements were magically eliminated, fitness would increase by perhaps 10%. Removing all other transposable element families would raise fitness only by another 2%. A *t* haplotype at 5% frequency also reduces mean fitness by about 2% (assuming heterozygous fitness is reduced 20%). Because frequencies vary widely among populations (0–70%), the average cost also varies widely. Similar calculations can be done for other selfish elements, and loads of a few percent are not unusual. The *VDE* HEG of yeast, on the other hand, imposes no detectable cost at all (<1% in the laboratory).

These effects on mean fitness are not to be confused with effects on population productivity. For elements that produce female-biased sex ratios (driving X chromosomes, CMS, PGL, hybridogenetic hemiclones), productivity may even be enhanced. By contrast, selfish genes that target female function, either directly (Y drive) or indirectly (androgenesis), seem most likely to reduce population productivity and perhaps even to cause population extinction. The androgenetic *Cupressus dupreziana* is down to 231 highly homozygous trees, with each hermaphrodite achieving only about 10% seed set.

In addition to their immediate costs for the host organism, selfish genetic elements can also have longer-term consequences for host adaptation. Transposable element insertions that are slightly deleterious may nonetheless become fixed in the population by chance or hitchhiking; likewise, nucleotide changes that are slightly deleterious may become fixed by biased gene conversion. In principle, such effects could lead to a slow evolutionary degeneration. Indeed, it has been suggested that the accumulation of transposable element insertions over millions of years has led to a bloating of the genome that may have driven some salamanders stupid and put many plants on the Red List, as well as some mammals and reptiles. Further analysis of these suggested effects, both empirical and theoretical, would be most welcome.

The long-term consequences of selfish genetic elements can also be beneficial to host adaptation. Again, transposable elements are the most relevant class. In our own species, insertions appear occasionally to produce beneficial mutations, which then (with the help of conventional natural selection) can raise the fitness of the entire population. We cannot rule out the possibility that most of the 4 million insertions in our genome that have accumulated over the last 200 million years have gone to fixation because they were beneficial to the host. A much larger set of insertions had deleterious effects and were promptly eliminated.

Occasionally, further long-term benefits to the host lineage can come from co-opting or domesticating the biochemical capabilities of selfish genetic elements. The best-understood example is the mechanism for mating type switching in yeast, but other likely examples include the adaptive immune system of vertebrates and telomere synthesis in both *Drosophila* and eukaryotes more generally. On a smaller scale, a *P* element insertion at the tip of the X chromosome of *D. melanogaster* appears to have been recruited by the host because it interferes with normal *P*-element activity.

Selfish elements can also play an important role in host evolution insofar as the host evolves mechanisms to suppress the elements, and then those mechanisms are co-opted for other purposes. DNA methylation is one oft-cited example of a gene regulatory mechanism that may originally have evolved to suppress transposable elements and subsequently became more widely used to control host gene expression during development. If so, methylation has also been co-opted back by selfish genes—possibly by transposable elements to regulate their own activity, and certainly by genes involved in genomic imprinting to manipulate expression levels across the generations. Methylation also introduces a mutation bias from C to T, which may select for a compensating bias in DNA repair from A and T to G and C, with knockon consequences for biased gene conversion. The long-term effects of selfish genes on host adaptation, like their effects on population productivity, may be varied and may be either positive or negative.

Selfish genetic elements are also expected to play an important role in the evolution of host genetic and sexual systems, although the data are as yet more suggestive than compelling (Table 12.3). Mutation, recombination, ploidy, sexual systems, uniparental inheritance, and germline development are all features that selfish elements may, one way or another, have influenced. Testing these ideas is critical to a deeper understanding of the subject.

Finally, phenotypic effects aside, selfish genetic elements can have profound long-term effects on the host genome. About half of our genome

Table 12.3   Host features that may have evolved as a result of selfish genetic activity*

Feature	Hypothesis / Scenario
GENETIC AND SEXUAL SYSTEMS	
Meiosis	Early evolution may have been influenced by the need to reduce the transmission of selfish genetic elements.
Mutation	Diverse effects possible. a. Transposable elements can greatly increase the frequency of certain classes of mutation (especially duplications, deletions, and rearrangements). Most will be harmful, but some may be beneficial. b. Lower mutation rates (e.g., fidelity of DNA polymerases, repair systems) can be selected to reduce the likelihood of selfish cell lineages arising. c. Methylation increases the mutation rate (C→T).
Recombination	Diverse effects possible. a. Distribution of hot- and coldspots in the genome and the molecular machinery that initiates recombination may evolve under the influence of biased gene conversion. b. *t* haplotype reduces recombination by the evolution of inversions. c. Maize neocentromeres increase recombination to increase the chance of a single crossover between themselves and the centromere. d. The *ENS2* HEG of yeast increases rates of recombination in the mitochondrial genome as a by-product of its imprecise recognition capabilities. e. Transposable elements may select for mechanisms preventing ectopic recombination between homologous sequences at different sites in the genome.
Diploid-on-haploid lifecycles	Paternally imprinted genes may be selected to cause diploid tissue to proliferate at the expense of the haploid mother (e.g., ascomycetes, mosses, and red algae).
Male haploidy	May evolve by maternal genes excluding paternal chromosomes in males.
Gynodioecy	A direct result of CMS genes.
Dioecy	CMS genes may select for nuclear genes that masculinize the hermaphrodite; at the limit there are separate males and females.
Sex-determining systems	X* of lemmings acts by converting males to females. More generally, any driving sex chromosome that produces skewed sex ratios may select for compensatory mutations that convert what would normally be a female into a male, or vice versa.
Sex chromosome systems	Some sex chromosome systems may be more prone to selfish sex ratio distorters than others, and so are less likely to persist over evolutionary time (e.g., female heterogamety with viviparity, XO female heterogamety).
Mating type switching	Evolved by the domestication of a HEG in saccharomycete yeasts.

473

Table 12.3   (continued)

Feature	Hypothesis / Scenario
Uniparental transmission of mitochondria and chloroplasts	Organelles with selfish mutations (e.g., that cause them to replicate fast but otherwise render them defective) may select for nuclear genes that destroy the organellar DNA from one parent, thus imposing uniparental inheritance.
Paternal transmission of mitochondria	CMS genes may select for nuclear genes that destroy maternal mitochondria and instead transmit paternal mitochondria.
DEVELOPMENT	
Methylation and RNAi	May have evolved to suppress transposable elements; now also used to control development. Methylation increases the C→T mutation rate, and so selects for biased gene conversion (A,T→G,C).
Germline development	Mutations that increase cell proliferation rates but are harmful if transmitted to the offspring will select for segregated germlines in which germ cells migrate through the body and whose replication depends on both cell-autonomous and cell-nonautonomous factors.
Somatic development	Many details of somatic development may have evolved to prevent cancer.
Early switch from maternal to zygotic control of development	Selection to suppress maternal effect killers may have contributed to the evolution of early switch in mammals.
Placentation	Evolution of mother-fetal interactions in mammals likely to have been influenced by maternal effect killers, gestational drive elements, and imprinted genes.
Adaptive immune system	Evolved by the domestication of a transposable element.
THE GENOME	
Genome size	Transposable elements, biased gene conversion, and drive at female meiosis all likely to play a role.
Karyotype shape	Chromosomal rearrangements can arise because of recombination between transposable elements and may spread because of drive at female meiosis.
Telomeres	Evolved by the domestication of retrotransposable elements.
Centromeres	Rapid evolution may be driven by competition for inclusion in the egg at female meiosis.
Holocentric chromosomes	Meiosis-specific centromeres may select for mitosis-specific centromeres that can then expand along the whole chromosome.
GC content	Evolves under the influence of biased gene conversion.
Dispersed repeat structure	Consists primarily of transposable elements.
Minisatellite repeat structure	Evolves under the influence of biased gene conversion.
Nuclear introns and spliceosomes	Evolved from group II introns; now often used to control gene expression.

* Many of these features are speculative; consult the relevant chapter for more details.

consists of transposable element fossils, and of which half are new since we diverged from mice 75mya. Differences in genome size in closely related species are usually due to the number of transposable elements or their fossils. And many chromosomal rearrangements that differ between species may also be the result of transposable elements, due either to ectopic recombination between insertions at different sites or to mistakes by the various transposition mechanisms. Biased gene conversion can also have a pervasive effect on the genome, molding its size, GC content, and minisatellite repeat structure. And selection at female meiosis to avoid polar bodies can lead to the accumulation of neocentromeric sequences throughout the genome (in maize and presumably other species); to the loss of "junk" sequences (in *Drosophila*); and probably to the reshaping of the karyotype, from metacentric to acrocentric and back again (in mammals). Non-Mendelian dynamics seem certain to play a leading role in the developing science of genome evolution.

The Hidden World of Selfish Genetic Elements

Perhaps nothing has impressed us more than the steady discovery of a vast, hidden world of selfish genetic elements inhabiting every species studied. From an evolutionary perspective, these elements are to a degree their own independent life forms, sometimes even with their own distinct evolutionary histories, always with genetic interests that diverge from those of the rest of the organism. They are often the simplest of life forms—one or a handful of genes and perhaps some noncoding repeats (if even that)—simple enough that we may soon be able to create our own, simple yet with complex, ramifying effects on the hosts they inhabit, both immediate and long-term. In our own species these effects include fundamental aspects of the genetic system itself—its size, organization, and degree of recombination—intense internal conflict over early development, and later internal conflict over juvenile and adult behavior. The unity of the organism is an approximation, undermined by these continuously emerging selfish elements with their alternative, narrowly self-benefiting means for boosting transmission to the next generation. The result: a parallel universe of (often intense) sociogenetic interactions *within* the individual organism—a world that evolves according to its own rules, as modulated by the sexual and social lives of the hosts and the Mendelian system that acts in part to suppress them.

References

Abbott AG, Hess JE, and Gerbi SA. 1981. Spermatogenesis in *Sciara coprophila*. I. Chromosome orientation on the monopolar spindle of meiosis. *Chromosoma* 83: 1–18.

Abraham S, Ames IH, and Smith HH. 1968. Autoradiographic studies of DNA synthesis in the B-chromosomes of *Crepis capillaris*. *J. Hered.* 59: 297–299.

Abrahamson S, Meyer HU, Himoe E, and Daniel G. 1966. Further evidence demonstrating germinal selection in early premeiotic germ cells of *Drosophila* males. *Genetics* 54: 687–696.

Acland GM, Halloran-Blanton S, Boughman JA, and Aguirre GD. 1990. Segregation distortion in inheritance of progressive rod cone degeneration (*prcd*) in miniature poodle dogs. *Am. J. Med. Genet.* 35: 354–359.

Adams KL, Qiu Y-L, Stoutemyer M, and Palmer JD. 2002. Punctuated evolution of mitochondrial gene content: high and variable rates of mitochondrial gene loss and transfer to the nucleus during angiosperm evolution. *Proc. Natl. Acad. Sci. USA* 99: 9905–9912.

Adams KL, Song K, Roessler PG, Nugent JM, Doyle JL, Doyle JJ, and Palmer JD. 1999. Intracellular gene transfer in action: dual transcription and multiple silencings of nuclear and mitochondrial *cox2* genes in legumes. *Proc. Natl. Acad. Sci. USA* 96: 13863–13868.

Adams S, Vinkenoog R, Spielman M, Dickinson HG, and Scott RJ. 2000. Parent-of-origin effects on seed development in *Arabidopsis thaliana* require DNA methylation. *Development* 127: 2493–2502.

Agrawal A, Eastman QM, and Schatz DG. 1998. Transposition mediated by RAG1 and RAG2 and its implications for the evolution of the immune system. *Nature* 394: 744–751.

Ågren G, Zhou Q, and Zhong W. 1989. Ecology and behavior of Mongolian gerbils, *Meriones unguiculatus*, at Xilinhot, Inner Mongolia, China. *Anim. Behav.* 37: 11–27.

Agulnik S, Adolph S, Winking H, and Traut W. 1993. Zoogeography of the chromosome 1 HSR in natural populations of the house mouse (*Mus musculus*). *Hereditas* 119: 39–46.

Ajioka JW, and Eanes WF. 1989. The accumulation of P-elements on the tip of the *X* chromosome in populations of *Drosophila melanogaster*. *Genet. Res.* 53: 1–6.

Akhverdyan M, and Fredga K. 2001. EM studies of female meiosis in wood lemmings with different sex chromosome constitutions. *J. Exp. Zool.* 290: 504–516.

Alberts B, Johnson A, Lewis J, Raff M, Roberts K, and Walter P. 2002. *Molecular Biology of the Cell*, 4th ed. New York: Garland Science.

Alberts SC. 1999. Paternal kin discrimination in wild baboons. *Proc. R. Soc. Lond. B* 266: 1501–1506.

Alexander MP. 1973. Investigation of the mode of action of the *X* locus in tomato. *Euphytica* 22: 344–350.

Alexander RD, and Borgia G. 1978. Group selection, altruism, and the levels of organization of life. *Annu. Rev. Ecol. Syst.* 9: 449–474.

Alfenito MR, and Birchler JA. 1990. Studies on B chromosome stability during development. *Maydica* 35: 359–366.

Alfenito MR, and Birchler JA. 1993. Molecular characterization of a maize B chromosome centric sequence. *Genetics* 135: 589–597.

Allard RW. 1963. An additional gametophyte factor in the lima bean. *Der Züchter* 33: 212–216.

Allen ND, Logan K, Lally G, Drage DJ, Norris ML, and Keverne EB. 1995. Distribution of parthenogenetic cells in the mouse brain and their influence on brain development and behavior. *Proc. Natl. Acad. Sci. USA* 92: 10782–10786.

Allen ND, and Mooslehner KA. 1992. Imprinting, transgene methylation and genotype-specific modification. *Semin. Dev. Biol.* 3: 87–98.

Allers T, and Lichten M. 2001. Differential timing and control of noncrossover and crossover recombination during meiosis. *Cell* 106: 47–57.

Alves MJ, Coelho MM, and Collares-Pereira MJ. 1998. Diversity of reproductive modes of females of the *Rutilus alburnoides* complex (Teleostei, Cyprinidae): a way to avoid the genetic constraints of uniparentalism. *Mol. Biol. Evol.* 15: 1233–1242.

Ames BN, Gold LS, and Willett WC. 1995. The causes and prevention of cancer. *Proc. Natl. Acad. Sci. USA* 92: 5258–5265.

Amikura R, Kashikawa M, Nakamura A, and Kobayashi S. 2001. Presence of mitochondria-type ribosomes outside mitochondria in germ plasm of *Drosophila* embryos. *Proc. Natl. Acad. Sci. USA* 98: 9133–9138.

Amor DJ, Bentley K, Ryan J, Perry J, Wong L, Slater H, and Choo KHA. 2004. Human centromere repositioning "in progress." *Proc. Natl. Acad. Sci. USA* 101: 6542–6547.

Amor DJ, and Choo KHA. 2002. Neocentromeres: role in human disease, evolution, and centromere study. *Am. J. Hum. Genet.* 71: 695–714.

Anagnou NP, O'Brien SJ, Shimada T, Nash WG, Chen MJ, and Nienhuis AW. 1984. Chromosomal organization of the human dihydrofolate-reductase genes— dispersion, selective amplification, and a novel form of polymorphism. *Proc. Natl. Acad. Sci. USA* 81: 5170–5174.

Ananiev EV, Phillips RL, and Rines HW. 1998. A knob-associated tandem repeat in maize capable of forming fold-back DNA segments: are chromosome knobs megatransposons? *Proc. Natl. Acad. Sci. USA* 95: 10785–10790.

Anderson WA. 1968. Structure and fate of the paternal mitochondrion during early embryogenesis of *Paracentrotus lividus. J. Ultrastruct. Res.* 24: 311–321.

Andervort HB, and Dunn TB. 1962. Occurrence of tumors in wild house mice. *J. Natl. Cancer Inst.* 28: 1153–1163.

Andrews JD, and Gloor GB. 1995. A role for the *KP* leucine zipper in regulating *P* element transposition in *Drosophila melanogaster. Genetics* 141: 587–594.

Angus RA. 1980. Geographical dispersal and clonal diversity in unisexual fish populations. *Am. Nat.* 115: 531–550.

Ankel-Simons F, and Cummins JM. 1996. Misconceptions about mitochondria and mammalian fertilization: implications for theories on human evolution. *Proc. Natl. Acad. Sci. USA* 93: 13859–13863.

Anxolabéhère D, Kidwell MG, and Periquet G. 1988. Molecular characteristics of diverse populations are consistent with the hypothesis of a recent invasion of *Drosophila melanogaster* by mobile *P* elements. *Mol. Biol. Evol.* 5: 252–269.

Araujo SMSR, Pompolo SG, Perfectti F, and Camacho JPM. 2001. Integration of a B chromosome into the A genome of a wasp. *Proc. Biol. Sci.* 268: 1127–1131.

Araujo SMSR, Pompolo SG, Perfectti F, and Camacho JPM. 2002. Integration of a B chromosome into the A genome of a wasp revisited. *Proc. Biol. Sci.* 269: 1475–1478.

Arcà B, Zabalou S, Loukeris TG, and Savakis C. 1997. Mobilization of a *Minos* transposon in *Drosophila melanogaster* chromosomes and chromatid repair by heteroduplex formation. *Genetics* 145: 267–279.

Ardlie KG. 1995. The frequency, distribution, and maintenance of *t* haplotypes in natural populations of mice. PhD thesis, Princeton University.

Ardlie KG. 1998. Putting the brake on drive: meiotic drive of *t* haplotypes in natural populations of mice. *Trends Genet.* 14: 189–193.

Ardlie KG, and Silver LM. 1996. Low frequency of mouse *t* haplotypes in wild populations is not explained by modifiers of meiotic drive. *Genetics* 144: 1787–1797.

Ardlie KG, and Silver LM. 1998. Low frequency of *t* haplotypes in natural populations of house mice (*Mus musculus domesticus*). *Evolution* 52: 1185–1196.

Arkhipova I, and Meselson M. 2000. Transposable elements in sexual and ancient asexual taxa. *Proc. Natl. Acad. Sci. USA* 97: 14473–14477.

Arkhipova IR, Pyatkov KI, Meselson M, and Evgen'ev MB. 2003. Retroelements containing introns in diverse invertebrate taxa. *Nat. Genet.* 33: 123–124.

Arnaud P, Monk D, Hitchens M, Gordon E, Dean W, Beechey CV, Peters J, Craigen W, Preece M, Stanier P, Moore GE, and Kelsey G. 2003. Conserved methylation imprints in the human and mouse GRB10 genes with divergent allelic expression suggests differential reading of the same mark. *Hum. Mol. Genet.* 12: 1005–1019.

Arney KL. 2003. H19 and Igf2—enhancing the confusion? *Trends Genet.* 19: 17–23.

Ashburner M. 1989. *Drosophila: A Laboratory Handbook.* Cold Spring Harbor: Cold Spring Harbor Laboratory Press.

Ashburner M, Misra S, Roote J, Lewis SE, Blazej R, Davis T, Doyle C, Galle R, George R, Harris N, Hartzell G, Harvey D, Hong L, Houston K, Hoskins R, Johnson G, Martin C, Moshrefi A, Palazzolo M, Reese MG, Spradling A, Tsang G, Wan K, Whitelaw K, Kimmel B, Celniker S, and Rubin GM. 1999. An exploration of the sequence of a 2.9-Mb region of the genome of *Drosophila melanogaster:* the *Adh* region. *Genetics* 153: 179–219.

Ashman T-L. 1999. Determinants of sex allocation in a gynodioecious wild strawberry: implications for the evolution of dioecy and sexual dimorphism. *J. Evol. Biol.* 12: 648–661.

Atlan A, Capillon C, Derome N, Couvet D, and Montchamp-Moreau C. 2003. The evolution of autosomal suppressors of *sex-ratio* drive in *Drosophila simulans.* *Genetica* 117: 47–58.

Atlan A, Joly D, Capillon C, and Montchamp-Moreau C. 2004. *Sex-ratio* distorter of *Drosophila simulans* reduces male productivity and sperm competition. *J. Evol. Biol.* 17: 744–751.

Atlan A, Merçot H, Landré C, and Montchamp-Moreau C. 1997. The *sex-ratio* trait in *Drosophila simulans:* geographical distribution of distortion and resistance. *Evolution* 51: 1886–1895.

Aulard S, Lemeunier F, Hoogland C, Chaminade N, Brookfield JF, and Biémont C. 1995. Chromosomal distribution and population dynamics of the 412 retrotransposon in a natural population of *Drosophila melanogaster*. *Chromosoma* 103: 693–699.

Avdulov N, and Titova N. 1933. Additional chromosomes in *Paspalum stoloniferum* Bosco. *Bull. Appl. Bot., Genet. Pl. Br.* 2: 165–172.

Avilés L, and Maddison W. 1991. When is the sex ratio biased in social spiders?

Chromosome studies of embryos and male meiosis in *Anelosimus* species (Araneae, Theridiidae). *J. Arachnol.* 19: 126–135.

Avilés L, McCormack J, Cutter A, and Bukowski T. 2000. Precise, highly female-biased sex ratios in a social spider. *Proc. Biol. Sci.* 267: 1445–1449.

Axelrod R, and Hamilton WD. 1981. The evolution of cooperation. *Science* 211: 1390–1396.

Ayonoadu U, and Rees H. 1971. The effects of B chromosomes on the nuclear phenotype in root meristems of maize. *Heredity* 27: 365–383.

Bailey JA, Carrel L, Chakravarti A, and Eichler EE. 2000. Molecular evidence for a relationship between LINE-1 elements and X chromosome inactivation: the Lyon repeat hypothesis. *Proc. Natl. Acad. Sci. USA* 97: 6634–6639.

Bailey MF, Delph LF, and Lively CM. 2003. Modeling gynodioecy: novel scenarios for maintaining polymorphism. *Am. Nat.* 161: 762–776.

Balk J, and Leaver CJ. 2001. The PET1-CMS mitochondrial mutation in sunflower is associated with premature programmed cell death and cytochrome *c* release. *Plant Cell* 13: 1803–1818.

Bantock C. 1961. Chromosome elimination in the Cecidomyidae. *Nature* 190: 466–467.

Barker JF. 1966. Climatological distribution of a grasshopper supernumerary chromosome. *Evolution* 20: 665–666.

Barlow DP, Stöger R, Herrmann BG, Saito K, and Schweifer N. 1991. The mouse insulin-like growth factor type-2 receptor is imprinted and closely linked to the *Tme* locus. *Nature* 349: 84–87.

Baroux C, Spillane C, and Grossniklaus U. 2002. Genomic imprinting during seed development. *Adv. Genet.* 46: 165–214.

Barr CM. 2004. Hybridization and regional sex ratios in *Nemophila menziesii*. *J. Evol. Biol.* 17: 786–794.

Barraclough TG, Birky CW, and Burt A. 2003. Diversification in sexual and asexual organisms. *Evolution* 57: 2166–2172.

Barrett SCH. 2002. The evolution of plant sexual diversity. *Nat. Rev. Genet.* 3: 274–284.

Bartolomé C, and Maside X. 2004. The lack of recombination drives the fixation of transposable elements on the fourth chromosome of *Drosophila melanogaster*. *Genet. Res.* 83: 91–100.

Bassett MJ, Lin-Bao X, and Hannah LC. 1990. Flower colors in common bean produced by interactions of the *Sal* and *V* loci and a gametophyte factor *Ga* linked to *Sal*. *J. Am. Soc. Hort.* 115: 1029–1033.

Battersby BJ, and Shoubridge EA. 2001. Selection of a mtDNA sequence variant in hepatocytes of heteroplasmic mice is not due to differences in respiratory chain function or efficiency of replication. *Hum. Mol. Genet.* 10: 2469–2479.

Beckenbach AT. 1978. The "sex-ratio" trait in *Drosophila pseudoobscura:* fertility relations of males and meiotic drive. *Am. Nat.* 112: 97–117.

Beckenbach AT. 1983. Fitness analysis of the "sex-ratio" polymorphism in experimental populations of *Drosophila pseudoobscura. Am. Nat.* 121: 630–648.

Beckenbach AT. 1996. Selection and the "sex-ratio" polymorphism in natural populations of *Drosophila pseudoobscura. Evolution* 50: 787–794.

Beckenbach AT, Curtsinger JW, and Policansky D. 1982. Fruitless experiments with fruit flies: the "sex ratio" chromosomes of *D. pseudoobscura. Dros. Inf. Serv.* 58: 22.

Beeman RW. 2003. Distribution of the *Medea* factor, *M4* in populations of *Tribolium castaneum* (Herbst) in the United States. *J. Stored Prod. Res.* 39: 45–51.

Beeman RW, and Friesen KS. 1999. Properties and natural occurrence of maternal-effect selfish genes (*"Medea"* factors) in the red flour beetle, *Tribolium castaneum. Heredity* 82: 529–534.

Beeman RW, Friesen KS, and Denell RE. 1992. Maternal-effect selfish genes in flour beetles. *Science* 256: 89–92.

Beerman S. 1977. The diminution of heterochromatic chromosomal segments in *Cyclops* (Crustacea; Copepoda). *Chromosoma* 60: 297–344.

Beladjal L, Vanderkerckhove TTM, Muyssen B, Heyrman J, de Caesemaeker J, and Mertens J. 2002. B-chromosomes and male-biased sex ratio with paternal inheritance in the fairy shrimp *Branchipus schaefferi* (Crustacea, Anostraca). *Heredity* 88: 356–360.

Belfort M, Derbyshire V, Parker MM, Cousineau B, and Lambowitz AM. 2002. Mobile introns: pathways and proteins. In: *Mobile DNA II*, eds. NL Craig, R Craigie, M Gellert, and AM Lambowitz. Washington, D.C.: ASM Press, pp. 761–783.

Bell G, and Burt A. 1990. *B*-chromosomes: germ-line parasites which induce changes in host recombination. *Parasitology* 100: S19-S26.

Belloni M, Tritto P, Bozzetti MP, Palumbo G, and Robbins LG. 2002. Does *Stellate* cause meiotic drive in *Drosophila melanogaster? Genetics* 161: 1551–1559.

Belshaw R, Katzourakis A, Paces J, Burt A, and Tristem M. 2005. High copy number in human endogenous retrovirus families is associated with copying mechanisms in addition to reinfection. *Mol. Biol. Evol.* 22: 814–817.

Belshaw R, Pereira V, Katzourakis A, Talbot G, Paces J, Burt A, and Tristem M. 2004. Long-term reinfection of the human genome by endogenous retroviruses. *Proc. Natl. Acad. Sci. USA* 101: 4894–4899.

Bemis WP. 1959. Selective fertilization in lima beans. *Genetics* 44: 555–562.

Ben-Aroya S, Mieczkowski PA, Petes TD, and Kupiec M. 2004. The compact chromatin structure of a Ty repeated sequence suppresses recombination hotspot activity in *Saccharomyces cerevisiae. Mol. Cell* 15: 221–231.

Bengtsson BO. 1977. Evolution of the sex ratio in the wood lemming, *Myopus schistocolor*. In: *Measuring Selection in Natural Populations. Lecture Notes in Biomathematics no. 19*, eds. FB Christiansen and TM Fenchel. Berlin: Springer-Verlag, pp. 333–343.

Bengtsson BO, and Uyenoyama MK. 1990. Evolution of the segregation ratio: modification of gene conversion and meiotic drive. *Theor. Popul. Biol.* 38: 192–218.

Bennett D. 1975. The T-locus of the mouse. *Cell* 6: 441–454.

Bennett D, Alton AK, and Artzt K. 1983. Genetic analysis of transmission ratio distortion by *t*-haplotypes in the mouse. *Genet. Res.* 41: 29–45.

Bennett ST, Wilson AJ, Esposito L, Bouzekri N, Undlien DE, Cucca F, Nistico L, Buzzetti R, IMDIAB Group, Bosi E, Pociot F, Nerup J, Cambon-Thomsen A, Pugliese A, Shield JPH, McKinney PA, Bain SC, Polychronakos C, and Todd JA. 1997. Insulin VNTR allele-specific effect in type 1 diabetes depends on identity of untransmitted paternal allele. *Nat. Genet.* 17: 350–352.

Ben-Shlomo R, Douek J, and Rinkevich B. 2001. Heterozygote deficiency and chimerism in remote populations of a colonial ascidian from New Zealand. *Mar. Ecol. Progr. Ser.* 209: 109–117.

Bentolila S, Alfonso AA, and Hanson MR. 2002. A pentatricopeptide repeat-containing gene restores fertility to cytoplasmic male-sterile plants. *Proc. Natl. Acad. Sci. USA* 99: 10887–10892.

Berend SA, Hale DW, Engstrom MD, and Greenbaum IF. 2001. Cytogenetics of collared lemmings (*Dicrostonyx groenlandicus*). II. Meiotic behavior of B chromosomes suggests a Y-chromosome origin of supernumerary chromosomes. *Cytogenet. Cell Genet.* 95: 85–91.

Bereterbide A, Hernould M, Farbos I, Glimelius K, and Mouras A. 2002. Restoration of stamen development and production of functional pollen in an alloplasmic CMS tobacco line by ectopic expression of the *Arabidopsis thaliana SUPERMAN* gene. *Plant J.* 29: 607–615.

Berger CA, and Witkus ER. 1954. The cytology of *Xanthisma texanum* D.C. I. Differences in the chromosome number of root and shoot. *Bull. Torrey Bot. Club* 82: 377–382.

Bergthorsson U, Adams KL, Thomason B, and Palmer JD. 2003. Widespread horizontal transfer of mitochondrial genes in flowering plants. *Nature* 424: 197–201.

Bestor TH. 2003. Cytosine methylation mediates sexual conflict. *Trends Genet.* 19: 185–190.

Beukeboom LW. 1994. Bewildering B's: an impression of the 1st B Chromosome Conference. *Heredity* 73: 328–336.

Beukeboom LW, Seif M, Plowman AB, de Ridder F, and Michiels NK. 1998. Phenotypic fitness effects of B chromosomes in the pseudogamous parthenogenetic planarian *Polycelis nigra. Heredity* 80: 594–603.

Beukeboom LW, and Werren JH. 1993a. Transmission and expression of the parasitic paternal sex ratio (PSR) chromosome. *Heredity* 70: 437–443.

Beukeboom LW, and Werren JH. 1993b. Population genetics of a parasitic chromosome: experimental analysis of PSR in subdivided populations. *Evolution* 46: 1257–1268.

Beukeboom LW, and Werren JH. 2000. The paternal-sex-ratio (PSR) chromosome in natural populations of *Nasonia* (Hymenoptera: Chalcidoidea). *J. Evol. Biol.* 13: 967–975.

Bhatnagar SP, and Johri BM. 1983. Embryology of Loranthaceae. In: *The Biology of Mistletoes*, eds. M Calder and P Bernhardt. Sydney: Academic Press, pp. 47–67.

Bianchi NO, Bianchi MS, Bailliet G, and de la Chapelle A. 1993. Characterization and sequencing of the sex determining region Y gene (*Sry*) in *Akodon* (Cricetidae) species with sex reversed females. *Chromosoma* 102: 389–395.

Bianchi NO, Reig OA, Molina OJ, and Dulout FN. 1971. Cytogenetics of the South American akodont rodents (Cricetidae). *Evolution* 25: 724–736.

Bibikova M, Carroll D, Segal DJ, Trautman JK, Smith J, Kim Y-G, and Chandrasegaran S. 2001. Stimulation of homologous recombination through targeted cleavage by chimeric nucleases. *Mol. Cell. Biol.* 21: 289–297.

Bidau CJ. 1986. Effects on cytokinesis and sperm formation of a B-isochromosome in *Metaleptea brevicornis adspersa. Caryologia* 39: 165–177.

Bidau CJ, and Marti DA. 2004. B chromosomes and Robertsonian fusions of *Dichroplus pratensis* (Acrididae): intraspecific support for the centromeric drive theory. *Cytogenet. Genome Res.* 106: 347–350.

Biémont C, Lemeunier F, Guerreiro MPG, Brookfield JF, Gautier C, Aulard S, and Pasyukova EG. 1994. Population dynamics of the *copia, mdg1, mdg3, gypsy*, and *P* transposable elements in natural populations of *Drosophila melanogaster. Genet. Res.* 63: 197–212.

Biémont C, Tsitrone A, Vieira C, and Hoogland C. 1997. Transposable element distribution in *Drosophila. Genetics* 147: 1997–1999.

Bigot Y, Hamelin M-H, and Periquet G. 1990. Heterochromatin condensation and evolution of unique satellite-DNA families in two parasitic wasp species: *Diadromus pulchellus* and *Eupelmus vuilleti* (Hymenoptera). *Mol. Biol. Evol.* 7: 351–364.

Bill CA, Taghian DG, Duran WA, and Nickoloff JA. 2001. Repair bias of large loop mismatches during recombination in mammalian cells depends on loop length and structure. *Mutat. Res. DNA Repair* 485: 255–265.

Birdsell JA. 2002. Integrating genomics, bioinformatics, and classical genetics to study the effects of recombination on genome evolution. *Mol. Biol. Evol.* 19: 1181–1197.

Birger Y, Shemer R, Perk J, and Razin A. 1999. The imprinting box of the mouse Igf2r gene. *Nature* 397: 84–88.

Birky Jr. CW. 1976. The inheritance of genes in mitochondria and chloroplasts. *BioScience* 26: 26–33.

Birky Jr. CW. 1995. Uniparental inheritance of mitochondrial and chloroplast genes: mechanisms and evolution. *Proc. Natl. Acad. Sci. USA* 92: 11331–11338.

Birky Jr. CW. 2001. The inheritance of genes in mitochondria and chloroplasts: laws, mechanisms, and models. *Annu. Rev. Genet.* 35: 125–148.

Bishop DVM, Canning E, Elgar K, Morris E, Jacobs PA, and Skuse DH. 2000. Distinctive patterns of memory function in subgroups of females with Turner syndrome: evidence for imprinted loci on the X-chromosome affecting neural development. *Neuropsychologia* 38: 712–721.

Black DM, Jackson MS, Kidwell MG, and Dover GA. 1987. KP elements repress P-induced hybrid dysgenesis in *Drosophila melanogaster. EMBO J.* 6: 4125–4135.

Blackman RL. 1995. Sex determination in insects. In: *Insect Reproduction,* eds. SL Leather and J Hardie. Boca Raton: CRC Press, pp. 57–94.

Blackstone NW, and Kirkwood TBL. 2003. Mitochondria and programmed cell death: "Slave revolt" or community homeostasis. In: *Genetic and Cultural Evolution of Cooperation,* ed. P Hammerstein. Cambridge, Mass.: MIT Press, pp. 309–325.

Boeke JD. 1997. LINEs and *Alu*s–the polyA connection. *Nat. Genet.* 16: 6–7.

Boeke JD, and Devine SE. 1998. Yeast retrotransposons: finding a nice quiet neighborhood. *Cell* 93: 1087–1089.

Boeke JD, and Pickeral OK. 1999. Retroshuffling the genomic deck. *Nature* 398: 108–111.

Boeke JD, and Stoye JP. 1997. Retrotransposons, endogenous retroviruses, and the evolution of retroelements. In: *Retroviruses,* eds. JM Coffin, SH Hughes, and HE Varmus. Cold Spring Harbor: Cold Spring Harbor Laboratory Press, pp. 343–435.

Bohn GW, and Tucker CM. 1940. Studies on *Fusarium* wilt of the tomato. I. Immunity in *Lycopersicon pimpinellifolium* Mill. and its inheritance in hybrids. *Missouri Agric. Exp. Stn. Res. Bull.* 311: 1–82.

Boissinot S, Chevret P, and Furano AV. 2000. L1 (LINE-1) retrotransposon evolution and amplification in recent human history. *Mol. Biol. Evol.* 17: 915–928.

Bolton EC, and Boeke JD. 2003. Transcriptional interactions between yeast tRNA genes, flanking genes and *Ty* elements: a genomic point of view. *Genome Res.* 13: 254–263.

Bonaccorsi S, Gatti M, Pisano C, and Lohe A. 1990. Transcription of a satellite DNA on two Y chromosome loops of *Drosophila melanogaster. Chromosoma* 99: 260–266.

Bondrup-Nielsen S, Ims RA, Fredriksson R, and Fredga K. 1993. Demography of the wood lemming (*Myopus schistocolor*). In: *The Biology of Lemmings,* eds. NC Stenseth and RA Ims. London: Academic Press, pp. 493–507.

Bongiorni S, Cintio O, and Prantera G. 1999. The relationship between DNA methylation and chromosome imprinting in the coccid *Planococcus citri. Genetics* 151: 1471–1478.

Bongiorni S, Fiorenzo P, Pippoletti D, and Prantera G. 2004. Inverted meiosis and meiotic drive in mealybugs. *Chromosoma* 112: 331–341.

Borie N, Maisonhaute C, Sarrazin S, Loevenbruck C, and Biémont C. 2002. Tissue-specificity of 412 retrotransposon expression in *Drosophila simulans* and *Drosophila melanogaster.* Heredity 89: 247–252.

Borsa P, and Kjellberg F. 1996. Experimental evidence for pseudoarrhenotoky in *Hypothenemus hampei* (Coleoptera: Scolytidae). *Heredity* 76: 130–135.

Bosemark NO. 1957. On accessory chromosomes in *Festuca pratensis.* V. Influence of accessory chromosomes on fertility and vegetative development. *Hereditas* 43: 211–235.

Bougourd SM, and Jones RN. 1997. B chromosomes: a physiological enigma. *New Phytol.* 137: 43–54.

Bougourd SM, and Parker JS. 1979a. The B-chromosome system of *Allium schoenoprasum.* III. An abrupt change in B-frequency. *Chromosoma* 75: 385–392.

Bougourd SM, and Parker JS. 1979b. The B-chromosome system of *Allium schoenoprasum.* II. Stability, inheritance and phenotypic effects. *Chromosoma* 75: 369–383.

Bougourd SM, and Plowman AB. 1996. The inheritance of B chromosomes in *Allium schoenoprasum* L. *Chromosome Res.* 4: 151–158.

Boulton A, Myers RS, and Redfield RJ. 1997. The hotspot conversion paradox and the evolution of meiotic recombination. *Proc. Natl. Acad. Sci. USA* 94: 8058–8063.

Boussy IA, and Daniels SB. 1991. *hobo* transposable elements in *Drosophila melanogaster* and *D. simulans. Genet. Res.* 58: 27–34.

Boveri T. 1887. Über Differenzierung der Zellkerne während der Furchung des Eies von *Ascaris megalocephala. Anat. Anz.* 2: 688–693.

Braden AWH. 1958. Influence of time of mating on the segregation ratio of alleles at the T locus in the house mouse. *Nature* 181: 786–787.

Brinkman JN, Sessions SK, Houben A, and Green DM. 2000. Structure and evolution of supernumerary chromosomes in the Pacific giant salamander, *Dicamptodon tenebrosus. Chromosome Res.* 8: 477–485.

Broach JR, and Volkert FC. 1991. Circular DNA plasmids of yeasts. In: *The Molecular and Cellular Biology of the Yeast Saccharomyces: Genome Dynamics, Protein Synthesis, and Energetics*, eds. JR Broach, EW Jones, and JR Pringle. Cold Spring Harbor: Cold Spring Harbor Laboratory Press, pp. 297–331.

Brookfield JFY. 1991. Models of repression of transposition in P-M hybrid dysgenesis by P cytotype and by zygotically encoded repressor proteins. *Genetics* 128: 471–486.

Brookfield JFY. 1995. Transposable elements as selfish DNA. In: *Mobile Genetic Elements*, ed. DJ Sherratt. Oxford: IRL Press, pp. 130–153.

Brookfield JFY. 1996. Models of the spread of non-autonomous selfish transposable elements when transposition and fitness are coupled. *Genet. Res.* 67: 199–209.

Brookman JJ, Toosy AT, Shashidhara LS, and White RAH. 1992. The 412 retrotransposon and the development of gonadal mesoderm in *Drosophila*. *Development* 116: 1185–1192.

Brosius J. 1999. Genomes were forged by massive bombardments with retroelements and retrosequences. *Genetica* 107: 209–238.

Brosseau GE. 1960. Genetic analysis of the male fertility factors on the Y chromosome of *Drosophila melanogaster*. *Genetics* 45: 257–274.

Brown J, Cebra-Thomas JA, Bleil JD, Wasserman PM, and Silver LM. 1989. A premature acrosome reaction is programmed by mouse *t* haplotypes during sperm differentiation and could play a role in transmission ratio distortion. *Development* 106: 769–773.

Brown LM, and Jones RN. 1976. B-chromosome effects at meiosis in *Crepis capillaris*. *Cytologia* 41: 493–506.

Brown SW. 1964. Automatic frequency response in the evolution of male haploidy and other coccid chromosome systems. *Genetics* 49: 797–817.

Brown SW. 1965. Chromosomal survey of the armored and palm scale insects (Coccoidea: Diaspididae and Phoenococcidae). *Hilgardia* 36: 189–294.

Brown SW, and Chandra HS. 1977. Chromosome imprinting and the differential regulation of homologous chromosomes. In: *Cell Biology: A Comprehensive Treatise*, eds. L Goldstein and DM Prescott. New York: Academic Press, pp. 109–189.

Brown SW, and Nelson-Rees WA. 1961. Radiation analysis of a lecanoid genetic system. *Genetics* 46: 983–1007.

Brown SW, and Nur U. 1964. Heterochromatic chromosomes in the coccids. *Science* 145: 130–136.

Brown TC, and Jiricny J. 1987. A specific mismatch repair event protects mammalian cells from loss of 5-methylcytosine. *Cell* 50: 945–950.

Brown TC, and Jiricny J. 1988. Different base/base mispairs are corrected with different efficiencies and specificities in monkey kidney cells. *Cell* 54: 705–711.

Bruck D. 1957. Male segregation advantage as a factor in maintaining lethal alleles in wild populations of house mice. *Proc. Natl. Acad. Sci. USA* 43: 152–158.

Brun LO, Borsa P, Gaudichon V, Stuart JJ, Aronstein K, Coustau C, and French-Constant RH. 1995a. Functional haplodiploidy. *Nature* 374: 506.

Brun LO, Stuart JJ, Gaudichon V, Aronstein K, and French-Constant RH. 1995b. Functional haplodiploidy: a mechanism for the spread of insecticide resistance in an important international insect pest. *Proc. Natl. Acad. Sci. USA* 92: 9861–9865.

Buard J, Shone AC, and Jeffreys AJ. 2000. Meiotic recombination and flanking marker exchange at the highly unstable human minisatellite CEB1 (*D2S90*). *Am. J. Hum. Genet.* 67: 333–344.

Bucheton A, Busseau I, and Teninges D. 2002. I elements in *Drosophila melanogaster*. In: *Mobile DNA II*, eds. NL Craig, R Craigie, M Gellert, and AM Lambowitz. Washington: ASM Press, pp. 796–812.

Buckler ES, Phelps-Durr TL, Keith Buckler CS, Dawe RK, Doebley JF, and Holtsford TP. 1999. Meiotic drive of chromosomal knobs reshaped the maize genome. *Genetics* 153: 415–426.

Budar F, and Pelletier G. 2001. Male sterility in plants: occurrence, determinism, significance and use. *C. R. Acad. Sci. Paris, Life Sci.* 324: 543–550.

Budar F, Touzet P, and De Paepe R. 2003. The nucleo-mitochondrial conflict in cytoplasmic male sterilities revisited. *Genetica* 117: 3–16.

Bull JJ. 1979. An advantage for the evolution of male haploidy and systems with similar genetic transmission. *Heredity* 43: 361–381.

Bull JJ. 1983. *Evolution of Sex Determining Mechanisms*. Menlo Park: Benjamin Cummings.

Bull JJ, and Bulmer MG. 1981. The evolution of *XY* females in mammals. *Heredity* 47: 347–365.

Bull JJ, and Charnov EL. 1977. Changes in the heterogametic mechanism of sex determination. *Heredity* 39: 1–14.

Bullejos M, Sanchez A, Burgos M, Jimenez R, and Diaz de la Guardia RD. 1999. Multiple mono- and polymorphic Y-linked copies of the SRY HMG-box in Microtidae. *Cytogenet. Cell Genet.* 86: 46–50.

Bulmer M. 1988. Sex ratio evolution in lemmings. *Heredity* 61: 231–233.

Bulmer MG. 1986. Genetic models of endosperm evolution in higher plants. In: *Evolutionary Processes and Theory*, eds. S Karlin and E Nevo. Orlando: Academic Press, pp. 743–763.

Bulmer MG, and Parker GA. 2002. The evolution of anisogamy: a game-theoretic approach. *Proc. Biol. Sci.* 269: 2381–2388.

Bureau TE, Ronald PC, and Wessler SR. 1996. A computer-based systematic survey reveals the predominance of small inverted-repeat elements in wild-type rice genes. *Proc. Natl. Acad. Sci. USA* 93: 8524–8529.

Bureau TE, White SE, and Wessler SR. 1994. Transduction of a cellular gene by a plant retroelement. *Cell* 77: 479–480.

Burger G, Forget L, Zhu Y, Gray MW, and Lang BF. 2003. Unique mitochondrial genome architecture in unicellular relatives of animals. *Proc. Natl. Acad. Sci. USA* 100: 892–897.

Burgos M, Jimenez R, and Diaz de la Guardia R. 1988. XY females in *Microtus cabrerae* (Rodentia, Microtidae)—a case of possibly Y-linked sex reversal. *Cytogenet. Genome Res.* 49: 275–277.

Burley N. 1981. Sex ratio manipulation and selection for attractiveness. *Science* 211: 721–722.

Burley N. 1986. Sex-ratio manipulation in color-banded populations of zebra finches. *Evolution* 40: 1191–1206.

Burt A. 1995. The evolution of fitness. *Evolution* 49: 1–8.

Burt A. 2000. Sex, recombination, and the efficacy of selection—was Weismann right? *Evolution* 54: 337–351.

Burt A. 2003. Site-specific selfish genes as tools for the control and genetic engineering of natural populations. *Proc. Biol. Sci.* 270: 921–928.

Burt A, and Koufopanou V. 2004. Homing endonuclease genes: the rise and fall and rise again of a selfish element. *Curr. Opin. Genet. Dev.* 14: 609–615.

Burt A, and Trivers R. 1998a. Genetic conflicts in genomic imprinting. *Proc. Biol. Sci.* 265: 2393–2397.

Burt A, and Trivers R. 1998b. Selfish DNA and breeding system in flowering plants. *Proc. R. Soc. Lond. B* 265: 141–146.

Burzynski A, Zbawicka M, Skibinski DOF, and Wenne R. 2003. Evidence for recombination of mtDNA in the marine mussel *Mytilus trossulus* from the Baltic. *Mol. Biol. Evol.* 20: 388–392.

Bushman F. 2002. *Lateral DNA Transfer: Mechanisms and Consequences.* Cold Spring Harbor: Cold Spring Harbor Press.

Bussières J, Lemieux C, Lee RW, and Turmel M. 1996. Optional elements in the chloroplast DNAs of *Chlamydomonas eugametos* and *C. moewusii:* unidirectional gene conversion and co-conversion of adjacent markers in high-viability crosses. *Curr. Genet.* 30: 356–365.

Buzek J, Ebert I, Ruffine-Catiglione M, Sirosky J, Vyskot B, and Greilhuber J. 1998. Structure and DNA methylation pattern of partially heterochromatinised endosperm nuclei in *Gagea lutea. Planta* 204: 506–514.

Byrne M, Phelps H, Church T, Adair V, Selvakumarswamy P, and Potts J. 2000. Reproduction and development of the freshwater clam *Corbicula australis* in southeast Australia. *Hydrobiologia* 418: 185–197.

Cabrero J, López-León MD, Bakkali M, and Camacho JPM. 1999. Common origin of B chromosome variants in the grasshopper *Eyprepocnemis plorans. Heredity* 83: 435–439.

Cabrero J, Perfectti F, Gómez R, Camacho JPM, and López-León MD. 2003. Population variation in the A chromosome distribution of satellite DNA and ribosomal DNA in the grasshopper *Eyprepocnemis plorans*. *Chromosome Res.* 11: 375–381.

Cabrero J, Viseras E, and Camacho JPM. 1984. The B-chromosomes of *Locusta migratoria*. I. Detection of negative correlation between mean chiasma frequency and the rate of accumulation of the B's: a reanalysis of the available data about the transmission of these B chromosomes. *Genetica* 64: 155–164.

Cáceres M, Ranz JM, Barbadilla A, Long M, and Ruiz A. 1999. Generation of a widespread *Drosophila* inversion by a transposable element. *Science* 285: 415–418.

Cairns J. 1975. Mutation selection and the natural history of cancer. *Nature* 255: 197–200.

Cairns J. 2002. Somatic stem cells and the kinetics of mutagenesis and carcinogenesis. *Proc. Natl. Acad. Sci. USA* 99: 10567–10570.

Calvi BR, and Gelbart WM. 1994. The basis for germline specificity of the *hobo* transposable element in *Drosophila melanogaster. EMBO J.* 13: 1636–1644.

Camacho JPM. 2005. B chromosomes. *The Evolution of the Genome,* ed. TR Gregory. Amsterdam: Elsevier, pp. 223–286.

Camacho JPM, ed. 2004. *B Chromosomes in the Eurkaryote Genome.* Basel: Karger.

Camacho JPM, Bakkali M, Corral JM, Cabrero J, López-León MD, Aranda I, Martin-Alganza A, and Perfectti F. 2002. Host recombination is dependent on the degree of parasitism. *Proc. Biol. Sci.* 269: 2173–2177.

Camacho JPM, Cabrero J, López-León MD, Bakkali M, and Perfectti F. 2003. The B chromosomes of the grasshopper *Eyprepocnemis plorans* and the intragenomic conflict. *Genetica* 117: 77–84.

Camacho JPM, Carballo AR, and Cabrero J. 1980. The B-chromosome system of the grasshopper *Eyprepocnemis plorans* subsp. *plorans* (Charpentier). *Chromosoma* 80: 163–166.

Camacho JPM, Perfectti F, Teruel M, López-León MD, and Cabrero J. 2004. The odd-even effect in mitotically unstable B chromosomes in grasshoppers. *Cytogenet. Genome Res.* 106: 325–331.

Camacho JPM, Sharbel TF, and Beukeboom LW. 2000. B-chromosome evolution. *Philos. Trans. R. Soc. Lond. B Biol. Sci.* 355: 163–178.

Camacho JPM, Shaw MW, López-León MD, Pardo MC, and Cabrero J. 1997. Population dynamics of a selfish B chromosome neutralized by the standard genome in the grasshopper *Eyprepocnemis plorans. Am. Nat.* 149: 1030–1050.

Cameron DR, and Moav RM. 1957. Inheritance in *Nicotiana tabacum* XXVII. Pollen killer, an alien genetic locus inducing abortion of microspores not carrying it. *Genetics* 42: 326–335.

Cameron FM, and Rees H. 1967. The influence of B chromosomes on meiosis in *Lolium. Heredity* 22: 446–450.

Cano MI, and Santos JL. 1989. Cytological basis of the B chromosome accumulation mechanism in the grasshopper *Heteracris littoralis* (Ramb). *Heredity* 62: 91–95.

Cao L, Kenchington E, and Zouros E. 2004. Differential segregation patterns of sperm mitochondria in embryos of the blue mussel (*Mytilus edulis*). *Genetics* 166: 883–894.

Capillon C, and Atlan A. 1999. Evolution of driving X chromosomes and resistance factors in experimental populations of *Drosophila simulans*. *Evolution* 53: 506–517.

Capy P, Vitalis R, Langin T, Higuet D, and Bazin C. 1996. Relationships between transposable elements based upon the integrase-transposase domains: is there a common ancestor? *J. Mol. Evol.* 42: 359–368.

Carlile MJ. 1987. Genetic exchange and gene flow: their promotion and prevention. In: *Evolutionary Biology of the Fungi*, eds. ADM Rayner, CM Brasier, and D Moore. Cambridge, England: Cambridge University Press: 203–214.

Carlson WR. 1969. Factors affecting preferential fertilization in maize. *Genetics* 62: 543–554.

Carlson WR. 1994. Crossover effects of B chromosomes may be "selfish." *Heredity* 72: 636–638.

Carlson WR, and Chou TS. 1981. B chromosome nondisjunction in corn: control by factors near the centromere. *Genetics* 97: 379–389.

Carlson WR, and Roseman R. 1992. A new property of the maize B chromosome. *Genetics* 131: 211–223.

Carmona JA, Sanjur OI, Doadrio I, Machordom A, and Vrijenhoek RC. 1997. Hybridogenetic reproduction and maternal ancestry of polyploid Iberian fish: the *Tropidophoxinellus alburnoides* complex. *Genetics* 146: 983–993.

Carpenter ATC. 1984. Meiotic roles of crossing-over and of gene conversion. *Cold Spring Harbor Symp. Quant. Biol.* 49: 23–29.

Carpenter ATC. 1987. Gene conversion, recombination nodules, and the initiation of meiotic synapsis. *BioEssays* 6: 232–236.

Carroll L, Meagher S, Morrisson L, Penn D, and Potts WK. 2004. Fitness effects of a selfish gene (the *Mus t* complex) are revealed in an ecological context. *Evolution* 58: 1318–1328.

Carson HL. 1983. Chromosomal sequences and interisland colonizations in Hawaiian *Drosophila*. *Genetics* 103: 465–482.

Carvalho AB. 2002. Origin and evolution of the *Drosophila* Y chromosome. *Curr. Opin. Genet. Dev.* 12: 664–668.

Carvalho AB, and Clark AG. 2004. Y chromosome of *D pseudoobscura* is not homologous to the ancestral *Drosophila* Y. *Science* 307: 108–110.

Carvalho AB, and Klaczko LB. 1994. Y-linked suppressors of the *sex-ratio* trait in *Drosophila mediopunctata*. *Heredity* 73: 573–579.

Carvalho AB, Sampaio MC, Varandas FR, and Klaczko LB. 1998. An experimental demonstration of Fisher's principle: evolution of sexual proportion by natural selection. *Genetics* 148: 719–731.

Carvalho AB, and Vaz SC. 1999. Are *Drosophila* SR drive chromosomes always balanced? *Heredity* 83: 221–228.

Carvalho AB, Vaz SC, and Klaczko LB. 1997. Polymorphism for Y-linked suppressors of *sex-ratio* in two natural populations of *Drosophila mediopunctata*. *Genetics* 146: 891–902.

Casane D, Dennebouy N, de Rochambeau H, Mounolou JC, and Monnerot M. 1994. Genetic analysis of systematic mitochondrial heteroplasmy in rabbits. *Genetics* 138: 471–480.

Casavant NC, Scott L, Cantrell MA, Wiggins LE, Baker RJ, and Wichman HA. 2000. The end of the LINE? Lack of recent L1 activity in a group of South American rodents. *Genetics* 154: 1809–1817.

Casselton LA. 2002. Mate recognition in fungi. *Heredity* 88: 142–147.

Castiglia R, and Capanna E. 2000. Contact zone between chromosomal races of *Mus musculus domesticus*. II. Fertility and segregation in laboratory-reared and wild mice heterozygous for multiple Robertsonian rearrangements. *Heredity* 85: 147–156.

Castillo-Davis CI, Mekhedov SL, Hartl DL, and Koonin EV. 2002. Selection for short introns in highly expressed genes. *Nat. Genet.* 31: 415–418.

Cattanach BM, and Beechey CV. 1990. Autosomal and X-chromosome imprinting. *Dev. Suppl.*: 63–72.

Caubet Y, Hatcher MJ, Mocquard J-P, and Rigaud T. 2000. Genetic conflict and changes in heterogametic mechanisms of sex determination. *J. Evol. Biol.* 13: 766–777.

Cazemajor M, Joly D, and Montchamp-Moreau C. 2000. *Sex-ratio* meiotic drive in *Drosophila simulans* is related to equational nondisjunction of the Y chromosome. *Genetics* 154: 229–236.

Cazemajor M, Landré C, and Montchamp-Moreau C. 1997. The *sex-ratio* trait in *Drosophila simulans*: genetic analysis of distortion and suppression. *Genetics* 147: 635–642.

Cervantes RB, Stringer JR, Shao C, Tischfield JA, and Stambrook PJ. 2002. Embryonic stem cells and somatic cells differ in mutation frequency and type. *Proc. Natl. Acad. Sci. USA* 99: 3586–3590.

Chaboissier M-C, Bucheton A, and Finnegan DJ. 1998. Copy number control of a transposable element, the *I* factor, a LINE-like element in *Drosophila. Proc. Natl. Acad. Sci. USA* 95: 11781–11785.

Chadwick-Furman NE, and Weissman IL. 1995. Life history plasticity in chimaeras of the colonial ascidian *Botryllus schlosseri. Proc. Biol. Sci.* 262: 157–162.

Chain AC, Zollman S, Tseng JC, and Laski FA. 1991. Identification of a *cis*-acting sequence required for germ line-specific splicing of the P-element ORF2-ORF3 intron. *Mol. Cell. Biol.* 11: 1538–1546.

Chandler VL, Stam M, and Sidorenko LV. 2002. Long-distance *cis* and *trans* interactions mediate paramutation. In: *Homology Effects*, eds. JC Dunlap and C-t Wu. San Diego: Academic Press, pp. 215–234.

Chandra HS, and Brown SW. 1975. Chromosome imprinting and the mammalian X chromosome. *Nature* 253: 165–168.

Chandrasegaran S, and Smith J. 1999. Chimeric restriction enzymes: what is next. *Biol. Chem.* 380: 841–848.

Charalambous M, Smith FM, Bennett WR, Crew TE, Mackenzie F, and Ward A. 2003. Disruption of the imprinted *Grb10* gene leads to disproportionate overgrowth by an *Igf2*-independent mechanism. *Proc. Natl. Acad. Sci. USA* 100: 8292–8297.

Charlat S, Hurst GDD, and Merçot H. 2003. Evolutionary consequences of *Wolbachia* infections. *Trends Genet.* 19: 217–223.

Charlesworth B. 1994. The evolution of lethals in the *t*-haplotype system of the mouse. *Proc. Biol. Sci.* 258: 101–107.

Charlesworth B, and Barton NH. 1996. Recombination load associated with selection for increased recombination. *Genet. Res.* 67: 27–41.

Charlesworth B, Borthwick H, Bartolomé C, and Pignatelli P. 2004. Estimates of the genomic mutation rate for detrimental alleles in *Drosophila melanogaster*. *Genetics* 167: 815–826.

Charlesworth B, and Charlesworth D. 1978. A model for the evolution of dioecy and gynodioecy. *Am. Nat.* 112: 975–997.

Charlesworth B, and Charlesworth D. 1983. The population dynamics of transposable elements. *Genet. Res.* 42: 1–27.

Charlesworth B, and Dempsey ND. 2001. A model of the evolution of the unusual sex chromosome system of *Microtus oregoni*. *Heredity* 86: 387–394.

Charlesworth B, and Hartl DL. 1978. Population dynamics of the segregation distorter polymorphism of *Drosophila melanogaster*. *Genetics* 89: 171–192.

Charlesworth B, and Langley CH. 1986. The evolution of self-regulated transposition of transposable elements. *Genetics* 112: 359–383.

Charlesworth B, and Langley CH. 1989. The population genetics of *Drosophila* transposable elements. *Annu. Rev. Genet.* 23: 251–287.

Charlesworth B, Langley CH, and Sniegowski PD. 1997. Transposable element distributions in *Drosophila*. *Genetics* 147: 1993–1995.

Charlesworth B, and Lapid A. 1989. A study of ten families of transposable elements on X chromosomes from a population of *Drosophila melanogaster*. *Genet. Res.* 54: 113–125.

Charlesworth B, Lapid A, and Canada D. 1992. The distribution of transposable elements within and between chromosomes in a population of *Drosophila melano-gaster*. I. Element frequencies and distribution. *Genet. Res.* 60: 103–114.

Charlesworth B, Sniegowski P, and Stephan W. 1994. The evolutionary dynamics of repetitive DNA in eukaryotes. *Nature* 371: 215–220.

Charlesworth D, and Ganders FR. 1979. The population genetics of gynodioecy with cytoplasmic-genic male-sterility. *Heredity* 43: 213–218.

Charlesworth D, and Laporte V. 1998. The male-sterility polymorphism of *Silene vulgaris:* analysis of genetic data from two populations and comparison with *Thymus vulgaris*. *Genetics* 150: 1267–1282.

Charlton WL, Keen CL, Merriman C, Lynch P, Greenland AJ, and Dickinson HG. 1995. Endosperm development in *Zea mays*, implication of gametic imprinting and paternal excess in regulation of transfer layer development. *Development* 121: 3089–3097.

Charnov EL. 1982. *The Theory of Sex Allocation*. Princeton: Princeton University Press.

Chen Q, Jahier J, and Cauderson Y. 1994. The B chromosome system of Inner Mongolian *Agropyron* Gaertn. I. Bs distribution, morphological and cytogenetic behaviour. *Caryologia* 46: 245–260.

Chen XJ, and Clark-Walker GD. 2000. The petite mutation in yeasts: 50 years on. *Int. Rev. Cytol.* 194: 197–238.

Chevalier BS, Kortemme T, Chadsey MS, Baker D, Monnat Jr. RJ, and Stoddard BL. 2002. Design, activity, and structure of a highly specific artificial endonuclease. *Mol. Cell* 10: 895–905.

Chevalier BS, and Stoddard BL. 2001. Homing endonucleases: structural and functional insight into the catalysts of intron/intein mobility. *Nucleic Acids Res.* 29(18): 3757–3774.

Chiavarino AM, González-Sánchez M, Poggio L, Puertas MJ, Rosato M, and Rosi P. 2001. Is maize B chromosome preferential fertilization controlled by a single gene? *Heredity* 86: 743–748.

Chiavarino AM, Rosato M, Manzanero S, Jiménez G, González-Sánchez M, and Puertas MJ. 2000. Chromosome nondisjunction and instabilities in tapetal cells are affected by B chromosomes in maize. *Genetics* 155: 889–897.

Chiavarino AM, Rosato M, Rosi P, Poggio L, and Naranjo CA. 1998. Localization of the genes controlling B chromosome transmission rate in maize (*Zea mays* ssp. *mays*, Poaceae). *Am. J. Bot.* 85: 1581–1585.

Chinnery PF, and Samuels DC. 1999. Relaxed replication of mtDNA: a model with implications for the expression of disease. *Am. J. Hum. Genet.* 64: 1158–1165.

Chinnery PF, Samuels DC, Elson J, and Turnbull DN. 2002. Accumulation of mitochondrial DNA mutations in ageing, cancer, and mitochondrial disease: is there a common mechanism? *Lancet* 360: 1323–1325.

Chinnery PF, Thorburn DR, Samuels DC, White SL, Dahl HM, Turnbull DM, Lightowlers RN, and Howell N. 2000. The inheritance of mitochondrial DNA heteroplasmy: random drift, selection or both? *Trends Genet.* 16: 500–505.

Chinnery PF, and Turnbull DN. 2000. Mitochondrial DNA mutations in the pathogenesis of human disease. *Mol. Med. Today* 6: 425–432.

Chippindale AK, Gibson JR, and Rice WR. 2001. Negative genetic correlation for adult fitness between sexes reveals ontogenetic conflict in *Drosophila. Proc. Natl. Acad. Sci. USA* 98: 1671–1675.

Cho Y, Qium Y-L, Kuhlman P, and Palmer JD. 1998. Explosive invasion of plant mitochondria by a group I intron. *Proc. Natl. Acad. Sci. USA* 95: 14244–14249.

Choi Y, Ishiguro N, Shinagawa M, Kim CJ, Okamoto Y, Minami S, and Ogihara K. 1999. Molecular structure of canine LINE-1 elements in canine transmissible venereal tumor. *Anim. Genet.* 30: 51–53.

Chorney MJ, Chorney K, Seese N, Owen MJ, Daniels J, McGuffin P, Thompson LA, Detterman DK, Benbow C, Lubinski D, Eley T, and Plomin R. 1998. A quantitative trait locus associated with cognitive ability in children. *Psychol. Sci.* 9: 159–166.

Christensen S, Pont-Kingdon G, and Carroll D. 2000. Target specificity of the endonuclease from the *Xenopus laevis* non-long terminal repeat retrotransposon, Tx1L. *Mol. Cell. Biol.* 20: 1219–1226.

Cimino MC. 1972. Egg-production, polyploidization and evolution in a diploid all-female fish of the genus *Poeciliopsis. Evolution* 26: 294–306.

Clark-Walker GD. 1992. Evolution of mitochondrial genomes in fungi. *Int. Rev. Cytol.* 141: 89–127.

Clark JB, Kim PC, and Kidwell MG. 1998. Molecular evolution of *P* transposable elements in the genus *Drosophila.* III. The *melanogaster* species group. *Mol. Biol. Evol.* 15: 746–755.

Clark MM, and Galef Jr. BG. 1988. Effect of uterine position on rate of sexual development in female Mongolian gerbils. *Physiol. Behav.* 42: 15–18.

Clark MM, and Galef Jr. BG. 1995. Prenatal influences on reproductive life history strategies. *Trends Ecol. Evol.* 10: 151–153.

Clark MM, Spencer CA, and Galef Jr. BG. 1986. Reproductive life history correlates of early and late sexual maturation in female Mongolian gerbils (*Meriones unguiculatus*). *Anim. Behav.* 34: 551–560.

Clark MM, Tucker L, and Galef Jr. BG. 1992. Stud males and dud males: intra-uterine position effects on the reproductive success of male gerbils. *Anim. Behav.* 43: 215–221.

Clayton DA. 1982. Replication of animal mitochondrial DNA. *Cell* 28: 693–705.

Cobbs G, Jewell L, and Gordon L. 1991. Male-sex-ratio trait in *Drosophila pseudoobscura*–frequency of autosomal aneuploid sperm. *Genetics* 127: 381–390.

Coen D, Deutch J, Netter P, Petrochillo E, and Slonimski PP. 1970. Mitochondrial genetics. I. Methodology and phenomenology. *Symp. Soc. Exp. Biol.* 24: 444–496.

Coffin JM, Hughes SM, and Varmus HE, eds. 1997. *Retroviruses.* Cold Spring Harbor: Cold Spring Harbor Laboratory Press.

Colleaux L, D'Auriol L, Betermier M, Cottarel G, Jacquier A, Galibert F, and Dujon B. 1986. Universal code equivalent of a yeast mitochondrial intron reading frame is expressed into *E. coli* as a specific double strand endonuclease. *Cell* 44: 521–533.

Comstock JH. 1940. *An Introduction to Entomology,* 9th ed. Ithaca, N.Y.: Comstock Publishing Associates.

Conley CA, and Hanson MR. 1995. How do alterations in plant mitochondrial genomes disrupt pollen development? *J. Bioenerg. Biomembr.* 27: 447–457.

Conn JS, and Blum U. 1981. Sex ratio of *Rumex hastatulus:* the effects of environmental factors and certation. *Evolution* 35: 1108–1116.

Cook LG. 2000. Extraordinary and extensive karyotypic variation: A 48-fold range in chromosome number in the gall-inducing scale insect *Apiomorpha* (Hemiptera: Coccoidea: Eriococcidae). *Genome* 43: 255–263.

Cooley L, Kelley R, and Spradling A. 1988. Insertional mutagenesis of the *Drosophila* genome with single P elements. *Science* 239: 1121–1128.

Coopersmith CB, and Lenington S. 1990. Preferences of female mice for males whose *t*-haplotype differs from their own. *Anim. Behav.* 40: 1179–1192.

Corish P, Black DM, Featherston DW, Merriam J, and Dover GA. 1996. Natural repressors of P-induced hybrid dysgenesis in *Drosophila melanogaster:* a model for repressor evolution. *Genet. Res.* 67: 109–121.

Correns C. 1906. Die vererbung der Geshlechstsformen bei den gynodiöcischen Pflanzen. *Ber. Dtsch. Bot. Ges.* 24: 459–474.

Cosmides LM, and Tooby J. 1981. Cytoplasmic inheritance and intragenomic conflict. *J. Theor. Biol.* 89: 83–129.

Costa HS, Toscano NC, and Henneberry TJ. 1996. Mycetocyte inclusion in the oocytes of *Bemisia argentifolii* (Homoptera, Aleyrodidae). *Ann. Entomol. Soc. Am.* 89: 694–699.

Cousineau B, Lawrence S, Smith D, and Belfort M. 2000. Retrotransposition of a bacterial group II intron. *Nature* 404: 1018–1021.

Couvet D, Atlan A, Belhassen E, Gliddon C, Gouyon PH, and Kjellberg F. 1990. Coevolution between two symbionts: the case of cytoplasmic male-sterility in higher plants. *Oxf. Surv. Evol. Biol.* 7: 225–249.

Covello PS, and Gray MW. 1992. Silent mitochondrial and active nuclear genes for subunit 2 of cytochrome c oxidase (cox2) in soybean: evidence for RNA-mediated gene transfer. *EMBO J.* 11: 3815–3820.

Coyne JA, and Orr HA. 1993. Further evidence against meiotic drive models of hybrid sterility. *Evolution* 47: 685–687.

Coyne JA, and Orr HA. 2004. *Speciation*. Sunderland, Mass.: Sinauer Associates.

Coyne RS, Chalker DL, and Yao M-C. 1966. Genome downsizing during ciliate development: nuclear division of labor through chromosome restructuring. *Annu. Rev. Genet.* 30: 557–578.

Coyne RS, Chalker DL, and Yao M-C. 1996. Genome downsizing during ciliate development: nuclear division of labor through chromosome restructuring. *Annu. Rev. Genet.* 30: 557–578.

Craig NL. 1995. Unity in transposition reactions. *Science* 270: 253–254.

Craig NL, Craigie R, Gellert M, and Lambowitz AM, eds. 2002. *Mobile DNA II*. Washington, D.C.: ASM Press.

Croteau S, Andrade MF, Huang F, Greenwood CMT, Morgan K, and Naumova AK. 2002. Inheritance patterns of maternal alleles in imprinted regions of the mouse genome at different stages of development. *Mamm. Genome* 13: 24–29.

Crouse HV. 1960. The controlling element in sex chromosome behavior in *Sciara*. *Genetics* 45: 1429–1443.

Crouse HV. 1979. X heterochromatin subdivision and cytogenetic analysis in *Sciara coprophila* (Diptera, Sciaridae). II. The controlling element. *Chromosoma* 74: 219–239.

Crow JF. 1957. Possible consequences of an increased mutation rate. *Eugen. Quart.* 4: 67–80.

Crow JF. 1979. Genes that violate Mendel's rules. *Sci. Am.* 240: 134–146.

Crow JF. 1982. Some perspectives from population-genetics. *Res. Publ.–Assoc. Res. Nerv. Ment. Dis.* 60: 93–104.

Crow JF. 1988. The ultraselfish gene. *Genetics* 118: 389–391.

Crow JF. 1991. Why is Mendelian segregation so exact? *BioEssays* 13: 305–312.

Crow JF, and Kimura M. 1970. *An Introduction to Population Genetics Theory*. New York: Harper and Row.

Cruickshank RH, and Thomas RH. 1999. Evolution of haplodiploidy in dermanyssine mites (Acari: Melostigmata). *Evolution* 53: 1796–1803.

Curcio MJ, and Belfort M. 1996. Retrohoming: cDNA-mediated mobility of group II introns requires a catalytic RNA. *Cell* 84: 9–12.

Curcio MJ, and Garfinkel DJ. 1991. Single-step selection for Ty*1* element retrotransposition. *Proc. Natl. Acad. Sci. USA* 88: 936–940.

Curley JP, Barton S, Surani A, and Keverne EB. 2004. Coadaptation in mother and infant regulated by a paternally expressed imprinted gene. *Proc. Biol. Sci.* 271: 1303–1309.

Curole JP, and Kocher TD. 2002. Ancient sex-specific extension of the cytochrome *c* oxidase II gene in bivalves and the fidelity of doubly-uniparental inheritance. *Mol. Biol. Evol.* 19: 1323–1328.

Curtis CF, Grover KK, Suguna SG, Uppal DK, Dietz K, Agarwal HV, and Kazmi SJ.

1976. Comparative field cage tests of the population suppressing efficiency of three genetic control systems for *Aedes aegypti. Heredity* 36: 11–29.

Cuypers H, Dash S, Peterson PA, Saedler H, and Gierl A. 1988. The defective En-I102 element encodes a product reducing the mutability of the En/Spm transposable element system of *Zea mays. EMBO J.* 7: 2953–2960.

D'Amours D, Stegmeier F, and Amon A. 2004. Cdc14 and condensin control the dissolution of cohesion-independent chromosome linkages at repeated DNA. *Cell* 117: 455–469.

Dahms NM, and Hancock MK. 2002. P-type lectins. *Biochimica et Biophysica Acta* 1572: 317–340.

Dallai R, Fanciulli PP, and Frati G. 1999. Chromosome elimination and sex determination in springtails (Insecta, Collembola). *J. Exp. Zool. (Mol. Dev. Evol.)* 285: 215–225.

Dallai R, Fanciulli PP, and Frati G. 2000. Aberrant spermatogenesis and the peculiar mechanism of sex determination in Symphypleonan Collembola (Insecta). *J. Hered.* 91: 351–358.

Dalstra HJP, Swart K, Debets AJM, Saupe SJ, and Hoekstra RF. 2003. Sexual transmission of the [Het-s] prion leads to meiotic drive in *Podospora anserina. Proc. Natl. Acad. Sci. USA* 100: 6616–6621.

Dalziel AC, and Stewart DT. 2002. Tissue-specific expression of male-transmitted mitochondrial DNA and its implications for rates of molecular evolution of *Mytilus* mussels (Bivalvia: Mytilidae). *Genome* 45: 348–355.

Daniel A. 2002. Distortion of female meiotic segregation and reduced male fertility in human Robertsonian translocations: consistent with the centromeric model of co-evolving centromere DNA/centromeric histone (CENP-A). *Am. J. Med. Genet.* 111: 450–452.

Daniels SB, Clark SH, Kidwell MG, and Chovnick A. 1987. Genetic transformation of *Drosophila melanogaster* with an autonomous *P* element: phenotypic and molecular analyses of long-established transformed lines. *Genetics* 115: 711–723.

Daniels SB, Peterson KR, Strausbaugh LD, Kidwell MG, and Chovnick A. 1990. Evidence for horizontal transmission of the *P* transposable element between *Drosophila* species. *Genetics* 124: 339–355.

Danilevskaya ON, Hermon P, Hantke S, Muszynski MG, Kollipara K, and Ananiev EV. 2003. Duplicated *fie* genes in maize: expression patterns and imprinting suggest distinct functions. *Plant Cell* 15: 425–438.

Darlington CD, and Wylie AP. 1956. *Chromosome Atlas of Flowering Plants.* New York: Macmillan.

Das U, and Das AK. 2000. Review of canine transmissible venereal sarcoma. *Vet. Res. Commun.* 24: 545–556.

Davies DR, Goryshin IY, Reznikoff WS, and Rayment I. 2000. Three-dimensional structure of the Tn5 synaptic complex transposition intermediate. *Science* 289: 77–85.

Davison A. 2000. Three-dimensional structure of the Tn5 synaptic complex transposition intermediate, *Cepea nemoralis. J. Mollusc. Stud.* 66: 143–147.

Dawe RK, and Cande WZ. 1996. Induction of centromeric activity in maize by *suppressor of meiotic drive 1. Proc. Natl. Acad. Sci. USA* 93: 8512–8517.

Dawe RK, and Hiatt EN. 2004. Plant neocentromeres: fast, focused, and driven. *Chromosome Res.* 12: 655–669.

Dawkins R. 1976. *The Selfish Gene.* Oxford: Oxford University Press.

Day T, and Taylor PD. 1998. Chromosomal drive and the evolution of meiotic nondisjunction and trisomy in humans. *Proc. Natl. Acad. Sci. USA* 95: 2361–2365.

de Haan AA, Koelewijn HP, Hundscheid MPJ, and Van Damme JMM. 1997a. The dynamics of gynodioecy in *Plantago lanceolata* L. II. Mode of action and frequencies of restorer alleles. *Genetics* 147: 1317–1328.

de Haan AA, Luyten RMJM, Bakx-Schotman TJMT, and Van Damme JMM. 1997b. The dynamics of gynodioecy in *Plantago lanceolata* L. I. Frequencies of male-steriles and their cytoplasmic male sterility types. *Heredity* 79: 453–462.

de Haan AA, Mateman AC, Van Dijk PJ, and Van Damme JMM. 1997c. New CMS types in *Plantago lanceolata* and their relatedness. *Theor. Appl. Genet.* 94: 539–548.

de Jesus CM, Galetti Jr. PM, Valentini SR, and Moreira-Filho O. 2003. Molecular characterization and chromosomal localization of two families of satellite DNA in *Prochilodus lineatus* (Pisces, Prochilodontidae), a species with B chromosomes. *Genetica* 118: 25–32.

de la Casa-Esperón E, and Sapienza C. 2003. Natural selection and the evolution of genome imprinting. *Annu. Rev. Genet.* 37: 349–370.

de Mazancourt C, Loreau M, and Abbadie L. 1998. Grazing optimization and nutrient cycling: when do herbivores enhance primary production? *Ecology* 79: 2242–2252.

de Saint Phalle B, and Sullivan W. 1996. Complete sister chromatid separation is the mechanism of programmed chromosome elimination during early *Sciara coprophila* embryogenesis. *Development* 122: 3775–3784.

de Stordeur E. 1997. Nonrandom partition of mitochondria in heteroplasmic *Drosophila. Heredity* 79: 615–623.

DeBerardinis RJ, Goodier JL, Ostertag EM, and Kazazian HH. 1998. Rapid amplification of a retrotransposon subfamily is evolving the mouse genome. *Nat. Genet.* 20: 288–290.

DeChiara TM, Robertson EJ, and Efstratiadis A. 1991. Parental imprinting of the mouse insulin-like growth factor II gene. *Cell* 64: 849–859.

Deininger PL, and Batzer MA. 1999. Alu repeats and human disease. *Mol. Genet. Metab.* 67: 183–193.

Delannay X, Gouyon PH, and Valdeyron G. 1981. Mathematical study of the evolution of gynodioecy with cytoplasmic inheritance under the effect of a nuclear restorer gene. *Genetics* 99: 169–181.

Delph LF. 1990. Sex-differential resource-allocation patterns in the subdioecious shrub *Hebe subalpina*. *Ecology* 71: 1342–1351.

Dermitzakis ET, Masly JP, Waldrip HM, and Clark AG. 2000. Non-Mendelian segregation of sex chromosomes in heterospecific *Drosophila* males. *Genetics* 154: 687–694.

Derome M, Métayer K, Montchamp-Moreau C, and Venille M. 2004. Signature of selective sweep associated with the evolution of *sex-ratio* drive in *Drosophila simulans*. *Genetics* 166: 1357–1366.

Derr LK, and Strathern JN. 1993. A role for reverse transcripts in gene conversion. *Nature* 361: 170–173.

Desfeux C, Maurice S, Henry J-P, Lejeune B, and Gouyon P-H. 1996. Evolution of reproductive systems in the genus *Silene*. *Proc. R. Soc. Lond. B* 263: 409–414.

Dhar MK, Friebe AK, Koul AK, and Gill BS. 2002. Origin of an apparent B chromosome by mutation, chromosome fragmentation and specific DNA-sequence amplification. *Chromosoma* 111: 332–340.

Diaz F, Bayona-Bafaluy MP, Rana M, Mora M, Hao H, and Moraes CT. 2002. Human mitochondrial DNA with large deletions repopulates organelles faster than full-length genomes under relaxed copy number control. *Nucleic Acids Res.* 30: 4626–4633.

Dill CL, Wise RP, and Schnable PS. 1997. Rf8 and Rf* mediate unique T-urf13-transcript accumulation, revealing a conserved motif associated with RNA processing and restoration of pollen fertility in T-cytoplasm maize. *Genetics* 147: 1367–1379.

Doak TG, Doerder FP, Jahn CL, and Herrick G. 1994. A proposed superfamily of transposase genes: transposon-like elements in ciliated protozoa and a common "D35E" motif. *Proc. Natl. Acad. Sci. USA* 91: 942–946.

Dobrovolskaïa-Zavadskaïa N, and Kobozieff N. 1927. Sur la reproduction des souris anoures. *C. R. Séanc. Soc. Biol.* 97: 116–119.

Dobson SL, and Tanouye MA. 1998. Evidence for a genomic imprinting sex determination mechanism in *Nasonia vitripennis* (Hymenoptera; Chalcidoidea). *Genetics* 149: 233–242.

Dod B, Litel C, Makoundou P, Orth A, and Boursot P. 2003. Identification and characterization of *t* haplotypes in wild populations using molecular markers. *Genet. Res.* 81: 103–114.

Domínguez A, and Albornoz J. 1996. Rates of movement of transposable elements in *Drosophila melanogaster*. *Mol. Gen. Genet.* 251: 130–138.

Doolittle WF, and Sapienza C. 1980. Selfish genes, the phenotype paradigm and genome evolution. *Nature* 284: 601–603.

Dornburg R, and Temin HM. 1990. cDNA genes formed after infection with retroviral vector particles lack the hallmarks of natural processed pseudogenes. *Mol. Cell. Biol.* 10: 68–74.

Dowe Jr. MF, Roman GW, and Klein AS. 1990. Excision and transposition of two *Ds* transposons from the *bronze mutable* 4 derivative 6856 allele of *Zea mays* L. *Mol. Gen. Genet.* 221: 475–485.

Drouin G, and Dover GA. 1987. A plant processed pseudogene. *Nature* 328: 557–558.

Dujon B. 1981. Mitochondrial genetics and functions. *The Molecular Biology of the Yeast Saccharomyces: Life Cycle and Inheritance.* JN Strathern, EW Jones, and JR Broach. Cold Spring Harbor, Cold Spring Harbor Laboratory: 505–635.

Dujon B. 1989. Group I introns as mobile genetic elements: facts and mechanistic speculations—a review. *Gene* 82: 91–114.

Dunn LC, Beasley AB, and Tinker H. 1958. Relative fitness of wild house mice heterozygous for a lethal allele. *Am. Nat.* 92: 215–220.

Dunn LC, and Levene H. 1961. Population dynamics of a variant *t*-allele in a confined population of wild house mice. *Evolution* 15: 385–393.

Durand D, Ardlie KG, Buttel L, Levin SA, and Silver LM. 1997. Impact of migration and fitness on the stability of lethal *t*-haplotype polymorphism in *Mus musculus:* a computer study. *Genetics* 145: 1093–1108.

Duvillié B, Bucchini D, Tang T, Jami J, and Pàldi A. 1998. Imprinting at the mouse *Ins2* locus: evidence for *cis-* and *trans-*allelic interactions. *Genomics* 47: 52–57.

Dvorak J. 1980. Homoeology between *Agropyron elongatum* chromosomes and *Triticum aestivum* chromosomes. *Can. J. Genet. Cytol.* 22: 237–259.

Dvorak J, and Appels R. 1986. Investigation of homologous crossing over and sister chromatid exchange in the wheat Nor-2 locus coding for rRNA and Gli-B2 locus coding for gliadins. *Genetics* 113: 1037–1056.

Eanes WF, Wesley C, Hey J, Houle D, and Ajioka JW. 1988. The fitness consequences of *P* element insertion in *Drosophila melanogaster. Genet. Res.* 52: 17–26.

Eaves IA, Bennett ST, Forster P, Ferber KM, Ehrmann D, Wilson AJ, Bhattacharyya S, Ziegler AG, Brinkmann B, and Todd JA. 1999. Transmission ratio distortion at the INS-IGF2 VNTR. *Nat. Genet.* 22: 324–325.

Eberhard WG. 1980. Evolutionary consequences of intracellular organelle competition. *Q. Rev. Biol.* 55: 231–249.

Eggleston WB, Johnson-Schlitz DM, and Engels WR. 1988. P-M hybrid dysgenesis does not mobilize other transposable element families in *D. melanogaster. Nature* 331: 368–370.

Ehara M, Watanabe KI, and Ohama T. 2000. Distribution of cognates of group II introns detected in mitochondrial cox1 genes of a diatom and a haptophyte. *Gene* 256: 157–167.

Eickbush DG, and Eickbush TH. 1995. Vertical transmission of the retrotransposable elements *R1* and *R2* during the evolution of the *Drosophila melanogaster* species subgroup. *Genetics* 139: 671–684.

Eickbush TH. 1997. Telomerase and retrotransposons: which came first? *Science* 277: 911–912.

Eickbush TH. 1999. Exon shuffling in retrospect. *Science* 283: 1465–1467.

Eickbush TH. 2002. R2 and related site-specific non-long terminal repeat retrotransposons. In: *Mobile DNA II*, eds. NL Craig, R Craigie, M Gellert, and AM Lambowitz. Washington, D.C.: ASM Press, pp. 813–835.

Eickbush TH, and Malik HS. 2002. Origins and evolution of retrotransposons. In: *Mobile DNA II*, eds. NL Craig, R Craigie, M Gellert, and AM Lambowitz. Washington, D.C.: ASM Press, pp. 1111–1144.

El Maâtaoui M, and Pichot C. 2001. Microsporogenesis in the endangered species *Cupressus dupreziana* A. Camus: evidence for meiotic defects yielding unreduced and abortive pollen. *Planta* 213: 543–549.

Embley TM, van der Giezen M, Horner DS, Dyal PL, and Foster P. 2003. Mitochondria and hydrogenosomes are two forms of the same fundamental organelle. *Philos. Trans. R. Soc. Lond. B Biol. Sci.* 358: 191–203.

Emerson JJ, Kaessmann H, Betrán E, and Long M. 2004. Extensive gene traffic on the mammalian X chromosome. *Science* 303: 537–540.

Endo TR. 1990. Gametocidal chromosomes and their induction of chromosome mutations in wheat. *Jap. J. Genet.* 65: 135–152.

Engeler B, and Reyer H-U. 2001. Choosy females and indiscriminate males: mate choice in mixed populations of sexual and hybridogenetic water frogs (*Rana lessonae, Rana esculenta*). *Behav. Ecol.* 12: 600–606.

Engels WR. 1989. P elements in *Drosophila melanogaster*. In: *Mobile DNA*, eds. D Berg and M Howe. Washington, D.C.: American Society for Microbiology, pp. 437–484.

Engels WR. 1992. The origin of P elements in *Drosophila melanogaster*. *BioEssays* 14: 681–686.

Engels WR, Johnson-Schlitz DM, Eggleston WB, and Sved J. 1990. High-frequency P element loss in *Drosophila* is homolog dependent. *Cell* 62: 515–525.

Engels WR, and Preston CR. 1984. Formation of chromosome rearrangements by P factors in *Drosophila*. *Genetics* 107: 657–678.

Ephrussi B, de Margerie-Hottinguer H, and Roman H. 1955. Suppressiveness: a new factor in the genetic determinism of the synthesis of respiratory enzymes in yeast. *Proc. Natl. Acad. Sci. USA* 41: 1065–1071.

Erickson RP. 2003. Somatic gene mutation and human disease other than cancer. *Mutat. Res.* 543: 125–136.

Eshel I. 1985. Evolutionary genetic stability of Mendelian segregation and the role of free recombination in the chromosomal system. *Am. Nat.* 125: 412–420.

Eskelinen O. 1997. On the population fluctuations and structure of the wood lemming *Myopus schistocolor. Z. Säugetierkunde* 62: 293–302.

Eskes R, Yang J, Lambowitz AM, and Perlman PS. 1997. Mobility of yeast mitochondrial group II introns: engineering a new site specificity and retrohoming via full reverse splicing. *Cell* 88: 865–874.

Esnault C, Maestre J, and Heidmann T. 2000. Human LINE retrotransposons generate processed pseudogenes. *Nat. Genet.* 24: 363–367.

Espinosa MB, and Vitullo AD. 1996. Offspring sex-ratio and reproductive performance in heterogametic females of the South American field mouse *Akodon azarae. Hereditas* 124: 57–62.

Espinosa MB, and Vitullo AD. 2001. Fast-developing preimplantation embryo progeny from heterogametic females in mammals. *Zygote* 9: 289–292.

Esteban MR, Campos MCC, Perondini ALP, and Goday C. 1997. Role of microtubules and microtubule organizing centers on meiotic chromosome elimination in *Sciara ocellaris. J. Cell Biol.* 110: 721–730.

Etter A, Bernard V, Kenzelmann M, Tobler H, and Müller F. 1994. Ribosomal heterogeneity from chromatin diminution in *Ascaris lumbricoides. Science* 265: 954–956.

Evans GM, Rees H, Snell CL, and Sun S. 1972. The relationship between nuclear DNA amount and the duration of the mitotic cycle. *Chromosomes Today* 3: 24–31.

Evans K, Fryer A, Inglehearn C, Duvall-Young J, Whittaker J, Gregory CY, Butler R, Ebenezer N, Hunt D, and Bhattacharya SS. 1994. Genetic linkage of cone-rod retinal dystrophy to chromosome 19q and evidence for segregation distortion. *Nat. Genet.* 6: 210–213.

Evgen'ev MB, Zelentsova H, Shostak N, Kozitsina M, Barskyi V, Lankenau D-H, and Corces VG. 1997. *Penelope,* a new family of transposable elements and its possible role in hybrid dysgenesis in *Drosophila virilis. Proc. Natl. Acad. Sci. USA* 94: 196–201.

Extavour C, and García-Bellido A. 2001. Germ cell selection in genetic mosaics in *Drosophila melanogaster. Proc. Natl. Acad. Sci. USA* 98: 11341–11346.

Faugeron-Fonty G, Kim CLV, de Zamaroczy M, Goursot R, and Bernardi G. 1984. A comparative study of the *ori* sequences from the mitochondrial genomes of twenty wild-type yeast strains. *Gene* 32: 459–473.

Fauré S, Noyer J-L, Carreel F, Horry J-P, Bakry F, and Lanaud C. 1994. Maternal inheritance of chloroplast genome and paternal inheritance of mitochondrial genome in bananas (*Musa acuminata*). *Curr. Genet.* 25: 265–269.

Fedoroff NV. 1983. Controlling elements in maize. In: *Mobile Genetic Elements*, ed. JA Shapiro. New York: Academic Press, pp. 1–63.

Fenocchio AS, Bertolo LAC, Takahashi CS, and Camacho JPM. 2000. B chromosomes in two fish species, genus *Rhamdia* (Siluriformes, Pimelodidae). *Folia Biol.-Krakow* 48: 105–109.

Fernandez R, Barragan MJL, Bullejos M, Marchal JA, Martinez S, Diaz de la Guardia RD, and Sanchez A. 2002. Mapping the SRY gene in *Microtus cabrerae:* a vole species with multiple SRY copies in males and females. *Genome* 45: 600–603.

Ferrigno O, Virolle T, Djabari Z, Ortonne J-P, White RJ, and Aberdam D. 2001. Transposable B2 SINE elements can provide mobile RNA polymerase II promoters. *Nat. Genet.* 28: 77–81.

Feschotte C, Zhang X, and Wessler SR. 2002. Miniature inverted-repeat transposable elements and their relationship to established DNA transposons. In: *Mobile DNA II*, eds. NL Craig, R Craigie, M Gellert, and AM Lambowitz. Washington, D.C.: ASM Press, pp. 1147–1158.

Finch RA, Miller TE, and Bennett MD. 1984. "Cuckoo" *Aegilops* addition chromosome in wheat ensures its transmission by causing chromosome breaks in meiospores lacking it. *Chromosoma* 90: 84–88.

Fincham JRS. 1983. *Genetics.* Boston: Jones and Bartlett.

Fink GR. 1987. Pseudogenes in yeast. *Cell* 49: 5–6.

Finnegan DJ. 1989. The I factor and I-R hybrid dysgenesis in *Drosophila melanogaster.* In: *Mobile DNA*, eds. D Berg and M Howe. Washington, D.C.: American Society for Microbiology, pp. 503–517.

Finnegan DJ. 1992. Transposable elements. In: *The Genome of Drosophila melanogaster,* eds. DL Lindsley and GG Zimm. San Diego: Academic Press, pp. 1096–1107.

Fischer SEJ, Weinholds E, and Plasterk RHA. 2001. Regulated transposition of a fish transposon in the mouse germ line. *Proc. Natl. Acad. Sci. USA* 98: 6759–6764.

Fisher RA. 1931. The evolution of dominance. *Biol. Rev.* 6: 345–368.

Fishman L, and Willis JH. 2005. A novel meiotic drive locus near-completely distorts segregation in *Mimulus* (monkeyflower). *Genetics* 169: 347–353.

Fluminhan A, and Kameya T. 1997. Involvement of knob heterochromatin in mitotic abnormalities in germinating aged seeds of maize. *Genome* 40: 91–98.

Fogel S, Mortimer R, Lusnak K, and Tavares F. 1979. Meiotic gene conversion: a signal of the basic recombination event in yeast. *Cold Spring Harbor Symp. Quant. Biol* 43: 1325–1341.

Forné T, Oswald J, Dean W, Saam JR, Bailleul B, Dandolo L, Tilghman SM, Walter J, and Reik W. 1997. Loss of the maternal *H19* gene induces changes in *Igf2* methylation in both *cis* and *trans. Proc. Natl. Acad. Sci. USA* 94: 10243–10248.

Fortunato A, Strassman JE, Santorelli L, and Queller DC. 2003. Co-occurrence in nature of different clones of the social amoeba, *Dictyostelium discoideum*. *Mol. Ecol.* 12: 1031–1038.

Foss HM, Roberts CJ, Claeys KM, and Selker EU. 1993. Abnormal chromosome behavior in *Neurospora* mutants defective in DNA methylation. *Science* 262: 1737–1741.

Foster GG, and Whitten MJ. 1991. Meiotic drive in *Lucilia cuprina* and chromosomal evolution. *Am. Nat.* 137: 403–415.

Foster JW, Brennan FE, Hampikian GK, Goodfellow PN, Sinclair AH, Lovell-Badge R, Selwood L, Renfree MB, Cooper DW, and Graves JAM. 1992. Evolution of sex determination and the Y chromosome: SRY-related sequences in marsupials. *Nature* 359: 531–533.

Francks C, DeLisi LE, Shaw SH, Fisher SE, Richardson AJ, Stein JF, and Monaco AP. 2003. Parent-of-origin effects on handedness and schizophrenia susceptibility on chromosome 2p12-q11. *Hum. Mol. Genet.* 12: 3225–3230.

Frank SA. 1989. The evolutionary dynamics of cytoplasmic male sterility. *Am. Nat.* 133: 345–376.

Frank SA. 1991. Divergence of meiotic drive-suppression systems as an explanation for sex-biased hybrid sterility and inviability. *Evolution* 45: 262–267.

Frank SA. 2000. Polymorphism of attack and defense. *Trends Ecol. Evol.* 15: 167–171.

Frank SA, and Barr CM. 2001. Spatial dynamics of cytoplasmic male sterility. In: *Integrating Ecology and Evolution in a Spatial Context,* eds. J Silvertown and J Antonovics. Oxford: Blackwell Science, pp. 219–243.

Frank SA, and Hurst LD. 1996. Mitochondria and male disease. *Nature* 383: 224.

Frank SA, Iwasa Y, and Nowak MA. 2003. Patterns of cell division and the risk of cancer. *Genetics* 163: 1527–1532.

Frankel WN, Stoye JP, Taylor BA, and Coffin JM. 1990. A linkage map of endogenous murine leukemia proviruses. *Genetics* 124: 221–236.

Frankham R, Torkamanzehi A, and Moran C. 1991. *P*-element transposon-induced quantitative genetic variation for inebriation time in *Drosophila melanogaster*. *Theor. Appl. Genet.* 81: 317–320.

Franks P, and Lenington S. 1986. Dominance and reproductive behavior of wild house mice in a seminatural environment correlated with T-locus genotype. *Behav. Ecol. Sociobiol.* 18: 395–404.

Franks TK, Houben A, Leach CR, and Timmis JN. 1996. The molecular organisation of a B chromosome tandem repeat sequence from *Brachycome dichromosomatica*. *Chromosoma* 105: 223–230.

Fraser MJ. 1986. Transposon-mediated mutagenesis of baculoviruses: transposon shuttling and implications for speciation. *Ann. Entomol. Soc. Am.* 79: 773–783.

Fredga K. 1988. Aberrant chromosomal sex-determining mechanisms in mammals, with special reference of species with XY females. *Philos. Trans. R. Soc. Lond. B Biol. Sci.* 322: 83–95.

Fredga K. 1994. Bizarre mammalian sex-determining mechanisms. In: *The Differences between the Sexes*, eds. RV Short and E Balaban. Cambridge, England: Cambridge University Press, pp. 419–431.

Fredga K, Fredriksson R, Bondrup-Nielsen S, and Ims RA. 1993. Sex ratio, chromosomes and isozymes in natural populations of the wood lemming (*Myopus schistocolor*). In: *The Biology of Lemmings*, eds. NC Stenseth and RA Ims. London: Academic Press, pp. 465–491.

Fredga K, Gropp A, Winking H, and Frank F. 1976. Fertile XX- and XY-type females in the wood lemming, *Myopus schistocolor. Nature* 261: 225–227.

Fredga K, Gropp A, Winking H, and Frank F. 1977. A hypothesis explaining the exceptional sex ratio in the wood lemming (*Myopus schistocolor*). *Hereditas* 85: 101–104.

Fredga K, Setterfield L, and Mittwoch U. 2000. Gonadal development and birth weight in X*X and X*Y females of the wood lemming, *Myopus schisticolor. Cytogenet. Cell Genet.* 91: 97–101.

Friebe B, Zhang P, Nasuda S, and Gill BS. 2003. Characterization of a knock-out mutation at the *Gc2* locus in wheat. *Chromosoma* 111: 509–517.

Fröst S. 1959. The cytological behaviour and mode of transmission of accessory chromosomes in *Plantago serraria. Hereditas* 46: 191–210.

Fry CL, and Wilkinson GS. 2004. Sperm survival in female stalk-eyed flies depends on seminal fluid and meiotic drive. *Evolution* 58: 1622–1626.

Fuge H. 1997. Nonrandom chromosome elimination in male meiosis of a sciarid fly: elimination of paternal chromosomes in first division is mediated by non-kinetochore microtubules. *Cell Motil. Cytoskeleton* 36: 84–94.

Fukuda MN, Sato T, Nakayama J, Klier G, Mikami M, Aoki D, and Nozawa S. 1995. Trophinin and tastin, a novel cell adhesion molecule complex with potential involvement in embryo implantation. *Genes Dev.* 9: 1199–1210.

Fundele R, Barton S, Christ B, Krause R, and Surani A. 1995a. Distribution of androgenetic cells in fetal mouse chimeras. *Roux's Arch. Dev. Biol.* 204: 484–493.

Fundele R, Bober E, Arnold H, Grim M, Bender R, Wilting J, and Christ B. 1994. Early skeletal muscle development proceeds normally in parthenogenetic mouse embryos. *Dev. Biol.* 161: 30–36.

Fundele R, Li LL, Herxfeld A, Barton S, and Surani A. 1995b. Proliferation and differentiation of androgenetic cells in fetal mouse chimeras. *Roux's Arch. Dev. Biol.* 204: 494–501.

Fundele R, and Surani A. 1994. Experimental embryological analysis of genetic imprinting in mouse development. *Dev. Genet.* 15: 515–522.

Fundele R, Surani A, and Allen ND. 1997. Consequences of genomic imprinting for fetal development. In: *Genomic Imprinting,* eds. W Reik and A Surani. Oxford: Oxford University Press, pp. 98–117.

Funnell DL, Matthews PS, and VanEtten HD. 2002. Identification of new pisatin demethylase genes (PDA5 and PDA7) in *Nectria haematococca* and non-Mendelian segregation of pisatin demethylating ability and virulence on pea due to loss of chromosomal elements. *Fungal Genet. Biol.* 37: 121–133.

Furano AV. 2000. The biological properties and evolutionary dynamics of mammalian LINE-1 retrotransposons. *Prog. Nucleic Acid Res. Mol. Biol.* 64: 255–294.

Futcher AB, and Cox BS. 1983. Maintenance of the 2-micron circle plasmid in populations of *Saccharomyces cerevisiae. J. Bacteriol.* 154: 612–622.

Futcher B, Reid E, and Hickey DA. 1988. Maintenance of the 2-micron circle plasmid of *Saccharomyces cerevisiae* by sexual transmission: an example of a selfish DNA. *Genetics* 118: 411–415.

Gabay-Laughnan S, and Laughnan JR. 1994. Male sterility and restorer genes in maize. In: *The Maize Handbook,* eds. M Freeling and V Walbot. New York: Springer-Verlag, pp. 418–423.

Gaillard-Sanchez M, Mattei G, Clauser E, and Corvol P. 1990. Assignment by in situ hybridization of the angiotensinogen gene to chromosome band 1q4, the same region as the human renin gene. *Hum. Genet.* 84: 341–343.

Galagan JE, and Selker EU. 2004. RIP: the evolutionary cost of genome defense. *Trends Genet.* 20: 417–423.

Galindo BE, Vacquier VD, and Swanson WJ. 2003. Positive selection in the egg receptor for abalone sperm lysin. *Proc. Natl. Acad. Sci. USA* 100: 4639–4643.

Galtier N. 2003. Gene conversion drives GC content evolution in mammalian histones. *Trends Genet.* 19: 65–68.

Galtier N, Piganeau G, Mouchiroud D, and Duret L. 2001. GC-content evolution in mammalian genomes: the biased gene conversion hypothesis. *Genetics* 159: 907–911.

Gao X, Rowley DJ, Gai XW, and Voytas DF. 2002. *Ty5 gag* mutations increase retrotransposition and suggest a role for hydrogen bonding in the function of the nucleocapsid zinc finger. *J. Virol.* 76: 3240–3247.

García-Bellido A, and Robbins LG. 1983. Viability of female germ-line cells homozygous for zygotic lethals in *Drosophila melanogaster. Genetics* 103: 235–247.

Garcillán-Barcia MP, Bernales I, Mendiola MV, and de la Cruz F. 2002. IS*91* rolling-circle transposition. In: *Mobile DNA II,* eds. NL Craig, R Craigie, M Gellert, and AM Lambowitz. Washington, D.C.: ASM Press, pp. 891–904.

Gardner MB, Kozak CA, and O'Brien SJ. 1991. The Lake Casitas wild mouse: evolving genetic resistance to retroviral disease. *Trends Genet.* 7: 22–27.

Garrido-Ramos MA, Stewart DT, Sutherland BW, and Zouros E. 1998. The distribution of male-transmitted mitochondrial DNA types in somatic tissues of blue mussels: implications for the operation of doubly uniparental inheritance of mitochondrial DNA. *Genome* 41: 818–824.

Georges M, Charlier C, and Cockett N. 2003. The callipyge locus: evidence for the *trans* interaction of reciprocally imprinted genes. *Trends Genet.* 19: 248–252.

Gerbi SA. 1986. Unusual chromosome movements in sciarid flies. In: *Results and Problems in Cell Differentiation 13: Germ Line-Soma Differentiation,* ed. W Hennig. Berlin: Springer-Verlag, pp. 71–104.

Gerdes K, Helin K, Christensen OW, and Lobner-Olesen A. 1988. Translational control and differential RNA decay are key elements regulating postsegregational expression of the killer protein encoded by the *parB* locus of plasmid R1. *J. Mol. Biol.* 203: 119–129.

Gerdes K, Nielsen A, Thorstead P, and Wagner EGH. 1992. Mechanism of killer gene activation. Antisense RNA-dependent RNase III cleavage ensures rapid turn-over of the stable Hok, SrnB and PndA effector messenger RNAs. *J. Mol. Biol.* 226: 637–649.

Gershenson S. 1928. A new sex-ratio abnormality in *Drosophila obscura. Genetics* 13: 488–507.

Gervais C, Roy G, Grandtner MM, and Desaulniers G. 1989. The B chromosomes of *Claytonia caroliniana* (Portulacaceae) and maple forest dieback. *Can. J. For. Res.* 19: 595–598.

Ghaffari SM, and Bidmeshpoor A. 2002. Presence and behaviour of B-chromosomes in *Acanthophyllum laxiusculum* (Caryophyllaceae). *Genetica* 115: 319–323.

Gierl A. 1990. How maize transposable elements escape negative selection. *Trends Genet.* 6: 155–158.

Gifford R, and Tristem M. 2003. The evolution, distribution and diversity of endogenous retroviruses. *Virus Genes* 26: 291–315.

Gileva EA. 1980. Chromosomal diversity and an aberrant genetic system of sex determination in the Arctic lemming, *Dicrostonyx torquatus* Pallas (177). *Genetica* 52: 99–103.

Gileva EA. 1987. Meiotic drive in the sex chromosome system of the varying lemming, *Dicrostonyx torquatus Pall.* (Rodentia, Microtinae). *Heredity* 59: 383–389.

Gileva EA, Benenson IE, Konopistseva LA, Puchkov VF, and Makaranets IA. 1982. XO females in the varying lemming, *Dicrostonyx torquatus:* reproductive performance and its evolutionary significance. *Evolution* 36: 601–609.

Gileva EA, and Chebotar NA. 1979. Fertile XO males and females in the varying lemming, *Dicrostonyx torquatus* Pall. (1979). A unique genetic system of sex determination. *Heredity* 42: 67–77.

Gileva EA, and Fedorov VB. 1991. Sex ratio, XY females and absence of inbreeding in a population of the wood lemming, *Myopus schistocolor* Lilljeborg, 1844. *Heredity* 66: 351–355.

Gill DE, Chao L, Perkins SL, and Wolf JB. 1995. Genetic mosaicism in plants and clonal animals. *Annu. Rev. Ecol. Syst.* 26: 423–444.

Gill JJB, Jones BMG, Marchant CJ, McLeish J, and Ochenden DJ. 1972. The distribution of chromosome races of *Ranunculus ficaria* L. in the British Isles. 36: 31–47.

Gillham NW. 1994. *Organelle Genes and Genomes.* New York: Oxford University Press.

Gimble FS. 2000. Invasion of a multitude of genetic niches by mobile endonuclease genes. *FEMS Microbiol. Lett.* 185: 99–107.

Gimble FS, and Thorner J. 1992. Homing of a DNA endonuclease gene by meiotic gene conversion in *Saccharomyces cerevisiae*. *Nature* 357: 301–306.

Gloor GB, Preston CR, Johnson-Schlitz DM, Nassif NA, Phillis RW, Benz WK, Robertson HM, and Engels WR. 1993. Type I repressors of *P* element mobility. *Genetics* 135: 81–95.

Goday C, and Pimpinelli S. 1989. Centromere organization in meiotic chromosomes of *Parascaris univalens*. *Chromosoma* 98: 160–166.

Goday C, and Ruiz MF. 2002. Differential acetylation of histones H3 and H4 in paternal and maternal germline chromosomes during development of sciarid flies. *J. Cell Sci.* 115: 4765–4775.

Goddard MR, and Burt A. 1999. Recurrent invasion and extinction of a selfish gene. *Proc. Natl. Acad. Sci. USA* 96: 13880–13885.

Goddard MR, Greig D, and Burt A. 2001. Outcrossed sex allows a selfish gene to invade yeast populations. *Proc. Biol. Sci.* 268: 2537–2542.

Goldsbrough PB, Ellis THN, and Cullis CA. 1981. Organisation of the 5S RNA genes in flax. *Nucleic Acids Res.* 9: 5895–5904.

González-Sánchez M, González-González E, Molina F, Chiavarino AM, Rosato M, and Puertas MJ. 2003. One gene determines maize B chromosome accumulation by preferential fertilisation; another gene(s) determines their meiotic loss. *Heredity* 90: 122–129.

Gonzalez P, and Lessios HA. 1999. Evolution of sea urchin retroviral-like (SURL) elements: evidence from 40 echinoid species. *Mol. Biol. Evol.* 16: 938–952.

Good AG, Meister GA, Brock HW, Grigliatti TA, and Hickey DA. 1989. Rapid spread of transposable P elements in experimental populations of *Drosophila melanogaster*. *Genetics* 122: 387–396.

Goodier JL, Ostertag EM, Du K, and Kazazian Jr. HH. 2001. A novel active L1 retrotransposon subfamily in the mouse. *Genome Res.* 11: 1677–1685.

Goodier JL, Ostertag EM, and Kazazian Jr. HH. 2000. Transduction of 3'-flanking sequences is common in L1 retrotransposition. *Hum. Mol. Genet.* 9: 653–657.

Goodwin TJD, and Poulter RTM. 2001. The DIRS1 group of retrotransposons. *Mol. Biol. Evol.* 18: 2067–2082.

Goodwin TJD, and Poulter RTM. 2004. A new group of tyrosine recombinase-encoding retrotransposons. *Mol. Biol. Evol.* 21: 746–759.

Goriely A, McVean GAT, Röjmyr M, Ingemarsson B, and Wilkie AOM. 2003. Evidence for selective advantage of pathogenic FGFR2 mutations in the male germline. *Science* 301: 643–646.

Gorlov IP, and Tsurusaki N. 2000a. Analysis of the phenotypic effects of B chromosomes in a natural population of *Metagagrella tenuipes* (Arachnida: Opiliones). *Heredity* 84: 209–217.

Gorlov IP, and Tsurusaki N. 2000b. Morphology and meiotic/mitotic behavior of B chromosomes in a Japanese harvestman, *Metagagrella tenuipes* (Arachnida: Opiliones): no evidence for B accumulation mechanisms. *Zoolog. Sci.* 17: 349–355.

Görtz H-D, Kuhlmann H-W, Möllenbeck M, Tiedtke A, Kusch J, Schmidt HJ, and Miyake A. 1999. Intra- and intercellular communication systems in ciliates. *Naturwissenschaften* 86: 422–434.

Goswami HK, and Khandelwal S. 1980. Chromosomal elimination in natural populations of *Ophioglossum. Cytologia* 45: 77–86.

Goulielmos G, and Zouros E. 1995. Incompatibility analysis of male hybrid sterility in 2 *Drosophila* species–lack of evidence for maternal, cytoplasmic, or transposable element effects. *Am. Nat.* 145: 1006–1014.

Gouyon P-H, Vichot F, and Van Damme JMM. 1991. Nuclear-cytoplasmic male-sterility–single-point equilibria versus limit-cycles. *Am. Nat.* 137: 498–514.

Graf J-D, and Polls-Pelaz MP. 1989. Evolutionary genetics of the *Rana esculenta* complex. In: *Evolution and Ecology of Unisexual Vertebrates*, eds. RM Dawley and JP Bogart. Albany, N.Y.: New York State Museum. Museum Bulletin 466: 289–301.

Graham J. 1992. *Cancer Selection: The New Theory of Evolution.* Lexington, Va.: Aculeus.

Grandbastien M-A. 1998. Activation of plant retrotransposons under stress conditions. *Trends Plant Sci.* 3: 181–187.

Graves JAM, and Shetty S. 2001. Sex from W to Z: evolution of vertebrate sex chromosomes and sex determining genes. *J. Exp. Zool.* 290: 449–462.

Gray CT, and Thomas SM. 1985. Germination and B chromosomes in *Allium porrum* L. *J. Plant Physiol.* 121: 281–285.

Gray MW. 2001. Speculations on the origin and evolution of editing. In: *RNA Editing*, ed. BL Bass. Oxford: Oxford University Press, pp. 160–184.

Gray YHM. 2000. It takes two transposons to tango. *Trends Genet.* 16: 461–468.

Grbić M, Nagy LM, and Strand MR. 1998. Development of polyembryonic insects: a major departure from typical insect embryogenesis. *Dev. Genes Evol.* 208: 69–81.

Green DM. 1988. Cytogenetics of the endemic New Zealand frog, *Leiopelma hochstetteri*: extraordinary supernumerary chromosome variation and a unique sex-chromosome system. *Chromosoma* 97: 55–70.

Green DM. 1990. Muller's Ratchet and the evolution of supernumerary chromosomes. *Genome* 33: 818–824.

Green DR, and Reed JC. 1998. Mitochondria and apoptosis. *Science* 281: 1309–1312.

Greenblatt IM. 1984. A chromosome replication pattern deduced from pericarp phenotypes resulting from movements of the transposable element, Modulator, in maize. *Genetics* 108: 471–485.

Gregg PC, Webb GC, and Adena MA. 1984. The dynamics of B chromosomes in populations of the Australian plague locust, *Chortoicetes terminifera* (Walker). *Can. J. Genet. Cytol.* 26: 194–208.

Gregory TR. 2000. Nucleotype effects without nuclei: genome size and erythrocyte size in mammals. *Genome* 43: 895–901.

Gregory TR. 2005. The C-value enigma in plants and animals: a review of parallels and an appeal for partnership. *Ann. Bot.* 95: 133–146.

Greihuber J, and Speta F. 1976. C-typed karyotypes in the *Scilla hohenackeri* group, *S. persica* and *Puschkinia* (Liliaceae). *Plant Syst. Evol.* 126: 149–188.

Gribnau J, Hochedlinger K, Hata K, Li E, and Jaenisch R. 2003. Asynchronous replication timing of imprinted loci is independent of DNA methylation, but consistent with differential subnuclear localization. *Genes Dev.* 17: 759–773.

Grohmann L, Brennicke A, and Schuster W. 1992. The mitochondrial gene encoding ribosomal protein S12 has been translocated to the nuclear genome in *Oenothera*. *Nucleic Acids Res.* 20: 5641–5646.

Gropp A, and Winking H. 1981. Robertsonian translocations: cytology, meiosis, segregation patterns and biological consequences of heterozygosity. *Symp. Zool. Soc. Lond.* 47: 141–181.

Grossniklaus U, Spillane C, Page DR, and Kohler C. 2000. Genomic imprinting and seed development: endosperm formation with and without sex. *Curr. Opin. Plant Biol.* 4: 21–27.

Grun P. 1976. *Cytoplasmic Genetics and Evolution*. New York: Columbia University Press.

Gueiros-Filho FJ, and Beverley SM. 1997. Trans-kingdom transposition of the *Drosophila* element *mariner* within the protozoan *Leishmania*. *Science* 276: 1716–1719.

Guerrini F, Bucci S, Ragghianti M, Mancino G, Hotz H, Uzzell T, and Berger L. 1997. Genomes of two water frog species resist germ line exclusion in interspecies hybrids. *J. Exp. Zool.* 297: 163–176.

Gummere GR, McCormick PJ, and Bennett D. 1986. The influence of genetic background and the homologous chromosome *17* on *t*-haplotype transmission ratio distortion in mice. *Genetics* 114: 235–245.

Guo H, Karberg M, Long M, Jones III JP, Sullenger B, and Lambowitz AM. 2000. Group II introns designed to insert into therapeutically relevant DNA target sites in human cells. *Science* 289: 452–457.

Gutknecht J, Sperlich D, and Bachmann L. 1995. A species specific satellite DNA family of *Drosophila subsilvestris* appearing predominantly in B chromosomes. *Chromosoma* 103: 539–544.

Gutz H, and Leslie JF. 1976. Gene conversion: a hitherto overlooked parameter in population genetics. *Genetics* 83: 861–866.

Hackstein JHP, Hackstein P, Hochstenbach R, Hauschteck-Jungen E, and Beukeboom LW. 1996. Is the Y chromosome of *Drosophila* an evolved supernumerary chromosome? *BioEssays* 18: 317–323.

Hagemann AT, and Craig NL. 1993. Tn7 transposition creates a hotspot for homologous recombination at the transposon donor site. *Genetics* 133: 9–16.

Hagemann S, Haring E, and Pinsker W. 1996. Repeated horizontal transfer of *P* transposons between *Scaptomyza pallida* and *Drosophila bifasciata*. *Genetica* 98: 43–51.

Hager R, and Johnstone RA. 2003. The genetic basis of family conflict resolution in mice. *Nature* 421: 533–535.

Haig D. 1992. Intragenomic conflict and the evolution of eusociality. *J. Theor. Biol.* 156: 401–403.

Haig D. 1993a. Alternatives to meiosis: the unusual genetics of red algae, microsporidia, and others. *J. Theor. Biol.* 163: 15–31.

Haig D. 1993b. Genetic conflicts in human pregnancy. *Q. Rev. Biol.* 68: 495–532.

Haig D. 1993c. The evolution of unusual chromosomal systems in coccoids: extraordinary sex ratios revisited. *J. Evol. Biol.* 6: 69–77.

Haig D. 1993d. The evolution of unusual chromosomal systems in sciarid flies: intragenomic conflict and the sex ratio. *J. Evol. Biol.* 6: 249–261.

Haig D. 1994. Cohabitation and pregnancy-induced hypertension. *Lancet* 344: 1633–1634.

Haig D. 1996a. Gestational drive and the green-bearded placenta. *Proc. Natl. Acad. Sci. USA* 93: 6547–6551.

Haig D. 1996b. Do imprinted genes have few and small introns? *BioEssays* 18: 351–353.

Haig D. 1996c. The altercation of generations: genetic conflicts of pregnancy. *Am. J. Reprod. Immunol.* 35: 226–232.

Haig D. 1996d. Placental hormones, genomic imprinting, and maternal-fetal communication. *J. Evol. Biol.* 9: 357–380.

Haig D. 1997. Parental antagonism, relatedness asymmetries, and genomic imprinting. *Proc. Biol. Sci.* 264: 1657–1662.

Haig D. 1999a. Asymmetric relations: internal conflicts and the horror of incest. *Evol. Hum. Behav.* 20: 83–98.

Haig D. 1999b. Genetic conflict and the private life of *Peromyscus polionotus*. *Nat. Genet.* 22: 131.

Haig D. 1999c. What is a marmoset? *Am. J. Primatol.* 49: 285–296.

Haig D. 2000a. The kinship theory of genomic imprinting. *Annu. Rev. Ecol. Syst.* 31: 9–32.

Haig D. 2000b. Genomic imprinting, sex-biased dispersal, and social behavior. *Ann. N.Y. Acad. Sci.* 907: 149–163.

Haig D. 2002. *Kinship and Genomic Imprinting.* New Brunswick: Rutgers University Press.

Haig D. 2003. On intrapersonal reciprocity. *Evol. Hum. Behav.* 24: 418–425.

Haig D. 2004a. Evolutionary conflicts in pregnancy and calcium metabolism—a review. *Placenta* 25 (Supplement A): S10-S15.

Haig D. 2004b. Genomic imprinting and kinship: how good is the evidence? *Annu. Rev. Genet.* 38: 553–585.

Haig D, and Grafen A. 1991. Genetic scrambling as a defence against meiotic drive. *J. Theor. Biol.* 153: 531–558.

Haig D, and Graham C. 1991. Genomic imprinting and the strange case of the insulin-like growth factor II receptor. *Cell* 64: 1045–1046.

Haig D, and Trivers R. 1995. The evolution of parental imprinting: a review of hypotheses. In: *Genomic Imprinting,* eds. R Ohlsson, K Hall, and M Ritzen. Cambridge, England: Cambridge University Press, pp. 17–28.

Haig D, and Westoby M. 1989. Parent-specific gene expression and the triploid endosperm. *Am. Nat.* 134: 147–155.

Haig D, and Westoby M. 1991. Genomic imprinting in endosperm: its effects on seed development in crosses between species and between different ploidies of the same species, and its implications for the evolution of apomixis. *Philos Trans. R. Soc. Lond. B Biol. Sci.* 333: 1–13.

Haig D, and Wharton R. 2003. Prader-Willi syndrome and the evolution of human childhood. *Am. J. Hum. Biol.* 15: 1–10.

Hall DW. 2004. Meiotic drive and sex chromosome cycling. *Evolution* 58: 925–931.

Hall JG. 1999. Human diseases and genomic imprinting. *Genomic Imprinting: An Interdisciplinary Approach,* ed. R Ohlsson. Berlin: Springer-Verlag, pp. 119–132.

Hamilton WD. 1964a. The genetical evolution of social behaviour, I. *J. Theor. Biol.* 7: 1–16.

Hamilton WD. 1964b. The genetical evolution of social behaviour, II. *J. Theor. Biol.* 7: 17–52.

Hamilton WD. 1967. Extraordinary sex ratios. *Science* 156: 477–488.

Hamilton WD. 1972. Altruism and related phenomena, mainly in social insects. *Annu. Rev. Ecol. Syst.* 3: 193–232.

Hamilton WD. 1979. Wingless and fighting males in fig wasps and other insects. In: *Reproductive Competition, Mate Choice and Sexual Selection in Insects,* eds. MS Blum and NA Blum. New York: Academic Press, pp. 167–220.

Hamilton WD. 1993. Inbreeding in Egypt and in this book: a childish perspective. In: *The Natural History of Inbreeding and Outcrossing*, ed. NW Thornhill. Chicago: University of Chicago Press, pp. 429–450.

Hammer MF, and Silver LM. 1993. Phylogenetic analysis of the alpha-globin pseudogene-4 (Hba-ps4) locus in the house mouse species complex reveals a stepwise evolution of *t* haplotypes. *Mol. Biol. Evol.* 10: 971–1001.

Hammerberg C, and Klein J. 1975. Linkage disequilibrium between H-2 and *t* complex in chromosome 17 of the mouse. *Nature* 258: 296–299.

Han JS, and Boeke JD. 2004. A highly active synthetic mammalian retrotransposon. *Nature* 429: 314–318.

Han YN, Liu XG, Benny U, Kistler HC, and VanEtten HD. 2001. Genes determining pathogenicity to pea are clustered on a supernumerary chromosome in the fungal plant pathogen *Nectria haematococca*. *Plant J.* 25: 305–314.

Harada K, Yukuhiro K, and Mukai T. 1990. Transposition rates of movable genetic elements in *Drosophila melanogaster*. *Proc. Natl. Acad. Sci. USA* 87: 3248–3252.

Harden N, and Ashburner M. 1990. Characterization of the *FB-NOF* transposable element of *Drosophila melanogaster*. *Genetics* 126: 387–400.

Hartl DL. 1972. Population dynamics of sperm and pollen killers. *Theor. Appl. Genet.* 42: 81–88.

Hartl DL, and Hiraizumi Y. 1976. Segregation distortion after fifteen years. In: *The Genetics and Biology of Drosophila*, vol. 1b, eds. M Ashburner and E Novitski. New York: Academic Press, pp. 615–666.

Hassold T, Abruzzo M, Adkins K, Griffin D, Merrill M, Millie E, Saker D, Shen J, and Zaragoza M. 1996. Human aneuploidy: incidence, origin and etiology. *Environ. Mol. Mutagen.* 28: 167–175.

Hastings IM. 1992. Population genetic aspects of deleterious cytoplasmic genomes and their effect on the evolution of sexual reproduction. *Genet. Res.* 59: 215–225.

Hastings IM. 1994. Manifestations of sexual selection may depend on the genetic basis of sex determination. *Proc. R. Soc. Lond. B* 258: 83–87.

Hatchett JH, and Gallun RL. 1970. Genetics of the ability of the Hessian fly, *Mayetiola destructor*, to survive on wheats having different genes for resistance. *Ann. Entomol. Soc. Am.* 63: 1400–1407.

Hatta R, Ito K, Hosaki Y, Tanaka T, Tanaka A, Yamamoto M, Akimitsu K, and Tsuge T. 2002. A conditionally dispensable chromosome controls host-specific pathogenicity in the fungal plant pathogen *Alternaria alternata*. *Genetics* 161: 59–70.

Haugen P, Reeb V, Lutzoni F, and Bhattacharya D. 2004. The evolution of homing endonuclease genes and group I introns in nuclear rDNA. *Mol. Biol. Evol.* 21: 129–140.

Hauschteck-Jungen E. 1990. Postmating reproductive isolation and modification of the "sex ratio" trait in *Drosophila subobscura* induced by the sex chromosome gene arrangement A2+3+5+7. *Genetica* 83: 31–44.

Hauschteck-Jungen E, and Hartl DL. 1978. DNA distribution in spermatid nuclei of normal and segregation distorter males of *Drosophila melanogaster*. *Genetics* 89: 15–35.

Hauschteck-Jungen E, and Maurer B. 1976. Sperm dysfunction in sex-ratio males of *Drosophila subobscura*. *Genetica* 46: 459–477.

Heemert C. 1977. Somatic pairing and meiotic nonrandom disjunction in a pericentric inversion of *Hylemya antiqua* (Meigen). *Chromosoma* 59: 193–206.

Heerdt BG, Houston MA, Rediske JJ, and Augenlicht LH. 1996. Steady-state levels of mitochondrial messenger RNA species characterize a predominant pathway culminating in apoptosis and shedding of HT29 human colonic carcinoma cells. *Cell Growth Diff.* 7: 101–106.

Heilbuth JC. 2000. Lower species richness in dioecious clades. *Am. Nat.* 156: 221–241.

Heinlein M, Brattig T, and Kunze R. 1994. *In vivo* aggregation of maize *Activator* (*Ac*) transposase in nuclei of maize endosperm and petunia protoplasts. *Plant J.* 5: 705–714.

Helle W, Bolland HR, van Arendonk R, de Boer R, Schulten GGM, and Russel VM. 1978. Genetic evidence for biparental males in haplo-diploid predator mites (Acarina: Phytoseiidae). *Genetica* 49: 165–171.

Henikoff S, Ahmad K, and Malik HS. 2001. The centromere paradox: stable inheritance with rapidly evolving DNA. *Science* 293: 1098–1102.

Henikoff S, and Malik HS. 2002. Selfish drivers. *Nature* 417: 227.

Henriques-Gil N, Santos JL, and Giráldez R. 1982. Genotype-dependent effect of B-chromosomes on chiasma frequency in *Eyprepocnemis plorans* (Acrididae: Orthoptera). *Genetica* 59: 223–227.

Herbst EW, Fredga K, Frank SA, and Winkling W. 1978. Cytological identification of two X-chromosome types in the wood lemming (*Myopus schisticolor*). *Chromosoma* 69: 185–191.

Herman H, Lu M, Anggraini M, Sikora A, Chang Y, and Yoon BJ. 2003. *Trans* allele methylation and paramutation-like effects in mice. *Nat. Genet.* 34: 199–202.

Herrera JA, López-León MD, Cabrero J, Shaw MW, and Camacho JPM. 1996. Evidence for B chromosome drive suppression in the grasshopper *Eyprepocnemis plorans*. *Heredity* 76: 633–639.

Herrick G, and Seger J. 1999. Imprinting and paternal genome elimination in insects. In: *Genomic Imprinting: An Interdisciplinary Approach*, ed. R Ohlsson. Heidelberg: Springer-Verlag, pp. 41–71.

Herrmann BG, Koschorz B, Wertz K, McLaughlin KJ, and Kispert A. 1999. A protein kinase encoded by the *t* complex responder gene causes non-Mendelian inheritance. *Nature* 402: 141–146.

Herzing LBK, Cook EH, and Ledbetter DH. 2002. Allele-specific expression analysis by RNA-FISH demonstrates preferential maternal expression of UBE3A and imprint maintenance within 15q11-q13 duplications. *Hum. Mol. Genet.* 11: 1707–1718.

Hess RT, and Hoy MA. 1982. Microorganisms associated with the spider mite predator *Metaseiulus* (= *Typhlodromus*) *occidentalis:* electron microscope observations. *J. Invertebr. Pathol.* 40: 98–106.

Hewitt GM. 1973. Evolution and maintenance of B-chromosomes. *Chromosomes Today* 4: 351–369.

Hewitt GM. 1976. Meiotic drive for B-chromosomes in the primary oocytes of *Myrmeleotettix maculatus* (Orthoptera: Acrididae). *Chromosoma* 56: 381–391.

Hewitt GM. 1979. *Orthoptera: Grasshoppers and Crickets.* Berlin: Gebrüder Borntraeger.

Hewitt GM, and Brown FM. 1970. The B chromosome system in *Myrmeleotettix maculatus.* V. A steep cline in East Anglia. *Heredity* 25: 363–371.

Hewitt GM, and John B. 1967. The B chromosome system of *Myrmeleotettix maculatus* (Thunb.) III. The statistics. *Chromosoma* 21: 140–162.

Hewitt GM, and John B. 1970. The B chromosome system of *Myrmeleotettix maculatus* (Thunb.). IV. The dynamics. *Evolution* 24: 169–180.

Hewitt GM, and Ruscoe CNE. 1971. Changes in microclimate correlated with a cline for B-chromosomes in the grasshopper *Myrmeleotettix maculatus. J. Anim. Ecol.* 40: 753–765.

Hiatt EN, and Dawe RK. 2003a. Four loci on abnormal chromosome 10 contribute to meiotic drive in maize. *Genetics* 164: 699–709.

Hiatt EN, and Dawe RK. 2003b. The meiotic drive system on maize abnormal chromosome 10 contains few essential genes. *Genetica* 117: 67–76.

Hiatt EN, Kentner EK, and Dawe RK. 2002. Independently regulated neocentromere activity of two classes of tandem repeat arrays. *Plant Cell* 14: 407–420.

Hickey DA. 1982. Selfish DNA: a sexually-transmitted nuclear parasite. *Genetics* 101: 519–531.

Hickey DA, Wang S, and Magoulas C. 1994. Gene duplication, gene conversion and codon bias. In: *Non-neutral Evolution: Theories and Molecular Data,* ed. B Golding. New York: Chapman & Hall, pp. 199–207.

Hickey WA. 1970. Factors influencing the distortion of sex ratio in *Aedes aegypti. J. Med. Entomol.* 7: 727–735.

Hickey WA, and Craig Jr. GB. 1966a. Distortion of sex ratio in populations of *Aedes aegypti. Can. J. Genet. Cytol.* 8: 260–278.

Hickey WA, and Craig Jr. GB. 1966b. Genetic distortion of sex ratio in a mosquito, *Aedes aegypti. Genetics* 53: 1177–1196.

Higashiyama T, Noutoshi Y, Fujie M, and Yamada T. 1997. Zepp, a LINE-like retrotransposon accumulated in the *Chlorella* telomeric region. *EMBO J.* 16: 3715–3723.

Hillis DM, Moritz C, Porter CA, and Baker RJ. 1991. Evidence for biased gene conversion in concerted evolution of ribosomal DNA. *Science* 251: 308–310.

Hiom K, Melek M, and Gellert M. 1998. DNA transposition by the RAG1 and RAG2 proteins: a possible source of oncogenic translocations. *Cell* 94: 463–470.

Hiraizumi Y, Albracht JM, and Albracht BC. 1994. X-linked elements associated with negative segregation distortion in the *SD* system of *Drosophila melanogaster. Genetics* 138: 145–152.

Hirschorn R, Yang DR, Puck JM, Huie ML, Jiang C-K, and Kurlandsky LE. 1996. Spontaneous *in vivo* reversion to normal of an inherited mutation in a patient with adenosine deaminase deficiency. *Nat. Genet.* 13: 290–295.

Hoeh WR, Stewart DT, and Guttman SI. 2002. High fidelity of mitochondrial genome transmission under the doubly uniparental mode of inheritance in freshwater mussels (Bivalvia: Unionoidae). *Evolution* 56: 2252–2261.

Hoeh WR, Stewart DT, Saavedra C, Sutherland BW, and Zouros E. 1997. Phylogenetic evidence for role-reversals of gender-associated mitochondrial DNA in *Mytilus* (Bivalvia: Mytilidae). *Mol. Biol. Evol.* 14: 959–967.

Hoeh WR, Stewart DT, Sutherland BW, and Zouros E. 1996. Multiple origins of gender-associated mitochondrial DNA lineages in bivalves (Mollusca: Bivalvia). *Evolution* 50: 2276–2286.

Hoekstra HE, and Edwards SV. 2000. Multiple origins of XY female mice (genus *Akodon*): phylogenetic and chromosomal evidence. *Proc. Biol. Sci.* 267: 1825–1831.

Hoekstra HE, and Hoekstra JM. 2001. An unusual sex-determination system in South American field mice (genus *Akodon*): the role of mutation, selection, and meiotic drive in maintaining XY females. *Evolution* 55: 190–197.

Hoekstra RF. 1990. Evolution of uniparental inheritance of cytoplasmic DNA. In: *Organizational Constraints on the Dynamics of Evolution*, eds. J Maynard Smith and J Vida. Manchester: Manchester University Press, pp. 269–278.

Hoffmann AA, and Turelli M. 1997. Cytoplasmic incompatibility in insects. In: *Influential Passengers: Inherited Microorganisms and Arthropod Reproduction*, eds. SL O'Neill, AA Hoffmann, and JH Werren. Oxford: Oxford University Press, pp. 42–80.

Hoffmann RJ, Boore JL, and Brown WM. 1992. A novel mitochondrial genome organization for the blue mussel, *Mytilus edulis. Genetics* 131: 397–412.

Holm PB, and Rasmussen SW. 1980. Chromosome pairing, recombination nodules and chiasma formation in diploid *Bombyx* males. *Carlsberg Res. Comm.* 45: 483–548.

Holmes DS, and Bougourd SM. 1989. B-chromosome selection in *Allium schoenoprasum*. I. Natural populations. *Heredity* 63: 83–87.

Holmes DS, and Bougourd SM. 1991. B-chromosome selection in *Allium schoenoprasum* II. Experimental populations. *Heredity* 67: 117–122.

Holmes WG, and Sherman PW. 1982. The ontogeny of kin recognition in two species of ground squirrels. *Am. Zool.* 22: 491–517.

Hoogland C, and Biémont C. 1996. Chromosomal distribution of transposable elements in *Drosophila melanogaster:* test of the ectopic recombination model for maintenance of insertion site number. *Genetics* 144: 197–204.

Hopkin SP. 1997. *Biology of the Springtails (Insecta: Collembola).* New York: Oxford University Press.

Hotz H, Semlitsch RD, Gutmann E, Guex G-D, and Beerli P. 1999. Spontaneous heterosis in larval life-history traits of hemiclonal frog hybrids. *Proc. Natl. Acad. Sci. USA* 96: 2171–2176.

Houben A, Thompson JN, Ahne R, Leach CR, Verlin D, and Timmis JN. 1999. A monophyletic origin of the B chromosomes of *Brachycome dichromosomatica* (Asteraceae). *Plant Syst. Evol.* 219: 127–135.

Houben A, Verlin D, Leach CR, and Timmis JN. 2001. The genomic complexity of micro B chromosomes of *Brachycome dichromosomatica*. *Chromosoma* 110: 451–459.

Houck MA, Clark JB, Peterson KR, and Kidwell MG. 1991. Possible horizontal transfer of *Drosophila* genes by the mite *Proctolaelaps regalis*. *Science* 253: 1125–1129.

Houtchens K, and Lyttle TW. 2003. Responder (Rsp) alleles in the Segregation Distorter (SD) system of meiotic drive in *Drosophila* may represent a complex family of satellite repeat sequences. *Genetica* 117: 291–302.

Hoy MA. 1979. Parahaploidy of the "arrhenotokous" predator, *Metaseiulus occidentalis* (Acarina, Phytoseiidae) demonstrated by X-irradiation of males. *Entomol. Exp. Appl.* 26: 97–104.

Hoy MA. 1985. Recent advances in genetics and genetic improvement of the Phytoseiidae. *Annu. Rev. Entomol.* 30: 345–370.

Hoy MA, and Smilanick JM. 1979. Sex-pheromone produced by immature and adult females of the predatory mite, *Metaseiulus occidentalis*, Acrina, Phytoseiidae. *Entomol. Exp. Appl.* 26: 291–300.

Hsu FC, Wang CJ, Chen CM, Hu HY, and Chen CC. 2003. Molecular characterization of a family of tandemly repeated DNA sequences, TR-1, in heterochromatic knobs of maize and its relatives. *Genetics* 164: 1087–1097.

Huang S-W, Ardlie KG, and Yu H-T. 2001. Frequency and distribution of *t*-haplotypes in the Southeast Asian house mouse (*Mus musculus castaneus*) in Taiwan. *Mol. Ecol.* 10: 2349–2354.

Hughes JF, and Coffin JM. 2001. Evidence for genomic rearrangements mediated by human endogenous retroviruses during primate evolution. *Nat. Genet.* 29: 487–489.

Hurst GDD, and Werren JH. 2001. The role of selfish genetic elements in eukaryotic evolution. *Nat. Genet.* 2: 597–606.

Hurst LD. 1992. Is *Stellate* a relict meiotic driver? *Genetics* 130: 229–230.

Hurst LD. 1993. *scat*$^+$ is a selfish gene analogous to *Medea* of *Tribolium castaneum*. *Cell* 75: 407–408.

Hurst LD. 1994. Embryonic growth and the evolution of the mammalian Y chromosome. I. The Y as an attractor for selfish growth factors. *Heredity* 73: 223–232.

Hurst LD. 1995. Selfish genetic elements and their role in evolution: the evolution of sex and some of what that entails. *Philos. Trans. R. Soc. Lond. B Biol. Sci.* 349: 321–332.

Hurst LD. 1996. Further evidence consistent with *Stellate*'s involvement in meiotic drive. *Genetics* 142:641–643.

Hurst LD. 1997. Evolutionary theories of imprinting. In: *Genomic Imprinting*, eds. W Reik and MA Surani. Oxford: Oxford University Press, pp. 211–237.

Hurst LD, Atlan A, and Bengtsson BO. 1996a. Genetic conflicts. *Q. Rev. Biol.* 71: 317–364.

Hurst LD, and Hamilton WD. 1992. Cytoplasmic fusion and the nature of sexes. *Proc. R. Soc. Lond. B Biol. Sci.* 247: 189–194.

Hurst LD, McVean G, and Moore T. 1996b. Imprinted genes have few and small introns. *Nat. Genet.* 12: 234–237.

Hurst LD, and Pomiankowski A. 1991a. Causes of sex ratio bias may account for unisexual sterility in hybrids: a new explanation of Haldane's rule and related phenomena. *Genetics* 128: 841–858.

Hurst LD, and Pomiankowski A. 1991b. Maintaining Mendelism: might prevention be better than cure? *BioEssays* 13: 489–490.

Hurst LD, and Randerson JP. 2000. Transitions in the evolution of meiosis. *J. Evol. Biol.* 13: 466–479.

Hutchinson J. 1975. Selection of B chromosomes in *Secale cereale* and *Lolium perenne*. *Heredity* 34: 39–52.

Iida S, Morita Y, Choi JD, Park KI, and Hoshino A. 2004. Genetics and epigenetics in flower pigmentation associated with transposable elements in morning glories. *Adv. Biophys.* 38: 141–159.

Ims RA. 1987. Determinants of reproductive success in *Clethrionomys rufocanus*. *Ecology* 68: 1812–1818.

Ims RA, Bondrup-Nielsen S, Fredga K, and Fredriksson R. 1993. Habitat use and spatial distribution of the wood lemming *Myopus schistocolor*. In: *The Biology of Lemmings,* eds. NC Stenseth and RA Ims. London: Academic Press: 509–518.

Inoue K, Nakada K, Ogura A, Isobe K, Goto Y-i, Nonaka I, and Hayashi J-I. 2000. Generation of mice with mitochondrial dysfunction by introducing mouse mtDNA carrying a deletion into zygotes. *Nat. Genet.* 26: 176–181.

Intano GW, McMahan CA, McCarrey JR, Walter RB, McKenna AE, Matsumoto Y, MacInnes MA, Chen DJ, and Walter CA. 2002. Base excision repair is limited by different proteins in male germ cell nuclear extracts prepared from young and old mice. *Mol. Cell. Biol.* 22: 2410–2418.

Intano GW, McMahan CA, Walter RB, McCarrey JR, and Walter CA. 2001. Mixed spermatogenic germ cell nuclear extracts exhibit high base excision repair activity. *Nucleic Acids Res.* 29: 1366–1372.

Isles AR, Baum MJ, Ma D, Keverne EB, and Allen ND. 2001. Urinary odour preferences in mice. *Nature* 409: 783–784.

Isles AR, Baum MJ, Ma D, Szeto A, Keverne EB, and Allen ND. 2002. A possible role for imprinted genes in inbreeding avoidance and dispersal from the natal area in mice. *Proc. Biol. Sci.* 269: 665–670.

Isles AR, and Wilkinson LS. 2000. Imprinted genes, cognition and behaviour. *Trends Cogn. Sci.* 4: 309–318.

Itier J-M, Tremp GL, Leonard J-F, Multon M-C, Ret G, Scweighoffer F, Tocque B, Bluet-Pajot M-T, Cormier V, and Dautry F. 1998. Imprinted gene in postnatal growth role. *Nature* 393: 125–126.

Ito J, Ghosh A, Moreira LA, Wimmer EA, and Jacobs-Lorena M. 2002. Transgenic anopheline mosquitoes impaired in transmission of a malaria parasite. *Nature* 417: 452–455.

Ivics Z, Izsvák Z, Minter A, and Hackett PB. 1996. Identification of functional domains and evolution of Tc1-like transposable elements. *Proc. Natl. Acad. Sci. USA* 93: 5008–5013.

Iwasa Y, and Pomiankowski A. 2001. The evolution of X-linked genomic imprinting. *Genetics* 158: 1801–1809.

Jackson RC. 1960. Supernumerary chromosomes in *Haplopappas gracilis. Evolution* 15: 135.

Jacob S, McClintock MK, Zelano B, and Ober C. 2002. Paternally inherited HLA alleles are associated with women's choice of male odor. *Nat. Genet.* 30: 175–179.

Jacobs HT. 1991. Structural similarities between a mitochondrially encoded polypeptide and a family of prokaryotic respiratory toxins involved in plasmid maintenance suggest a novel mechanism for the evolutionary maintenance of mitochondrial DNA. *J. Mol. Evol.* 32: 333–339.

Jacobs MS, and Wade MJ. 2003. A synthetic review of the theory of gynodioecy. *Am. Nat.* 161: 837–851.

Jacquier A, and Dujon B. 1985. An intron-encoded protein is active in a gene conversion process that spreads an intron into a mitochondrial gene. *Cell* 41: 383–394.

Jaenike J. 1996. Sex-ratio meiotic drive in the *Drosophila quinaria* group. *Am. Nat.* 148: 237–254.

Jaenike J. 2001. Sex chromosome meiotic drive. *Annu. Rev. Ecol. Syst.* 32: 25–49.

Jahn CL, and Klobutcher LA. 2002. Genome remodeling in ciliated protozoa. *Annu. Rev. Microbiol.* 56: 489–520.

Jakob SS, Meister A, and Blattner FR. 2004. The considerable genome size variation of *Hordeum* species (Poaceae) is linked to phylogeny, life form, ecology and speciation rates. *Mol. Biol. Evol.* 21: 860–869.

James AC. 1992. "Sex ratio" meiotic drive in *Drosophila neotestacea.* PhD thesis, University of Rochester.

James AC, and Ballard JWO. 2003. Mitochondrial genotype affects fitness in *Drosophila simulans. Genetics* 164: 187–194.

James AC, and Jaenike J. 1990. "Sex ratio" meiotic drive in *Drosophila testacea. Genetics* 125: 651–656.

Jamilena M, Ruiz Rejón C, and Ruiz Rejón M. 1994. A molecular analysis of the origin of the *Crepis capillaris* B chromosome. *J. Cell Sci.* 107: 703–708.

Jansen J, Olsen AK, Wiger R, Naegeli H, de Boer P, van der Hoeven F, Holme JA, Brunborg G, and Mullenders L. 2001. Nucleotide excision repair in rat male germ cells: low level of repair in intact cells contrasts with high dual incision activity *in vitro. Nucleic Acids Res.* 29: 1791–1800.

Jarrell GH. 1995. A male-biased natal sex-ratio in inbred collared lemmings, *Dicrostonyx groenlandicus. Hereditas* 123: 31–37.

Jayalakshmi K, and Pantulu JV. 1982. Selective advantage of B-chromosomes in pearl millet. *Bionature* 2: 53–57.

Jayaram M, Grainge I, and Tribble G. 2002. Site-specific recombination by the Flp protein of *Saccharomyces cerevisiae.* In: *Mobile DNA II,* eds. NL Craig, R Craigie, M Gellert, and AM Lambowitz. Washington, D.C.: ASM Press, pp. 192–218.

Jeffreys AJ, Barber R, Bois P, Buard J, Dubrova YE, Grant G, Hollies CRH, May CA, Neumann R, Panayi M, Ritchie AE, Shone AC, Signer E, Stead JDH, and Tamaki K. 1999. Human minisatellites, repeat DNA instability and meiotic recombination. *Electrophoresis* 20: 1665–1675.

Jeffreys AJ, and Neumann R. 2002. Reciprocal crossover asymmetry and meiotic drive in a human recombination hot spot. *Nat. Genet.* 31: 267–271.

Jeffs P, and Ashburner M. 1991. Processed pseudogenes in *Drosophila. Proc. Biol. Sci.* 244: 151–159.

Jenkins G. 1986. Synaptonemal complex formation in hybrids of *Lolium temulentum* X *Lolium perenne* (L.) III. Tetraploid. *Chromosoma* 93: 413–419.

Jensen RE, Aiken Hobbs AE, Cerveny KL, and Sesaki H. 2000. Yeast mitochondrial dynamics: fusion, division, segregation, and shape. *Microsc. Res. Tech.* 51: 573–583.

Jenuth JP, Peterson AC, Fu K, and Shoubridge EA. 1996. Random genetic drift in the female germline explains the rapid segregation of mammalian mitochondrial DNA. *Nat. Genet.* 14: 146–151.

Jenuth JP, Peterson AC, and Shoubridge EA. 1997. Tissue-specific selection for different mtDNA genotypes in heteroplasmic mice. *Nat. Genet.* 16: 93–95.

Jiggins FM, Hurst GDD, and Majerus MEN. 1999. How common are meiotically driving sex chromosomes in insects? *Am. Nat.* 154: 481–483.

Jiménez MM, Romera F, González-Sánchez M, and Puertas MJ. 1997. Genetic control of the rate of transmission of rye B chromosomes. III. Male meiosis and gametogenesis. *Heredity* 78: 636–644.

Jockusch E. 1997. An evolutionary correlate of genome size change in plethodontid salamanders. *Proc. R. Soc. Lond. B* 264: 597–604.

John B. 1990. *Meiosis.* Cambridge, England: Cambridge University Press.

John B, and Freeman M. 1975. The cytogenetic structure of Tasmanian populations of *Phaulacridium vittatum*. *Chromosoma* 53: 283–293.

John B, and Hewitt GM. 1965. The B chromosome system of *Myrmeleotettix maculatus* (Thunb.), I. The mechanics. *Chromosoma* 16: 548–578.

John RM, and Surani MS. 2000. Genomic imprinting, mammalian evolution, and the mystery of egg-laying mammals. *Cell* 101: 585–588.

Johnson DT, and Croft BA. 1976. Laboratory study of the dispersal behavior of *Amblyseius fallacis* (Acarina: Phytoseiidae). *Ann. Entomol. Soc. Am.* 69: 1019–1023.

Johnson LJ, Koufopanou V, Goddard MR, Hetherington R, Schäfer SM, and Burt A. 2004. Population genetics of the wild yeast *Saccharomyces paradoxus*. *Genetics* 166: 43–52.

Johnson NA, and Wu C-I. 1992. An empirical test of the meiotic drive models of hybrid sterility: sex-ratio data from hybrids between *Drosophila simulans* and *D. sechellia*. *Genetics* 130: 507–511.

Johnson PG, and Brown GH. 1969. A comparison of the relative fitness of genotypes segregating for the *tw2* allele in laboratory stock and its possible effect on gene frequency in mouse populations. *Am. Nat.* 103: 5–21.

Johnston PG, and Cattanach BM. 1981. Controlling elements in the mouse. IV. Evidence of non-random X-inactivation. *Genet. Res.* 37: 151–160.

Johnston PG, Watson CM, Adams M, and Paull DJ. 2002. Sex chromosome elimination, X chromosome inactivation and reactivation in the southern brown bandicoot *Isodon obeselius* (Marsupialia: Peramelidae). *Cytogenet. Genome Res.* 99: 119–124.

Johri BM. 1963. Female gametophyte. In: *Recent Advances in the Embryology of Angiosperms,* ed. P Maheshwari. Delhi: University of Delhi, pp. 69–103.

Jones N, and Houben A. 2003. B chromosomes in plants: escapees from the A chromosome genome? *Trends Plant Sci.* 8: 417–423.

Jones RN. 1975. B chromosome systems in flowering plants and animal species. *Int. Rev. Cytol.* 40: 1–100.

Jones RN. 1985. Are B chromosomes "selfish"? In: *The Evolution of Genome Size,* ed. T Cavalier-Smith. Chichester: John Wiley & Sons, pp. 397–425.

Jones RN. 1991. B-chromosome drive. *Am. Nat.* 137: 430–442.

Jones RN. 1995. B chromosomes in plants. *New Phytol.* 131: 411–434.

Jones RN, and Puertas MJ. 1993. The B-chromosomes of rye (*Secale cereale* L.). In: *Frontiers in Plant Science Research,* eds. KK Dhir and TS Sareen. Delhi: Bhagwati Enterprises, pp. 81–112.

Jones RN, and Rees H. 1967. Genotypic control of chromosome behaviour in rye. XI. The influence of B chromosomes on meiosis. *Heredity* 22: 333–347.

Jones RN, and Rees H. 1982. *B Chromosomes.* London: Academic Press.

Jordan IK, Matyunina LV, and McDonald JF. 1999. Evidence for the recent horizontal transfer of long terminal repeat retrotransposon. *Proc. Natl. Acad. Sci. USA* 96: 12621–12625.

Jordan IK, Rogozin IB, Glazko GV, and Koonin EV. 2003. Origin of a substantial fraction of human regulatory sequences from transposable elements. *Trends Genet.* 19: 68–72.

Just W, Rau W, Vogel W, Akhverdyan M, Fredga K, Graves JAM, and Lyapunova E. 1995. Absence of SRY in species of the vole *Ellobius*. *Nat. Genet.* 11: 117–118.

Kalendar R, Tanskanen J, Immonen S, Nevo E, and Schulman A. 2000. Genome evolution of wild barley (*Hordeum spontaneum*) by *BARE*-1 retrotransposon dynamics in response to sharp microclimate divergence. *Proc. Natl. Acad. Sci. USA* 97: 6603–6607.

Kalendar R, Vicient CM, Peleg O, Anamthawat-Jonsson K, Bolshoy A, and Schulman AH. 2004. Large retrotransposon derivatives: abundant, conserved but nonautonomous retroelements of barley and related genomes. *Genetics* 166: 1437–1450.

Kamps TL, McCarty DR, and Chase CD. 1996. Gametophyte genetics in *Zea mays* L.: dominance of a restoration-of-fertility allele (*Rf3*) in diploid pollen. *Genetics* 142: 1001–1007.

Kaneda H, Hayashi J-I, Takahama S, Taya C, Fischer Lindahl K, and Yonekawa H.

1995. Elimination of paternal mitochondrial DNA in intraspecific crosses during early mouse embryogenesis. *Proc. Natl. Acad. Sci. USA* 92: 4542–4546.

Kaneko-Ishino T, Kohda T, and Ishino F. 2003. The regulation and biological significance of genomic imprinting in mammals. *J. Biochem.* 133: 699–711.

Kapitonov VV, and Jurka J. 2001. Rolling-circle transposons in eukaryotes. *Proc. Natl. Acad. Sci. USA* 98: 8714–8719.

Kaplan N, Darden T, and Langley CH. 1985. Evolution and extinction of transposable elements in Mendelian populations. *Genetics* 109: 459–480.

Karakashian S, and Milkman R. 1967. Colony fusion compatibility types in *Botryllus schlosseri*. *Biol. Bull.* 133: 473.

Karamysheva TV, Andreenkova OV, Bochkaerev MN, Borissov YM, Bogdanchikova N, Borodin PM, and Rubtsov NB. 2002. B chromosomes of Korean field mouse *Apodemus peninsulae* (Rodentia. Murinae) analysed by microdissection and FISH. *Cytogenet. Genome Res.* 96: 154–160.

Kaszas E, and Birchler JA. 1998. Meiotic transmission rates correlate with physical features of rearranged centromeres in maize. *Genetics* 150: 1683–1692.

Kathariou S, and Spieth PT. 1982. Spore killer polymorphism in *Fusarium moniliforme*. *Genetics* 102: 19–24.

Kato YTA. 1976. Cytological studies of maize (*Zea mays* L.) and teosinte (*Zea mexicana* Schrader Kuntze) in relation to their origin and evolution. *Mass. Agric. Exp. Stn. Bull.* 635: 1–185.

Kaul MLH. 1988. *Male Sterility in Higher Plants*. Berlin: Springer-Verlag.

Kawano S, and Kuroiwa T. 1989. Transmission pattern of mitochondrial DNA during plasmodium formation in *Physarum polycephalum*. *J. Gen. Microbiol.* 135: 1559–1566.

Kawano S, Takano H, Mori K, and Kuroiwa T. 1991. A mitochondrial plasmid that promotes mitochondrial fusion in *Physarum polycephalum*. *Protoplasma* 160: 167–169.

Kayano H. 1957. Cytogenetic studies in *Lilium callosum* III. Preferential segregation of a supernumerary chromosome. *Proc. Jap. Acad.* 33: 553–558.

Kayano H. 1971. Accumulation of B chromosomes in the germ-line of *Locusta migratoria*. *Heredity* 27: 119–123.

Kazazian Jr. HH. 1999. An estimated frequency of endogenous insertional mutations in humans. *Nat. Genet.* 22: 130.

Kazazian Jr. HH, and Moran JV. 1998. The impact of L1 retrotransposons on the human genome. *Nat. Genet.* 19: 19–24.

Ke N, and Voytas DF. 1997. High frequency cDNA recombination of the *Saccharomyces* retrotransposon Ty5: the LTR mediates formation of tandem elements. *Genetics* 147: 545–556.

Keegan-Rogers V. 1984. Unfamiliar-female advantage among clones of unisexual fish (*Poeciliopsis*, Poeciliidae). *Copeia* 1984: 169–174.

Kempken F, and Kück U. 1998. Transposons in filamentous fungi–facts and perspectives. *BioEssays* 20: 652–659.

Kenchington E, MacDonald B, Cao L, Tsagkarakis D, and Zouros E. 2002. Genetics of mother-dependent sex ratio in blue mussels (*Mytilus* spp) and implications for double uniparental inheritance of mitochondrial DNA. *Genetics* 161: 1579–1588.

Kennison JA, and Southworth JW. 2002. Transvection in *Drosophila*. *Adv. Genet.* 46: 399–420.

Kermicle JL. 1970. Dependence of the R-mottled aleurone phenotype in maize on mode of sexual transmission. *Genetics* 66: 69–85.

Kesler SR, Blasey CM, Brown WE, Yankowitz J, Zeng SM, Bender BG, and Reiss AL. 2003. Effects of X-monosomy and X-linked imprinting on superior temporal gyrus morphology in Turner syndrome. *Biol. Psychiatry* 54: 636–646.

Ketting RF, Haverkamp THA, van Luenen HGAM, and Plasterk RHA. 1999. *mut-7* of *C. elegans*, required for transposon silencing and RNA interference, is a homolog of Werner syndrome helicase and RNaseD. *Cell* 99: 133–141.

Keverne EB, Fundele R, Narasimha M, Barton SC, and Surani MA. 1996. Genomic imprinting and the differential roles of parental genomes in brain development. *Dev. Brain Res.* 92: 91–100.

Khajavi M, Tari AM, Patel NB, Tsuji K, Siwak DR, Meistrich ML, Terry NHA, and Ashizawa T. 2001. "Mitotic drive" of expanded CTG repeats in myotonic dystrophy type 1 (DM1). *Hum. Mol. Genet.* 10: 855–863.

Kidwell MG. 1983. Evolution of hybrid dysgenesis determinants in *Drosophila melanogaster*. *Proc. Natl. Acad. Sci. USA* 80: 1655–1659.

Kidwell MG, Kimura K, and Black DM. 1988. Evolution of hybrid dysgenesis potential following *P* element contamination in *Drosophila melanogaster*. *Genetics* 119: 815–828.

Kidwell MG, and Lisch DR. 2001. Transposable elements, parasitic DNA, and genome evolution. *Evolution* 55: 1–24.

Kikudome GY. 1959. Studies on the phenomenon of preferential segregation in maize. *Genetics* 44: 815–831.

Killian JK, Nolan CM, Stewart N, Munday BL, Andersen NA, Nicol S, and Jirtle RL. 2001. Monotreme *IGF2* expression and ancestral origin of genomic imprinting. *J. Exp. Zool.* 291: 201–212.

Killian JK, Nolan CM, Wylie AA, Li T, Vu TH, Hoffman AR, and Jirtle RL. 2001. Divergent evolution in *M6P/IGF2R* imprinting from the Jurassic to the Quaternary. *Hum. Mol. Genet.* 10: 1721–1728.

Kim A, Terzian C, Santamaria P, Pélisson A, Prud'homme N, and Bucheton A.

1994. Retroviruses in invertebrates: the *gypsy* retrotransposon is apparently an infectious retrovirus of *Drosophila melanogaster. Proc. Natl. Acad. Sci. USA* 91: 1285–1289.

Kim HG, Meinhardt LW, Benny U, and Kistler HC. 1995. Nrs1, a repetitive element linked to pisatin demethylase genes on a dispensable chromosome of *Nectria haematococca. Mol. Plant Microbe Interact.* 8: 524–531.

Kim JM, Vanguri S, Boeke JD, Gabriel A, and Voytas DF. 1998. Transposable elements and genome organization: a comprehensive survey of retrotransposons revealed by the complete *Saccharomyces cerevisiae* genome sequence. *Genome Res.* 8: 464–478.

Kimura M, and Kayano H. 1961. The maintenance of supernumerary chromosomes in wild populations of *Lilium callosum* by preferential segregation. *Genetics* 46: 1699–1712.

Kirk D, and Jones RN. 1970. Nuclear genetic activity in B chromosomes rye, in terms of quantitative interrelationships between nuclear protein, nuclear RNA and histone. *Chromosoma* 31: 241–254.

Kitchin RM. 1975. Intranuclear destruction of heterochromatin in two species of armored scale insects. *Genetica* 45: 227–235.

Klein J, Sipos P, and Figueroa F. 1984. Polymorphism of *t*-complex genes in European wild mice. *Genet. Res.* 44: 39–46.

Kloeckener-Gruissem B, and Freeling M. 1995. Transposon-induced promoter scrambling: a mechanism for the evolution of new alleles. *Proc. Natl. Acad. Sci. USA* 92: 1836–1840.

Knipling EP. 1979. *The Basic Principles of Insect Population Suppression and Management.* Agriculture Handbook No. 512. Washington, D.C.: U.S. Department of Agriculture.

Kobayashi I. 1998. Selfishness and death: raison d'être of restriction, recombination and mitochondria. *Trends Genet.* 14: 368–374.

Kohn JR. 1989. Sex ratio, seed production, biomass allocation, and the cost of male function in *Cucurbita foetidissima* HBK (Cucurbitaceae). *Evolution* 43: 1424–1434.

Kohn JR, and Biardi JE. 1995. Outcrossing rates and inferred levels of inbreeding depression in gynodioecious *Cucurbita foetidissima* (Cucurbitaceae). *Heredity* 75: 77–83.

Koizuka N, Imai R, Fujimoto H, Hayakawa T, Kimura Y, Kohno-Murase J, Sakai T, Kawasaki S, and Imamura J. 2003. Genetic characterization of a pentatricopeptide repeat protein gene, *orf687,* that restores fertility in the cytoplasmic male-sterile Kosena radish. *Plant J.* 34: 407–415.

Komaru A, Kawagishi T, and Konishi K. 1998. Cytological evidence of spontaneous androgenesis in the freshwater clam *Corbicula leana* Prime. *Dev. Genes Evol.* 208: 46–50.

Komaru A, and Konishi K. 1999. Non-reductional spermatozoa in three shell color types of the freshwater clam *Corbicula fluminea* in Taiwan. *Zoolog. Sci.* 16: 105–108.

Komaru A, Ookubo K, and Kiyomoto M. 2000. All meiotic chromosomes and both centrosomes at spindle pole in the zygotes discarded as two polar bodies in clam *Corbicula leana:* unusual polar body formation observed by antitubulin immunofluorescence. *Dev. Genes Evol.* 210: 263–269.

Komdeur J. 1996. Facultative sex ratio bias in the offspring of Seychelles warblers. *Proc. R. Soc. Lond. B* 263: 661–666.

Komdeur J, Daan S, Tinbergen J, and Mateman C. 1997. Extreme adaptive modification in sex ratio of the Seychelles warbler's eggs. *Nature* 385: 522–525.

Konovalov AA. 1995. Two gametophytic genes linked to a gene controlling recombination in sugar beet (*Beta vulgaris* L.). *Hereditas* 123: 245–250.

Koop BF, Baker RJ, and Genoways HH. 1983. Numerous chromosomal polymorphisms in a natural population of rice rats (*Oryzomys*, Cricetidae). *Cytogenet. Cell Genet.* 35: 131–135.

Koufopanou V, and Burt A. 2005. Degeneration and domestication of a selfish gene in yeast: molecular evolution versus site-directed mutagenesis. *Mol. Biol. Evol.:* (in press).

Koufopanou V, Goddard MR, and Burt A. 2002. Adaptation for horizontal transfer in a homing endonuclease. *Mol. Biol. Evol.* 19: 239–246.

Kovacevic M, and Schaeffer SW. 2000. Molecular population genetics of *X*-linked genes in *Drosophila pseudoobscura*. *Genetics* 156: 155–172.

Krackow S. 1995. Potential mechanisms for sex-ratio adjustment in mammals and birds. *Biol. Rev.* 70: 225–241.

Kraytsberg Y, Schwartz M, Brown TA, Ebralidse K, Kunz WS, Clayton DA, Vissing J, and Khrapko K. 2004. Recombination of human mitochondrial DNA. *Science* 304: 981.

Krebs CJ, and Myers JH. 1974. Population cycles in small mammals. *Adv. Ecol. Res.* 8: 267–399.

Kremer H, Hennig W, and Dijkhof R. 1986. Chromatin organization in the male germ line of *Drosophila hydei*. *Chromosoma* 94: 147–161.

Kruuk LEB, Clutton-Brock TH, Albon SD, Pemberton JM, and Guinness FE. 1999. Population density affects sex ratio variation in red deer. *Nature* 399: 459–461.

Kubai DF. 1982. Meiosis in *Sciara coprophila:* structure of the spindle and chromosome behavior during the first meiotic division. *J. Cell Biol.* 93: 655–669.

Kubota S, Takano J-i, Tsuneishi R, Kobayakawa S, Fujikawa N, Nabeyama M, and

Kohno S-i. 2001. Highly repetitive DNA families restricted to germ cells in a Japanese hagfish (*Eptatretus burgeri*): a hierarchical and mosaic structure in eliminated chromosomes. *Genetica* 111: 319–328.

Kunkel TA. 1985. The mutational specificity of DNA polymerases-α and -γ during *in vitro* DNA synthesis. *J. Biol. Chem.* 260: 12866–12874.

Kunz BA, and Haynes RH. 1981. Phenomenology and genetic control of mitotic recombination in yeast. *Annu. Rev. Genet.* 15: 57–89.

Kunze R, Starlinger P, and Schwartz D. 1988. DNA methylation of the maize transposable element *Ac* interferes with its transcription. *Mol. Gen. Genet.* 214: 325–327.

Kunze R, and Weil CF. 2002. The *hAT* and CACTA superfamilies of plant transposons. In: *Mobile DNA II*, eds. NL Craig, R Craigie, M Gellert, and AM Lambowitz. Washington, D.C.: ASM Press, pp. 565–610.

Kusano A, Staber C, Chan HYE, and Ganetzky B. 2003. Closing the (Ran)GAP on segregation distortion in *Drosophila*. *BioEssays* 25: 108–115.

Kusano A, Staber C, and Ganetzky B. 2001. Nuclear mislocalization of enzymatically active RanGAP causes segregation distortion in *Drosophila*. *Dev. Cell* 1: 351–361.

Kusano A, Staber C, and Ganetzky B. 2002. Segregation distortion induced by wild-type RanGAP in *Drosophila*. *Proc. Natl. Acad. Sci. USA* 99: 6866–6870.

Ladoukakis ED, Saavedra C, Magoulas A, and Zouros E. 2002. Mitochondrial DNA variation in a species with two mitochondrial genomes: the case of *Mytilus galloprovincialis* from the Atlantic, the Mediterranean and the Black Sea. *Mol. Ecol.* 11: 755–769.

Ladoukakis ED, and Zouros E. 2001. Direct evidence for homologous recombination in mussel (*Mytilus galloprovincialis*) mitochondrial DNA. *Mol. Biol. Evol.* 18: 1168–1175.

Lahn BT, Pearson NM, and Jegalian K. 2001. The human Y chromosome, in the light of evolution. *Nat. Rev. Genet.* 2: 207–216.

Lal SK, and Hannah LC. 1999. Maize transposable element *Ds* is differentially spliced from primary transcripts in endosperm and suspension cells. *Biochem. Biophys. Res. Commun.* 261: 798–801.

Lamb BC. 1984. The properties of meiotic gene conversion important in its effects on evolution. *Heredity* 53: 113–138.

Lamb BC. 1998. Gene conversion disparity in yeast: its extent, multiple origins, and effects on allele frequencies. *Heredity* 80: 538–552.

Lampe DJ, Akerley BJ, Rubin EJ, Mekalanos JJ, and Robertson HM. 1999. Hyperactive transposase mutants of the *Himar1 mariner* transposon. *Proc. Natl. Acad. Sci. USA* 96: 11428–11433.

Lampe DJ, Grant TE, and Robertson HM. 1998. Factors affecting transposition of the *Himar mariner* transposon *in vitro*. *Genetics* 149: 179–187.

LaMunyon CW, and Ward S. 1997. Increased competitiveness of nematode sperm bearing the male X chromosome. *Proc. Natl. Acad. Sci. USA* 94: 185–189.

Lander ES, Linton LM, Birren B, Nusbaum C, and (+ 239 others). 2001. Initial sequencing and analysis of the human genome. *Nature* 409: 860–921.

Lang BF, Gray MW, and Burger G. 1999. Mitochondrial genome evolution and the origin of eukaryotes. *Annu. Rev. Genet.* 33: 351–397.

Langdon T, Seago C, Jones RN, Ougham H, Thomas H, Forster JW, and Jenkins G. 2000. *De novo* evolution of satellite DNA on the rye B chromosome. *Genetics* 154: 869–884.

Langin T, Capy P, and Daboussi MJ. 1995. The transposable element *Impala,* a fungal member of the *Tc1-mariner* superfamily. *Mol. Gen. Genet.* 246: 19–28.

Langley CH, Montgomery EA, Hudson R, Kaplan N, and Charlesworth B. 1988. On the role of unequal exchange in the containment of transposable element copy number. *Genet. Res.* 52: 223–235.

Laporte V, Viard F, Bena G, Valero M, and Cuguen J. 2001. The spatial structure of sexual and cytonuclear polymorphism in the gynodioecious *Beta vulgaris* ssp. *maritima:* I/ at a local scale. *Genetics* 157: 1699–1710.

LaSalle JM, and Lalande M. 1996. Homologous association of oppositely imprinted chromosomal domains. *Science* 272: 725–728.

Laser KD, and Lersten NR. 1972. Anatomy and cytology of microsporogenesis in cytoplasmic male sterile angiosperms. *Bot. Rev.* 38: 425–454.

Lathe III WC, and Eickbush TH. 1997. A single lineage of R2 retrotransposable elements is an active, evolutionarily stable component of the *Drosophila* rDNA locus. *Mol. Biol. Evol.* 14: 1232–1241.

Le Bras S, Cohen-Tannoudji M, Kress C, Vandormael-Pournin S, Babinet C, and Baldacci P. 2000. BALB/c alleles at modifier loci increase the severity of the maternal effect of the "DDK syndrome." *Genetics* 154: 803–811.

Leach CR, Donald TM, Franks TK, Spiniello SS, and Hanrahan CF. 1995. Organisation and origin of a B chromosome centromeric sequence from *Brachycome dichromosomatica. Chromosoma* 103: 708–714.

Leamy LJ, Meagher S, Taylor S, Carroll L, and Potts WK. 2001. Size and morphometric characters in mice: their association with inbreeding and *t*-haplotype. *Evolution* 55: 2333–2341.

Leaver MJ. 2001. A family of *Tc1*-like transposons from the genomes of fishes and frogs: evidence for horizontal transmission. *Gene* 271: 203–214.

Leblanc O, Pointe C, and Hernandez M. 2002. Cell cycle progression during endosperm development in *Zea mays* depends on parental dosage effects. *Plant J.* 32: 1057–1086.

Leblanc P, Desset S, Giorgi F, Taddei AR, Fausto AM, Mazzini M, Dastugue B, and Vaury C. 2000. Life cycle of an endogenous retrovirus, *ZAM,* in *Drosophila melanogaster. J. Virol.* 74: 10658–10669.

Leblon A. 1972a. Mechanism of gene conversion in *Ascobolus immersus* I. Existence of a correlation between the origin of mutants induced by different mutagens and their conversion spectrum. *Mol. Gen. Genet.* 115: 36–48.

Leblon A. 1972b. Mechanism of gene conversion in *Ascobolus immersus* II. The relationships between the genetic alterations in b_1 or b_2 mutants and their conversion spectrum. *Mol. Gen. Genet.* 116: 322–335.

Leblon G. 1979. Intragenic suppression at the *b2* locus in *Ascobolus immersus*. II. Characteristics of the mutations in groups *A* and *E*. *Genetics* 92: 1093–1106.

Lee CC, Beall EL, and Rio DC. 1998. DNA binding by the KP repressor protein inhibits P-element transposase activity *in vitro*. *EMBO J.* 17: 4166–4174.

Lee CC, Mul YM, and Rio DC. 1996. The *Drosophila* P-element KP repressor protein dimerizes and interacts with multiple sites on P-element DNA. *Mol. Cell. Biol.* 16: 5616–5622.

Lee J, and Jaenisch R. 1997. The (epi)genetic control of mammalian X-chromosome inactivation. *Curr. Opin. Genet. Dev.* 7: 274–289.

Lee S-H, Motomura T, and Ichimura T. 2002. Light and electron microscopic observations of preferential destruction of chloroplast and mitochondrial DNA at early male gametogenesis of the anisogamous green alga *Derbesia tenuissima* (Chlorophyta). *J. Phycol.* 38: 534–542.

Lee SJ, and Warmke HE. 1979. Organelle size and number in fertile and T-cytoplasmic male-sterile corn. *Am. J. Bot.* 66: 141–148.

Lefebvre L, Viville S, Barton SC, Ishino F, Keverne EB, and Surani MA. 1998. Abnormal maternal behaviour and growth retardation associated with loss of the imprinted gene *Mest*. *Nat. Genet.* 20: 163–169.

Lefevre G, and Jonsson UB. 1962. Sperm transfer, storage, displacement, and utilization in *Drosophila melanogaster*. *Genetics* 47: 1719–1736.

Leigh Brown AJ, and Moss JE. 1987. Transposition of the I element and *copia* in a natural population of *Drosophila melanogaster*. *Genet. Res.* 49: 121–128.

Leigh Jr. EG. 1970. Natural selection and mutability. *Am. Nat.* 104: 301–305.

Leigh Jr. EG. 1973. The evolution of mutation rates. *Genetics* 73 (Suppl.): 1–18.

LeMaire-Adkins R, and Hunt PA. 2000. Nonrandom segregation of the mouse univalent X chromosome: evidence of spindle-mediated meiotic drive. *Genetics* 156: 775–783.

Lenington S. 1991. The *t* complex: a story of genes, behavior, and populations. *Adv. Study Behav.* 20: 51–86.

Lenington S, Coopersmith C, and Williams J. 1992. Genetic basis of mating preferences in wild house mice. *Am. Zool.* 32: 40–47.

Lenington S, Coopersmith CB, and Erhart M. 1994. Female preference and variability among t-haplotypes in wild house mice. *Am. Nat.* 143: 766–784.

Lenington SL, Drickamer LC, Erhart ME, and Robinson AS. 1996. Genetic basis for male aggression and survivorship in wild house mice (*Mus domesticus*). *Aggr. Behav.* 22: 135–145.

Leroi AM, Koufopanou V, and Burt A. 2003. Cancer selection. *Nat. Rev. Cancer* 3: 226–231.

Leslie JF, and Vrijenhoek RC. 1978. Genetic dissection of clonally inherited genomes of *Poeciliopsis*. I. Linkage analysis and preliminary assessment of deleterious gene loads. *Genetics* 90: 801–811.

Leslie JF, and Vrijenhoek RC. 1980. Consideration of Müller's ratchet mechanism through studies of genetic linkage and genomic compatibilities in clonally reproducing *Poeciliopsis*. *Evolution* 34: 1105–1115.

Levin DA, Palestis BG, Jones RN, and Trivers R. 2005. Phyletic hot spots for B chromosomes in angiosperms. *Evolution* 59: 962–969.

Levin HL. 2002. Newly identified retrotransposons of the Ty3/gypsy class in fungi, plants, and vertebrates. In: *Mobile DNA II,* eds. NL Craig, R Craigie, M Gellert, and AM Lambowitz. Washington, D.C.: ASM Press, pp. 684–701.

Levin JG, and Rosenak MJ. 1976. Synthesis of murine leukemia virus proteins associated with virions assembled in actinomycin D-treated cells: evidence for persistence of viral messenger RNA. *Proc. Natl. Acad. Sci. USA* 73: 1154–1158.

Levings III CS. 1990. The Texas cytoplasm of maize: cytoplasmic male sterility and disease susceptibility. *Science* 250: 942–947.

Levis RW, Ganesan R, Houtchens K, Tolar LA, and Sheen FM. 1993. Transposons in place of telomeric repeats at a *Drosophila* telomere. *Cell* 75: 1083–1093.

Lewis D. 1941. Male sterility in natural populations of hermaphrodite plants: the equilibrium between females and hermaphrodites to be expected with different types of inheritance. *New Phytol.* 40: 56–63.

Li LL, Keverne EB, Aparicio SA, Ishino F, Barton SC, and Surani MA. 1999. Regulation of maternal behavior and offspring growth by paternally expressed *Peg3*. *Science* 284: 330–333.

Li X-Q, Jean M, Landry BS, and Brown GG. 1998. Restorer genes for different forms of *Brassica* cytoplasmic male sterility map to a single nuclear locus that modifies transcripts of several mitochondrial genes. *Proc. Natl. Acad. Sci. USA* 95: 10032–10037.

Lifschytz E, and Lindsley DL. 1972. The role of X-chromosome inactivation during spermatogenesis. *Proc. Natl. Acad. Sci. USA* 69: 182–186.

Lim JK, and Simmons MJ. 1994. Gross chromosome rearrangements mediated by transposable elements in *Drosophila melanogaster. BioEssays* 16: 269–275.

Lin B-Y. 1978. Regional control of nondisjunction of B chromosome in maize. *Genetics* 90: 613–627.

Lin B-Y. 1979. Two new *B-10* translocations involved in the control of nondisjunction of the B chromosome in maize. *Genetics* 92: 931–945.

Liu H-P, Mitton JB, and Wu S-K. 1996. Paternal mitochondrial DNA differentiation far exceeds maternal mitochondrial DNA and allozyme differentiation in the freshwater mussel, *Anodonta grandis grandis*. *Evolution* 50: 952–957.

Liu W-S, Eriksson L, and Fredga K. 1998. XY sex reversal in the wood lemming is associated with deletion of Xp[21–23] as revealed by chromosome microdissection and fluorescence *in situ* hybridization. *Chromosome Res.* 6: 379–383.

Liu W-S, Nordqvist K, Lau Y-FC, and Fredga K. 2001. Characterization of the Xp21-23 region in the wood lemming, a region involved in XY sex reversal. *J. Exp. Zool.* 290: 551–557.

Lively CM, and Johnson SG. 1994. Brooding and the evolution of parthenogenesis: strategy models and evidence from aquatic invertebrates. *Proc. Biol. Sci.* 256: 89–95.

Lizarralde M, Bianchi NO, and Merani M. 1982. Cytogenetics of South American akodon rodents (Cricetidae). VII. Origin of sex chromosome polymorphism in *Akodon azarae*. *Cytologia* 47: 183–193.

Lloyd DG. 1976. The transmission of genes via pollen and ovules in gynodioecious angiosperms. *Theor. Popul. Biol.* 9: 299–316.

Loegering WQ, and Sears ER. 1963. Distorted inheritance of stem-rust resistance of timstein wheat caused by a pollen-killing gene. *Can. J. Genet. Cytol.* 5: 65–72.

Lohe AR, and Hartl DL. 1996. Autoregulation of *mariner* transposase activity by overproduction and dominant-negative complementation. *Mol. Biol. Evol.* 13: 549–555.

Lohe AR, Hilliker AJ, and Roberts PA. 1993. Mapping simple repeated DNA sequences in heterochromatin of *Drosophila melanogaster*. *Genetics* 134: 1149–1174.

Lohe AR, Moriyama EN, Lidholm D-A, and Hartl DL. 1995. Horizontal transmission, vertical inactivation, and stochastic loss of *mariner*-like transposable elements. *Mol. Biol. Evol.* 12: 62–72.

Longley AE. 1945. Abnormal segregation during megasporogenesis in maize. *Genetics* 30: 100–113.

López-León MD, Cabrero J, and Camacho JPM. 1991. Meiotic drive against an autosomal supernumerary segment promoted by the presence of a B chromosome in females of the grasshopper *Eyprepocnemis plorans*. *Chromosoma* 100: 282–287.

López-León MD, Cabrero J, and Camacho JPM. 1992a. Male and female segregation distortion for heterochromatic supernumerary segments on the S(8) chromosome of the grasshopper *Chorthippus jacobsi*. *Chromosoma* 101: 511–516.

López-León MD, Cabrero J, Camacho JPM, Cano MI, and Santos JL. 1992b. A widespread B chromosome polymorphism maintained without apparent drive. *Evolution* 46: 529–539.

López-León MD, Cabrero J, Pardo MC, Viseras E, Camacho JPM, and Santos JL.

1993. Generating high variability of B chromosomes in *Eyprepocnemis plorans* (grasshopper). *Heredity* 71: 352–362.

López-León MD, Pardo MC, Cabrero J, and Camacho JPM. 1992c. Random mating and absence of sexual selection for B chromosomes in two natural populations of the grasshopper *Eyprepocnemis plorans*. *Heredity* 69: 558–561.

López S, and Domínguez CA. 2003. Sex choice in plants: facultative adjustment of the sex ratio in the perennial herb *Begonia gracilis*. *J. Evol. Biol.* 16: 1177–1185.

Loppin B, Berger F, and Couble P. 2001. Paternal chromosome incorporation into the zygote nucleus is controlled by *maternal haploid* in Drosophila. *Dev. Biol.* 231: 383–396.

Loppin B, Docquier M, Bonneton F, and Couble P. 2000. The maternal effect mutation *sésame* affects the formation of the male pronucleus in *Drosophila melanogaster*. *Dev. Biol.* 222: 392–404.

Lovering R, Harden N, and Ashburner M. 1991. The molecular structure of *TE146* and its derivatives in *Drosophila melanogaster*. *Genetics* 128: 357–372.

Luan DD, Korman MH, Jakubczak JL, and Eickbush TH. 1993. Reverse transcription of R2Bm RNA is primed by a nick at the chromosomal target site: a mechanism for non-LTR retrotransposition. *Cell* 72: 595–605.

Lucov Z, and Nur U. 1973. Accumulation of B-chromosomes by preferential segregation in females of the grasshopper *Melanoplus femur-rubrum*. *Chromosoma* 42: 289–306.

Ludwig T, Eggenschwiler J, Fisher P, D'Ercole AJ, Davenport ML, and Efstratiadis A. 1996. Mouse mutants lacking the type 2 IGF receptor (IGF2R) are rescued from perinatal lethality in *Igf2* and *Igf1r* null backgrounds. *Dev. Biol.* 177: 517–535.

Lundrigan BL, and Tucker PK. 1997. Evidence for multiple functional copies of the mole sex-determining locus, *Sry*, in African murine rodents. *J. Mol. Evol.* 45: 60–65.

Lynch M, and Conery JS. 2003. The origins of genome complexity. *Science* 302: 1401–1404.

Lyon MF. 1961. Gene action in X chromosome of mouse (*Mus musculus* L). *Nature* 190: 372–373.

Lyon MF. 1984. Transmission ratio distortion in mouse *t* haplotypes is due to multiple distorter genes acting on a responder locus. *Cell* 37: 621–628.

Lyon MF. 1991. The genetic basis of transmission-ratio distortion and male sterility due to the *t* complex. *Am. Nat.* 137: 349–358.

Lyon MF. 1992. Deletion of mouse *t*-complex distorter-1 produces an effect like that of the *t*-form of the distorter. *Genet. Res.* 59: 27–33.

Lyon MF. 1998. X-chromosome inactivation: a repeat hypothesis. *Cytogenet. Cell Genet.* 80: 133–137.

Lyon MF. 2000. An answer to a complex problem: cloning the mouse *t*-complex responder. *Mamm. Genome* 11: 817–819.

Lyon MF. 2003. Transmission ratio distortion in mice. *Annu. Rev. Genet.* 37: 393–408.

Lyon MF, Schimenti JC, and Evans EP. 2000. Narrowing the critical regions for mouse *t* complex transmission ratio distortion factors by use of deletions. *Genetics* 155: 793–801.

Lyttle TW. 1977. Experimental population genetics of meiotic drive systems I. Pseudo-Y chromosomal drive as a means of eliminating cage populations of *Drosophila melanogaster*. *Genetics* 86: 413–445.

Lyttle TW. 1981. Experimental population genetics of meiotic drive systems. III. Neutralization of sex-ratio distortion in *Drosophila* through sex-chromosome aneuploidy. *Genetics* 98: 317–334.

Lyttle TW. 1991. Segregation distorters. *Annu. Rev. Genet.* 25: 511–557.

Lyttle TW, and Haymer DS. 1992. The role of the transposable element *hobo* in the origin of endemic inversions in wild populations of *Drosophila melanogaster*. *Genetica* 86: 113–126.

Lyttle TW, Sandler LM, Prout T, and Perkins DD, eds. 1991. The genetics and evolutionary biology of meiotic drive. *Am. Nat.* 37:281–456.

MacAlpine DM, Kolesar J, Okamoto K, Butow RA, and Perlman PS. 2001. Replication and preferential inheritance of hypersuppressive petite mitochondrial DNA. *EMBO J.* 20: 1807–1817.

MacAlpine DM, Perlman PS, and Butow RA. 2000. The numbers of individual mitochondrial DNA molecules and mitochondrial DNA nucleoids in yeast are co-regulated by the general amino acid control pathway. *EMBO J.* 19: 767–775.

Mackay TFC, Lyman RF, and Jackson MS. 1992. Effects of *P* element insertions on quantitative traits in *Drosophila melanogaster*. *Genetics* 130: 315–332.

Macleod D, Clark VH, and Bird A. 1999. Absence of genome-wide changes in DNA methylation during development of the zebrafish. *Nat. Genet.* 23: 139–140.

Maheshwari P. 1950. *An Introduction to the Embryology of Angiosperms*. New York: McGraw-Hill.

Malik HS, Burke WD, and Eickbush TH. 1999. The age and evolution of non-LTR retrotransposable elements. *Mol. Biol. Evol.* 16: 793–805.

Malik HS, and Henikoff S. 2001. Adaptive evolution of Cid, a centromere-specific histone in *Drosophila*. *Genetics* 157: 1293–1298.

Manicacci D, Atlan A, Rossello JAE, and Couvet D. 1998. Gynodioecy and reproductive trait variation in three *Thymus* species (Lamiaceae). *Int. J. Plant Sci.* 159: 948–957.

Mantovani B, Passamonti M, and Scali V. 2001. The mitochondrial cytochrome oxidase II gene in *Bacillus* stick insects: ancestry of hybrids, androgenesis, and phylogenetic relationships. *Mol. Phylogenet. Evol.* 19: 157–163.

Mantovani B, and Scali V. 1992. Hybridogenesis and androgenesis in the stick-insect *Bacillus rossius-grandii benazzii* (Insecta, Phasmatodea). *Evolution* 46: 783–796.

Mantovani B, Tinti F, Bachmann L, and Scali V. 1997. The *Bag*320 satellite DNA family in *Bacillus* stick insects (Phasmatodea): different rates of molecular evolution of highly repetitive DNA in bisexual and parthenogenetic taxa. *Mol. Biol. Evol.* 14: 1197–1205.

Marais G. 2003. Biased gene conversion: implications for genome and sex evolution. *Trends Genet.* 19: 330–338.

Marescalchi O, and Scali V. 2001. New DAPI and FISH findings on egg maturation processes in related hybridogenetic and parthenogenetic *Bacillus* hybrids (Insecta, Phasmatodea). *Mol. Reprod. Dev.* 60: 270–276.

Marín I, and Baker BS. 1998. The evolutionary dynamics of sex determination. *Science* 281: 1990–1994.

Martin-Alganza A, Cabrero J, López-León MD, and Perfectti F. 1997. Supernumerary heterochromatin does not affect several morphological and physiological traits in the grasshopper *Eyprepocnemis plorans*. *Hereditas* 126: 187–189.

Martin S, Arana P, and Henriques-Gil N. 1996. The effect of B chromosomes on mating success of the grasshopper *Eyprepocnemis plorans*. *Genetica* 97: 197–203.

Maruyama K, and Hartl DL. 1991. Evidence for interspecific transfer of the transposable element *mariner* between *Drosophila* and *Zaprionus*. *J. Mol. Evol.* 33: 514–524.

Masand P, and Kapil RN. 1966. Nutrition of the embryo sac and embryo—a morphological approach. *Phytomorphology* 16: 158–175.

Maside X, Assimacopoulos S, and Charlesworth B. 2000. Rates of movement of transposable elements on the second chromosome of *Drosophila melanogaster*. *Genet. Res.* 75: 275–284.

Mason JM, and Biessmann H. 1993. Transposition as a mechanism for maintaining telomere length in *Drosophila*. In: *Chromosome Segregation and Aneuploidy*, ed. BK Vig. Berlin: Springer-Verlag, pp. 143–149.

Mathews LM, Chi SY, Greenberg N, Ovchinnikov I, and Swergold GD. 2003. Large differences between LINE-1 amplification rates in the human and chimpanzee lineages. *Am. J. Hum. Genet.* 72: 739–748.

Mathiopoulos KD, della Torre A, Predazzi V, Petrarca V, and Coluzzi M. 1998. Cloning of inversion breakpoints in the *Anopheles gambiae* complex traces a transposable element at the inversion junction. *Proc. Natl. Acad. Sci. USA* 95: 12444–12449.

Matthews RB, and Jones RN. 1983. Dynamics of the B-chromosome polymorphism in rye. I. Simulated populations. *Heredity* 48: 347–371.

Maurice S, Belhassen E, Couvet D, and Gouyon P-H. 1994. Evolution of dioecy: can nuclear-cytoplasmic interactions select for maleness? *Heredity* 73: 346–354.

Maurice S, Charlesworth D, Desfeux C, Couvet D, and Gouyon P-H. 1993. The evolution of gender in hermaphrodites of gynodioecious populations with nucleo-cytoplasmic male-sterility. *Proc. R. Soc. Lond. Ser. B Biol. Sci.* 251: 253–261.

Mayer W, Niveleau A, Walter J, Fundele R, and Haaf T. 2000. Embryogenesis: demethylation of the zygotic paternal genome. *Nature* 403: 501–502.

Maynard Smith J, and Stenseth NC. 1978. On the evolutionary stability of the female-biased sex ratio in the wood lemming (*Myopus schisticolor*). *Heredity* 41: 205–214.

McAllister BF, and Werren JH. 1997. Hybrid origin of a B chromosome (PSR) in the parasitic wasp *Nasonia vitripennis*. *Chromosoma* 106: 243–253.

McCarrey JR, and Thomas K. 1987. Human testis-specific PGK gene lacks introns and possesses characteristics of a processed gene. *Nature* 326: 501–505.

McCauley DE, and Brock MT. 1998. Frequency-dependent fitness in *Silene vulgaris*, a gynodioecious plant. *Evolution* 52: 30–36.

McCauley DE, Olson MS, Emery SN, and Taylor DR. 2000. Population structure influences sex ratio evolution in a gynodioecious plant. *Am. Nat.* 155: 814–819.

McCauley DE, and Taylor DR. 1997. Local population structure and sex ratio: evolution of gynodioecious plants. *Am. Nat.* 150: 406–419.

McCauley DE, and Wade MJ. 1980. Group selection–the genetic and demographic basis for the phenotypic differentiation of small populations of *Tribolium castaneum*. *Evolution* 34: 813–821.

McClintock B. 1952. Chromosome organization and genic expression. *Cold Spring Harbor Symp. Quant. Biol.* 16: 13–47.

McClintock B. 1956. Intranuclear systems controlling gene action and mutation. *Brookhaven Symp. Biol.* 8: 58–74.

McClintock B. 1984. The significance of responses of the genome to challenge. *Science* 226: 792–801.

McGrath J, and Hillman N. 1980. The *in vitro* transmission frequency of the *t12* mutation in the mouse. *J. Embryol. Exp. Morphol.* 60: 141–151.

McKay FE. 1971. Behavioral aspects of population dynamics in unisexual-bisexual *Poeciliopsis* (Pisces: Poeciliidae). *Ecology* 52: 778–790.

McKee BD, and Handel MA. 1993. Sex chromosomes, recombination, and chromatin conformation. *Chromosoma* 102: 71–80.

McKee BD, Hong C-s, and Das S. 2000. On the roles of heterochromatin and euchromatin in meiosis in *Drosophila*: mapping chromosomal pairing sites and testing candidate mutations for effects on X-Y nondisjunction and meiotic drive in male meiosis. *Genetica* 109: 77–93.

McKone MJ, and Halpern SL. 2003. The evolution of androgenesis. *Am. Nat.* 161: 641–656.

McLean C, Bucheton A, and Finnegan DJ. 1993. The 5' untranslated region of the I factor, a long interspersed nuclear element-like retrotransposon of *Drosophila melanogaster*, contains an internal promoter and sequences that regulate expression. *Mol. Cell. Biol.* 13: 1042–1050.

McVean GAT, Myers SR, Hunt S, Deloukas P, Bentley DR, and Donnelly P. 2004. The fine-scale structure of recombination rate variation in the human genome. *Science* 304: 581–584.

McVean GT, and Hurst LD. 1997. Molecular evolution of imprinted genes: no evidence for antagonistic coevolution. *Proc. Biol. Sci.* 264: 739–746.

McVean GT, Hurst LD, and Moore T. 1996. Genomic evolution in mice and men: imprinted genes have little intronic content. *Bioessays* 18: 773–775.

Mead DJ, Gardner DCJ, and Oliver SG. 1986. The yeast 2 micron plasmid: strategies for the survival of a selfish DNA. *Mol. Gen. Genet.* 205: 417–421.

Meffe GK, and Snelson Jr. FF, eds. 1989. *Ecology and Evolution of Livebearing Fishes (Poeciliidae)*. Englewood Cliffs: Prentice Hall.

Meland S, Johansen S, Johansen T, Haugli K, and Haugli F. 1991. Rapid disappearance of one parental mitochondrial genotype after isogamous mating in the myxomycete *Physarum polycephalum*. *Curr. Genet.* 19: 55–60.

Merrill C, Bayraktaroglu L, Kusano A, and Ganetzky B. 1999. Truncated RanGAP encoded by the *Segregation Distorter* locus of *Drosophila*. *Science* 283: 1742–1745.

Messing J, and Grossniklaus U. 1999. Genomic imprinting in plants. In: *Genomic Imprinting: An Interdisciplinary Approach*, ed. R Ohlsson. Berlin: Springer-Verlag, pp. 23–40.

Metz CW. 1926. Genetic evidence of a selective segregation of chromosomes in *Sciara* (Diptera). *Proc. Natl. Acad. Sci. USA* 12: 690–692.

Metz CW. 1938. Chromosome behavior, inheritance and sex determination in *Sciara*. *Am. Nat.* 72: 485–520.

Metz CW, and Lawrence EG. 1938. Preliminary observations on *Sciara* hybrids. *J. Hered.* 29: 179–186.

Meunier J, and Duret L. 2004. Recombination drives the evolution of GC-content in the human genome. *Mol. Biol. Evol.* 21: 984–990.

Miao VP, Covert SF, and VanEtten HD. 1991a. A fungal gene for antibiotic resistance on a dispensable ("B") chromosome. *Science* 254: 1773–1776.

Miao VP, Matthews DE, and VanEtten HD. 1991b. Identification and chromosomal location of a family of cytochrome P-450 genes for pisatin detoxification in the fungus *Nectria haematococca*. *Mol. Gen. Genet.* 226: 214–223.

Miklos GLG, and Rubin GM. 1996. The role of the genome project in determining gene function: insights from model organisms. *Cell* 86: 521–529.

Miller DD, and Roy R. 1964. Further data on Y chromosome types in *Drosophila athabasca*. *Can. J. Genet. Cytol.* 6: 334–348.

Miller DD, and Stone LE. 1962. A reinvestigation of karyotype in *Drosophila affinis* and related species. *J. Hered.* 53: 12–24.

Miller DD, and Voelker RA. 1969. Salivary gland chromosome variation in the *Drosophila affinis* subgroup. IV. The short arm of the X chromosome in "western" and "eastern" *Drosophila athabasca*. *J. Hered.* 60: 307–311.

Miller DR, and Kosztarab M. 1979. Recent advances in the study of scale insects. *Annu. Rev. Entomol.* 24: 1–27.

Miller DW, and Miller LK. 1982. A virus mutant with an insertion of a *copia*-like transposable element. *Nature* 299: 562–564.

Miller WJ, McDonald JF, Nouaud D, and Anxolabéhère D. 1999. Molecular domestication—more than a sporadic episode in evolution. *Genetica* 107: 197–207.

Mills KI, Woodgate LJ, Gilkes AF, Walsh V, Sweeney MC, Brown G, and Burnett AK. 1999. Inhibition of mitochondrial function in HL 60 cells is associated with an increased apoptosis and expression of CD14. *Biochem. Biophys. Res. Commun.* 263: 294–300.

Mittal RD, Wileman G, and Hillman N. 1989. Acrosin activity in spermatozoa from sterile t6/tw32 and fertile control mice. *Genet. Res.* 54: 143–148.

Miyamura S, Kuroiwa T, and Nagata T. 1987. Disappearance of plastid and mitochondrial nucleoids during the formation of generative cells of higher plants revealed by fluorescence microscopy. *Protoplasma* 128: 1–13.

Mizumura H, Shibata T, and Morishima N. 1999. Stable association of 70-kDa heat shock protein induces latent multisite specificity of a unisite-specific endonuclease in yeast mitochondria. *J. Biol. Chem.* 274: 25682–25690.

Mizumura H, Shibata T, and Morishima N. 2002. Association of HSP70 with endonucleases allows the expression of otherwise silent mutations. *FEBS Lett.* 522: 177–182.

Mochizuki A, Takeda Y, and Iwasa Y. 1996. The evolution of genomic imprinting. *Genetics* 144: 1283–1295.

Mogensen HL. 1996. The hows and whys of cytoplasmic inheritance in seed plants. *Am. J. Bot.* 83: 383–404.

Moloney DM, Slaney SR, Oldridge M, Wall SA, Sahlin P, Stenman G, and Wilkie AOM. 1996. Exclusive paternal origin of new mutations in Apert syndrome. *Nat. Genet.* 13: 48–53.

Monro K, and Poore AGB. 2004. Selection in modular organisms: is intraclonal variation in macroalgae evolutionarily important? *Am. Nat.* 163: 564–578.

Montchamp-Moreau C, and Atlan A, eds. 2003. *Intragenomic Conflict. Genetica* 117:1–110.

Montchamp-Moreau C, and Cazemajor M. 2002. Sex-ratio drive in *Drosophila simulans:* variation in segregation ratio of *X* chromosomes from a natural population. *Genetics* 162: 1221–1231.

Montchamp-Moreau C, Ginhoux V, and Atlan A. 2001. The Y chromosomes of *Drosophila simulans* are highly polymorphic for their ability to suppress sex-ratio drive. *Evolution* 55: 728–737.

Montchamp-Moreau C, and Joly D. 1997. Abnormal spermiogenesis is associated with the X-linked sex-ratio trait in *Drosophila simulans. Heredity* 79: 24–30.

Montoya-Burgos JI, Boursot P, and Galtier N. 2003. Recombination explains isochores in mammalian genomes. *Trends Genet.* 19: 128–130.

Moore T, and Haig D. 1991. Genomic imprinting in mammalian development: a parental tug-of-war. *Trends Genet.* 7: 45–49.

Moore T, and Reik W. 1996. Genetic conflict in early development: parental imprinting in normal and abnormal growth. *Rev. Reprod.* 1: 73–77.

Moore WS. 1976. Components of fitness in the unisexual fish *Poeciliopsis monacha-occidentalis*. *Evolution* 30: 564–578.

Moran JV, DeBerardinis RJ, and Kazazian Jr. HH. 1999. Exon shuffling by L1 retrotransposition. *Science* 283: 1530–1534.

Moran JV, and Gilbert N. 2002. Mammalian LINE-1 retrotransposons and related elements. In: *Mobile DNA II*, eds. NL Craig, R Craigie, M Gellert, and AM Lambowitz. Washington, D.C.: ASM Press, pp. 836–869.

Morgan MT. 2001. Transposable element number in mixed mating populations. *Genet. Res.* 77: 261–275.

Morita T, Kubota H, Murata K, Nozaki M, Delarbre C, Willison K, Satta Y, Sakaizumi M, Takahata N, Gachelin G, and Matsushiro A. 1992. Evolution of the mouse *t* haplotype: recent and worldwide introgression to *Mus musculus*. *Proc. Natl. Acad. Sci. USA* 89: 6851–6855.

Morris JS, Ohman A, and Dolan RJ. 1998. Conscious and unconscious emotional learning in the human amygdala. *Nature* 393: 467–470.

Moss JP. 1969. B-chromosomes and breeding systems. *Chromosomes Today* 2: 268 (abstract).

Moure CM, Gimble FS, and Quiocho FA. 2002. Crystal structure of the intein homing endonuclease PI-*Sce*I bound to its recognition sequence. *Nat. Struct. Biol.* 9: 764–770.

Mueller MW, Allmaier M, Eskes R, and Schweyen RJ. 1993. Transposition of group II intron *al1* in yeast and invasion of mitochondrial genes at new locations. *Nature* 366: 174–176.

Müller F, Bernard V, and Tobler H. 1996. Chromatin diminution in nematodes. *Bioessays* 18: 133–138.

Müller F, and Tobler H. 2000. Chromatin diminution in the parasitic nematodes *Ascaris suum* and *Parascaris univalens*. *Int. J. Parasit.* 30: 391–399.

Munier F, Spence MA, Pescia G, Balmer A, Gailloud C, Thonney F, van Melle G, and Rutz HP. 1992. Paternal selection favoring mutant alleles of the retinoblastoma susceptibility gene. *Hum. Genet.* 89: 508–512.

Müntzing A. 1945. Cytological studies of extra fragment chromosomes in rye. II. Transmission and multiplication of standard fragments and iso-fragments. *Hereditas* 31: 457–477.

Müntzing A. 1949. Accessory chromosomes in *Secale* and *Poa*. *Proc. 8th Int. Cong. Genet. (Stockholm):* 402–411.

Müntzing A. 1954. Cytogenetics of accessory chromosomes (B chromosomes). *Caryologia* S6: 282–301.

Müntzing A. 1963. Effects of accessory chromosomes in diploid and tetraplid rye. *Hereditas* 49: 371–426.

Müntzing A. 1970. Chromosomal variation in the Lindstrom strain of wheat carrying accessory chromosomes of rye. *Hereditas* 66: 279–286.

Müntzing A, Jaworska H, and Carlibom C. 1969. Studies of meiosis in the Lindstrom strain of wheat carrying accessory chromosomes of rye. *Hereditas* 61: 179–207.

Müntzing A, and Nygren A. 1955. A new diploid variety of *Poa alpina* with two accessory chromosomes at meiosis. *Hereditas* 41: 405–422.

Murai K, Takumi S, Koga H, and Ogihara Y. 2002. Pistillody, homeotic transformation of stamens into pistil-like structures, caused by nuclear-cytoplasm interaction in wheat. *Plant J.* 29: 169–181.

Murphy SK, and Jirtle RL. 2003. Imprinting evolution and the price of silence. *Bioessays* 25: 577–588.

Murphy TD, and Karpen GH. 1995. Localization of centromere function in a *Drosophila* minichromosome. *Cell* 82: 599–609.

Nachman MW, and Myers P. 1989. Exceptional chromosomal mutations in a rodent population are not strongly underdominant. *Proc. Natl. Acad. Sci. USA* 86: 6666–6670.

Nachman MW, and Searle JB. 1995. Why is the house mouse karyotype so variable? *Trends Ecol. Evol.* 10: 397–402.

Nadeau JH, Britton-Davidian J, Bonhomme F, and Thaler L. 1988. *H-2* polymorphisms are more uniformly distributed than allozyme polymorphisms in natural populations of house mice. *Genetics* 118: 131–140.

Nagamine CM, and Carlisle C. 1994. Duplication and amplification of the SRY locus in Muridae. *Cytogenet. Cell Genet.* 67: 393.

Nagelkerke CJ, and Sabelis MW. 1998. Precise control of sex allocation in pseudo-arrhenotokous phytoseiid mites. *J. Evol. Biol.* 11: 649–684.

Nakagawa K-i, Morishima N, and Shibata T. 1991. A maturase-like subunit of the sequence-specific endonuclease Endo.*Sce*I from yeast mitochondria. *J. Biol. Chem.* 266: 1977–1984.

Nakagawa K-i, Morishima N, and Shibata T. 1992. An endonuclease with multiple cutting sites, Endo.*Sce*I, initiates genetic recombination at its cutting site in yeast mitochondria. *EMBO J.* 11: 2707–2715.

Nakajima Y, Yamamoto T, Muranaka T, and Oeda K. 1999. Genetic variation of petaloid male-sterile cytoplasm of carrots revealed by sequence-tagged sites (STSs). *Theor. Appl. Genet.* 99: 837–843.

Nakajima Y, Yamamoto T, Muranaka T, and Oeda K. 2001. A novel *orfB*-related gene of carrot mitochondrial genomes that is associated with homeotic cytoplasmic male sterility (CMS). *Plant Mol. Biol.* 46: 99–107.

Nassif N, Penney J, Pal S, Engels WR, and Gloor GB. 1994. Efficient copying of nonhomologous sequences from ectopic sites via P-element-induced gap repair. *Mol. Cell. Biol.* 14: 1613–1625.

Nasuda S, Friebe B, and Gill BS. 1998. Gametocidal genes induce chromosome breakage in the interphase prior to the first mitotic cell division of the male gametophyte in wheat. *Genetics* 149: 1115–1124.

Naumova AK, Greenwood CMT, and Morgan K. 2001. Imprinting and deviation from Mendelian transmission ratios. *Genome* 44: 311–320.

Nauta MJ, and Hoekstra RF. 1993. Evolutionary dynamics of spore killers. *Genetics* 135: 923–930.

Neale DB, Marshall KA, and Sederoff RR. 1989. Chloroplast and mitochondrial DNA are paternally inherited in *Sequoia sempervirens* D. Don Endl. *Proc. Natl. Acad. Sci. USA* 86: 9347–9349.

Nee S, and Maynard Smith J. 1990. The evolutionary biology of molecular parasites. *Parasitology* 100: S5-S18.

Negovetic S, Anholt BR, Semlitsch RD, and Reyer H-U. 2001. Specific responses of sexual and hybridogenetic European waterfrog tadpoles to temperature. *Ecology* 82: 766–774.

Nei M. 1971. Total number of individuals affected by a single deleterious mutation in large populations. *Theor. Popul. Biol.* 2: 426–430.

Neitzel H, Kalscheuer V, Singh AP, Henschel S, and Sperling K. 2002. Copy and paste: the impact of a new non-L1 retroposon on the gonosomal heterochromatin of *Microtus agrestis*. *Cytogenet. Genome Res.* 96: 179–185.

Nelson-Rees WA, Hoy MA, and Roush RT. 1980. Heterochromatinization, chromatin elimination and haploidization in the parahaploid mite *Metaseiulus occidentalis* (Nesbitt) (Acarina: Phytoseiidae). *Chromosoma* 77: 263–275.

Nelson OE. 1994. The gametophyte factors of maize. In: *The Maize Handbook*, eds. M Freeling and V Walbot. New York: Springer-Verlag, pp. 496–503.

Néo D, Moreira Filho O, and Camacho JPM. 2000. Altitudinal variation for B chromosome frequency in the characid fish *Astyanax scabripinnis*. *Heredity* 85: 136–141.

Nesse RM, and Williams GC. 1994. *Evolution and Healing: The New Science of Darwinian Medicine*. London: Phoenix.

Newton ME, Wood RJ, and Southern DI. 1976. A cytogenetic analysis of meiotic drive in the mosquito, *Aedes aegypti* (L.). *Genetica* 46: 297–318.

Newton ME, Wood RJ, and Southern DI. 1978. Cytological mapping of the M and D loci in the mosquito, *Aedes aegypti* (L.). *Genetica* 48: 137–143.

Nicolas A, Treco D, Schultes NP, and Szostak JW. 1989. An initiation site for meiotic gene conversion in the yeast *Saccharomyces cerevisiae*. *Nature* 338: 35–39.

Nigumann P, Redik K, Mätlik K, and Speek M. 2002. Many human genes are transcribed from the antisense promoter of L1 retrotransposon. *Genomics* 79: 628–634.

Nikitin AG, and Woodruff RC. 1995. Somatic movement of the *mariner* transposable element and lifespan of *Drosophila* species. *Mutat. Res.* 338: 43–49.

Nishimura Y, Misumi O, Kato K, Inada N, Higashiyama T, Momoyama Y, and Kuroiwa T. 2002. An *mt*⁺ gamete-specific nuclease that targets *mt*⁻ chloroplasts during sexual reproduction in *C. reinhardtii*. *Genes Dev.* 16: 1116–1128.

Niwa K, and Sakamoto S. 1995. Origin of B-chromosomes in temperate rye. *Genome* 38: 307–312.

Nizetic D, Figueroa F, and Klein J. 1984. Evolutionary relationships between the *t* and *H-2* haplotypes in the house mice. *Immunogenetics* 19: 311–320.

Nokkala S, Kuznetsova V, and Maryanska-Nadachowska A. 2000. Achiasmatic segregation of a B chromosome from the X chromosome in two species of psyllids (Psylloidea, Homoptera). *Genetica* 108: 181–189.

Nokkala S, and Nokkala C. 2004. Interaction of B chromosomes with A or B chromosomes in segregation in insects. *Cytogenet. Genome Res.* 106: 394–397.

Nolan CM, Killian JK, Pettite JN, and Jirtle RL. 2001. Imprint status of *M6P/IGF2R* and *IGF2* in chickens. *Dev. Genes Evol.* 211: 179–183.

Normark BB. 2004. Haplodiploidy as an outcome of coevolution between male-killing cytoplasmic elements and their hosts. *Evolution* 58: 790–798.

Normark BB, Jordal BH, and Farrell BD. 1999. Origin of a beetle haplodiploid lineage. *Proc. R. Soc. Lond. B Biol. Sci.* 266: 2253–2259.

Novitski E. 1951. Non-random disjunction in *Drosophila*. *Genetics* 36: 267–280.

Novitski E. 1967. Nonrandom disjunction in *Drosophila*. *Annu. Rev. Genet.* 1: 71–86.

Novitski E, Peacock WJ, and Engel J. 1965. Cytological basis of "sex ratio" in *Drosophila pseudoobscura*. *Science* 148: 516–517.

Nowell PC. 1976. The clonal evolution of tumor cell populations. *Science* 194: 23–28.

Nunney L. 1999. Lineage selection and the evolution of multistage carcinogenesis. *Proc. Biol. Sci.* 266: 493–498.

Nur U. 1962. A supernumerary chromosome with an accumulation mechanism in the lecanoid genetic system. *Chromosoma* 13: 249–271.

Nur U. 1963. A mitotically unstable supernumerary chromosome with an accumulation mechanism in a grasshopper. *Chromosoma* 14: 407–422.

Nur U. 1969a. Mitotic instability leading to an accumulation of B-chromosomes in grasshoppers. *Chromosoma* 27: 1–19.

Nur U. 1969b. Harmful B-chromosomes in a mealy bug: additional evidence. *Chromosoma* 28: 280–297.

Nur U. 1977. Maintenance of a "parasitic" B chromosome in the grasshopper *Melanoplus femur-rubrum*. *Genetics* 87: 499–512.

Nur U. 1980. Evolution of unusual chromosome systems in scale insects (Coccoidea: Homoptera). In: *Insect Cytogenetics*, eds. RL Blackman, GM Hewitt, and M Ashburner. London: Blackwell Scientific Publications, pp. 97–117.

Nur U. 1982. Destruction of specific heterochromatic chromosomes during spermatogenesis in the Comstockiella chromosome system (Coccoidea: Homoptera). *Chromosoma* 85: 519–530.

Nur U. 1990a. Chromosomes, sex-ratios, and sex determination. In: *Armored Scale Insects: Their Biology, Natural Enemies and Control*, vol. A, ed. D Rosen. Amsterdam: Elsevier, pp. 179–190.

Nur U. 1990b. Heterochromatization and euchromatization of whole genomes in scale insects (Coccoidea: Homoptera). *Dev. Suppl.*: 29–34.

Nur U, and Brett BLH. 1985. Genotypes suppressing meiotic drive of a B chromosome in the mealybug, *Pseudococcus obscurus*. *Genetics* 110: 73–92.

Nur U, and Brett BLH. 1987. Control of meiotic drive of B chromosomes in the mealybug, *Pseudococcus affinis (obscurus)*. *Genetics* 115: 499–510.

Nur U, and Brett BLH. 1988. Genotypes affecting the condensation and transmission of heterochromatic B chromosomes in the mealybug *Pseudococcus affinis*. *Chromosoma* 96: 205–212.

Nur U, Werren JH, Eickbush DG, Burke WD, and Eickbush TH. 1988. A "selfish" B chromosome that enhances its transmission by eliminating the paternal genome. *Science* 240: 512–514.

Nuzhdin SV. 1995. The distribution of transposable elements on X chromosomes from a natural population of *Drosophila simulans*. *Genet. Res.* 66: 159–166.

Nuzhdin SV. 1999. Sure facts, speculations, and open questions about the evolution of transposable element copy number. *Genetica* 107: 129–137.

Nuzhdin SV, and Mackay TFC. 1995. The genomic rate of transposable element movement in *Drosophila melanogaster*. *Mol. Biol. Evol.* 12: 180–181.

Nyquist WE. 1962. Differential fertilization in the inheritance of stem rust resistance in hybrids involving a common wheat strain derived from *Triticum timopheevi*. *Genetics* 47: 1109–1124.

O'Brochta DA, and Handler AM. 1988. Mobility of *P* elements in drosophilids and nondrosophilids. *Proc. Natl. Acad. Sci. USA* 85: 6052–6056.

O'Neill SL, Hoffmann AA, and Werren JH, eds. 1997. *Influential Passengers: Inherited Microorganisms and Arthropod Reproduction*. Oxford: Oxford University Press.

Obata Y, and Kono T. 2002. Maternal primary imprinting is established at a specific time for each gene throughout oocyte growth. *J. Biol. Chem.* 277: 5285–5289.

Ogawa O, Becroft DM, Morison IM, Eccles MR, Skeen JE, Mauger DC, and Reeve AE. 1993. Constitutional relaxation of insulin-like growth factor II gene imprinting associated with Wilms' tumour and gigantism. *Nat. Genet.* 5: 408–412.

Ogielska M, and Bartmanska J. 1999. Development of testes and differentiation of germ cells in water frogs of the *Rana esculenta*-complex (Amphibia, Anura). *Amphibia-Reptilia* 20: 251–263.

Ogiwara I, Miya M, Ohshima K, and Okada N. 1999. Retropositional parasitism of SINEs on LINEs: identification of SINEs and LINEs in elasmobranchs. *Mol. Biol. Evol.* 16: 1238–1250.

Ohno S, Jainchill J, and Stenius C. 1963. The creeping vole (*Microtus oregoni*) as a gonosomic mosaic. I. The OY/XY constitution of the male. *Cytogenetics* 2: 232–239.

Ohsako T, Hirai K, and Yamamoto M-T. 2003. The *Drosophila misfire* gene has an essential role in sperm activation during fertilization. *Genes Genet. Syst.* 78: 253–266.

Okada N, Hamada M, Ogiwara I, and Ohshima K. 1997. SINEs and LINEs share common 3' sequences: a review. *Gene* 205: 229–243.

Olds-Clarke P. 1997. Models for male infertility: the *t* haplotypes. *Rev. Reprod.* 2: 157–164.

Olds-Clarke P, and Johnson LR. 1993. *t* haplotypes in the mouse compromise sperm flagellar function. *Dev. Biol.* 155: 14–25.

Olds-Clarke P, and Peitz B. 1986. Fertility of sperm from *t*/+ mice: evidence that +–bearing sperm are dysfunctional. *Genet. Res.* 47: 49–52.

Oliver Jr. JH. 1983. Chromosomes, genetic variance and reproductive strategies among mites and ticks. *Bull. Entomol. Soc. Am.* 29: 8–17.

Olsen A-K, Bjørtuft H, Holme J, Seeberg E, Bjørås M, and Brunborg G. 2001. Highly efficient base excision repair (BER) in human and rat male germ cells. *Nucleic Acids Res.* 29: 1781–1790.

Orgel LE, and Crick FHC. 1980. Selfish DNA: the ultimate parasite. *Nature* 284: 604–607.

Orr-Weaver TL, and Szostak JW. 1985. Fungal recombination. *Microbiol. Rev.* 49: 33–58.

Orr HA, and Irving S. 2005. Segregation distortion in hybrids between the Bogota and USA subspecies of *Drosophila pseudoobscura*. *Genetics* 169: 671–682.

Östergren G. 1945. Parasitic nature of extra fragment chromosomes. *Bot. Notiser* 2: 157–163.

Otto SP, and Hastings IM. 1998. Mutation and selection within the individual. *Genetica* 102/103: 507–524.

Owusu-Daaku KO, Wood RJ, and Butler RD. 1997. Selected lines of *Aedes aegypti* with persistently distorted sex ratios. *Heredity* 79: 388–393.

Özen RS, Baysal FE, Devlin B, Farr JE, Gorry M, Ehrlich GD, Richard CW. 1999. Fine mapping of the split-hand/split-foot locus (SHFM3) at 10q24: evidence for anticipation and segregation distortion. *Am. J. Hum. Genet.* 64: 1646–1654.

Pagano A, Joly P, Plénet S, Lehman A, and Grolet O. 2001. Breeding habitat partitioning in the *Rana esculenta* complex: the intermediate niche hypothesis supported. *Ecoscience* 8: 294–300.

Pagel M, and Johnstone RF. 1992. Variation across species in the size of the nuclear genome supports the junk-DNA explanation for the C-value paradox. *Proc. R. Soc. Lond. B Biol. Sci.* 249: 119–124.

Paldi A, Gyapay G, and Jami J. 1995. Imprinted chromosomal regions of the human genome display sex-specific meiotic recombination frequencies. *Curr. Biol.* 5: 1030–1035.

Palestis BG, Burt A, Jones RN, and Trivers R. 2004a. B chromosomes are more frequent in mammals with acrocentric karyotypes: support for the theory of centromeric drive. *Proc. R. Soc. Lond. B Biol. Sci.* 271 Suppl.: S22-S24.

Palestis BG, Jones RN, and Trivers R. 2004b. Drive and phenotypic effects of B chromosomes, analyzed by sex and mode of drive. Second B Chromosome Conference, 26–29 June, 2004, Bubion, Spain. Available at http://www.wagner.edu/faculty/users/bpalesti/B%20poster.ppt.

Palestis BG, Trivers R, Jones RN, and Burt A. 2004c. The distribution of B chromosomes across species. *Cytogenet. Genome Res.* 106: 151–158.

Paliwal RL, and Hyde BB. 1959. The association of a single B chromosome with male sterility in *Plantago coronopus. Am. J. Bot.* 46: 460–466.

Palmer JD. 1990. Contrasting modes and tempos of genome evolution in land plant organelles. *Trends Genet.* 6: 115–120.

Palopoli MF, and Wu C-I. 1996. Rapid evolution of a coadapted gene complex: evidence from the *Segregation Distorter (SD)* system of meiotic drive in *Drosophila melanogaster. Genetics* 143: 1675–1688.

Palumbi SR. 2001. Humans as the world's greatest evolutionary force. *Science* 293: 1786–1790.

Pantulu JV, and Manga V. 1975. Influence of B-chromosomes on meiosis in pearl millet. *Genetica* 45: 237–251.

Pâques F, and Haber JE. 1999. Multiple pathways of recombination induced by double-strand breaks in *Saccharomyces cerevisiae. Microbiol. Mol. Biol. Rev.* 63: 349–404.

Pardo-Manuel de Villena F, de la Casa-Esperón E, Briscoe TL, and Sapienza C. 2000. A genetic test to determine the origin of maternal transmission ratio distortion: meiotic drive at the mouse *Om* locus. *Genetics* 154: 333–342.

Pardo-Manuel de Villena F, and Sapienza C. 2001a. Female meiosis drives karyotypic evolution in mammals. *Genetics* 159: 1179–1189.

Pardo-Manuel de Villena F, and Sapienza C. 2001b. Nonrandom segregation during meiosis: the unfairness of females. *Mamm. Genome* 12: 331–339.

Pardo-Manuel de Villena F, and Sapienza C. 2001c. Transmission ratio distortion in offspring of heterozygous female carriers of Robertsonian translocations. *Hum. Genet.* 108: 31–36.

Pardo MC, López-León MD, Cabrero J, and Camacho JPM. 1994. Transmission analysis of mitotically unstable B chromosomes in *Locusta migratoria. Genome* 37: 1027–1034.

Pardue M-L, and DeBaryshe PG. 2002. Telomeres and transposable elements. In: *Mobile DNA II,* eds. NL Craig, R Craigie, M Gellert, and AM Lambowitz. Washington, D.C.: ASM Press, pp. 870–887.

Parker JS, Jones GH, Edgar L, and Whitehouse C. 1989. The population cytogenetics of *Crepis capillaris.* II. The stability and inheritance of B chromosomes. *Heredity* 63: 19–27.

Parker JS, Jones GH, Edgar LA, and Whitehouse C. 1991. The population cytogenetics of *Crepis capillaris.* IV. The distribution of B chromosomes in British populations. *Heredity* 66: 211–218.

Parker JS, Taylor S, and Ainsworth CC. 1982. The B-chromosome system of *Hypochoeris maculata.* III. Variation in B-chromosome transmission rates. *Chromosoma* 85: 299–310.

Pascual L, and Periquet G. 1991. Distribution of *hobo* transposable elements in natural populations of *Drosophila melanogaster. Mol. Biol. Evol.* 8: 282–296.

Passamonti M, Boore JL, and Scali V. 2003. Molecular evolution and recombination in gender-associated mitochondrial DNAs of the Manila clam *Tapes philippinarum. Genetics* 164: 603–611.

Passamonti M, and Scali V. 2001. Gender-associated mitochondrial DNA heteroplasmy in the venerid clam *Tapes philippinarum* (Mollusca Bivalvia). *Curr. Genet.* 39: 117–124.

Patterson J, and Stone W. 1952. *Evolution in the Genus Drosophila.* New York: Macmillan.

Patton JL. 1977. B chromosome systems in the pocket mouse, *Perognathus baileyi:* meiosis and C-band studies. *Chromosoma* 60: 1–14.

Peacock WJ, Dennis ES, Rhoades MM, and Pryor AJ. 1981. Highly repeated DNA-sequence limited to knob heterochromatin in maize. *Proc. Natl. Acad. Sci. USA* 78: 4490–4494.

Perfectti F, Corral JM, Mesa JA, Cabrero J, Bakkali M, López-León MD, and Camacho JPM. 2004. Rapid suppression of drive for a parasitic B chromosome. *Cytogenet. Genome Res.* 106: 338–343.

Perfectti F, and Werren JH. 2001. The interspecific origin of B chromosomes: experimental evidence. *Evolution* 55: 1069–1073.

Perkins DD. 2003. A fratricidal fungal prion. *Proc. Natl. Acad. Sci. USA* 100: 6292–6294.

Perkins DD, Lande R, and Stahl FW. 1993. Estimates of the proportion of recombination intermediates that are resolved with crossing over in *Neurospora crassa*. *Genetics* 133: 690–691.

Perkins HD, and Howells AJ. 1992. Genomic sequences with homology to the *P* element of *Drosophila melanogaster* occur in the blowfly *Lucilia cuprina*. *Proc. Natl. Acad. Sci. USA* 89: 10753–10757.

Perondini ALP, and Ribeiro AF. 1997. Chromosome elimination in germ cells of *Sciara* embryos: involvement of the nuclear envelope. *Invertebr. Reprod. Dev.* 32: 131–141.

Perry J, and Ashworth A. 1999. Evolutionary rate of a gene affected by chromosomal position. *Curr. Biol.* 9: 987–989.

Peter AKH. 2001. Dispersal rates and distances in adult water frogs, *Rana lessonae, R. ridibunda*, and their hybridogenetic associate *R. esculenta*. *Herpetologica* 57: 449–460.

Peter AKH, Reyer H-U, and Tietje GA. 2002. Species and sex ratio differences in mixed populations of hybridogenetic water frogs: the influence of pond features. *Ecoscience* 9: 1–11.

Peters LL, and Barker JE. 1993. Novel inheritance of the murine severe combined anemia and thrombocytopenia (scat) phenotype. *Cell* 74: 135–142.

Petersen RF, Langkjær RB, Hvidtfeldt J, Gartner J, Palmen W, Ussery DW, and Piškur J. 2002. Inheritance and organisation of the mitochondrial genome differ between two *Saccharomyces* yeasts. *J. Mol. Biol.* 318: 627–636.

Peto R, Roe FJ, Lee PN, Levy L, and Clack J. 1975. Cancer and ageing in mice and men. *Br. J. Cancer* 32: 411–442.

Petras ML. 1967. Studies of natural populations of *Mus*. II. Polymorphism at the *T* locus. *Evolution* 21: 466–478.

Petrov DA. 2002. Mutational equilibrium model of genome size evolution. *Theor. Popul. Biol.* 61: 533–546.

Petrov DA, Aminetzach YT, Davis JC, Bensasson D, and Hirsh AE. 2003. Size matters: non-LTR retrotransposable elements and ectopic recombination in *Drosophila*. *Mol. Biol. Evol.* 20: 880–892.

Petrov DA, Sangster TA, Johnston JS, Hartl DL, and Shaw KL. 2000. Evidence for DNA loss as a determinant of genome size. *Science* 287: 1060–1062.

Pichot C, Borrut A, and El Maâtaoui M. 1998. Unexpected DNA content in the endosperm of *Cupressus dupreziana* A. Camus seeds and its implications in the reproductive process. *Sex. Plant Reprod.* 11: 148–152.

Pichot C, and El Maâtaoui M. 2000. Unreduced diploid nuclei in *Cupressus dupreziana* A. Camus pollen. *Theor. Appl. Genet.* 101: 574–579.

Pichot C, El Maâtaoui M, Raddi S, and Raddi P. 2001. Surrogate mother for endangered *Cupressus*. *Nature* 412: 39.

Pickeral OK, Makalowski W, Boguski MS, and Boeke JD. 2000. Frequent human genomic DNA transduction driven by LINE-1 retrotransposition. *Genome Res.* 10: 411–415.

Pigozzi MI, and Solari AJ. 1998. Germ cell restriction and regular transmission of an accessory chromosome that mimics a sex body in the zebra finch. *Chromosome Res.* 6: 105–113.

Pike TW, and Petrie M. 2003. Potential mechanisms of avian sex manipulation. *Biol. Rev.* 78: 553–574.

Pimpinelli S, and Dimitri P. 1989. Cytogenetic analysis of segregation distortion in *Drosophila melanogaster:* the cytological organization of the Responder (*Rsp*) locus. *Genetics* 121: 765–772.

Pineda-Krch M, and Lehtilä K. 2002. Cell lineage dynamics in stratified apical meristems. *J. Theor. Biol.* 219: 495–505.

Plagge A, Gordon E, Dean W, Boiani R, Cinti S, Peters J, and Kelsey G. 2004. The imprinted signaling protein XL$^{\alpha}$s is required for postnatal adaptation to feeding. *Nat. Genet.* 36: 818–826.

Planchart A, You Y, and Schimenti JC. 2000. Physical mapping of male fertility and meiotic drive quantitative trait loci in the mouse *t* complex using chromosome deficiencies. *Genetics* 155: 803–812.

Plasterk RHA. 1991. The origin of footprints of the Tc1 transposon of *Caenorhabditis elegans. EMBO J.* 10: 1919–1925.

Plasterk RHA, and Groenen JTM. 1992. Targeted alterations of the *Caenorhabditis elegans* genome by transgene instructed DNA double strand break repair following Tc1 excision. *EMBO J.* 11: 287–290.

Plasterk RHA, Izsvák Z, and Ivics Z. 1999. Resident aliens: the Tc1/*mariner* super-family of transposable elements. *Trends Genet.* 15: 326–332.

Plenet S, Hervant F, and Joly P. 2000. Ecology of the hybridogenetic *Rana esculenta* complex: differential oxygen requirements of tadpoles. *Evol. Ecol.* 14: 13–23.

Plenge RM, Percec I, Nadeau JH, and Willard HF. 2000. Expression-based assay of an X-linked gene to examine effects of the X-controlling element (*Xce*) locus. *Mamm. Genome* 11: 405–408.

Plowman AB, and Bougourd SM. 1994. Selectively advantageous effects of B chromosomes on germination behaviour in *Allium schoenoprasum* L. *Heredity* 72: 587–593.

Policansky D. 1974. "Sex ratio," meiotic drive, and group selection in *Drosophila pseudoobscura. Am. Nat.* 108: 75–90.

Policansky D, and Ellison J. 1970. Sex ratio in *Drosophila pseudoobscura:* spermiogenic failure. *Science* 169: 888–889.

Polukonova NV, and Belianina SI. 2002. About the possibility of hybridogenesis in the species origin of the midge *Chironomus usenicus* Loginova et Beljanina (Chironomidae, Diptera). *Genetika* 38: 1635–1640.

Ponte P, Gunning P, and Kedes L. 1983. Human actin genes are single copy for alpha-skeletal and alpha-cardiac actin but multicopy for beta-cytoskeletal and gamma-cytoskeletal genes–3′ untranslated regions are isotype specific but are conserved in evolution. *Mol. Cell. Biol.* 3: 1783–1791.

Poot P. 1997. Reproductive allocation and resource compensation in male-sterile and hermaphroditic plants of *Plantago lanceolata* (Plantaginaceae). *Am. J. Bot.* 84: 1256–1265.

Posey KL, Koufopanou V, Burt A, and Gimble FS. 2004. Evolution of divergent DNA recognition specificities in VDE homing endonucleases from two yeast species. *Nucleic Acids Res.* 32: 3947–3956.

Potten CS, Owen G, and Booth D. 2002. Intestinal stem cells protect their genome by selective segregation of template DNA strands. *J. Cell Sci.* 115: 2381–2388.

Poulter RTM, Goodwin TJD, and Butler MI. 2003. Vertebrate helentrons and other novel *Helitrons*. *Gene* 313: 201–212.

Prescott DM. 1994. The DNA of ciliated protozoa. *Microbiol. Rev.* 58: 233–267.

Presgraves DC, Severance E, and Wilkinson GS. 1997. Sex chromosome meiotic drive in stalk-eyed flies. *Genetics* 147: 1169–1180.

Procunier WS. 1975. The B chromosomes of *Cnephia dacotensis* and *Cnephia ornithophilia* (Diptera, Simulidae). *Can. J. Zool.* 53: 1638–1647.

Prokopowich CD, Gregory TR, and Crease TJ. 2003. The correlation between rDNA copy number and genome size in eukaryotes. *Genome* 46: 48–50.

Pryor A, Faulkner K, Rhoades MM, and Peacock WJ. 1980. Asynchronous replication of heterochromatin in maize. *Proc. Natl. Acad. Sci. USA* 77: 6705–6709.

Puertas MJ. 2002. Nature and evolution of B chromosomes in plants: a non-coding but information-rich part of plant genomes. *Cytogenet. Genome Res.* 96: 198–205.

Puertas MJ, and Carmona R. 1976. Greater ability of pollen tube growth in rye plants with 2B chromosomes. *Theor. Appl. Genet.* 47: 41–43.

Puertas MJ, González-Sánchez M, Manzanero S, Romera F, and Jiménez MM. 1998. Genetic control of the rate of transmission of rye B chromosomes. IV. Localization of the genes controlling B transmission rate. *Heredity* 80: 209–213.

Puertas MJ, Jiménez MM, Romera F, Vega JM, and Diez M. 1990. Maternal imprinting effect on B chromosome transmission in rye. *Heredity* 64: 197–204.

Puertas MJ, Ramirez A, and Baeza F. 1987. The transmission of B chromosomes in *Secale cereale* and *Secale vavilovii* populations. II. Dynamics of populations. *Heredity* 58: 81–85.

Qi ZX, Zeng H, Li HL, Chen CB, Song WQ, and Chen RY. 2002. The molecular characterization of maize B chromosome specific AFLPs. *Cell Res.* 12: 63–68.

Qiu AD, Shi AJ, and Komaru A. 2001. Yellow and brown shell color morphs of *Corbicula fluminea* (Bivalvia: Corbiculidae) from Sichuan Province, China, are triploids and tetraploids. *J. Shellfish Res.* 20: 323–328.

Quattro JM, Avise JC, and Vrijenhoek RC. 1991. Molecular evidence for multiple origins of hybridogenetic fish clones (Poeciliidae: *Poeciliopsis*). *Genetics* 127: 391–398.

Quattro JM, Avise JC, and Vrijenhoek RC. 1992. An ancient clonal lineage in the fish genus *Poeciliopsis* (Antheriniformes: Poeciliidae). *Proc. Natl. Acad. Sci. USA* 89: 348–352.

Queller DC. 2003. Theory of genomic imprinting conflict in social insects. *BMC Evol. Biol.* 3: 15.

Queller DC, Ponte E, Bozzaro S, and Strassmann JE. 2003. Single-gene greenbeard effects in the social amoeba *Dictyostelium discoideum*. *Science* 299: 105–106.

Quesada H, Stuckas H, and Skibinski DOF. 2003. Heteroplasmy suggests paternal co-transmission of multiple genomes and pervasive reversion of maternally into paternally transmitted genomes of mussel *Mytilus* mitochondrial DNA. *J. Mol. Evol.* 57: S138-S147.

Quesada H, Wenne R, and Skibinski DOF. 1999. Interspecies transfer of female mitochondrial DNA is coupled with role-reversals and departure from neutrality in the mussel *Mytilus trossulus*. *Mol. Biol. Evol.* 16: 655–665.

Quesneville H, and Anxolabéhère D. 1998. Dynamics of transposable elements in metapopulations: a model of *P* element invasion in *Drosophila*. *Theor. Popul. Biol.* 54: 175–193.

Quicke DLJ. 1997. *Parasitic Wasps*. London: Chapman & Hall.

Raghuvanshi SS, and Kumar G. 1983. No male sterility gene on B chromosome of *Plantago coronopus*. *Heredity* 51: 429–433.

Rainier S, Johnson LA, Dobry CJ, Ping AJ, Grundy PE, and Feinberg AP. 1993. Relaxation of imprinted genes in human cancer. *Nature* 362: 747–749.

Raju NB. 1979. Cytogenetic behavior of spore killer genes in *Neurospora*. *Genetics* 93: 607–623.

Raju NB. 1980. Meiosis and ascospore genesis in *Neurospora*. *Eur. J. Cell Biol.* 23: 208–223.

Raju NB. 1994. Ascomycete spore killers: chromosomal elements that distort genetic ratios among the products of meiosis. *Mycologia* 86: 461–473.

Raju NB, and Perkins DD. 1991. Expression of meiotic drive elements *Spore killer-2* and *Spore killer-3* in asci of *Neurospora tetrasperma*. *Genetics* 129: 25–37.

Ramachandran V. 1996. The evolutionary biology of self-deception, laughter, dreaming and depression: some clues from anosognosia. *Med. Hypotheses* 47: 347–362.

Ramsdale M. 1999. Genomic conflict in fungal mycelia: a subcellular population biology. In: *Structure and Dynamics of Fungal Populations*, ed. JJ Worrall. Dordrecht: Kluwer Academic: 139–174.

Rand E, and Cedar H. 2003. Regulation of imprinting: a multi-tiered process. *J. Cell. Biochem.* 88: 400–407.

Randerson JP, and Hurst LD. 1999. Small sperm, uniparental inheritance and selfish cytoplasmic elements: a comparison of two models. *J. Evol. Biol.* 12: 1110–1124.

Randerson JP, and Hurst LD. 2001. The uncertain evolution of the sexes. *Trends Ecol. Evol.* 16: 571–579.

Rao SRV, and Padmaja M. 1992. Mammalian-type dosage compensation mechanism in an insect–*Gryllotalpa fossor* (Scudder) Orthoptera. *J. Biosci.* 17: 253–273.

Rawson PD, and Hilbish TJ. 1995. Evolutionary relationships among the male and female mitochondrial DNA lineages in the *Mytilus edulis* species complex. *Mol. Biol. Evol.* 12: 893–901.

Reed KM, and Werren JH. 1995. Induction of paternal genome loss by the paternal-sex-ratio chromosome and cytoplasmic incompatibility bacteria (*Wolbachia*): a comparative study of early embryonic events. *Mol. Reprod. Dev.* 40: 408–418.

Rees H, and Ayonoadu U. 1973. B chromosome selection in rye. *Theor. Appl. Genet.* 43: 162–166.

Rees H, and Hutchinson J. 1973. Nuclear DNA variation due to B chromosomes. *Cold Spring Harbor Symp. Quant. Biol.* 38: 175–182.

Reik W, and Walter J. 2001a. Genomic imprinting: parental influence on the genome. *Nat. Rev. Genet.* 2: 21–32.

Reik W, and Walter J. 2001b. Evolution of imprinting mechanisms: the battle of the two sexes begins in the zygote. *Nat. Genet.* 27: 255–256.

Renard JP, and Babinet C. 1986. Identification of a paternal development effect on the cytoplasm of one-cell-stage mouse embryos. *Proc. Natl. Acad. Sci. USA* 83: 6883–6886.

Renard JP, Baldacci P, Richoux-Duranthon V, Pournin S, and Babinet C. 1994. A maternal factor affecting mouse blastocyst formation. *Development* 120: 797–802.

Reyer H-U, Frei G, and Som C. 1999. Cryptic female choice: frogs reduce clutch size when amplexed by undesired males. *Proc. Biol. Sci.* 266: 2101–2107.

Reyer H-U, Niederer B, and Hettyey A. 2003. Variation in fertilisation abilities between hemiclonal hybrid and sexual parental males of sympatric water frogs (*Rana lessonae, R. esculenta, R. ridibunda*). *Behav. Ecol. Sociobiol.* 54: 274–284.

Rhoades MM. 1942. Preferential segregation in maize. *Genetics* 27: 395–407.

Rhoades MM. 1952. Preferential segregation in maize. *Heterosis,* ed. JW Gowen. Ames: Iowa College Press, pp. 66–80.

Rhoades MM, and Dempsey E. 1966. The effect of abnormal chromosome 10 on preferential segregation and crossing over in maize. *Genetics* 53: 989–1020.

Rhoades MM, and Dempsey E. 1972. On the mechanism of chromatin loss induced by the B chromosome of maize. *Genetics* 71: 73–96.

Rhoades MM, and Dempsey E. 1973. Chromatin elimination induced by B chromosome of maize. I. Mechanism of loss and the pattern of endosperm variegation. *J. Hered.* 64: 12–18.

Rhoades MM, and Dempsey E. 1988. Structure of K10-II chromosome and comparison with K10-I. *Maize Genet. Coop. Newsl.* 62: 33.

Ribeiro JMC, and Kidwell MG. 1994. Transposable elements as population drive mechanisms: specification of critical parameter values. *J. Med. Entomol.* 31: 10–16.

Rice WR. 1992. Sexually antagonistic genes: experimental evidence. *Science* 256: 1436–1439.

Rice WR, and Chippindale AK. 2001. Intersexual ontogenetic conflict. *J. Evol. Biol.* 14: 685–693.

Richards AJ. 1986. *Plant Breeding Systems.* London: George Allen & Unwin.

Rick CM. 1966. Abortion of male and female gametes in the tomato determined by allelic interaction. *Genetics* 53: 85–96.

Rick CM. 1971. The tomato *Ge* locus: linkage relations and geographic distribution of alleles. *Genetics* 67: 75–85.

Ridley M, and Grafen A. 1981. Are green beard genes outlaws? *Anim. Behav.* 29: 954–955.

Rigaud T, Juchault P, and Mocquard JP. 1997. The evolution of sex determination in isopod crustaceans. *Bioessays* 19: 409–416.

Rinkevich B. 1996. Bi- versus multichimerism in colonial urochordates: a hypothesis for links between natural tissue transplantation, allogenetics and evolutionary ecology. *Exp. Clin. Immunogenet.* 13: 61–69.

Rinkevich B. 2001. Human natural chimerism: an acquired character or a vestige of evolution. *Hum. Immunol.* 62: 651–657.

Rinkevich B. 2002. The colonial urochordate *Botryllus schlosseri:* from stem cells and natural tissue transplantation to issues in evolutionary ecology. *BioEssays* 24: 730–740.

Rinkevich B, Porat R, and Goren M. 1995. Allorecognition elements on a urochordate histocompatibility locus indicate unprecedented extensive polymorphism. *Proc. R. Soc. Lond. B* 259: 319–324.

Rist L, Semlitsch RD, Hotz H, and Reyer H-U. 1997. Feeding behaviour, food consumption, and growth efficiency of hemiclonal and parental tadpoles of the *Rana esculenta* complex. *Funct. Ecol.* 11: 735–742.

Robbins LG, Palumbo G, Bonaccorsi S, and Pimpinelli S. 1996. Measuring meiotic drive. *Genetics* 142: 645–647.

Robertson HM. 2002. Evolution of DNA transposons in eukaryotes. In: *Mobile DNA II*, eds. NL Craig, R Craigie, M Gellert, and AM Lambowitz. Washington, D.C.: ASM Press, pp. 1093–1110.

Robertson HM, and Lampe DJ. 1995. Recent horizontal transfer of a *mariner* transposable element among and between Diptera and Neuroptera. *Mol. Biol. Evol.* 12: 850–862.

Roeder GS, and Fink GR. 1983. Transposable elements in yeast. In: *Mobile Genetic Elements*, ed. JA Shapiro. New York: Academic Press, pp. 299–328.

Roesli M, and Reyer H-U. 2000. Male vocalization and female choice in the hybridogenetic *Rana lessonae/Rana esculenta* complex. *Anim. Behav.* 60: 745–755.

Roman H. 1948. Directed fertilisation in maize. *Proc. Natl. Acad. Sci. USA* 34: 36–42.

Romera F, Jimenez MM, and Puertas MJ. 1991. Factors controlling the dynamics of the B chromosome polymorphism in Korean rye. *Heredity* 67: 189–195.

Rong YS, and Golic KG. 2003. The homologous chromosome is an effective template for the repair of mitotic DNA double-strand breaks in *Drosophila*. *Genetics* 165: 1831–1842.

Ronsseray S, and Anxolabéhère D. 1986. Chromosomal distribution of P and I transposable elements in a natural population of *Drosophila melanogaster*. *Chromosoma* 94: 433–440.

Ronsseray S, Lehmann M, and Anxolabéhère D. 1991. The maternally inherited regulation of *P* elements in *Drosophila melanogaster* can be elicited by two *P* copies at cytological site 1A on the *X* chromosome. *Genetics* 129: 501–512.

Roth G, Blanke J, and Ohle M. 1995. Brain size and morphology in miniaturized plethodontid salamanders. *Brain Behav. Evol.* 45: 84–95.

Roth G, Blanke J, and Wake DB. 1994. Cell size predicts morphological complexity in the brains of frogs and salamanders. *Proc. Natl. Acad. Sci. USA* 91: 4796–4800.

Roth G, Nishikawa KC, Naujoks-Manteuffel C, Schmidt A, and Wake DB. 1993. Paedomorphosis and simplification in the nervous system of salamanders. *Brain Behav. Evol.* 42: 137–170.

Roth G, Nishikawa KC, and Wake DB. 1997. Genome size, secondary simplification, and the evolution of the brain in salamanders. *Brain Behav. Evol.* 50: 50–59.

Rubin EJ, Akerley BJ, Novik VN, Lampe DJ, Husson RN, and Mekalanos JJ. 1999. *In vivo* transposition of *mariner*-based elements in enteric bacteria and mycobacteria. *Proc. Natl. Acad. Sci. USA* 96: 1645–1650.

Rubin GM. 1983. Dispersed repetitive DNAs in *Drosophila*. In: *Mobile Genetic Elements*, ed. JA Shapiro. New York: Academic Press, pp. 329–361.

Rubin GM, and Spradling AC. 1982. Genetic transformation of *Drosophila* with transposable element vectors. *Science* 218: 348–353.

Ruiz-Pesini E, Lapeña A-C, Díez-Sánchez C, Pérez-Martos A, Montoya J, Alvarez E, Díaz M, Urriés A, Montoro L, López-Pérez MJ, and Enríquez JA. 2000. Human mtDNA haplogroups associated with high or reduced spermatozoa motility. *Am. J. Hum. Genet.* 67: 682–696.

Runge S, Nielsen FC, Nielsen J, Lykke-Andersen J, Werer UM, and Christiansen J. 2000. H19 RNA binds four molecules of insulin-like growth factor II mRNA-binding protein. *J. Biol. Chem.* 275: 29562–29569.

Rushforth AM, and Anderson P. 1996. Splicing removes the *Caenorhabditis elegans* transposon Tc1 from most mutant pre-mRNAs. *Mol. Cell. Biol.* 16: 422–429.

Rutishauser A, and Röthlisberger E. 1966. Boosting mechanism of B chromosomes in *Crepis capillaris*. *Chromosomes Today* 1: 28–30.

Rychlewski J, and Zarzycki K. 1981. Sex ratio in seeds of *Rumex acetosa* L. as a result of sparse or abundant pollination. *Acta Biol. Cracov. (Series: Botanica)* 18: 101–114.

Saavedra C, Reyero MI, and Zouros E. 1997. Male-dependent doubly uniparental inheritance of mitochondrial DNA and female-dependent sex-ratio in the mussel *Mytilus galloprovincialis*. *Genetics* 145: 1073–1082.

Sabba Rao MV, and Pantulu JV. 1978. The effects of derived B-chromosomes on meiosis in pearl millet *Pennisitum typhoides*. *Chromosoma* 69: 121–130.

Sabelis MW, and Nagelkerke CJ. 1987. Sex allocation strategies of pseudo-arrhenotokous phytoseiid mites. *Neth. J. Zool.* 37: 117–136.

Sabelis MW, and van der Meer J. 1986. Local dynamics of the interaction between predatory mites and two-spotted spider mites. In: *The Dynamics of Physiologically Structured Populations. Lecture Notes in Biomathematics 68*, eds. JAJ Metz and D Diekmann. Pp. 322–344.

Saffman EE, and Lasko P. 1999. Germline development in vertebrates and invertebrates. *Cell. Mol. Life Sci.* 55: 1141–1163.

Sanchez L, and Perondini ALP. 1999. Sex determination in sciarid flies: a model for the control of X-chromosome elimination. *J. Theor. Biol.* 197: 247–259.

Sandell LL, Guan X-J, Ingram R, and Tilghman SM. 2003. Gatm, a creatine synthesis enzyme, is imprinted in mouse placenta. *Proc. Natl. Acad. Sci. USA* 100: 4622–4627.

Sandler L, Hiraizumi Y, and Sandler I. 1959. Meiotic drive in natural populations of *Drosophila melanogaster*. I. The cytogenetic basis of segregation-distortion. *Genetics* 44: 233–250.

Sandler L, and Novitski E. 1957. Meiotic drive as an evolutionary force. *Am. Nat.* 857: 105–110.

SanMiguel P, Gaut BS, Tikhonov A, Nakajima Y, and Bennetzen JL. 1998. The paleontology of intergene retrotransposons of maize. *Nat. Genet.* 20: 43–45.

Sano Y. 1983. A new gene controlling sterility in F1 hybrids of two cultivated rice species. *J. Hered.* 74: 435–439.

Sano Y. 1990. The genic nature of gamete eliminator in rice. *Genetics* 125: 183–191.

Sano Y, Chu Y-E, and Oka H-I. 1979. Genetic studies of speciation in cultivated rice. I. Genic analysis for the F1 sterility between *O. sativa* L. and *O. glaberrima* Steud. *Jap. J. Genet.* 54: 121–132.

Santelices B, Correa JA, Hormazábal M, and Flores V. 2003. Contact responses between spores and sporelings of different species, karyological phases and cysto-carps of coalescing Rhodophyta. *Mar. Biol.* 143: 381–392.

Santos JL, del Cerro AL, Fernández A, and Díez M. 1993. Meiotic behaviour of B chromosomes in the grasshopper *Omocestus burri:* a case of drive in females. *Hereditas* 118: 139–143.

Sapre AB, and Deshpande DS. 1987. Origin of B chromosomes in *Coix* L. through spontaneous interspecific hybridization. *J. Hered.* 78: 191–196.

Sarkar KR, and Coe EH. 1971. Analysis of events leading to heterofertilization in maize. *J. Hered.* 62: 118–120.

Sassaman DM, Dombroski BA, Moran JV, Kimberland ML, Naas TP, DeBerardinis RJ, Gabriel A, Swergold GD, and Kazazian Jr. HH. 1997. Many human L1 elements are capable of retrotransposition. *Nat. Genet.* 16: 37–43.

Scali V, Tinti F, Mantovani B, and Marescalchi O. 1995. Mate recognition and gamete cytology features allow hybrid species production and evolution in *Bacillus* stick insects. *Bollettino di Zoologia* 62: 59–70.

Schartl M, Nanda I, Schlupp I, Wilde B, Epplen JT, Schmid M, and Parzefall J. 1995. Incorporation of subgenomic amounts of DNA as compensation for mutational load in a gynogenetic fish. *Nature* 373: 68–71.

Schimenti JC, and Hammer MF. 1990. Rapid identification of mouse *t* haplotypes by PCR polymorphism (PCRP). *Mouse Genome* 87: 108.

Schlupp I, Nanda I, Dobler M, Lamatsch DK, Epplen JT, Parzefall J, Schmid M, and Schartl M. 1998. Dispensable and indispensable genes in an ameiotic fish, the Amazon molly *Poecilia formosa. Cytogenet. Cell. Genet.* 80: 193–198.

Schmid M, Ziegler CG, Steinlein C, Nanda I, and Haaf T. 2002. Chromosome banding in Amphibia. XXIV. The B chromosomes of *Gastrotheca espeletia* (Anura, Hylidae). *Cytogenet. Genome Res.* 97: 205–218.

Schmidt BR. 1993. Are hybridogenetic frogs cyclical parthenogens? *Trends Ecol. Evol.* 8: 271–273.

Schnable PS, and Wise RP. 1998. The molecular basis of cytoplasmic male sterility and fertility restoration. *Trends Plant Sci.* 3: 175–180.

Schneeberger RG, and Cullis CA. 1992. Intraspecific 5S rRNA gene variation in flax, *Linum usitatissimum* (Linaceae). *Plant Syst. Evol.* 183: 265–280.

Schrader F, and Hughes-Schrader S. 1931. Haploidy in Metazoa. *Q. Rev. Biol.* 6: 411–438.

Schramke V, and Allshire R. 2004. Those interfering little RNAs! Silencing and eliminating chromosomes. *Curr. Opin. Genet. Dev.* 14: 174–180.

Schulten GGM, Arendonk RCM, Russell VM, and Roorda FA. 1978. Copulation, egg production and sex-ratio in *Phytoseiulus persimilis* and *Amblyseius bibens* (Acari: Phytoseiidae). *Entomol. Exp. Appl.* 24: 145–153.

Schultes NP, and Szostak JW. 1991. A poly(dA · dT) tract is a component of the re-combination initiation site at the *ARG4* locus in *Saccharomyces cerevisiae. Mol. Cell. Biol.* 11: 322–328.

Schultz RJ. 1961. Reproductive mechanisms of unisexual and bisexual strains of the viviparous fish *Poeciliopsis. Evolution* 15: 302–325.

Schultz RJ. 1973. Unisexual fish: laboratory synthesis of a "species." *Science* 179: 180–181.

Schultz RJ. 1977. Evolution and ecology of unisexual fishes. *Evol. Biol.* 10: 277–331.

Schultz ST. 1994. Nucleo-cytoplasmic male sterility and alternative routes to dioecy. *Evolution* 48: 1933–1945.

Schultz ST, and Ganders FR. 1996. Evolution of unisexuality in the Hawaiian flora: a test of microevolutionary theory. *Evolution* 50: 842–855.

Schütt C, and Nöthiger R. 2000. Structure, function and evolution of sex-determin-ing systems in Dipteran insects. *Development* 127: 667–677.

Schwacha A, and Kleckner N. 1994. Identification of joint molecules that form fre-quently between homologs but rarely between sister chromatids during yeast mei-osis. *Cell* 76: 51–63.

Schwartz A, Chan DC, Brown LG, Alagappan R, Pettay D, Disteche C, McGillivray B, de la Chapelle A, and Page DC. 1998. Reconstructing hominid Y evolution: X-homologous block, created by X-Y transposition, was disrupted by Yp inversion through LINE-LINE recombination. *Hum. Mol. Genet.* 7: 1–11.

Schwartz D. 1950. The analysis of a case of cross-sterility in maize. *Proc. Natl. Acad. Sci. USA* 36: 719–724.

Schwartz M, and Vissing J. 2002. Paternal inheritance of mitochondrial DNA. *N. Engl. J. Med.* 347: 576–580.

Schwarz-Sommer Z, Gierl A, Cuypers H, Peterson PA, and Saedler H. 1985. Plant transposable elements generate the DNA sequence diversity needed in evolution. *EMBO J.* 4: 591–597.

Scoles GJ, and Kibirge-Sebunya IN. 1982. Preferential abortion of gametes in wheat induced by an *Agropyron* chromosome. *Can. J. Genet. Cytol.* 25: 1–6.

Scott RJ, Spielman M, Bailey J, and Dickinson HG. 1998. Parent-of-origin effects on seed development in *Arabidopsis thaliana. Development* 125: 3329–3341.

Selander RK. 1970. Behavior and genetic variation in natural populations. *Am. Zool.* 10: 53–66.

Seleme M-d-C, Busseau I, Malinsky S, Bucheton A, and Teninges D. 1999. High-fre-quency retrotransposition of a marked *I* factor in *Drosophila melanogaster* corre-lates with a dynamic expression pattern of the ORF1 protein in the cytoplasm of oocytes. *Genetics* 151: 761–771.

Seligman LM, Chisholm KM, Chevalier BS, Chadsey MS, Edwards ST, Savage JH, and Veillet AL. 2002. Mutations altering the cleavage specificity of a homing endonuclease. *Nucleic Acids Res.* 30: 3870–3879.

Sellem CH, Lecellier G, and Belcour L. 1993. Transposition of a group II intron. *Nature* 366: 176–178.

Semlitsch RD, Hotz H, and Guex GD. 1997. Competition among tadpoles of coexisting hemiclones of hybridogenetic *Rana esculenta:* support for the frozen niche variation model. *Evolution* 51: 1249–1261.

Semlitsch RD, Schmiedehausen S, Hotz H, and Beerli P. 1996. Genetic compatibility between sexual and clonal genomes in local populations of the hybridogenetic *Rana esculenta* complex. *Evol. Ecol.* 10: 531–543.

Semple JC. 1976. The cytogenetics of *Xanthisma texanum* DC. (Asteraceae) and its B chromosomes. *Am. J. Bot.* 63: 388–398.

Seperack PK, Strobel MC, Corrow DJ, Jenkins NA, and Copeland NG. 1988. Somatic and germ-line reverse mutation rates of the retrovirus-induced dilute coat-color mutation of DBA mice. *Proc. Natl. Acad. Sci. USA* 85: 189–192.

Sessions SK, and Larson A. 1987. Developmental correlates of genome size in plethodontid salamanders and their implications for genome evolution. *Evolution* 41: 1239–1251.

Sharbel TF, Green DM, and Houben A. 1998. B-chromosome origin in the endemic New Zealand frog *Leiopelma hockstetteri* through sex chromosome evolution. *Genome* 41: 14–22.

Sharon G, Burkett TJ, and Garfinkel DJ. 1994. Efficient homologous recombination of Ty1 element cDNA when integration is blocked. *Mol. Cell. Biol.* 14: 6540–6551.

Sharp CB, Hilliker AJ, and Holm DG. 1985. Further characterization of genetic elements associated with the segregation distorter phenomenon in *Drosophila melanogaster. Genetics* 110: 671–688.

Shaw MW. 1984. The population genetics of the B-chromosome polymorphism of *Myrmeleotettix maculatus* (Orthoptera: Acrididae). *Biol. J. Linn. Soc.* 23: 77–100.

Shaw MW, and Hewitt GM. 1984. The effect of temperature on meiotic transmission rates of the B chromosome of *Myrmeleotettix maculates* (Orthoptera: Acrididae). *Heredity* 53: 259–268.

Shaw MW, and Hewitt GM. 1985. The genetic control of meiotic drive acting on the B-chromosome of *Myrmeleotettix maculatus. Heredity* 54: 187–194.

Shaw MW, and Hewitt GM. 1990. B chromosomes, selfish DNA and theoretical models: where next? *Oxf. Surv. Evol. Biol.* 7: 197–223.

Shaw MW, Hewitt GM, and Anderson DA. 1985. Polymorphism in the rates of meiotic drive on the B-chromosome of *Myrmeleotettix maculatus. Heredity* 55: 61–68.

Sheldon BC, Andersson S, Griffith SC, Örnborg J, and Sendecka J. 1999. Ultraviolet colour variation influences blue tit sex ratios. *Nature* 402: 874–877.

Sherman PW. 1977. Nepotism and the evolution of alarm calls. *Science* 197: 1246–1253.

Sherman PW. 1980. The limits of ground squirrel nepotism. In: *Sociobiology: Beyond*

Nature/Nurture? eds. GB Barlow and J Silverberg. Boulder, Colorado: Westview Press, pp. 505–544.

Shin H-S, Stavnezer J, Artzt K, and Bennett D. 1982. Genetic structure and origin of *t* haplotypes of mice, analysed with *H-2* cDNA probes. *Cell* 29: 969–976.

Shirafuji N, Takahashi S, Metsuda S, and Asano S. 1997. Mitochondrial antisense RNA for cytochrome c oxidase (MARCO) can induce morphologic changes and cell death in human hematopoietic cell lines. *Blood* 90: 4567–4577.

Shitara H, Kaneda H, Sato A, Inoue K, Ogura A, Yonekawa H, and Hayashi J-I. 2000. Selective and continuous elimination of mitochondria microinjected into mouse eggs from spermatids, but not from liver cells, occurs throughout embryogenesis. *Genetics* 156: 1277–1284.

Siebel CW, Admon A, and Rio DC. 1995. Soma-specific expression and cloning of PSI, a negative regulator of P element pre-mRNA splicing. *Genes Dev.* 9: 269–283.

Silver LM. 1982. Genomic analysis of the *H-2* complex region associated with mouse *t* haplotypes. *Cell* 29: 961–968.

Silver LM. 1993. The peculiar journey of a selfish chromosome: mouse t haplotypes and meiotic drive. *Trends Genet.* 9: 250–254.

Simmons GM. 1992. Horizontal transfer of hobo transposable elements within the *Drosophila melanogaster* species complex: evidence from DNA sequencing. *Mol. Biol. Evol.* 9: 1050–1060.

Skibinski DOF, Gallagher C, and Beynon CM. 1994. Sex-limited mitochondrial DNA transmission in the marine mussel *Mytilus edulis. Genetics* 138: 801–809.

Skuse D, James RS, Bishop DVM, Coppin B, Dalton P, Aamodt-Leeper G, Bacarese-Hamilton M, Creswell C, McGurk R, and Jacobs PA. 1997. Evidence from Turner's syndrome of an imprinted X-linked locus affecting cognitive function. *Nature* 387: 705–708.

Sleutels F, and Barlow DP. 2002. The origins of genomic imprinting in mammals. *Adv. Genet.* 46: 119–163.

Small ID, and Peeters N. 2000. The PPR motif–a TPR-related motif prevalent in plant organellar proteins. *Trends Biochem. Sci.* 25: 46–47.

Smit AFA. 1996. The origin of interspersed repeats in the human genome. *Curr. Opin. Genet. Dev.* 6: 743–748.

Smit AFA. 1999. Interspersed repeats and other mementos of transposable elements in mammalian genomes. *Curr. Opin. Genet. Dev.* 9: 657–663.

Smit AFA, Tóth G, Riggs AD, and Jurka J. 1995. Ancestral, mammalian-wide subfamilies of LINE-1 repetitive sequences. *J. Mol. Biol.* 246: 401–417.

Smith-White S, and Carter CR. 1970. The cytology of *Brachycome lineariloba*. II. The chromosome species and their relationships. *Chromosoma* 30: 129–153.

Smith DG, and Rolfs B. 1984. Segregation distortion and differential fitness at the albumin locus in rhesus monkeys *(Macaca mulatta). Am. J. Primatol.* 7: 285–289.

Smith NGC. 1998. The dynamics of maternal-effect selfish genetic elements. *J. Theor. Biol.* 191: 173–180.

Smithies O, and Powers PA. 1986. Gene conversions and their relationship to homologous pairing. *Phil. Trans. R. Soc. Lond. B Biol. Sci.* 291–302.

Solari AJ, Espinosa MB, Vitullo AD, and Merani MS. 1989. Meiotic behavior of gonosomically variant females of *Akodon azarae* (Rodentia, Cricetidae). *Cytogenet. Cell Genet.* 52: 57–61.

Sommerfeldt AD, Bishop JDD, and Wood CA. 2003. Chimerism following fusion in a clonal ascidian (Urochordata). *Biol. J. Linn. Soc.* 79: 183–192.

Song SU, Kurkulos M, Boeke JD, and Corces VG. 1997. Infection of the germ line by retroviral particles produced in the follicle cells: a possible mechanism for the mobilization of the *gypsy* retroelement of *Drosophila. Development* 124: 2789–2798.

Sonstegard TS, and Hackett PB. 1996. Autogenous regulation of RNA translation and packaging by Rous sarcoma virus Pr76^{gag}. *J. Virol.* 70: 6642–6652.

Soullier S, Hanni C, Catzeflis F, Berta P, and Laudet U. 1998. Male sex determination in the spiny rat *Tokudaia osimensis* (Rodentia: Muridae) is not *Sry* dependent. *Mamm. Genome* 9: 590–592.

Spahn L, and Barlow DP. 2003. An ICE pattern crystallizes. *Nat. Genet.* 35: 11–12.

Spakulova M, Kralova-Hromadova I, Dudinak V, and Reddy PV. 2002. Karyotype of *Acanthocephalus lucii:* the first record of supernumerary chromosomes in thorny-headed worms. *Parasitol. Res.* 88: 778–780.

Spinella DG, and Vrijenhoek RC. 1982. Genetic dissection of clonally inherited genomes of *Poeciliopsis.* II. Investigation of a silent carboxylesterase allele. *Genetics* 100: 279–286.

Sreenivasan TV. 1981. Cytogenetical studies in *Erianthus:* meiosis and behaviour of chromosomes in 2n=20 forms. *Genetica* 55: 129–132.

Srivastava M, Frolova E, Rottinghaus B, Boe SP, Grinberg A, Lee E, Love PE, and Pfeifer K. 2003. Imprint control element-mediated secondary methylation imprints at the *Igf2/H19* locus. *J. Biol. Chem.* 278: 5977–5983.

Stahl FW, and Lande R. 1995. Estimating interference and linkage map distance from two-factor tetrad data. *Genetics* 139: 1449–1454.

Stalker HD. 1961. The genetic systems modifying meiotic drive in *Drosophila peramelanica. Genetics* 46: 177–202.

Stark EA, Connerton I, Bennett ST, Barnes SR, Parker JS, and Forster JW. 1996. Molecular analysis of the structure of the maize B-chromosome. *Chromosome Res.* 4: 15–23.

Stebbins GL. 1971. *Chromosomal Evolution in Higher Plants.* London: Edward Arnold.

Stenseth NC. 1978. Is female biased sex-ratio in wood lemming *Myopus schisticolor* maintained by cyclic inbreeding? *Oikos* 30: 83–89.

Stenseth NC, and Ims RA. 1993. Population dynamics of lemmings: temporal and spatial variation: an introduction. In: *The Biology of Lemmings,* eds. NC Stenseth and RA Ims. London: Academic Press, pp. 61–96.

Stenseth NC, Kirkendall LR, and Moran N. 1985. On the evolution of pseudogamy. *Evolution* 39: 294–307.

Stevens JP, and Bougourd SM. 1988. Inbreeding depression and the outcrossing rate in natural populations of *Allium schoenoprasum* L (wild chives). *Heredity* 60: 257–261.

Stewart CEH, and Rotwein P. 1996. Growth, differentiation and survival: multiple physiological functions for insulin-like growth factors. *Physiol. Rev.* 76: 1005–1026.

Stewart DT, Kenchington ER, Singh RK, and Zouros E. 1996. Degree of selective constraint as an explanation of the different rates of evolution of gender-specific mitochondrial DNA lineages in the mussel *Mytilus. Genetics* 143: 1349–1357.

Stewart DT, Saavedra C, Stanwood RR, Ball AO, and Zouros E. 1995. Male and female mitochondrial DNA lineages in the blue mussel (*Mytilus edulis*) species group. *Mol. Biol. Evol.* 12: 735–747.

Stitou S, Diaz de la Guardia RD, Jiménez R, and Burgos M. 2000. Inactive ribosomal cistrons are spread throughout the B chromosomes of *Rattus rattus* (Rodentia, Muridae). Implications for their origin and evolution. *Chromosome Res.* 8: 305–311.

Stitou S, Zurita F, and Diaz de la Guardia RD. 2004. Transmission analysis of B chromosomes in *Rattus rattus* from Northern Africa. *Cytogenet. Genome Res.* 106: 344–346.

Stoner DS, Rinkevich B, and Weissman IL. 1999. Heritable germ and somatic cell lineage competitions in chimeric colonial protochordates. *Proc. Natl. Acad. Sci. USA* 96: 9148–9153.

Stoner DS, and Weissman IL. 1996. Somatic and germ cell parasitism in a colonial ascidian: possible role for a highly polymorphic allorecognition system. *Proc. Natl. Acad. Sci. USA* 93: 15254–15259.

Stouthamer R. 1997. *Wolbachia*-induced parthenogenesis. In: *Influential Passengers, Inherited Microorganisms and Arthropod Reproduction,* eds. SC O'Neill, JH Werren, and AA Hoffman. Oxford: Oxford University Press, pp. 102–124.

Stouthamer R, and Kazmer DJ. 1994. Cytogenetics of microbe-associated parthenogenesis and its consequences for gene flow in *Trichogramma* wasps. *Heredity* 73: 317–327.

Stouthamer R, van Tilborg M, de Jong JH, Nunney L, and Luck RF. 2001. Selfish element maintains sex in natural populations of a parasitoid wasp. *Proc. Biol. Sci.* 268: 617–622.

Stuart JJ, and Hatchett JH. 1988. Cytogenetics of the Hessian fly. II. Inheritance and behavior of somatic and germ-line-limited chromosomes. *J. Hered.* 79: 190–199.

Stuart JR, Haley KJ, Swedzinski D, Lockner S, Kocian PE, Merriman PJ, and Simmons MJ. 2002. Telomeric *P* elements associated with cytotype regulation of the *P* transposon family in *Drosophila melanogaster. Genetics* 162: 1641–1654.

Sturm S, Figueroa F, and Klein J. 1982. The relationship between *t* and *H-2* complexes in wild mice. The *H-2* haplotypes of 20 *t*-bearing strains. *Genet. Res.* 40: 73–88.

Sturtevant AH, and Dobzhansky T. 1936. Geographical distribution and cytology of "sex ratio" in *Drosophila pseudoobscura* and related species. *Genetics* 21: 473–490.

Subramanyam K. 1967. Some aspects of the embryology of *Sedum chrysanthum* (Boissier) Raymond-Hamlet with a discussion on its systematic position. *Phytomorphology* 17: 240–247.

Suguna SG, Wood RJ, Curtis CF, Whitelaw A, and Kazmi SJ. 1977. Resistance to meiotic drive at the M^D locus in an Indian wild population of *Aedes aegypti. Genet. Res.* 29: 123–132.

Suh D-S, Choi E-H, Yamazaki T, and Harada K. 1995. Studies on the transposition rates of mobile genetic elements in a natural population of *Drosophila melanogaster. Mol. Biol. Evol.* 12: 748–758.

Sullivan M, Higuchi T, Katis VL, and Uhlmann F. 2004. Cdc14 phosphatase induces rDNA condensation and resolves cohesin-independent cohesion during budding yeast anaphase. *Cell* 117: 471–482.

Summers DK. 1996. *The Biology of Plasmids.* Oxford: Blackwell Science.

Summers K, and Crespi B. 2005. Cadherins in maternal-foetal interactions: red queen with a green beard? *Proc. Biol. Sci.* 272: 643–649.

Sun H, Treco D, Schultes NP, and Szostak JW. 1989. Double-strand breaks at an initiation site for meiotic gene conversion. *Nature* 338: 87–90.

Sun H, Treco D, and Szostak JW. 1991. Extensive 3′-overhanging, single-stranded DNA associated with the meiosis-specific double-strand breaks at the *ARG4* recombination initiation site. *Cell* 64: 1155–1161.

Sun M. 1987. Genetics of gynodioecy in Hawaiian *Bidens* (Asteraceae). *Heredity* 59: 327–336.

Sun X, Wahlstrom J, and Karpen G. 1997. Molecular structure of a functional *Drosophila* centromere. *Cell* 91: 1007–1019.

Suomalainen E, Saura A, and Lokki J. 1987. *Cytology and Evolution in Parthenogenesis.* Boca Raton: CRC Press.

Susin SA, Lorenzo HK, Zamzami N, Marzo I, Snow BE, Brothers GM, Mangion J,

Jacotot E, Costantini P, Loeffler M, Larochette N, Goodlett DR, Aebersold R, Siderovski DP, Penninger JM, and Kroemer G. 1999. Molecular characterization of mitochondrial apoptosis-inducing factor. *Nature* 397: 441–446.

Sutovsky P, Moreno RD, Ramalho-Santos J, Dominko T, Simerly C, and Schatten G. 2000. Ubiquitinated sperm mitochondria, selective proteolysis, and the regulation of mitochondrial inheritance in mammalian embryos. *Biol. Reprod.* 63: 582–590.

Svensson E, and Nilsson J-Å. 1996. Mate quality affects offspring sex ratio in blue tits. *Proc. R. Soc. Lond. B* 263: 357–361.

Sweeny TL, and Barr AR. 1978. Sex ratio distortion caused by meiotic drive in a mosquito, *Culex pipiens* L. *Genetics* 88: 427–446.

Szklarzewicz T, and Moskal A. 2001. Ultrastructure, distribution, and transmission of endosymbionts in the whitefly *Aleurochiton aceris* Modeer (Insecta, Hemiptera, Aleyrodinea). *Protoplasma* 218: 45–53.

Tabara H, Sarkissian M, Kelly WG, Fleenor J, Grishok A, Timmons L, Fire A, and Mello CC. 1999. The *rde-1* gene, RNA interference, and transposon silencing in *C. elegans*. *Cell* 99: 123–132.

Tabata M. 1961. Studies of a gametophyte factor in barley. *Jap. J. Genet.* 36: 157–167.

Takeda S, Sugimoto K, Otsuki H, and Hirochika H. 1999. A 13-bp *cis*-regulatory element in the LTR promoter of the tobacco retrotransposon *Tto1* is involved in responsiveness to tissue culture, wounding, methyl jasmonate and fungal elicitors. *Plant J.* 18: 383–393.

Talbert PB, Bryson TD, and Henikoff S. 2004. Adaptive evolution of centromere proteins in plants and animals. *J. Biol.* 3: 1–17.

Talbert PB, Masuelli R, Tyagi AP, Comai L, and Henikoff S. 2002. Centromeric localization and adaptive evolution of an *Arabidopsis* histone H3 variant. *Plant Cell* 14: 1053–1066.

Tan S-S, Faulkner-Jones B, Breen S, Walsh M, Bertram J, and Reese B. 1995. Cell dispersion patterns in different cortical regions studied with an X-inactivated transgenic marker. *Development* 121: 1029–1039.

Tang Y, Schon EA, Wilichowski E, Vazquez-Memije ME, Davidson E, and King MP. 2000. Rearrangements of human mitochondrial DNA (mtDNA): new insights into the regulation of mtDNA copy number and gene expression. *Mol. Biol. Cell* 11: 1471–1485.

Tao Y, Hartl DL, and Laurie CC. 2001. Sex-ratio segregation distortion associated with reproductive isolation in *Drosophila*. *Proc. Natl. Acad. Sci. USA* 98: 13183–13188.

Tarayre M, Saumitou-Laprade P, Cuguen J, Couvet D, and Thompson JD. 1997. The spatial genetic structure of cytoplasmic (cpDNA) and nuclear (allozyme) markers within and among populations of the gynodioecious *Thymus vulgaris* (Labiatae) in southern France. *Am. J. Bot.* 84: 1675–1684.

Tatum LA. 1971. The southern corn blight epidemic. *Science* 171: 1113–1116.

Taylor DR, Zeyl C, and Cooke E. 2002. Conflicting levels of selection in the accumulation of mitochondrial defects in *Saccharomyces cerevisiae. Proc. Natl. Acad. Sci. USA* 99: 3690–3694.

Taylor JE, and Jaenike J. 2002. Sperm competition and the dynamics of X chromosome drive: stability and extinction. *Genetics* 160: 1721–1723.

Taylor JE, and Jaenike J. 2003. Sperm competition and the dynamics of X chromosome drive in finite and structured populations. *Annales Zoologici Fennica* 40: 195–206.

Taylor PD, and Bulmer MG. 1980. Local mate competition and the sex ratio. *J. Theor. Biol.* 86: 409–419.

Tchénio T, Segal-Bendirdjian E, and Heidmann T. 1993. Generation of processed pseudogenes in murine cells. *EMBO J.* 12: 1487–1497.

Teague JW, Morton NE, Dennis NR, Curtis G, McKechnie N, Macpherson JN, Murray A, Pound MC, Sharrock AJ, Youings SA, and Jacobs PA. 1998. FRAXA and FRAXE: Evidence against segregation distortion and for an effect of intermediate alleles on learning disability. *Proc. Natl. Acad. Sci. USA* 95: 719–724.

Telesnitsky A, and Goff SP. 1997. Reverse transcriptase and the generation of retroviral DNA. In: *Retroviruses,* eds. JM Coffin, SH Hughes, and HE Varmus. Cold Spring Harbor: Cold Spring Harbor Laboratory Press, pp. 121–160.

Temin HM. 1991. Sex and recombination in retroviruses. *Trends Genet.* 7: 71–74.

Temin RG, Ganetzky B, Powers PA, Lyttle TW, Pimpinelli S, Dimitri P, Wu C-I, and Hiraizumi Y. 1991. Segregation distortion in *Drosophila melanogaster:* genetic and molecular analyses. *Am. Nat.* 137: 287–331.

Temin RG, and Marthas M. 1984. Factors influencing the effect of segregation distortion in natural populations of *Drosophila melanogaster. Genetics* 107: 375–393.

Teoh SB, Rees H, and Hutchinson J. 1976. B chromosome selection in *Lolium. Heredity* 37: 207–213.

Terzian C, Ferraz C, Demaille J, and Bucheton A. 2000. Evolution of the *Gypsy* endogenous retrovirus in the *Drosophila melanogaster* subgroup. *Mol. Biol. Evol.* 17: 908–914.

Thomas DD, Donnelly CA, Wood RJ, and Alphey LS. 2000. Insect population control using a dominant, repressible, lethal genetic system. *Science* 287: 2474–2476.

Thomson MS, and Beeman RW. 1999. Assisted suicide of a selfish gene. *J. Hered.* 90: 191–194.

Thomson RL. 1984. *B* chromosomes in *Rattus fuscipes.* II. The transmission of *B* chromosomes to offspring and population studies: support for the "parasitic" model. *Heredity* 52: 363–372.

Thornberry NA, and Lazebnik Y. 1998. Caspases: enemies within. *Science* 281: 1312–1316.

Tikhonov AP, SanMiguel PJ, Nakajima Y, Gorenstein NM, Bennetzen JL, and Avramova Z. 1999. Colinearity and its exceptions in orthologous *adh* regions of maize and sorghum. *Proc. Natl. Acad. Sci. USA* 96: 7409–7414.

Tinti F, Mantovani B, and Scali V. 1995. Reproductive features of homospecific hybridogenetically-derived stick insects suggest how unisexuals can evolve. *J. Evol. Biol.* 8: 81–92.

Tinti F, and Scali V. 1992. Genome exclusion and gametic DAPI DNA content in the hybridogenetic *Bacillus rossius-grandii benazzii* complex (Insecta Phasmatodea). *Mol. Reprod. Dev.* 33: 235–242.

Tinti F, and Scali V. 1993. Chromosomal evidence of hemiclonal and all-paternal offspring production in *Bacillus rossius-grandii benazzii* (Insecta, Phasmatodea). *Chromosoma* 102: 403–414.

Tinti F, and Scali V. 1995. Allozymic and cytological evidence for hemiclonal, all-paternal, and mosaic offspring of the hybridogenetic stick insect *Bacillus rossius-grandii. J. Exp. Zool.* 273: 149–159.

Tokuyasu KT, Peacock WJ, and Hardy RW. 1977. Dynamics of spermiogenesis in *Drosophila melanogaster.* VII. Effects of *Segregation Distorter* (*SD*) chromosome. *J. Ultrastruct. Res.* 58: 96–107.

Topp CN, Zhong CX, and Dawe RK. 2004. Centromere-encoded RNAs are integral components of the maize kinetochore. *Proc. Natl. Acad. Sci. USA* 101: 15986–15991.

Torkamanzehi A, Moran C, and Nicholas FW. 1992. *P* element transposition contributes substantial new variation for a quantitative trait in *Drosophila melanogaster. Genetics* 131: 73–78.

Tosi LRO, and Beverley SM. 2000. *cis* and *trans* factors affecting *Mos1 mariner* evolution and transposition *in vitro,* and its potential for functional genomics. *Nucleic Acids Res.* 28: 784–790.

Tower J, Karpen GH, Craig NL, and Spradling AC. 1993. Preferential transposition of *Drosophila P* elements to nearby chromosomal sites. *Genetics* 133: 347–359.

Tram U, and Sullivan W. 2002. Role of delayed nuclear envelope breakdown and mitosis in *Wolbachia*-induced cytoplasmic incompatibility. *Science* 296: 1124–1126.

Traut W. 1994. Sex determination in the fly *Megaselia scalaris,* a model system for primary steps of sex chromosome evolution. *Genetics* 136: 1097–1104.

Traut W, and Marec F. 1996. Sex chromatin in lepidoptera. *Q. Rev. Biol.* 71: 239–256.

Traut W, Rahn IM, Winking H, Kunze B, and Weichenhan D. 2001. Evolution of a 6–200 Mb long-range repeat cluster in the genus *Mus. Chromosoma* 110: 247–252.

Traut W, and Willhoeft U. 1990. A jumping sex determining factor in the fly *Megaselia scalaris. Chromosoma* 99: 407–412.

Tremblay E, and Caltagirone LE. 1973. Fate of polar bodies in insects. *Annu. Rev. Entomol.* 18: 421–444.

Tristem M. 2000. Identification and characterization of novel human endogenous retrovirus families by phylogenetic screening of the human genome mapping project database. *J. Virol.* 74: 3715–3730.

Trivers R. 1974. Parent-offspring conflict. *Am. Zool.* 14: 249–264.

Trivers R. 1985. *Social Evolution.* Menlo Park: Benjamin Cummings.

Trivers R. 2000. The elements of a scientific theory of self-deception. *Ann. N.Y. Acad. Sci.* 907: 114–131.

Trivers R. 2002. *Natural Selection and Social Theory.* Oxford: Oxford University Press.

Trivers R, and Burt A. 1999. Kinship and genomic imprinting. In: *Genomic Imprinting: An Interdisciplinary Approach,* ed. R Ohlsson. Heidelberg: Springer-Verlag, pp. 1–21.

Trivers R, Burt A, and Palestis BG. 2004. B chromosomes and genome size in flowering plants. *Genome* 47: 1–8.

Trivers R, and Hare H. 1976. Haplodiploidy and the evolution of the social insects. *Science* 191: 249–263.

Tucker PK, and Lundrigan BL. 1993. Rapid evolution of the sex determining locus in Old World mice and rats. *Nature* 364: 715–717.

Tufarelli C, Stanley JAS, Garrick D, Sharpe JA, Ayyub H, Wood WG, and Higgs DR. 2003. Transcription of antisense RNA leading to gene silencing and methylation as a novel cause of human genetic disease. *Nat. Genet.* 34: 157–165.

Turelli M, and Hoffmann AA. 1999. Microbe-induced cytoplasmic incompatibility as a mechanism for introducing transgenes into arthropod populations. *Insect Mol. Biol.* 8: 243–255.

Turner BC. 2001. Geographical distribution of *Neurospora* spore killer strains and strains resistant to killing. *Fungal Genet. Biol.* 32: 93–104.

Turner BC, and Perkins DD. 1991. Meiotic drive in *Neurospora* and other fungi. *Am. Nat.* 137: 416–429.

Tycko B. 1999. Genomic imprinting and cancer. In: *Genomic Imprinting: An Interdisciplinary Approach,* ed. R Ohlsson. Berlin: Springer-Verlag, pp. 133–169.

Tycko B, and Morison IM. 2002. Physiological functions of imprinted genes. *J. Cell. Physiol.* 192: 245–258.

Udomkit A, Forbes S, McLean C, Arkhipova I, and Finnegan DJ. 1996. Control of expression of the *I* factor, a LINE-like transposable element in *Drosophila melanogaster. EMBO J.* 15: 3174–3181.

Ullstrup AJ. 1972. The impacts of the southern corn leaf blight epidemics of 1970–1971. *Annu. Rev. Phytopath.* 10: 37–50.

van Boven M, and Weissing FJ. 1999. Segregation distortion in a deme-structured population: opposing demands of gene, individual and group selection. *J. Evol. Biol.* 12: 80–93.

van Boven M, and Weissing FJ. 2001. Competition at the mouse *t* complex: rare alleles are inherently favored. *Theor. Popul. Biol.* 60: 343–358.

Van Damme JMM. 1983. Gynodioecy in *Plantago lanceolata* L. II. Inheritance of three male sterility types. *Heredity* 50: 253–273.

Van Damme JMM. 1984. Gynodioecy in *Plantago lanceolata* L. III. Sexual reproduction and the maintenance of male steriles. *Heredity* 52: 77–93.

Van Damme JMM, and Van Delden W. 1982. Gynodioecy in *Plantago lanceolata* L. I. Polymorphism for plasmon type. *Heredity* 49: 303–318.

van de Lagemaat LN, Landry J-R, Mager DL, and Medstrand P. 2003. Transposable elements in mammals promote regulatory variation and diversification of genes with specialized functions. *Trends Genet.* 19: 530–536.

van der Gaag M, Debets AJM, Oosterhof J, Slakhorst M, Thijssen JAGM, and Hoekstra RF. 2000. Spore-killing meiotic drive factors in a natural population of the fungus *Podospora anserina*. *Genetics* 156: 593–605.

van Vugt JFA, Salverda M, de Jong JH, and Stouthamer R. 2003. The paternal sex chromosome in the parasitic wasp *Trichogramma kaykai* condenses the paternal chromosomes into a dense chromatin mass. *Genome* 46: 580–587.

VandeBerg JL, Robinson ES, Samollow PB, and Johnston PG. 1987. X-linked gene expression and X-chromosome inactivation: marsupials, mouse, and man compared. In: *Isozymes: Current Topics in Biological and Medical Research,* vol. 15, eds. MC Rattazzi, JG Scandalios, and GS Whitt. New York: Allan R. Liss, pp. 225–253.

van Dijk BA, Boomsma DI, and de Man AJM. 1996. Spore-killing meiotic drive factors in a natural population of the fungus *Podospora anserina*. *Am. J. Med. Genet.* 61: 264–268.

VanEtten HD, Funnell-Baerg D, Wasmann C, and McCluskey K. 1994. Location of pathogenicity genes on dispensable chromosomes in *Nectria haematococca* MPVI. *Antonie van Leeuwenhoek Int. J. Gen. Mol. Microbiol.* 65: 263–267.

Vanin EF. 1985. Processed pseudogenes: characteristics and evolution. *Annu. Rev. Genet.* 19: 253–272.

Varandas FR, Sampaio MC, and Carvalho AB. 1997. Heritability of sexual proportion in experimental sex-ratio populations of *Drosophila mediopunctata*. *Heredity* 79: 104–112.

Varndell NP, and Godfray HCJ. 1996. Facultative adjustment of the sex ratio in an insect (*Plannococcus citri*, Pseudococcidae) with paternal genome loss. *Evolution* 50: 2100–2105.

Vassiliadis C, Saumitou-Laprade P, Lepart J, and Viard F. 2002. High male reproductive success of hermaphrodites in the androdioecious *Phillyrea angustifolia. Evolution* 56: 1362–1373.

Vastenhouw NL, and Plasterk RHA. 2004. RNAi protects the *Caenorhabditis elegans* germline against transposition. *Trends Genet.* 20: 314–319.

Vaury C, Pélisson A, Abad P, and Bucheton A. 1993. Properties of transgenic strains of *Drosophila melanogaster* containing I transposable elements from *Drosophila teissieri. Genet. Res.* 61: 81–90.

Vaz SC, and Carvalho AB. 2004. Evolution of autosomal suppression of the *Sex-Ratio* trait in *Drosophila. Genetics* 166: 265–277.

Vázquez-Manrique RP, Hernández M, Martínez-Sebastián MJ, and de Frutos R. 2000. Evolution of *gypsy* endogenous retrovirus in the *Drosophila obscura* species group. *Mol. Biol. Evol.* 17: 1185–1193.

Vázquez M, Ben-Dov C, Lorenzi H, Moore T, Schijman A, and Levin MJ. 2000. The short interspersed repetitive element of *Trypanosoma cruzi,* SIRE, is part of VIPER, an unusual retroelement related to long terminal repeat retrotransposons. *Proc. Natl. Acad. Sci. USA* 97: 2128–2133.

Venturelli M. 1981. Embryologia *Struthianthus vulgaris* (Loranthaceae–Loranthoideae). *Kurtziana* 14: 73–100.

Verona RI, Mann MRW, and Bartolomei MS. 2003. Genomic imprinting: intricacies of epigenetic regulation in clusters. *Annu. Rev. Cell Dev. Biol.* 19: 237–259.

Vicente VE, Moreira-Filho O, and Camacho JPM. 1996. Sex-ratio distortion associated with the presence of a B chromosome in *Astyanax scabripinnis* (Teleostei, Characidae). *Cytogenet. Cell Genet.* 74: 70–75.

Vicient CM, Suoniemi A, Anamthawat-Jonsson K, Tanskanen J, Beharav A, Nevo E, and Schulman A. 1999. Retrotransposon *BARE-1* and its role in genome evolution in the genus *Hordeum. Plant Cell* 11: 1769–1784.

Vieira C, and Biémont C. 1996. Selection against transposable elements in *D. simulans* and *D. melanogaster. Genet. Res.* 68: 9–15.

Vieira C, and Biémont C. 1997. Transposition rate of the 412 retrotransposable element is independent of copy number in natural populations of *Drosophila simulans. Mol. Biol. Evol.* 14: 185–188.

Vijayaraghavan MR, and Prabhakar K. 1984. The endosperm. In: *Embryology of Angiosperms,* ed. BM Johri. Berlin: Springer-Verlag, pp. 319–376.

Vincent A, and Petes TD. 1989. Mitotic and meiotic gene conversion of Ty elements and other insertions in *Saccharomyces cerevisiae. Genetics* 122: 759–772.

Vinogradov AE. 2003. Selfish DNA is maladaptive: evidence from the plant Red List. *Trends Genet.* 19: 609–614.

Vinogradov AE. 2004. Genome size and extinction risk in vertebrates. *Proc. Biol. Sci.* 271: 1701–1705.

Vinogradov AE, Borkin LJ, Gunther R, and Rosanov JM. 1990. Genome elimination in diploid and triploid *Rana esculenta* males: cytological evidence from DNA flow cytometry. *Genome* 33: 619–627.

Vinogradov AE, and Chubinishvili AT. 1999. Genome reduction in a hemiclonal frog *Rana esculenta* from radioactively contaminated areas. *Genetics* 151: 1123–1125.

Viseras E, Camacho JPM, Cano MI, and Santos JL. 1990. Relationship between mitotic instability and accumulation of B chromosomes in males and females of *Locusta migratoria*. *Genome* 33: 23–29.

Voelker RA. 1972. Preliminary characterization of "sex ratio" and rediscovery and reinterpretation of "male sex ratio" in *Drosophila affinis*. *Genetics* 71: 597–606.

Vogels A, Matthijs G, Legius E, Devriendt K, and Fryns JP. 2003. Chromosome 15 maternal uniparental disomy and psychosis in Prader-Willi syndrome. *J. Med. Genet.* 40: 72–73.

Vogelstein B, and Kinzler KW. 1993. The multistep nature of cancer. *Trends Genet.* 9: 138–141.

Vogt VM. 1997. Retroviral virions and genomes. In: *Retroviruses*, eds. JM Coffin, SH Hughes, and HE Varmus. Cold Spring Harbor: Cold Spring Harbor Laboratory Press, pp. 27–69.

Volobujev VT. 1980. The B-chromosome system of mammals. *Genetica* 52/53: 333–337.

vom Saal FS. 1981. Variation in phenotype due to random intrauterine positioning of male and female fetuses in rodents. *J. Reprod. Fertil.* 62: 633–650.

vom Saal FS. 1989. Sexual differentiation in litter-bearing mammals: influence of adjacent fetuses *in utero*. *J. Anim. Sci.* 67: 1824–1840.

vom Saal FS, and Bronson FH. 1978. *In utero* proximity of female mouse fetuses to males: effect on reproductive performance during later life. *Biol. Reprod.* 19: 842–853.

Vorburger C. 2001a. Heterozygous fitness effects of clonally transmitted genomes. *J. Evol. Biol.* 14: 602–610.

Vorburger C. 2001b. Fixation of deleterious mutations in clonal lineages: evidence from hybridogenetic frogs. *Evolution* 55: 2319–2332.

Vorburger C. 2001c. Non-hybrid offspring from matings between hemiclonal hybrid waterfrogs suggest occasional recombination between clonal genomes. *Ecol. Lett.* 4: 628–636.

Vorburger C, and Reyer H-U. 2003. A genetic mechanism of species replacement in European waterfrogs. *Conserv. Genet.* 4: 141–155.

Voytas DF, and Boeke JD. 2002. Ty1 and Ty5 of *Saccharomyces cerevisiae*. In: *Mobile DNA II*, eds. NL Craig, R Craigie, M Gellert, and AM Lambowitz. Washington, D.C.: ASM Press, pp. 631–662.

Vrana PB, Fossella JA, Matteson P, del Rio T, O'Neill MJ, and Tilghman SM. 2000.

Genetic and epigenetic incompatibilities underlie hybrid dysgenesis in *Peromyscus. Nat. Genet.* 25: 120–124.

Vrana PB, Guan X-J, Ingram RS, and Tilghman SM. 1998. Genomic imprinting is disrupted in interspecific *Peromyscus* hybrids. *Nat. Genet.* 20: 362–365.

Vrana PB, Matteson PG, Schmidt JV, Ingram RS, Joyce A, Prince KL, Dewey MJ, and Tilghman SM. 2001. Genomic imprinting of a placental lactogen gene in *Peromyscus. Dev. Genes Evol.* 211: 523–532.

Vrijenhoek RC. 1979. Factors affecting clonal diversity and coexistence. *Am. Zool.* 19: 787–797.

Vrijenhoek RC. 1984. The evolution of clonal diversity in *Poeciliopsis*. In: *Evolutionary Genetics of Fishes*, ed. BJ Turner. New York: Plenum Press, pp. 399–429.

Wade MJ, and Beeman RW. 1994. The population dynamics of maternal-effect selfish genes. *Genetics* 138: 1309–1314.

Walbot V, and Rudenko GN. 2002. *MuDR/Mu* transposable elements of maize. In: *Mobile DNA II,* eds. NL Craig, R Craigie, M Gellert, and AM Lambowitz. Washington, D.C.: ASM Press, pp. 533–564.

Walter CA, Intano GW, McCarrey JR, McMahan CA, and Walter RB. 1998. Mutation frequency declines during spermatogenesis in young mice but increases in old mice. *Proc. Natl. Acad. Sci. USA* 95: 10015–10019.

Walter J, and Paulsen M. 2003. Imprinting and disease. *Semin. Cell. Dev. Biol.* 14: 101–110.

Wang L, and Kunze R. 1998. Transposase binding site methylation in the epigenetically inactivated *Ac* derivative *Ds-cy. Plant J.* 13: 577–582.

Wasserman M. 1982. Evolution of the *repleta* group. In: *The Genetics and Biology of Drosophila,* eds. M Ashburner, HL Carson, and JN Thompson. London: Academic Press, pp. 61–139.

Waterston RH, Lindblad-Toh K, Birney E, Rogers J (plus 218 others). 2002. Initial sequencing and comparative analysis of the mouse genome. *Nature* 420: 520–562.

Watson JD. 1972. Origin of concatemeric T7 DNA. *Nat. New Biol.* 239: 197–201.

Weeks SC, Gaggiotti OE, Schenck RA, Spindler KP, and Vrijenhoek RC. 1992. Feeding behavior in sexual and clonal strains of *Poeciliopsis. Behav. Ecol. Sociobiol.* 30: 1–6.

Weichenhan D, Kunze B, Winking H, van Geel M, Osoegawa K, de Jong PJ, and Traut W. 2001. Source and component genes of a 6–200 Mb gene cluster in the house mouse. *Mamm. Genome* 12: 590–594.

Weichenhan D, Traut W, Kunze B, and Winking H. 1996. Distortion of Mendelian recovery ratio for a mouse HSR is caused by maternal and zygotic effects. *Genet. Res.* 68: 125–129.

Weil CF, and Kunze R. 2000. Transposition of maize *Ac/Ds* transposable elements in the yeast *Saccharomyces cerevisiae. Nat. Genet.* 26: 187–190.

Weiner AM, Deininger PL, and Efstratiadis A. 1986. Nonviral retroposons: genes, pseudogenes, and transposable elements generated by the reverse flow of genetic information. *Annu. Rev. Biochem.* 55: 631–661.

Weiss U, and Wilson JH. 1987. Repair of single-stranded loops in heteroduplex DNA transfected into mammalian cells. *Proc. Natl. Acad. Sci. USA* 84: 1619–1623.

Wendel JF, Cronn RC, Johnston JS, and Price HJ. 2002. Feast and famine in plant genomes. *Genetica* 115: 37–47.

Wenzlau JM, Saldanha RJ, Butow RA, and Perlman PS. 1989. A latent intron-encoded maturase is also an endonuclease needed for intron mobility. *Cell* 56: 421–430.

Werren JH, and Beukeboom LW. 1993. Population genetics of a parasitic chromosome: theoretical analysis of PSR in subdivided populations. *Am. Nat.* 142: 224–241.

Werren JH, and Beukeboom LW. 1998. Sex determination, sex ratios, and genetic conflict. *Annu. Rev. Ecol. Syst.* 29: 233–261.

Werren JH, and Charnov EL. 1978. Facultative sex ratios and population dynamics. *Nature* 272: 348–350.

Werren JH, and Hatcher MJ. 2000. Maternal-zygotic gene conflict over sex determination: effects of inbreeding. *Genetics* 155: 1469–1479.

Werren JH, Hatcher MJ, and Godfray HCJ. 2002. Maternal-offspring conflict leads to the evolution of dominant zygotic sex determination. *Heredity* 88: 102–111.

Werren JH, Nur U, and Wu C-I. 1988. Selfish genetic elements. *Trends Ecol. Evol.* 3: 297–302.

Werren JH, and Stouthamer R. 2003. PSR (paternal sex ratio) chromosomes: the ultimate selfish genetic elements. *Genetica* 117: 85–101.

Wessler SR. 1996. Plant retrotransposons: turned on by stress. *Curr. Biol.* 6: 959–961.

Westerman M, and Dempsey J. 1977. Population cytology of the genus *Phaulacridium*. VI. Seasonal changes in the frequency of the B-chromosomes in a population of *P. vittatum. Austr. J. Biol. Sci.* 30: 329–336.

Wetherington JD, Kotora KE, and Vrijenhoek RC. 1987. A test of the spontaneous heterosis hypothesis for unisexual vertebrates. *Evolution* 41: 721–731.

White MJD. 1973. *Animal Cytology and Evolution,* 3rd ed. Cambridge, England: Cambridge University Press.

White NJ, Nosten F, Looareesuwan S, Watkins WM, Marsh K, Snow RW, Kokwaro G, Ouma J, Hien TT, Molyneux ME, Taylor TE, Newbold CI, Ruebush II TK,

Danis M, Greenwood BM, Anderson RM, and Olliaro P. 1999. Averting a malaria disaster. *Lancet* 353: 1965–1967.

Whitfield LS, Lovell-Badge R, and Goodfellow PN. 1993. Rapid-sequence evolution of the mammalian sex-determining gene SRY. *Nature* 364: 713–715.

Whitham TG, and Slobodchikoff CN. 1981. Evolution by individuals, plant-herbivore interactions, and mosaics of genetic variability: the adaptive significance of somatic mutations in plants. *Oecologia* 49: 287–292.

Wickner RB. 1996. Double-stranded RNA viruses of *Saccharomyces cerevisiae*. *Microbiol. Rev.* 60: 250–265.

Widdig A, Nürnburg P, Krawczak M, Streich WJ, and Bercovitch FB. 2001. Paternal relatedness and age proximity regulate social relationships among adult female rhesus macaques. *Proc. Natl. Acad. Sci. USA* 98: 13769–13773.

Wilby AS, and Parker JS. 1988. The supernumerary segment systems of *Rumex acetosa*. *Heredity* 60: 109–117.

Wilde CD. 1986. Pseudogenes. *CRC Crit. Rev. Biochem.* 19: 323–352.

Wilkins JF, and Haig D. 2001. Genomic imprinting of two antagonistic loci. *Proc. Biol. Sci.* 268: 1861–1867.

Wilkins JF, and Haig D. 2002. Parental modifiers, antisense transcripts and loss of imprinting. *Proc. Biol. Sci.* 269: 1841–1846.

Wilkins JF, and Haig D. 2003a. What good is genomic imprinting: the function of parent-specific gene expression. *Nat. Rev. Genet.* 4: 359–368.

Wilkins JF, and Haig D. 2003b. Inbreeding, maternal care and genomic imprinting. *J. Theor. Biol.* 221: 559–564.

Wilkinson GS, and Sanchez MI. 2001. Sperm development, age and sex chromosome meiotic drive in the stalk-eyed fly, *Cyrtodiopsis whitei*. *Heredity* 87: 17–24.

Williams GC. 1988. Retrospect on sex and kindred topics. *The Evolution of Sex: An Examination of Current Ideas,* eds. RE Michod and BR Levin. Sunderland: Sinauer, pp. 287–298.

Williams GC, and Nesse RM. 1991. The dawn of Darwinian medicine. *Q. Rev. Biol.* 66: 1–22.

Williams J, and Lenington S. 1993. Environmental and genetic factors affecting preferences of female house mice (*Mus musculus*) for males that differ in *t*-complex genotype. *Behav. Genet.* 23: 51–58.

Wilson EB. 1906. Studies on chromosomes. V. The chromosomes of *Metapodius*. A contribution to the hypothesis of genetic continuity of chromosomes. *J. Exp. Zool.* 6: 147–205.

Wilson EB. 1907. The supernumerary chromosomes of Hemiptera. *Science* 26: 870–871.

Wilson EB. 1928. *The Cell in Development and Heredity*. New York: Macmillan.

Winking H, Gropp A, and Fredga K. 1981. Sex determination and phenotype in wood lemmings with XXY and related karyotypic anomalies. *Hum. Genet.* 58: 98–104.

Winston F, Chaleff DT, Valent B, and Fink GR. 1984. Mutations affecting Ty-mediated expression of the *HIS4* gene of *Saccharomyces cerevisiae*. *Genetics* 107: 179–197.

Wischmann C, and Schuster W. 1995. Transfer of *rps10* from the mitochondrion to the nucleus in *Arabidopsis thaliana:* evidence for RNA-mediated transfer and exon shuffling at the integration site. *FEBS Lett.* 374: 152–156.

Wise RP, and Pring DR. 2002. Nuclear-mediated mitochondrial gene regulation and male fertility in higher plants: light at the end of the tunnel? *Proc. Natl. Acad. Sci. USA* 99: 10240–10242.

Witte CP, Le QH, Bureau T, and Kumar A. 2001. Terminal-repeat retrotransposons in miniature (TRIM) are involved in restructuring plant genomes. *Proc. Natl. Acad. Sci. USA* 98: 13778–13783.

Wolf KW, Mertl HG, and Traut W. 1991. Structure, mitotic and meiotic behaviour, and stability of centromere-like elements devoid of chromosome arms in the fly *Megaselia scalaris* (Phoridae). *Chromosoma* 101: 99–108.

Wolf KW, Mitchell A, and Liu GC. 1996. Centromere-like elements in *Megaselia spiracularis* (Diptera: Phoridae): a fine structure and cytogenetic study. *Hereditas* 124: 203–209.

Wolfe KH, Morden CW, Ems SC, and Palmer JD. 1992. Rapid evolution of the translational apparatus in a nonphotosynthetic plant: loss or accelerated evolution of tRNA and ribosomal protein genes. *J. Mol. Evol.* 35: 304–317.

Wood RJ. 1976. Between-family variation in sex ratio in the Trinidad (T-30) strain of *Aedes aegypti* (L.) indicating differences in sensitivity to the meiotic drive gene M^D. *Genetica* 46: 345–361.

Wood RJ, and Newton ME. 1991. Sex-ratio distortion caused by meiotic drive in mosquitoes. *Am. Nat.* 137: 379–391.

Woodruff RC, and Nikitin AG. 1995. P DNA element movement in somatic cells reduces lifespan in *Drosophila melanogaster:* evidence in support of the somatic mutation theory of aging. *Mutat. Res.* 338: 35–42.

Wright A-DG, and Lynn DH. 1997. Maximum ages of ciliate lineages estimated using a small subunit rRNA molecular clock: crown eukaryotes date back to the paleoproterozoic. *Arch. Protistenkd.* 148: 329–341.

Wright SI, and Schoen DJ. 1999. Transposon dynamics and the breeding system. *Genetica* 107: 139–148.

Wu-Scharf D, Jeong BR, Zhang CM, and Cerutti H. 2000. Transgene and transposon silencing in *Chlamydomonas reinhardtii* by a DEAH-Box RNA helicase. *Science* 290: 1159–1162.

Wu C-I. 1983a. Virility deficiency and the sex-ratio trait in *Drosophila pseudoobscura*. I. Sperm displacement and sexual selection. *Genetics* 105: 651–662.

Wu C-I. 1983b. Virility deficiency and the sex-ratio trait in *Drosophila pseudoobscura*. II. Multiple mating and overall virility selection. *Genetics* 105: 663–679.

Wu C-I, and Beckenbach AT. 1983. Evidence for extensive genetic differentiation between the sex-ratio and standard arrangement of *Drosophila pseudoobscura* and *D. persimilis* and identification of hybrid sterility factors. *Genetics* 105: 71–86.

Wu C-I, Lyttle TW, Wu M-L, and Lin G-F. 1988. Association between a satellite DNA sequence and the *Responder* of *Segregation Distorter* in *D. melanogaster*. *Cell* 54: 179–189.

Wu C-I, True JR, and Johnson N. 1989. Fitness reduction associated with the deletion of a satellite DNA array. *Nature* 341: 248–251.

Wu T-C, and Lichten M. 1994. Meiosis-induced double-strand break sites determined by yeast chromatin structure. *Science* 263: 515–518.

Wu T. 1992. B chromosomes in *Sorghum stipoideum*. *Heredity* 68: 457–463.

Wyngaard GA, and Gregory TR. 2001. Temporal control of DNA replication and the adaptive value of chromatin diminution in copepods. *J. Exp. Zool.* 291: 310–316.

Xu H, and Boeke JD. 1990. Localization of sequences required in *cis* for yeast Ty1 element transposition near the long terminal repeats: analysis of mini-Ty1 elements. *Mol. Cell. Biol.* 10: 2695–2702.

Yamashita S, Takano-Shimizu T, Kitamura K, Mikami T, and Kishima Y. 1999. Resistance to gap repair of the transposon Tam3 in *Antirrhinum majus*: a role of the end regions. *Genetics* 153: 1899–1908.

Yang H-P, and Nuzhdin SV. 2003. Fitness costs of *Doc* expression are insufficient to stabilize its copy number in *Drosophila melanogaster*. *Mol. Biol. Evol.* 20: 800–804.

Yang J, Malik HS, and Eickbush TH. 1999. Identification of the endonuclease domain encoded by R2 and other site-specific, non-long terminal repeat retrotransposable elements. *Proc. Natl. Acad. Sci. USA* 96: 7847–7852.

Yang ZH, and Bielawski JP. 2000. Statistical methods for detecting molecular adaptation. *Trends Ecol. Evol.* 15: 496–503.

Yokomine T, Kuroiwa A, Tanaka K, Tsudzuki Y, and Sasaki H. 2001. Sequence polymorphisms, allelic expression status and chromosome locations of the chicken *IGF2* and *MPR1* genes. *Cytogenet. Cell Genet.* 93: 109–113.

Yoshiyama M, Tu Z, Kainoh Y, Honda H, Shono T, and Kimura K. 2001. Possible horizontal transfer of a transposable element from host to parasitoid. *Mol. Biol. Evol.* 18: 1952–1958.

Yosida TH. 1978. Some genetic analysis of supernumerary chromosomes in the black rat in laboratory matings. *Proc. Jap. Acad. Ser. B Phys. Biol. Sci.* 54: 440–445.

Yu-Sun CC, Wickramaratne MRT, and Whitehouse HLK. 1977. Mutagen specificity in conversion pattern in *Sordaria brevicollis. Genet. Res.* 29: 65–81.

Yu H-G, Hiatt EN, Chan A, Sweeney M, and Dawe RK. 1997. Neocentromere-mediated chromosome movement in maize. *J. Cell Biol.* 139: 831–840.

Zamzami N, Susin SA, Marchetti P, Hirsch T, Gomez-Monterrey I, Castedo M, and Kroemer G. 1996. Mitochondrial control of nuclear apoptosis. *J. Exp. Med.* 183: 1533–1544.

Zbawicka M, Skibiniski DOF, and Wenne R. 2003. Doubly uniparental transmission of mitochondrial DNA length variants in the mussel *Mytilus trossolus. Mar. Biol.* 142: 455–460.

Zečević L, and Paunović D. 1969. The effect of B chromosomes on chiasma frequency. *Chromosoma* 27: 198–200.

Zeh DW, and Zeh JA. 1999. Transmission distortion at a minisatellite locus in the harlequin beetle riding pseudoscorpion. *J. Hered.* 90: 320–323.

Zeyl C, and Bell G. 1996. Symbiotic DNA in eukaryotic genomes. *Trends Ecol. Evol.* 11: 10–15.

Zhang J, and Peterson T. 1999. Genome rearrangements by nonlinear transposons in maize. *Genetics* 153: 1403–1410.

Zhang P, and Spradling AC. 1993. Efficient and dispersed local *P* element transposition from *Drosophila* females. *Genetics* 133: 361–373.

Zhang Z, and Gerstein M. 2004. Large-scale analysis of pseudogenes in the human genome. *Curr. Opin. Genet. Dev.* 14: 328–335.

Zheng Y, Deng X, and Martin-DeLeon PA. 2001. Lack of sharing of Spam1 (Ph-20) among mouse spermatids and transmission ratio distortion. *Biol. Reprod.* 64: 1730–1738.

Zhou L, Mitra R, Atkinson PW, Hickman AB, Dyda F, and Craig NL. 2004. Transposition of *hAT* elements links transposable elements and V(D)J recombination. *Nature* 432: 995–1001.

Ziegler CG, Lamatsch DK, Steinlein C, Engel W, Scharl M, and Schmid M. 2003. A giant B chromosome of the cyprinid fish *Alburnus alburnus* harbours a retrotransposon-derived DNA sequence. *Chromosome Res.* 11: 23–35.

Zielinski WJ, vom Saal FS, and Vandenbergh JG. 1992. The effect of intrauterine position on the survival, reproduction and home range size of female house mice (*Mus musculus*). *Behav. Ecol. Sociobiol.* 30: 185–191.

Zimmerly S, Hausner G, and Wu XC. 2001. Phylogenetic relationships among group II intron ORFs. *Nucleic Acids Res.* 29: 1238–1250.

Zinic SD, Ugarkovic D, Cornudella L, and Plohl M. 2000. A novel interspersed type of organisation of satellite DNAs in *Tribolium madens* heterochromatin. *Chromosome Res.* 8: 201–212.

Zipora L, and Nur U. 1973. Accumulation of B-chromosomes by preferential segregation in females of the grasshopper *Melanoplus femur-rubrum*. *Chromosoma* 42: 289–306.

Zouros E. 2000. The exceptional mitochondrial DNA system of the mussel family Mytilidae. *Genes Genet. Syst.* 75: 313–318.

Zouros E, Ball AO, Saavedra C, and Freeman KR. 1994a. An unusual type of mitochondrial DNA inheritance in the blue mussel *Mytilus*. *Proc. Natl. Acad. Sci. USA* 91: 7463–7467.

Zouros E, Ball AO, Saavedra C, and Freeman KR. 1994b. Mitochondrial DNA inheritance. *Nature* 368: 819–820.

Zouros E, Freeman KR, Ball AO, and Pogson GH. 1992. Direct evidence for extensive paternal mitochondrial DNA inheritance in the marine mussel *Mytilus*. *Nature* 359: 412–414.

Župunski V, Gubenšek F, and Kordiš D. 2001. Evolutionary dynamics and evolutionary history in the RTE clade of non-LTR retrotransposons. *Mol. Biol. Evol.* 18: 1849–1863.

Zurita S, Cabrero J, López-León MD, and Camacho JPM. 1998. Polymorphism regeneration for a neutralized selfish B chromosome. *Evolution* 52: 274–277.

Zwick ME, Salstrom JL, and Langley CH. 1999. Genetic variation in rates of nondisjunction: association of two naturally occurring polymorphisms in the chromokinesin *nod* with increased rates of nondisjunction in *Drosophila melanogaster*. *Genetics* 152: 1605–1614.

Glossary

A chromosome anything but a B chromosome, i.e., the usual chromosomes

acrocentric a chromosome with the centromere near one end

allele one of a set of alternative forms of a gene

altruistic beneficial to others, costly to self

androgenetic developing from 2 sets of paternal chromosomes due to the fusion of 2 sperm pronuclei

aneuploidy having other than the basic complement of chromosomes (too many or too few)

antisense RNA RNA complementary to a particular RNA transcript to which it can bind and block function

apoptosis programmed cell death; usually beneficial to the larger organism

autosome a typical chromosome other than a sex chromosome

B chromosome extranumerary chromosome found in some individuals but not all (hence, unnecessary for normal development)

BGC biased gene conversion, that is, gene conversion (see later) that occurs more frequently in one direction than in the other

bivalent the figure produced by pairing of homologous chromosomes at the beginning of meiosis

bp (base pair) abbreviation for the shortest unit length of DNA

cDNA complementary DNA, formed by reverse transcription from an RNA template

centriole a small body in animal cells that lies next to the nucleus. Usually there are 2 centrioles that serve as centers for spindle fiber formation

centromere the (constricted) region of the chromosome that becomes associated with spindle fibers during mitosis and meiosis and is important in movement of chromosomes to the poles

chiasma (pl. chiasmata) the point at which homologous chromosomes remain attached after pairing has ceased in meiosis. These are sites at which there has been crossing-over

chimera a multicellular individual in which different cells are derived from genetically different progenitor cells (e.g., 2 or more zygotes)

chromatid either of 2 structures that comprise the duplicated chromosome arms prior to the division of the centromere to form daughter chromosomes. Each chromatid becomes a chromosome after division of the centromere

chromosome a linear array of genes physically linked together (along with associated proteins)

cis-**acting locus** a locus that affects the activity only of DNA sequences on its own molecule of DNA; usually noncoding

CMS cytoplasmic male sterility in which a cytoplasmic factor (mtDNA) converts hermaphrodites into females

compensation the replacement of nonsurviving offspring with additional offspring

cooperative beneficial to both parties

CpG island A region of DNA with a relatively high density of CpG dinucleotides (i.e., C followed by G); often associated with the promoter region of genes

crossing-over a process in which homologous chromosomes exchange distal segments of a chromatid. Crossing-over involves breakage and physical exchange of segments, followed by repair of the breaks. Crossing-over is a regular event in meiosis but occurs only rarely in mitosis

cytoplasm contents of a cell inside the plasma membrane but outside the nucleus

deletion loss of a segment of a chromosome

dioecious species with 2 separate sexes

diploid having 2 sets of chromosomes

direct repeats identical (or related) sequences present in 2 or more copies in the same orientation on the same DNA molecule; they are not necessarily adjacent

disjunction movement of members of a chromosome pair to opposite poles during cell division

dosage compensation mechanisms used to compensate for the discrepancy between the presence of 2 sex chromosomes in 1 sex, but only 1 in the other

drag less than 50% (sub-Mendelian) transmission into the next generation

drift random change in gene frequencies in a population over time

drive greater than 50% (super-Mendelian) transmission into the next generation

ectopic recombination recombination between homologous sequences at nonhomologous sites in the genome, such as between 2 copies of a transposable element

endonuclease an enzyme that cleaves DNA or RNA

endosymbiont a species (or descendent of one) living inside another species

enhancer a *cis*-acting sequence that increases the utilization of a promoter, and so increases gene expression

equational division the separation of a chromosome into daughter cells with complements similar to the parent cell. The term is used especially for the second division of meiosis

euchromatin region of an interphase chromosome that stains diffusely; "normal" chromatin, as opposed to the more condensed heterochromatin; associated with high gene number and expression

eukaryote an organism composed of cells with nuclei. Eukaryotes include all organisms except viruses, bacteria, and archaea

exon segment of a gene (usually coding) that is represented in the mature RNA product, after splicing out of introns

F_1 the first generation formed by crossing 2 parental lines

fitness of a genotype, which equals the average number of offspring produced by organisms with that genotype

fixation of a gene = 100% frequency

gene a section of DNA coding for an RNA, which usually in turn codes for a protein

gene conversion process by which DNA sequence information is transferred from one DNA helix (that remains unchanged) to another (whose sequence is altered)

gene pool all the genes (or DNA) in a sexual population

genome all the genes (or DNA) in a cell or organism

germline any lineage of cells potentially leading to gametes and the next generation (as opposed to somatic cells)

gynodioecious population consisting of females and hermaphrodites

haplodiploid males haploid, females diploid

haploid having a single set of chromosomes, that is, only 1 of each type of chromosome

HEG homing endonuclease gene, a class of selfish genes

heterochromatin regions of the genome that are highly condensed and (mostly) transcriptionally inactive throughout the cell cycle. May be constitutive or facultative

heterogametic producing dissimilar gametes with respect to the sex chromosomes (e.g., male mammals producing X and Y sperm). In contrast to **homogametic**, producing only one kind of gamete with respect to the sex chromosomes (e.g., egg cells in mammals)

heterokaryon a cell having 2 or more genetically nonidentical nuclei

heteroplasmy presence of genetically different types of mitochondria or chloroplasts in a single organism

inbreeding mating between relatives

intron segment of DNA that is transcribed but then removed from the transcript by splicing together the sequences (exons) on either side of it; typically noncoding

inversion flipping of a segment of DNA within a chromosome 180° so that gene order is exactly reversed compared to wildtype

inverted repeats two copies of the same DNA sequence in opposite orientation on the same DNA molecule

karyotype the entire chromosomal complement of a cell or species, typically as visualized during mitosis

kb (kilobase) abbreviation for 1000 base pairs (bp) of DNA or 1000 bases of RNA

knockout a mutation that deletes function of a gene entirely

LINE abbreviation for long interspersed nuclear element; 1 of the 2 major classes of retrotransposable elements

linkage loci are linked if they are located on the same chromosome; in general, the closer they are located, the less recombination there is between them (= tighter linkage)

locus (pl. loci) the position of a gene on a chromosome

LTR abbreviation for long terminal repeat; LTR retrotransposable elements are 1 of the 2 major classes of retrotransposable elements

Mb (megabase) abbreviation for 10^6 bp of DNA

meiosis division of diploid germ cells leading to the formation of haploid gametes

metacentric chromosome having a centromere in the middle; more generally, biarmed

metaphase the stage of mitosis and meiosis at which the chromosomes are maximally condensed and arrayed in a plane between the spindle poles

metaphase plate the plane on which metaphase chromosomes are aligned. Equatorial plate

methylation modification (e.g., of DNA) by the addition of methyl groups ($-CH_3$)

mitotype mitochondrial genotype (or genome)

mtDNA mitochondrial DNA

my/mya abbreviation for million years (ago)

neutral having negligible selective effect

nondisjunction the failure of chromosomes to separate properly in cell division

NOR nucleolus organizing region, section of some chromosomes carrying rDNA devoted to production of rRNA and ribosomes (sometimes associated with a constriction)

nucleolus structure in the nucleus where rRNA is synthesized and ribosomal subunits are assembled

ORF abbreviation for open reading frame; segment of DNA without stop codons and translatable into a protein

outbreeding, outcrossing mating between unrelated individuals

pachytene the stage in prophase of meiosis during which chromosomes are paired and somewhat shortened

parthenogenesis formation of an embryo from an unfertilized egg

PGL paternal genome loss; failure of paternal chromosomes to be transmitted to the next generation

plasmid small circular DNA molecule that replicates independently of the main chromosomes

prokaryote an organism that lacks a nucleus. Prokaryotes include bacteria and archaea

promoter region of DNA to which RNA polymerase binds to begin transcription

pseudogene a nonfunctional member of a gene family. **Processed pseudogenes** arise by reverse transcription of mRNA into DNA and subsequent ligation into the genome

r **(relatedness)** the chance for any given gene that an identical copy is found in another individual by direct descent from a common ancestor

rDNA ribosomal DNA; the genes encoding rRNA

recombination the formation of new combinations of alleles, typically during meiosis. This may include the process by which DNA molecules are broken and the fragments rejoined into new combinations

reduction division first meiotic division in which the number of chromosomes is reduced by one-half

retro- prefix denoting the use of reverse transcription from RNA into DNA. Retroelements include retrohoming group II introns, retrotransposable elements, and retroviruses

ribosome particle composed of RNAs and proteins that translates mRNAs into proteins

rRNA ribosomal RNA; any one of a number of RNA molecules that form part of the structure of ribosomes

selfing union of 2 gametes produced by the same organism

selfish beneficial to self, costly to another

sex-antagonistic gene a gene whose effect on fitness in one sex is positive and in the other, negative

sex chromosome an X or Y chromosome (or Z and W)

SINE abbreviation for short interspersed nuclear element; a nonautonomous retrotransposable element that is mobilized by LINEs

somatic refers to all cells of the body other than germ cells or their precursors

spindle (apparatus) the structure formed during nuclear division on which chromosomes first align and then move to opposite poles

spiteful costly to self, costly to another

tandem repeats multiple copies of the same sequence lying end to end

telomere the terminal structure of each chromosome arm (consisting of small tandem repeats)

trans-**acting locus** a locus that affects the activity of DNA sequences regardless of whether they are on its own molecule of DNA; usually protein coding

transcription the production of a strand of RNA from a DNA template

translation the production of a protein from an RNA template

translocation change in location of a chromosome segment by becoming attached to another chromosome

transposable element segment of DNA capable of moving to a new chromosomal location

univalent a single, unpaired chromosome

wildtype the typical allele at a locus in contrast to a mutation or a selfish element; conventionally denoted by a + sign

zygote the diploid cell formed from fusion of haploid gametes (e.g., sperm and egg)

Expanded Contents

Taxonomic Index

Salamanders: genome size, 290–291; B chromosomes in *Dicamptodon,* 347

Reptiles: X chromosomes in snakes, 95, and viviparity, 456–457; transposable elements in snakes, 270; genome size, 289

Birds: sex chromosomes, 62, 95; sex ratio adjustments in *Acrocephalus, Taeniopygia, Parus,* 93–95; mate choice, 94–95; imprinting, 101, 139; transposable elements, 244, 271; genome size, 288, 289; B chromosomes in *Taeniopygia,* 379

Mammals: maternal effect killers, 52–55; zygotic expression, 55; gestational drive, 55–57; sex chromosomes, 74–75, 94–95, 133–137, 293, 433; sex ratio adjustments in ungulates, 94; mate choice, 94–95; imprinting, 99–137, in sheep, 119; mitochondrial inheritance, 148, 156–157; GC content, 193–194; transposable elements, 244, 271, 277–279, 297; processed pseudogenes, 285; karyotype evolution, 315–318; transmissible tumors in dogs, 424–425; B chromosomes, 372–373

Rodents: sex chromosomes 78, 81, 88, in *Dicrostonyx, Myopus,* 82–87, 455, 459, 463, 469, 473, *Microtus,* 87–88, 90–91, *Akodon,* 88–90, 470, *Ellobius,* 88, *Tokudaia,* 88; intrauterine effects in mice and gerbils, 86, 134–135; imprinting in *Peromyscus,* 117–118; relatedness in Belding's ground squirrel, 124–125; X inactivation in *Peromyscus,* 137; transposable elements in *Holochilus, Oryzomys,* 265, *Rattus,* 278, 279; B chromosomes in *Rattus,* 340, 363, 373, *Apodemus,* 344, 365, *Dicrostonyx,* 379

Mus musculus: t-haplotype 19, 20, 21–39, 463; maternal effect killers, 52–54, 318–319; imprinting, 99–111, 113, 116–117, 119–120, 121–122, 125–126, 129–132, 134–137; X-inactivation, 136; mitochondrial inheritance in hybrids, 150; mitochondrial selection, 154–155; pseudoautosomal region, 194, 195; transposable elements, 244, 256, 262, 276, 278–279; meiotic chromosomes, 312; Robertsonian translocations, 315–318; cancer, 422; germline, 426, 430, 431; versus humans, 117, 470

Primates: possible drive in *Macaca,* 48; *Sry,* 90; *Igf2r,* 100; sex-biased dispersal, 128; biased gene conversion, 190, 191, 197–198; transposable elements, 244, 278; chimerism in callitrichids, 420, 437–438

Humans: disease, 15, 47, 99, 100, 105, 120, 121, 125, 135, 149, 154, 156, 161–162, 217, 222, 229, 244, 248–249, 314, 315, 325, 422–424, 426–427, 428, 429; associated species, 46; pregnancy, 56–57, 139; sex chromosomes and mate choice, 95; imprinting, 96, 99, 100–106, 108–109, 113, 117, 120, 121–122, 125, 128–129, 132–133, 135, 136, 460; X-inactivation, 136; mtDNA, 144; mitochondrial inheritance, 149; mitochondrial selection, 154–156; gene therapy, 186; GC content, 193, 194; minisatellites, 196; recombinational hotspot, 197; transposable elements, 239, 248–249, 254, 259, 262, 263–264, 265, 266, 267, 273, 276, 277–279, 281, 285, 286, 287, 295, 300, 471–472; female meiosis, 314; Robertsonian translocations, 315; neocentromeres, 366; mosaics, 420, 421; cell lineage selection, 421–424, 426–428; germline, 425–428; chimerism, 437; versus mice, 117, 470

General Index